Optimal Coordination of Power Protective Devices with Illustrative Examples

IEEE Press
445 Hoes Lane
Piscataway, NJ 08854

Optimal Coordination of Power Protective Devices with Illustrative Examples

Ali R. Al-Roomi
Dalhousie University
Canada

WILEY

Published by John Wiley & Sons, Inc., Hoboken, New Jersey.
Published simultaneously in Canada.

Library of Congress Cataloging-in-Publication Data

Names: Al-Roomi, Ali R., author. | John Wiley & Sons, publisher.
Title: Optimal coordination of power protective devices with illustrative examples / Ali R. Al-Roomi.
Description: Hoboken, New Jersey : Wiley, [2022] | Includes bibliographical references and index.
Identifiers: LCCN 2021047950 (print) | LCCN 2021047951 (ebook) | ISBN 9781119794851 (hardback) | ISBN 9781119794905 (adobe pdf) | ISBN 9781119794912 (epub)
Subjects: LCSH: Electric power systems–Protection. | Electric power system stability. | Protective relays.
Classification: LCC TK1010 .A52 2022 (print) | LCC TK1010 (ebook) | DDC 621.31–dc23/eng/20211104
LC record available at https://lccn.loc.gov/2021047950
LC ebook record available at https://lccn.loc.gov/2021047951

Cover Design: Wiley
Cover Images: (center) © PozitivStudija/Shutterstock, (top) © Sam Robinson/Getty Images

Set in 9.5/12.5pt STIXTwoText by Straive, Chennai, India

10 9 8 7 6 5 4 3 2 1

I dedicate this book to the memory of my father "A.Ali", my tender mother "Fatima", my patient wife "Alya", and my two little angels "Ridha" and "Murtadha".

Contents

Author Biography *xvi*
Preface *xvii*
Acknowledgments *xviii*
Acronyms *xix*
About The Companion Website *xxiii*
Introduction *xxv*

1 Fundamental Steps in Optimization Algorithms *1*
1.1 Overview *1*
1.1.1 Design Variables *4*
1.1.2 Design Parameters *4*
1.1.3 Design Function *5*
1.1.4 Objective Function(s) *5*
1.1.5 Design Constraints *7*
1.1.5.1 Mathematical Constraints *8*
1.1.5.2 Inequality Constraints *8*
1.1.5.3 Side Constraints *9*
1.1.6 General Principles *10*
1.1.6.1 Feasible Space vs. Search Space *10*
1.1.6.2 Global Optimum vs. Local Optimum *11*
1.1.6.3 Types of Problem *12*
1.1.7 Standard Format *12*
1.1.8 Constraint-Handling Techniques *13*
1.1.8.1 Random Search Method *17*
1.1.8.2 Constant Penalty Function *17*
1.1.8.3 Binary Static Penalty Function *18*
1.1.8.4 Superiority of Feasible Points (SFPs) – Type I *18*
1.1.8.5 Superiority of Feasible Points (SFP) – Type II *18*
1.1.8.6 Eclectic Evolutionary Algorithm *18*
1.1.8.7 Typical Dynamic Penalty Function *19*
1.1.8.8 Exponential Dynamic Penalty Function *19*
1.1.8.9 Adaptive Multiplication Penalty Function *19*
1.1.8.10 Self-Adaptive Penalty Function (SAPF) *20*
1.1.9 Performance Criteria Used to Evaluate Algorithms *21*
1.1.10 Types of Optimization Techniques *23*
1.2 Classical Optimization Algorithms *23*

1.2.1 Linear Programming 25
1.2.1.1 Historical Time-Line 25
1.2.1.2 Mathematical Formulation of LP Problems 26
1.2.1.3 Linear Programming Solvers 26
1.2.2 Global-Local Optimization Strategy 28
1.2.2.1 Multi-Start Linear Programming 29
1.2.2.2 Hybridizing LP with Meta-Heuristic Optimization Algorithms as a Fine-Tuning Unit 31
1.3 Meta-Heuristic Algorithms 33
1.3.1 Biogeography-Based Optimization 34
1.3.1.1 Migration Stage 40
1.3.1.2 Mutation Stage 41
1.3.1.3 Clear Duplication Stage 43
1.3.1.4 Elitism Stage 44
1.3.1.5 The Overall BBO Algorithm 45
1.3.2 Differential Evolution 45
1.4 Hybrid Optimization Algorithms 46
1.4.1 BBO-LP 48
1.4.2 BBO/DE 51
Problems 51
Written Exercises 51
Computer Exercises 53

2 Fundamentals of Power System Protection 57
2.1 Faults Classification 57
2.2 Protection System 61
2.3 Zones of Protection 65
2.4 Primary and Backup Protection 66
2.5 Performance and Design Criteria 66
2.5.1 Reliability 66
2.5.1.1 Dependability 66
2.5.1.2 Security 66
2.5.2 Sensitivity 67
2.5.3 Speed 67
2.5.4 Selectivity 67
2.5.5 Performance versus Economics 67
2.5.6 Adequateness 67
2.5.7 Simplicity 67
2.6 Overcurrent Protective Devices 67
2.6.1 Fuses 68
2.6.2 Bimetallic Relays 69
2.6.3 Overcurrent Protective Relays 69
2.6.4 Instantaneous OCR (IOCR) 70
2.6.5 Definite Time OCR (DTOCR) 71
2.6.6 Inverse Time OCR (ITOCR) 72
2.6.7 Mixed Characteristic Curves 73
2.6.7.1 Definite-Time Plus Instantaneous 73

2.6.7.2 Inverse-Time Plus Instantaneous *74*
2.6.7.3 Inverse-Time Plus Definite-Time Plus Instantaneous *74*
2.6.7.4 Inverse-Time Plus Definite-Time *75*
2.6.7.5 Inverse Definite Minimum Time (IDMT) *76*
Problems *76*
Written Exercises *76*
Computer Exercises *77*

3 Mathematical Modeling of Inverse-Time Overcurrent Relay Characteristics *79*
3.1 Computer Representation of Inverse-Time Overcurrent Relay Characteristics *79*
3.1.1 Direct Data Storage *79*
3.1.2 Curve Fitting Formulas *82*
3.1.2.1 Polynomial Equations *82*
3.1.2.2 Exponential Equations *89*
3.1.2.3 Artificial Intelligence *93*
3.1.3 Special Models *94*
3.1.3.1 RI-Type Characteristic *94*
3.1.3.2 RD-Type Characteristic *95*
3.1.3.3 FR Short Time Inverse *95*
3.1.3.4 UK Rectifier Protection *95*
3.1.3.5 BNP-Type Characteristic *95*
3.1.3.6 Standard CO Series Characteristics *95*
3.1.3.7 IAC and ANSI Special Equations *96*
3.1.4 User-Defined Curves *98*
3.2 Dealing with All the Standard Characteristic Curves Together *99*
3.2.1 Differentiating Between Time Dial Setting and Time Multiplier Setting *99*
3.2.2 Dealing with Time Dial Setting and Time Multiplier Setting as One Variable *104*
3.2.2.1 Fixed Divisor *106*
3.2.2.2 Linear Interpolation *108*
3.2.3 General Guidelines Before Conducting Researches and Studies *111*
Problems *113*
Written Exercises *113*
Computer Exercises *114*

4 Upper Limit of Relay Operating Time *117*
4.1 Do We Need to Define T^{max}? *117*
4.2 How to Define T^{max}? *118*
4.2.1 Thermal Equations *118*
4.2.1.1 Thermal Overload Protection for 3ϕ Overhead Lines and Cables *118*
4.2.1.2 Thermal Overload Protection for Motors *122*
4.2.1.3 Thermal Overload Protection for Transformers *124*
4.2.2 Stability Analysis *126*
Problems *136*
Written Exercises *136*
Computer Exercises *138*

5 Directional Overcurrent Relays and the Importance of Relay Coordination *139*
5.1 Relay Grading in Radial Systems *139*
5.1.1 Time Grading *140*
5.1.2 Current Grading *140*
5.1.3 Inverse-Time Grading *143*
5.2 Directional Overcurrent Relays *146*
5.3 Coordination of DOCRs *148*
5.4 Is the Coordination of DOCRs an Iterative Problem? *148*
5.5 Minimum Break-Point Set *161*
5.6 Summary *163*
Problems *164*
Written Exercises *164*
Computer Exercises *166*

6 General Mechanism to Optimally Coordinate Directional Overcurrent Relays *169*
6.1 Constructing Power Network *169*
6.2 Power Flow Analysis *170*
6.2.1 Per-Unit System and Three-to-One-Phase Conversion *172*
6.2.2 Power Flow Solvers *173*
6.2.3 How to Apply the Newton–Raphson Method *175*
6.2.4 Sparsity Effect *179*
6.3 P/B Pairs Identification *186*
6.3.1 Inspection Method *186*
6.3.2 Graph Theory Methods *186*
6.3.3 Special Software *188*
6.4 Short-Circuit Analysis *189*
6.4.1 Short-Circuit Calculations *189*
6.4.2 Electric Power Engineering Software Tools *190*
6.4.2.1 Minimum Short-Circuit Current *190*
6.4.2.2 Maximum Short-Circuit Current *192*
6.4.3 Most Popular Standards *193*
6.4.3.1 ANSI/IEEE Standards C37 & UL 489 *193*
6.4.3.2 IEC 61363 Standard *194*
6.4.3.3 IEC 60909 Standard *194*
6.5 Applying Optimization Techniques *201*
Problems *202*
Written Exercises *202*
Computer Exercises *205*

7 Optimal Coordination of Inverse-Time DOCRs with Unified TCCC *207*
7.1 Mathematical Problem Formulation *207*
7.1.1 Objective Function *208*
7.1.1.1 Other Possible Objective Functions *210*
7.1.2 Inequality Constraints on Relay Operating Times *211*
7.1.3 Side Constraints on Relay Time Multiplier Settings *211*

7.1.4 Side Constraints on Relay Plug Settings *211*
7.1.5 Selectivity Constraint Among Primary and Backup Relay Pairs *212*
7.1.5.1 Transient Selectivity Constraint *213*
7.1.6 Standard Optimization Model *216*
7.2 Optimal Coordination of DOCRs Using Meta-Heuristic Optimization Algorithms *217*
7.2.1 Algorithm Implementation *217*
7.2.2 Constraint-Handling Techniques *218*
7.2.3 Solving the Infeasibility Condition *222*
7.3 Results Tester *228*
Problems *229*
Written Exercises *229*
Computer Exercises *231*

8 Incorporating LP and Hybridizing It with Meta-heuristic Algorithms *235*
8.1 Model Linearization *235*
8.1.1 Classical Linearization Approach *236*
8.1.1.1 IEC Curves: Fixing Plug Settings and Varying Time Multiplier Settings *236*
8.1.1.2 IEEE Curves: Fixing Current Tap Settings and Varying Time Dial Settings *237*
8.1.2 Transformation-Based Linearization Approach *237*
8.1.2.1 IEC Curves: Fixing Time Multiplier Settings and Varying Plug Settings *238*
8.1.2.2 IEEE Curves: Fixing Time Dial Settings and Varying Current Tap Settings *238*
8.2 Multi-start Linear Programming *242*
8.3 Hybridizing Linear Programming with Population-Based Meta-heuristic Optimization Algorithms *245*
8.3.1 Classical Linearization Approach: Fixing PS/CTS and Varying TMS/TDS *245*
8.3.2 Transformation-Based Linearization Approach: Fixing TMS/TDS and Varying PS/CTS *245*
8.3.3 Innovative Linearization Approach: Fixing/Varying TMS/TDS and PS/CTS *250*
Problems *250*
Written Exercises *250*
Computer Exercises *251*

9 Optimal Coordination of DOCRs With OCRs and Fuses *253*
9.1 Simple Networks *253*
9.1.1 Protecting Radial Networks by Just OCRs *253*
9.1.2 Protecting Double-Line Networks by OCRs and DOCRs *255*
9.2 Little Harder Networks *257*
9.2.1 Combination of OCRs and DOCRs *258*
9.2.2 Combination of Fuses, OCRs, and DOCRs *261*
9.3 Complex Networks *264*
Problems *265*
Written Exercises *265*
Computer Exercises *266*

10 Optimal Coordination with Considering Multiple Characteristic Curves *271*
10.1 Introduction *271*
10.2 Optimal Coordination of DOCRs with Multiple TCCCs *273*

10.3 Optimal Coordination of OCRs/DOCRs with Multiple TCCCs *278*
10.4 Inherent Weaknesses of the Multi-TCCCs Approach *279*
Problems *280*
Written Exercises *280*
Computer Exercises *281*

11 Optimal Coordination with Considering the Best TCCC *283*
11.1 Introduction *283*
11.2 Possible Structures of the Optimizer *284*
11.3 Technical Issue *287*
Problems *290*
Written Exercises *290*
Computer Exercises *291*

12 Considering the Actual Settings of Different Relay Technologies in the Same Network *293*
12.1 Introduction *293*
12.2 Mathematical Formulation *294*
12.2.1 Objective Function *294*
12.2.2 Selectivity Constraint Among Primary and Backup Relay Pairs *295*
12.2.3 Inequality Constraints on Relay Operating Times *296*
12.2.4 Side Constraints on Relay Time Multiplier Settings *296*
12.2.5 Side Constraints on Relay Plug Settings *296*
12.3 Biogeography-Based Optimization Algorithm *297*
12.3.1 Clear Duplication Stage *297*
12.3.2 Avoiding Facing Infeasible Selectivity Constraints *297*
12.3.2.1 Linear Programming Stage *297*
12.3.3 Linking $PS_i^{y_i}$ and $TMS_i^{y_i}$ with y_i *298*
12.4 Further Discussion *299*
Problems *300*
Written Exercises *300*
Computer Exercises *301*

13 Considering Double Primary Relay Strategy *303*
13.1 Introduction *303*
13.2 Mathematical Formulation *306*
13.2.1 Objective Function *307*
13.2.2 Selectivity Constraint *308*
13.2.3 Inequality Constraints on Relay Operating Times *308*
13.2.4 Side Constraints on Relay Time Multiplier Settings *308*
13.2.5 Side Constraints on Relay Plug Settings *309*
13.3 Possible Configurations of Double Primary ORC Problems *309*
Problems *315*
Written Exercises *315*
Computer Exercises *316*

14 Adaptive ORC Solver *319*
14.1 Introduction *319*
14.2 Types of Network Changes *320*
14.2.1 Operational Changes *321*
14.2.2 Topological Changes *321*
14.3 AI-Based Adaptive ORC Solver *322*
14.3.1 Generating Datasets *323*
14.3.2 Applying ANN to Solve ORC Problems *324*
Problems *328*
Written Exercises *328*
Computer Exercises *329*

15 Multi-objective Coordination *333*
15.1 Basic Principles *333*
15.1.1 Conventional Aggregation Method *334*
15.2 Multi-objective Formulation of ORC Problems *335*
15.2.1 Operating Time vs. System Reliability *336*
15.2.2 Operating Time vs. System Cost *336*
15.2.3 Operating Time vs. System Reliability vs. System Cost *342*
15.3 Further Discussions *342*
Problems *345*
Written Exercises *345*
Computer Exercises *345*

16 Optimal Coordination of Distance and Overcurrent Relays *347*
16.1 Introduction *347*
16.2 Basic Mathematical Modeling *348*
16.3 Mathematical Modeling with Considering Multiple TCCCs *350*
16.3.1 Inequality Constraints *351*
16.3.2 Objective Function *352*
16.4 Mathematical Modeling with Considering Different Fault Locations *353*
16.4.1 Objective Function *353*
16.4.2 Inequality Constraints *354*
16.4.2.1 Near-End Faults *354*
16.4.2.2 Middle-Point Faults *354*
16.4.2.3 Far-End Faults *355*

17 Trending Topics and Existing Issues *357*
17.1 New Inverse-Time Characteristics *357*
17.1.1 Scaled Standard TCCC Models *357*
17.1.2 Stepwise TCCCs *358*
17.1.3 New Customized TCCCs *359*
17.2 Smart Grid *359*
17.2.1 Distributed Generation *359*
17.2.2 Series Compensation and Flexible Alternating Current Transmission System *360*

17.2.3 Fault Current Limiters 360
17.3 Economic Operation 360
17.4 Power System Realization 361
17.4.1 Power Lines 361
17.4.2 Economic Operation 363
17.4.2.1 Combined-Cycle Power Plants 364
17.4.2.2 Degraded Efficiency Phenomenon 364
17.4.2.3 Unaccounted Losses in Power Stations 365
17.5 Locating Faults in Mesh Networks by DOCRs 367
17.5.1 Mechanism of the Proposed Fault Location Algorithm 370
17.5.1.1 Approach No. 1: Classical Linear Interpolation 373
17.5.1.2 Approach No. 2: Logarithmic/Nonlinear Interpolation 374
17.5.1.3 Approach No. 3: Polynomial Regression 375
17.5.1.4 Approach No. 4: Asymptotic Regression 375
17.5.1.5 Approach No. 5: DTCC-Based Regression 375
17.5.2 Final Structure of the Proposed Fault Locator 377
17.5.3 Overall Accuracy vs. Uncertainty 379
17.5.4 Further Discussion 380

Appendix A Some Important Data Used in Power System Protection 381
A.1 Standard Current Transformer Ratios 381
A.2 Standard Device/Function Number and Function Acronym Descriptions 382
A.2.1 Standard Device/Function Numbers 382
A.2.2 Device/Function Acronyms 383
A.2.3 Suffix Letters 383
A.2.3.1 Auxiliary Devices 383
A.2.3.2 Actuating Quantities 383
A.2.3.3 Main Device 384
A.2.3.4 Main Device Parts 384
A.2.3.5 Other Suffix Letters 384

Appendix B How to Install PowerWorld Simulator (Education Version) 387

Appendix C Single-Machine Infinite Bus 391

Appendix D Linearizing Relay Operating Time Models 393
D.1 Linearizing the IEC/BS Model of DOCRs by Fixing Time Multiplier Settings 393
D.2 Linearizing the ANSI/IEEE Model of DOCRs by Fixing Time Multiplier Settings 394

Appendix E Derivation of the First Order Thermal Differential Equation 397

Appendix F List of ORC Test Systems 399
F.1 Three-Bus Test Systems 399
F.1.1 System No. 1 399
F.1.2 System No. 2 399
F.2 Four-Bus Test Systems 403
F.2.1 System No. 1 403

F.2.2 System No. 2 *403*
F.3 Five-Bus Test System *408*
F.4 Six-Bus Test Systems *410*
F.4.1 System No. 1 *410*
F.4.2 System No. 2 *410*
F.4.3 System No. 3 *411*
F.4.4 System No. 4 *413*
F.5 Eight-Bus Test Systems *418*
F.5.1 System No. 1 *418*
F.5.2 System No. 2 *422*
F.5.3 System No. 3 *423*
F.5.4 System No. 4 *424*
F.5.5 System No. 5 *425*
F.6 Nine-Bus Test System *427*
F.7 14-Bus Test Systems *430*
F.7.1 System No. 1 *431*
F.7.2 System No. 2 *433*
F.8 15-Bus Test System *437*
F.9 30-Bus Test Systems *441*
F.9.1 System No. 1 *441*
F.9.2 System No. 2 *444*
F.10 42-Bus Test System *448*
F.11 118-Bus Test System *453*

References *457*
Index *479*

Author Biography

Ali R. Al-Roomi (ali.ridha@dal.ca) earned his B.Sc. degree in Process Instrumentation and Control Engineering and his M.Sc. degree (Hons.) in Electrical and Electronics Engineering from the University of Bahrain, Sakhir, Bahrain, in 2006 and 2014, respectively. He was awarded the best B.Sc. project, two times on Dean's List, two times on President's List, and two silver medals. He earned his Ph.D. degree (Hons. with a full GPA of 4.3/4.3) in Electrical and Computer Engineering from Dalhousie University, Halifax, Nova Scotia, Canada, in 2020. For four years, he worked as a research assistant with Professor Mohamed E. El-Hawary for many projects funded by the Natural Sciences and Engineering Research Council (NSERC) of Canada. In addition to NSERC research assistantship funds, he received two scholarships from the Faculty of Graduate Studies (FGS) and Bruce and Dorothy Rosetti Engineering Research Scholarship. In April 2018, he has been awarded the best Ph.D. presentation at the conference conducted annually by his engineering department. Ali has more than seven years of experience spent in different automation companies and power stations. In addition to many superior hybrid meta-heuristic optimization algorithms that he designed in the past ten years, Ali also invented two novel symbolic-based machine learning computing systems and named them "universal functions originator" and "universal linear regression." He has different soft skills; including programming languages, math and science software, engineering software, and office and multimedia software. Recently, Al-Roomi has successfully accomplished a big engineering/programming project, which is about an enterprise application software (EAS). The software is known as the Meta-Tuner, which is specialized in finding optimal parameters of PID controllers with many advanced tools and features. His research interests include machine learning systems, meta-heuristic optimization algorithms, hybrid computing systems, power operation, power protection, state estimation, smart grid, power system design and analysis, system realization, load forecasting, process instruments and sensors, control systems, and some interdisciplinary engineering fields. He is a member of IEEE and his personal website is www.al-roomi.org.

Preface

It has been more than three decades since the first attempt to optimally coordinate protective devices was suggested in the literature. That framework was so simple and applicable to coordinate just directional overcurrent relays of small mesh networks with lots of simplifications. During this long period, many other frameworks were presented, which are more advanced and capable to adaptively coordinate different protective devices for both dynamic configurations; i.e. operational and topological changes. Nowadays, the grids are smarter and equipped with state-of-the-art numerical relays that can communicate with each other and be adjusted remotely. Also, there are many significant advancements in the fields of optimization algorithms, machine learning algorithms, numerical programming languages, and computing machines. Further, this hot research area is always active because of the continuous research and development in many branches of modern electric power systems.

All the aforementioned barriers make understanding the topic of optimal coordination so hard. It is an interdisciplinary research area where general electrical engineers require to grasp load flow analysis, fault analysis, stability analysis, protection systems, communication and networking, smart grid and renewable energy, classical and modern optimization algorithms, classical and modern machine learning algorithms, and numerical analysis in programming languages. Add to these challenges and obstacles, there are many thousands of conference and journal papers in the literature, which make the topic highly dispersed and very complicated! This is why many new researchers in this field deeply suffer and don't know where they should start.

With more than fifteen years in optimization algorithms, ten years in optimal coordination, and seven years in different power stations and automation companies, I have developed many innovative optimizers that can solve this highly constrained nonlinear non-convex mixed-integer engineering problem. I still remember the time when I faced all the preceding challenges. I was forced to spend days and nights without enough sleeping and consumed most of my daily energy just to find the correct way to program my early very primitive optimizers. This hard way of understanding exposed me to many known and hidden aspects that need to be taken into account during designing optimal coordinators for getting feasible solutions. To avoid facing this again, I wrote this book where all my knowledge, skills, and experiences in this field are put. This is the first book of its kind. It can be used as a reference and a quick start to expedite reaching professionalism. The readers will find many helpful examples and ready-made MATLAB codes and programs. The book also helps the instructors where two different kinds of supporting materials are provided for the exercises presented at the end of each chapter. The first one is a solution manual for all the written exercises, while the other is a compressed folder that contains the programming codes and electronic files of the computer exercises. Both the student and instructor materials can be downloaded from the companion website hosted by Wiley's server.

Acknowledgments

I have to start by thanking God who created the universe and continually maintains its existence. Thank God for giving meaning to our lives through Your love.

I would like to thank my awesome wife, Alya, for her continuous support, encouragement, understanding, patient, and constant love. Thank you so much, dear.

My special thanks to the commissioning editor, Mary Hatcher, for studying and approving the idea, researching the potential market, searching for editors, preparing the product proposal, publishing the contract, taking care of sales and marketing activities, and providing me with some complimentary copies. I would also like to thank the senior editorial assistant, Victoria Bradshaw, for taking her time in getting the legal documents done and following through on the book cover design. Additionally, my special thanks to the managing editor, Teresa Netzler, for regular check-ins, answering questions, submission of materials, manuscript assessment, permissions assessment, copyediting, typesetting, proof of corrections, indexing, printing, binding, and distribution. Last but not least, I would like to thank the content refinement specialist, Devi Ignasi, and the rest of Wiley's production team for overseeing all the production processes from an unedited manuscript to final files for print and electronic publication. I could not have done this without your help and passion for networking.

Acronyms

ACO	ant colony optimization
AI	artificial intelligence
ANN	artificial neural network
ANSI	American National Standards Institute
ASI	actuator sensor interface
ASOTPBR	average sum of the operating times of primary and backup relays
BBO	biogeography-based optimization
BCO	bee colony optimization
BFR	breaker failure relay
BNPF	branch-to-node matrix-based power flow
BPS	break-point set
BS	British Standards
CA	contingency analysis
CAN	controller area network
CB	circuit breaker
CCPP	combined-cycle power plant
CDT	coordination delay time
COx	oxides of carbon
CPU	central processing unit
CSM	current setting multiplier
CT	current transformer
CTCC	current–time characteristic curve
CTI	coordination time interval
CTR	CT-ratio
CTS	current tap setting
CWA	conventional weight aggregation
DCOCR	definite current over-current relay
DCS	distributed control system
DE	differential evolution
DEE	differential equation editor
DER	distributed energy resources
DFR	digital fault recorder
DG	distributed generation
DH+	data highway plus
DOCR	directional over-current relay

DM discrimination margin
DNP distributed network protocol
DSP digital signal processor
DTCC distance–time characteristic curve
DTOCR definite time over-current relay
DWA dynamic weight aggregation
EA evolutionary algorithm
EAS enterprise application software
ELD economic load dispatch
EMS energy management system
EOH equivalent operating hours
EPF exterior penalty function
FACTS flexible alternating current transmission system
FC feasibility checker
FCC fault current controller
FCL fault current limiter
FDLF fast-decoupled load flow
FL fuzzy logic
FME Fourier–Motzkin elimination
FPZ fault probability zone
FSA fast simulated annealing
GA genetic algorithm
GD gradient descent
GIC geomagnetically induced current
GM grading margin
GN Gauss–Newton
GPS global positioning system
GS Gauss–Seidel
GSA gravitational search algorithm
GT gas turbine
HART highway addressable remote transducer
HEM holomorphic embedding method
HMI human–machine interface
HRSG heat recovery steam generator
HSI habitat suitability index
IBT inter-bus transformer
IDMT inverse definite minimum time
IEC International Electrotechnical Commission
IED intelligent electronic device
IEEE Institute of Electrical and Electronics Engineers
ILP integer linear programming
INLP integer non-linear programming
IOCR instantaneous over-current relay
IPF interior penalty function
ISI island suitability index
ITOCR inverse time over-current relay
JFNK Jacobian-free Newton–Krylov
JM Jacobi method

Jr	jumping rate
KCL	Kirchhoff's current law
KKT	Karush–Kuhn–Tucker
LCR	local control room
LES	least error squares
LF	load flow
LM	Levenberg–Marquardt
LP	linear programming
LPP	linear programming problem
LR	linear regression
MCR	main control room
MINLP	mixed-integer non-linear programming
MLR	multiple linear regression
MMS	manufacturing message specification
MOEA	multi-objective evolutionary algorithm
MOP	multi-objective problem
MpBBO	metropolis biogeography-based optimization
NC	normally-closed
NFE	number of function evaluations
NFT	near feasibility threshold
NLP	non-linear programming
NLR	non-linear regression
NO	normally-open
NOx	oxides of nitrogen
NR	Newton–Raphson
NRLF	Newton–Raphson load flow
OBL	Opposition-Based Learning
OCPD	over-current protective device
OCR	over-current relay
ODE	ordinary differential equation
OEM	original equipment manufacturer
OMIB	one-machine infinite bus
OPC	open platform communications
OPF	optimal power flow
OR	operations research
ORC	optimal relay coordination
PB	population-based
P/B	primary/backup
PCS	pickup current setting
PF	power flow
PLC	programmable logic controller
PM	periodic maintenance
PMB	partial migration-based
PS	plug setting
PSA	possible solution area
PSM	plug setting multiplier
PSO	particle swarm optimization
PT	potential transformer

pu	per-unit
RAM	random-access memory
R&D	Research and Development
RO	reverse osmosis
ROM	read-only memory
RSA	random search algorithm
RTD	resistance-temperature detector
RTU	remote terminal unit
RWA	random weight aggregation
SA	simulated annealing
SAPF	self-adaptive penalty function
SC	super-capacitor
SC	series compensator
SCADA	supervisory control and data acquisition
SE	state estimation
S–E	complex power–complex voltage
SFP	superiority of feasible points
SIV	suitability index variable
SMB	single migration-based
SMIB	single-machine infinite bus
SOx	oxides of sulfur
SPMB	simplified partial migration-based
SR	symbolic regression
SSMB	simplified single migration-based
ST	steam turbine
STI	selective time interval
SVD	singular value decomposition
SVM	support vector machine
SVR	support vector regression
TCCC	time–current characteristic curve
TDCC	time–distance characteristic curve
TDR	time domain reflectometer
TDS	time dial setting
TFB	temperature/frequency-based
TLS	time lever setting
TMS	time multiplier setting
TS	tabu search
TSM	time setting multiplier
UCA2	Utility Communication Architecture - version 2
UFO	universal functions originator
UHCs	unburned hydro-carbons
UPFC	unified power flow controller
UPS	uninterruptible power source
VT	voltage transformer
WSCC	Western System Coordinating Council
WSM	weighted sum method

About The Companion Website

This book is accompanied by a companion website:

www.wiley.com/go/al-roomi/optimalcoordination

The website has Solution Manual & Exercises and Computer Codes for Exercises (MATLAB).

Introduction

To protect electric power system equipment and elements, different protective devices could be used. These devices could be categorized based on their technology (electromechanical, solid-state, digital, and numerical), signal (electrical and non-electrical), measurement (voltage, current, impedance, frequency, temperature, pressure, and force), speed, mechanism, and working principle.

Overcurrent protection is one of the most popular techniques that is used for a long time. Overcurrent protective devices (OCPDs) could be fuses, bi-metallic contacts, relays, or even resistance-temperature detectors (RTDs) and thermistors. The protection can cover small house appliances to large power equipment, such as generators, step-up transformers, switchgear, transmission lines, sub-transmission lines, inter-bus transformers (IBTs), distribution systems, motors, etc.

As compared with the other OCPDs, overcurrent relays (OCRs) are preferred in many applications because of their available features, settings, and relations between their operating times and fault currents. This special type of OCPDs appeared at the beginning of the last century when the electromechanical relays were presented. Solid-state, digital (hardware-based), and numerical relays all make this research area of electric power systems engineering very active with many advanced techniques appearing from time to time where different disciplines are involved; such as optimization algorithms, signal processing, hardware development, communication, automation, etc.

To be more specific, directional overcurrent relays (DOCRs) are the main part of this book.[1] These special types of OCRs (and sub-types of OCPDs) are applied in multi-loop systems, ring feeder systems, double-end fed power systems, or even single-end fed power systems of parallel feeders. Because DOCRs are cheaper than other protective relays and because DOCRs can compromise between different design criteria, they are commonly used as primary relays for interconnected sub-transmission and distribution systems. Also, for transmission systems, DOCRs are used as local backup relays where distance relays are the primary protective devices (Urdaneta et al., 1997; Gers and Holmes, 2004; Paithankar and Bhide, 2003; Rao, 2008).

The coordination of DOCRs is a very important stage for any protection design. Correct relay coordination involves selecting the suitable relay setting that assures faults in the protected zone are cleared first by the corresponding primary relays and if they fail, the corresponding backup relays act after a coordination time delay. This delay is known as the coordination time interval[2]

1 Of course, as stated in the book's title, the other protective devices will also be covered. Knowing how DOCRs being coordinated will allow the reader to capture this principle for any other protective device.

2 In some references, it is also known as the selective time interval (*STI*) (Paithankar and Bhide, 2003) and the discrimination margin (*DM*) (Gers and Holmes, 2004).

(*CTI*) (Albasri et al., 2015). The relays have two settings, namely, *plug setting* (*PS*) and *time multiplier setting* (*TMS*). The *operating time* (*T*) of each relay is a function of these two settings.

The coordination problem can be briefly defined as "**the quality of selectivity among protective devices** (Anthony, 1995)." It is considered as a highly constrained mixed-integer non-linear programming (MINLP) problem where *TMS* is continuous and *PS* is discrete. The feasible solution should satisfy the design constraints associated with the problem. Such constraints are the side constraints of *TMS* and *PS*, the variable bounds of *T*, and the selectivity between the primary and backup relays. It requires an expert protection engineer to solve it analytically where all the fault possibilities, system contingencies, and abnormalities are analyzed and predetermined. Alternatively, it can be easily solved by employing optimization algorithms (Noghabi et al., 2009).

The objective of *optimal coordination* is to find the best relay settings, i.e. *PS* and *TMS*, so that the sum of their operating times when they act as primary relays is minimized and all the design constraints are satisfied (Urdaneta et al., 1988). That is, optimal relay coordination (ORC) of OCRs/-DOCRs can be done by achieving two goals:

1. Correctly coordinating primary/backup (P/B) relay pairs, and
2. Effectively minimizing the sum of operating times of all relays when they act as primary protective devices.

Therefore, ORC is the core of this book. From the 1960s until now, this subject becomes very popular where many advanced and innovative techniques are presented with a further realization of its mathematical modeling. This is a normal consequence of the advanced technologies that have been employed in different areas of electric power systems engineering.[3]

The adaptive optimal coordination and optimal coordination of mixed types of protective devices (including non-directional OCRs, fuses, distance relays, etc.) will be studied in this book as well.

The remaining parts of this introduction will give a brief literature review and then followed by the main outlines of this book.

Literature Review

Solving the coordination problem of protective devices is an old tedious task. In the past, based on the availability of techniques and technologies at that time, getting optimal coordination was impossible. Instead, acceptable non-optimal coordination was implemented to meet the protection requirements of the simple electrical networks that were existing at that time.

There are different approaches to solve the coordination problem. They can be classified into two main categories (Hui, 2006):

1. Topological methods; which include graph theory and functional dependencies, and
2. Optimization methods; which include adaptive techniques.

In the 1960s, computers were used to overcome the laborious calculation (Knable, 1961; Radke, 1963; Albrecht et al., 1964; Tsien, 1964). It was assumed that the relative sequence of computation of relay settings is predetermined. A trial and error approach is considered in Albrecht et al. (1964). Knable (1969) is the first one who considered both parts of the coordination problem; namely, the relay settings computation and the relative sequence of that computation by using a heuristic scheme. Dwarakanath and Nowitz (1980) used the concept of linear graph theory to determine the

3 This includes the generation, transmission, and distribution parts.

relative sequence in a multi-loop network based on a minimum break-point set (MBPS). Jenkins et al. (1992) used the concept of functional dependency as a flexible and powerful alternative to identify BPS.

In 1988, the Gauss-Seidel (GS) iterative technique was successfully used (Urdaneta et al., 1988). However, because the coordination problem is nonlinear, predetermined values of *PS* are considered to convert it to a linear problem. Although this approach reduces the CPU time[4] and makes the problem easier to solve, it needs an expert protection engineer for setting the initial values of *PS*. In addition, there is no guarantee for converging to the global optima.

The linear programming (LP) techniques[5] are commonly used if the coordination problem is constructed in a linear form (Moravej et al., 2012). However, when both *PS* and *TMS* are considered as variables, the problem becomes nonlinear, and thus nonlinear programming (NLP) is used. Also, to increase the chance for getting feasible solutions, the values of *PS* are taken as continuous and then rounded-off at the end of the program[6] for getting a discretized *PS*. As a result, the final solution may drift away from its previous optimum state. LP is a fast and simple algorithm, but it is highly vulnerable to trap into local minima. NLP gives better performance, but it also could converge to local minima, especially with wrong selected initial values of *PS* and *TMS*. Also, NLP is harder than LP and it consumes more CPU time. In addition to the previously mentioned traditional algorithms, evolutionary algorithms (EAs)[7] have been successfully applied to solve the ORC problem (Ezzeddine and Kaczmarek, 2011b). Most EAs are multi-point search techniques[8,9] that can be used to ensure covering the search space and thus reaching the area where the global optimum is located in. However, these optimization algorithms are so slow and thus they are unsuitable for adaptive ORC problems where the processing speed is the winning factor. Based on that, different designs have been considered to hybridize EAs with LP/NLP so optimal or near-optimal settings can be attained within a very short time (Noghabi et al., 2009; Bedekar and Bhide, 2011a; Albasri et al., 2015; Al-Roomi and El-Hawary, 2019a). However, the programming structures of these optimization algorithms are more complex and thus only experienced engineers or programmers can implement them to solve real-world problems.

To effectively minimize the operating time of protective relays, some studies propose to use multiple time–current characteristic curves (TCCCs) for OCRs. However, this approach significantly increases the dimension of the optimization problem and could violate the selectivity constraint for some fault points. The other more practical approach is to stick with only one TCCC for all the protective relays. The idea here is to optimize both the relay settings and TCCC. Also, some studies take into account the difference in relay settings and their internal delays due to the relay versions and the technology embedded inside them. More realized studies take into account the phenomenon where both OCRs and fuses are used. The others cover the combination of overcurrent and distance relays. Also, the phenomenon of using old OCRs as local backup protective devices is covered in some recent studies.

---

4 Due to using few iteration steps and less complicated algorithms.
5 Such as simplex, two-phase simplex, and dual simplex methods.
6 The modern approach is to use MINLP directly instead of rounding-off *PS*. Also, if the problem is linearized, then integer linear programming (ILP) can be used to optimize discrete *TMS*. Further, if the problem is not linearized and both settings are discrete, then integer nonlinear programming (INLP) can be used. These ILP and INLP are applicable for electromechanical relays.
7 Such as ant colony optimization (ACO), bee colony optimization (BCO), genetic algorithm (GA), differential evolution (DE), and particle swarm optimization (PSO).
8 In other words, they are called population-based (PB) algorithms.
9 Some examples of single-point searchers are simulated annealing (SA) and tabu search (TA).

Because the dimension of practical coordination problems is generally high, so various artificial intelligence (AI) tools have been implemented to hit multiple birds with one stone. First, the optimal settings can be quickly obtained. Second, the dynamic changes in the network can be easily accounted for, which means that the AI-based approaches are very good for adaptive optimal coordination problems. Third, the uncertainty due to many variations of surrounding weather, system frequency, and load conditions can be effectively minimized, especially with the big data provided by real-time measurement devices mounted in modern smart grids.

Book Organization

The book is organized as follows: Chapter 1 paves the way to fully understand the optimization algorithms which are the heart of any ORC solver. Chapter 2 is an essential step to go forward in understanding the ORC subject. It helps the readers to easily follow the contents of this book and accelerating their ability to grasp the materials. Protection engineers can skip this chapter if they know all these fundamental materials. Chapter 3 gives a comprehensive overview of the mathematical models used to emulate inverse time OCRs in computers. The content of this chapter is important as it explains all the vague points behind the different models used in North America and Europe. Chapter 4 is responsible to answer the question of why there is an upper limit on relay operating times. It explains it in terms of thermal limit and system stability. Chapter 5 gives a brief idea about relay coordination and why it is important in power system protection. Further, this chapter explains the main drawback of conventional OCRs and how to overcome this technical problem by involving DOCRs. It also explains the current and time grading and shows how to optimally set inverse-time DOCRs using the GS method. Moreover, it covers the concept of the MBPS. Chapter 6 shows the general mechanism required to design an applicable optimal relay coordinator. Although the mechanism described in this chapter is created based on just DOCRs, it builds the foundation required to understand the main concept of ORC problems and how to model them mathematically. Chapter 7 can be considered as the first step to enter the ORC subject. The general mathematical expressions and optimization models are presented in this chapter with detailed descriptions of many concepts and coordination criteria. In this chapter, the IEC standard inverse TCCC is unified for all DOCRs before a meta-heuristic optimization algorithm is involved to solve different ORC problems. Also, this chapter explains how to validate the optimal results reported in the literature. Therefore, this chapter explains the core of the subject and thus more weight has to be given to it because the next chapters depend on the contents of this chapter. Chapter 8 solves the processing speed issue of conventional ORC solvers where population-based meta-heuristic optimization algorithms are implemented. In this chapter, the reader will know how to solve different problems by using conventional and multi-start LP techniques. Such hybrid algorithms can converge to very good results within a very limited number of iterations and a small population size. Thus, a significant amount of CPU time can be saved. The classical way is to linearize the objective function and design constraints by fixing *PS* and varying *TMS*. The other way, which is so tricky and requires to employ the transformation technique from regression analysis, is to fix *TMS* and vary *PS*. The most advanced scheme is to add two independent LP stages; one to fine-tune *PS* and the other for *TMS*. All these steps are covered. The linearized models presented in this chapter are modified to include North American relays. In Chapter 9, the ORC problem is solved by considering DOCRs with OCRs and fuses. This chapter shows the complexity of practical protection designs. Chapter 10 takes the reader to a more advanced level where multiple TCCCs are considered at the same time during optimizing the settings of DOCRs. The practical problems and weaknesses of this coordination approach are addressed and then solved in Chapter 11 by selecting the best TCCC for all DOCRs. Chapter 12 considers the actual settings and their limits when different relay technologies are imposed in the same network. It studies the effect of relay technology in solving ORC problems. For example, when electromechanical, static, digital, and numerical relays are installed in one network. The chapter incorporates both directional and non-directional inverse-time overcurrent relays (ITOCRs). There is a fact that some old relays are kept as local backup protective devices when the network is upgraded with state-of-the-art numerical relays. Chapter 13 deals with the double primary relay strategy (DPRS) where both main-1 and main-2 (or local backup) relays are involved. In this chapter, a new mathematical model is derived for this special ORC problem and then solved for different test systems. It covers both the case when just some primary relays are equipped with DPRS and the extreme case where all the primary relays are equipped with DPRS. The study is conducted based on different assumptions. For example, only electromechanical, only static, or a mixture of them are used as local backup relays. Also, the mathematical model presented in this chapter is solved when both directional and non-directional ITOCRs are involved in the ORC problem. In real life, no electrical network can preserve a steady-state operation forever. The changes could happen in the components' status (shutting down generating machines, taking lines out of service, opening busbars, disconnecting loads, etc.) or settings (increasing/decreasing the output of generators, contributions of storage elements, variations in the demand side, etc.). These two dynamic configurations are, respectively, called the topological and operational changes. An introduction to the adaptive optimal coordination strategy is covered in Chapter 14. One way to effectively solve this adaptive ORC problem is to hybridize meta-heuristic optimization algorithms with LP. The other way is to use machine learning (ML) tools such as artificial neural networks (ANNs) and support vector machines (SVMs). This chapter covers all these tools plus linear regression (LR) and nonlinear regression (NLR). One of the most recent researches is to use multi-objective problems (MOPs), which is highlighted in Chapter 15. It will cover the basic principles of multi-objective optimization, translating the ORC problems into MOPs, and then solving them by modifying the conventional meta-heuristic optimization algorithms designed in the earlier chapters. For transmission and sub-transmission systems, protection engineers could see both distance and OCRs. Chapter 16 gives the basic idea behind this special coordination problem and how to model it in several ways. Finally, Chapter 17 is dedicated to highlighting the trending optimal coordination topics in the literature. It can be used as a starter to know your position and where would you like to reach.

Optimization Tools

Everyone knows that there are many types of traditional and non-traditional optimization techniques. The professionalism on ORC is not affected by knowing this or that type of optimization techniques. From their names, they are just tools that help us to solve highly constrained coordination problems.

The main optimizer used in this book is the biogeography-based optimization (BBO) algorithm. There are three reasons behind this selection. Firstly, BBO still is a new meta-heuristic population-based EA and it has lots of opportunities to conduct many studies. Secondly, BBO proved itself as a very competitive optimization algorithm (Simon, 2008a). Thirdly, the author has good skills and experience in this particular optimization algorithm. The book also covers DE and LP.

If someone is an expert in GA or PSO, then he/she can easily follow the book contents without facing any problem. Furthermore, the modifications and hybridizations applied to BBO, in this book, can also be applied to other algorithms.

Examples

We try to maximize the understanding level by showing different examples. In addition to illustrative examples, we provide MATLAB codes as weapons to accomplish this task. The book contains more than 180 files of .m and .slx types. Also, we provide some examples done in some professional software used in electric power systems engineering.

The illustrative examples will handle the hard to understand mathematical expressions, while the programming examples will lift the readers to a professional level and will let them conduct their own researches, independently.

All these codes and soft files can be downloaded through the link provided by Wiley or the author's website [url: `www.al-roomi.org`].

Exercises

At the end of each chapter,[10] we give a variety of illustrative and computer exercises. Solving them will expand the understanding and grasping the whole ideas behind the topics. The book has a companion website where the instructors can download the solution manual for the illustrative exercises and the codes for the computer exercises. There are more than 550 files ready to download. They are organized according to their chapters as well as the exercise number. This solution manual and the soft coding files can save a significant amount of time that could be spent by instructors to solve them. Also, they can be used to modify or write new illustrative and computer exercises.

10 Except for the last two chapters, which are given just to introduce some advanced topics.

1

Fundamental Steps in Optimization Algorithms

Aim

One of the main barriers that make ORC very difficult to understand is its dependency on one or multiple optimization stages. This chapter gives a quick overview. It explains the meaning of optimization algorithms, the main types and subtypes, the design constraints, the differences between classical and meta-heuristic algorithms, etc. Although this chapter belongs to applied mathematics and data science, it is very essential to be added here before starting our journey into ORC. Failing in understanding the core of this chapter will have direct impacts and multiple consequences on understanding the rest of the book.

1.1 Overview

The term **mathematical optimization**, or just **optimization**, is frequently heard in mathematics, computer science, engineering, and even in economic and management science. Also, it can be found in proceedings, journals, books, encyclopedias, websites, etc., under different sections and names, like **soft computing**, **applied mathematics and optimization**, **evolutionary computation**, **numerical analysis**, etc.

From the basic of mathematics, suppose that there is a function (f) and it changes as the **independent variable** (x) changes, then f becomes the **dependent variable** of x and known shortly as $f(x)$.

Based on the system requirements, or in another word the **objective function**, the best solution to such a problem is called the **optimum** (or **optimal**) **solution**. This solution is located at a specified value of the **design variable** "x." The **optima** could be either **maxima** or **minima**, and the tool used to find this point is called an **optimization algorithm**.

Example 1.1 Consider the following arbitrary function:

$$f(x) = 30 - x + \sin(x) + \frac{5}{20x} + e^{0.3x} \quad ; \quad 0 \leqslant x \leqslant 10$$

Design a MATLAB code to graphically represent this function when x moves from 0 to 10 with steps of $\Delta x = 0.01$. Then, use the command `fsolve` to find the global minimum and its solution. The best point on the plot can be used as the initial point for the minimization process.

Optimal Coordination of Power Protective Devices with Illustrative Examples, First Edition. Ali R. Al-Roomi.

Companion website: www.wiley.com/go/al-roomi/optimalcoordination

Solution

The following is the code shown in `Minimization_1dimProb.m`:

```
1	clc
2	clear
3	format long
4	syms x
5	fx = '30 - x + sin(x) + 5./(20*x) + exp(0.3*x)';
6	% Finding good initial point
7	x = [0:0.01:10];
8	f = eval(fx);
9	plot(x, f, 'b', 'LineWidth', 2)
10	grid on; grid minor
11	xlabel('x'); ylabel('f(x)');
12	set(gca, 'FontSize', 13)
13	[f0, ind] = min(f);
14	x0 = x(ind)
15	% Finding optimal point
16	xopt = fsolve(fx, x0)
17	x = xopt; fmin = eval(fx)
```

If we run this simple program, the following result will be found in the command window:

```
x0 =
    4.550000000000000
xopt =
    4.549282360076904
fmin =
  28.433823902224816
```

Figure 1.1 graphically represents the optimized solution of the arbitrary minimization problem given in the question, which can also be accessed by opening `Minimization_1dimProb.fig`.

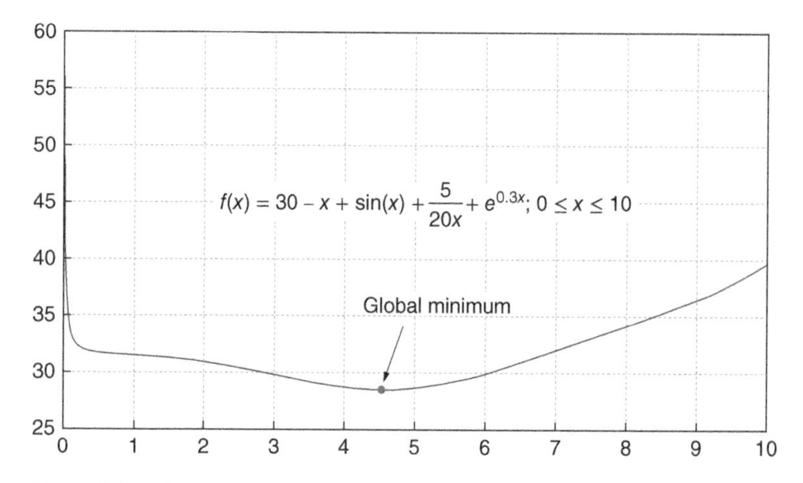

Figure 1.1 Minimization problem with the optimum point.

When dealing with optimization, there are so many types of algorithms. In general, they can be classified into three main categories:

- Classical "Traditional" Optimization Algorithms
- Modern "Non-Traditional" Optimization Algorithms
- Hybrid Optimization Algorithms

The last category contains algorithms that are constructed by combining or merging multi-algorithms into one final algorithm. The algorithms of these combinations could be taken from one category as well as from different categories. The main reason behind going to the hybrid approach is to accumulate the strengths of different techniques and at the same time to prevent, or at least to reduce, the associated weaknesses of each technique.[1].

The general optimal design formulation of any problem can be depicted by the flowchart shown in Figure 1.2 (Deb, 2010), which is explained in Sections 1.1.1, 1.1.2, 1.1.3, 1.1.4, and 1.1.5:

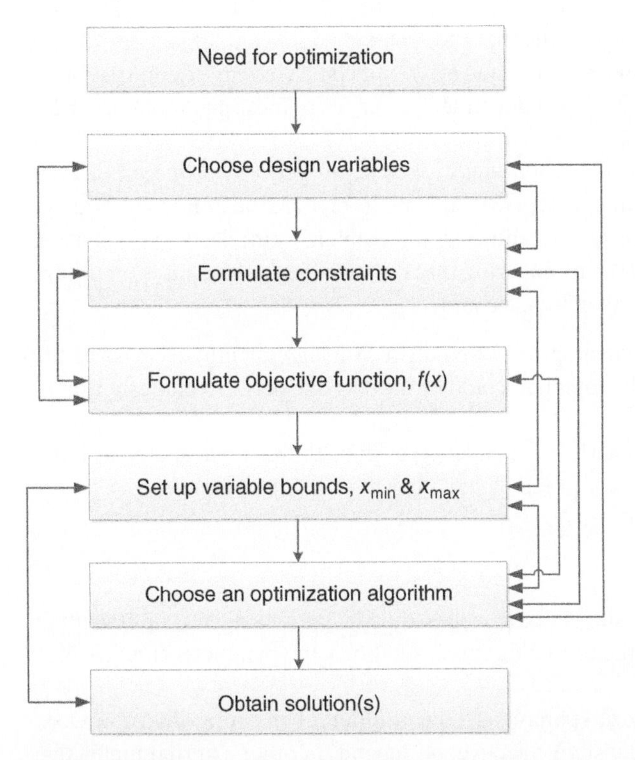

Figure 1.2 Flowchart of the optimal design procedure. Source: Deb (2010).

1 Also, if a system problem needs an algorithm having some abilities of self-learning, logical thinking and decision making, then the techniques built based on **artificial intelligence** (AI) are the correct choice here. For example, finding the optimum solution of a numerical problem is one task that can be assigned to **artificial neural networks (ANNs)** Thus, by enough training, ANNs can solve that problem smartly without expressing any mathematical model (i.e. **modeling-free techniques**). Thus, ANNs act as black-boxes. Actually, this is a double-edged sword. It can be seen as a source of strength because ANNs can estimate the solution directly without trying to express or optimize any weary mathematical models. However, this approach makes the whole process secret and nobody knows what is going on inside these mysterious networks (Rudin, 2019).

1.1.1 Design Variables

These variables are also known as **decision variables** or **solution features**, which are the independent variables of the optimization problem. If f depends on two or more variables, then it can be mathematically expressed as $f(x_1, x_2, \ldots, x_n)$. The subscript n denotes the total number of independent variables, and it is called the **dimension** of the problem. In optimization, if the given problem contains n decision variables, then it can be expressed in vector notation as $f(X)$ where $X = [x_1, x_2, \ldots, x_i, \ldots, x_n]$ and $i = 1, 2, \ldots, n$. $[X]$ is called the set of design variables or simply the **design vector**, and x_i is the ith element of $[X]$ (Venkataraman, 2009; Simon, 2013; Deb, 2010). For more understanding, let's take the following simple functions:

$$f(x) = ax^2 + bx + c = 0 \tag{1.1}$$

$$f(x_1, x_2, x_3) = (a - x_1)^2 + (b - x_2)^2 + (c - x_3)^2 \geqslant 4.5 \tag{1.2}$$

$$f(t) = \frac{\sigma}{\sqrt{8 - t} + \sqrt{t - 1}}, \qquad 1 \leqslant t \leqslant 8 \tag{1.3}$$

It can be concluded from these three equations that the problem could contain just one design variable (i.e. ***one*-dimensional problem**) as in (1.1) and (1.3), or it could contain n design variables (i.e. ***n*-dimensional problem**) as in (1.2). The independent variables of these equations are: x for (1.1), $\{x_1, x_2, x_3\}$ for (1.2), and t for (1.3).

These design variables could be defined as *continuous*, *discrete*, *integer*, *mixed-integer*, or even *binary*; which is a subset of discrete and integer types. The elements of X are the unknown variables that need to be determined by solving the given n-dimensional problem numerically (i.e. via using optimization algorithms). This can be done by ensuring that the variable f settled on its optimum value. Two important points have to be taken into account:

- The speed and efficiency of any proposed optimization algorithm significantly decrease as the problem dimension increases. This phenomenon could be used as one performance criterion in the evaluation stage.
- The set of design variables must be linearly independent; i.e. since they are considered as independent variables, then x_i should not be affected by the other $n - 1$ elements of X.

1.1.2 Design Parameters

The **parameters** are taken as **constants** or fixed values during initializing the optimization algorithm. For example, by referring to the preceding three equations, the parameters are: $\{a, b, c\}$ for (1.1) and (1.2), and σ for (1.3).

Note that, from mathematics, the word "parameter" has a different meaning than the word "constant." If some variables are held constant, inactive, or depend on other external ineffective variables during optimizing the function, then they are treated as constants at some given conditions of those external variables. For example, if the parameters $\{a, b, c\}$ of (1.1) are varied, then a family or set of quadratic functions can be generated.[2] As a real example from physics, the weight (W) of a body is equal to its mass (m) multiplied by the gravity acceleration (g):

$$W(m) = m \times g \tag{1.4}$$

The standard gravity acceleration is $g = 9.80\ 665\ \text{m/s}^2$. It is taken as a constant value since all the simulations are done within the Earth's surface. But, if there is a significant difference in the

2 This part is covered in **fuzzy optimization** where both the goals and parameters are **fuzzified**.

altitude, then the value of g in (1.4) will definitely change too. To clarify it more, let's take the following arbitrary example:

$$f(x) = 3x + \frac{\beta(t)}{\sqrt{(x)}} \tag{1.5}$$

In (1.5), f depends only on x, but β is a function of t that does not have any effect on f if it is held constant. Thus, β has to be defined first at a predetermined value of t, then it can be treated as a constant in f.

In addition to these design parameters, there are other parameters that need to be defined before starting the optimizer. Such parameters are the total number of **iterations**,[3] **tolerance**,[4], etc.

1.1.3 Design Function

To solve any faced numerical problem, first it has to be expressed mathematically so the variability and behavior of that problem can be translated into a meaningful and measurable format.

This mathematical model can be created arbitrary for virtual/non-real problems, such as (1.1)–(1.3). Alternatively, the model can be created based on some data collected from records, readings, analysis, surveys, and/or inspections of a specific machine, system, factory, goods, market, etc., at different conditions and times. One of the ways to do that is to employ regression analysis. This includes both, linear regression (LR) and nonlinear regression (NLR). The other innovative ways include the techniques of symbolic regression (SR) (Uy et al., 2011; Wang et al., 2019) and universal functions originator (UFO) (Al-Roomi and El-Hawary, 2020h).

Therefore, all the required information, such as objective (maximization or minimization mode) and the associated constrains (less than, equal to, greater than, etc.) should be identified and then defined in the design function.

1.1.4 Objective Function(s)

It has been seen that f depends on X, and it varies with any change in any element of X. The variable f could be a *linear* or *nonlinear* function, in a *continuous* or *non-continuous* domain, and its value could be a *continuous, discrete, integer, mixed-integer* or even *binary*.

The term **objective function**, which is also known as **criterion** or **merit** (Rao, 2009), means finding the optimum value of f. Based the objective function, the term "optimum," as covered before, could mean minimum or maximum. For example, minimizing the cost of a specific product or minimizing the time consumed to manufacture a product could be considered as the objective of such a problem. Also, that objective could be formulated as finding the maximum profit/revenue of a project suggested in a risky market or finding the maximum electric power delivered within different material characteristics of transmission lines. In general, the objective function is called the **cost function** when dealing with minimization problems and is called the **fitness** when dealing with maximization problems (Simon, 2013):

$$\min_{x} f(x) \Rightarrow f(x) \text{ is called "objective" or "cost"}$$

$$\max_{x} f(x) \Rightarrow f(x) \text{ is called "objective" or "fitness"} \tag{1.6}$$

3 Also known as **loops** or, in modern optimization, as **generations.**
4 Also known as the **minimum acceptable error** "ε" or the **early stopping criterion.**

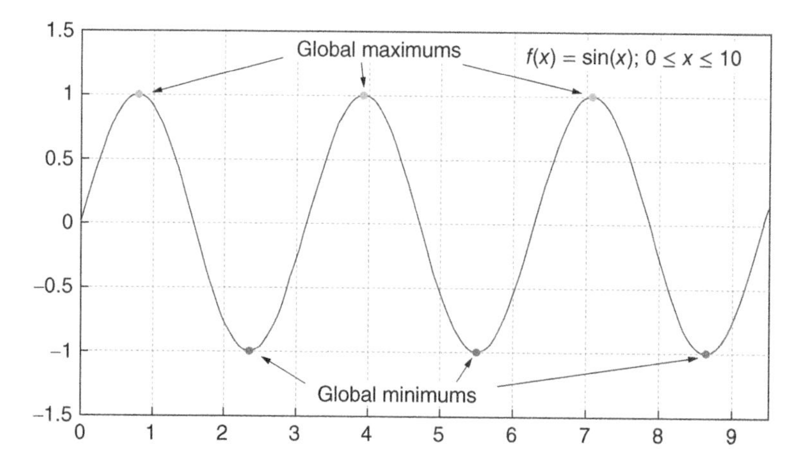

Figure 1.3 Multi-optimum points of a single-objective function.

If a **single-objective function** has multi-similar peaks and valleys, then its multi-optimum points mean either multi-minimum or multi-maximum points as illustrated in Figure 1.3. Thus, finding multi-optimum points does not mean finding a mixture of minimum and maximum points at the same time. Based on this, if the objective is to find the maximum value of a function, then the minimum value is considered as the worst solution, and vice versa if the objective is to minimize that function.

Suppose that the programming code of an optimization algorithm is designed to act as a minimizer. If the objective of a new design function is to find the maximum point, then the algorithm designer has to modify the structure of his/her code to act as a maximizer. Practically, this approach is totally not preferred and impractical, simply because this primitive correction is a time-consuming approach and it requires unnecessary effort and new codes to accomplish that task; especially when that program was coded in the past without embedding enough comments or supporting documents from the programmer. Alternatively, the **duality principle** can solve this technical issue directly by just reforming the objective.[5] One possible way to do that is by adding a negative sign as (Venkataraman, 2009; Simon, 2013):

$$\min_x f(x) \Leftrightarrow \max_x \left[-f(x)\right]$$

$$\max_x f(x) \Leftrightarrow \min_x \left[-f(x)\right] \tag{1.7}$$

Thus, the optimization problem can be switched easily from the maximization mode to the minimization mode, and vice versa, by just changing the sign. This is a very useful trick, especially when the researcher wants to find both the minimum and maximum points of f. That is, this trick can convert such optimization algorithms into general purpose optimizers where the design function can be plugged-in externally by a user and its objective function can be defined via a separate command coded somewhere in the program.

Example 1.2 Consider Bartels-Conn's function (Al-Roomi, 2015a):

$$f(X) = |x_1^2 + x_2^2 + x_1x_2| + |\sin(x_1)| + |\cos(x_2)| \quad ; \quad 500 \leqslant x_1, x_2 \leqslant 500$$

where the global minima is 1 and happens at $X^* = [0, 0]^T$.

5 This is valid with modern optimization algorithms. In linear programming, the duality approach is not an easy task.

If you already have a maximization algorithm, what should you do before being able to minimize the preceding function in that algorithm?

Solution

We have a maximization algorithm and the question wants to use it for a minimization problem. Thus, based on the duality principle, we have to just modify the objective function using the following expression:

$$\min_{x_1,x_2} f(X) \Leftrightarrow \max_{x_1,x_2} [-f(X)]$$

This means that we have to modify the objective function by multiplying it by a "−ve" sign:

$$\begin{aligned} f(X) &= -1 \times [|x_1^2 + x_2^2 + x_1x_2| + |\sin(x_1)| + |\cos(x_2)|] \\ &= -|x_1^2 + x_2^2 + x_1x_2| - |\sin(x_1)| - |\cos(x_2)| \end{aligned}$$

Then, the final fitness should multiplied by −1 again for getting the correct value.

1.1.5 Design Constraints

If f exists in all the points of X, then optimizing $f(X)$ becomes a relatively easy task. Such function is known as **unconstrained function**. If the objective is to minimize $f(X)$, then the formulation of the optimization problem can be expressed as:

$$\text{Find } X = \begin{Bmatrix} x_1 \\ x_2 \\ \vdots \\ x_3 \end{Bmatrix} \text{ which minimizes } f(X) \tag{1.8}$$

The large expression of (1.8) can be replaced by the following small expression:

$$\min_x f(x_1, x_2, \ldots, x_n) \tag{1.9}$$

Unfortunately, most of the problems faced in real-life applications have many obstacles and restrictions called **design constraints**. They could exist on f as **behavior**,[6] or could exist on each element of [X] as **side constraints** (Rao, 2009). The functional constraints can be further classified into:

- **Equality Constraints**
- **Inequality Constraints**

Since the side constraints are assigned to the design features (independent variables), so they are defined as just constant values (lower and upper limits). On the opposite side, the functional constraints could be defined as linear or nonlinear equations (Marriott and Stuckey, 1998).

If all these design constraints are satisfied, then the obtained solution becomes useful, and thus it can be effectively utilized and implemented. Such solution is called a **feasible solution**. If any one of these constraints is **violated**, then that solution is called an **infeasible solution**. It is a worthless solution,[7] which might not have any logical representation.[8]

---

6 Also called **functional constraints**.

7 This statement is applied in conventional optimization. In fuzzy optimization, the violated constraints and parameters are accepted with a certain degree, as well be seen later.

8 To be more precise, these infeasible solutions could contain good information that can guide the optimization algorithm to settle on the optimum value. This is the main reason why some **constraint-handling techniques** have better performance than others.

1.1.5.1 Mathematical Constraints

These constraints force the solution to be equal to only one value, which is equal to zero in (1.1). If (1.1), for instance, equals 30 instead of zero, then this right-hand side value should be subtracted in the left-hand side for getting a standard equality constraint format as follows:

$$f(x) = ax^2 + bx + c = 30 \quad (1.10)$$

$$f(x) = ax^2 + bx + c - 30 = 0 \quad (1.11)$$

If (1.1) or (1.11) is optimized, then the numerical solution should approach the analytical one. Unfortunately, satisfying equality constraint is a very hard task and it needs a large **number of function evaluations** (**NFE**)[9] in the optimization algorithm, and the desired answer may not be reached at all. The reason behind this is that when the equality constraint is assigned to a function, then the algorithm will accept the obtained solution as a feasible solution if and only if the value of the function is equal to the value of the equality constraint; it is zero in the standard format.

Practically, if the equality constraints exist in the design function, then they could be satisfied by a certain amount of tolerance ($\pm\varepsilon$) rather than setting them to zero. This approach can save a significant amount of CPU time and it can avoid getting infinite loops if the algorithm stopping criterion is activated with zero value.

The symbol h is frequently used in references to represent one equality constraint. If l equality constraints exist in the design function, then they could be represented as $[h_1, h_2, \ldots, h_l]$, or h_p : $p = 1, 2, \ldots, l$, or simply by using the vector notation $[H]$.

It is important to check if these l equality constraints are linearly independent or not. If not, then the design function has a nonlogical expression and thus the entire model needs to be reformulated again. If the given problem has a vector $[X]$ with a length n and a vector $[H]$ with a length l, then the optimization problem can be mathematically expressed as follows:

$$\min_{x} f(x_1, x_2, \ldots, x_n)$$

Subjected to:

$$h_1(x_1, x_2, \ldots, x_n) = 0$$
$$h_2(x_1, x_2, \ldots, x_n) = 0$$
$$h_l(x_1, x_2, \ldots, x_n) = 0 \quad (1.12)$$

By referring to linear algebra, if (1.12) is a linear model, then there are three possibilities (Venkataraman, 2009):

- $n > l$ → *under-determined case*: the problem has many solutions and thus the optimization technique is applicable.
- $n = l$ → *critical case*: the problem has one unique solution, so it is a solvable problem.
- $n < l$ → *over-determined case*: the problem has no solution, so the design function needs some corrections.

1.1.5.2 Inequality Constraints

Rather than the previous very intensive type of constraints that accepts only one solution as a feasible solution, this inequality constraint is more flexible. It requires fewer efforts from optimization algorithms, and it can be satisfied by many possible feasible vectors of X.

---

9 NFE = population size × number of iterations × number of evaluations per iteration.

Let's take (1.2) as an example. If the objective here is to minimize $f(X)$, then the smallest point is zero. This point occurs when the three design variables equal their corresponding parameters $\{x_1 = a, x_2 = b, x_3 = c\}$. But, because of its inequality constraint, the answer zero is considered as an infeasible solution. That is, the feasible solutions start when $f(X)$ equals 4.5 or above. If (1.2) is subjected to an equality constraint with zero on the right-hand side (i.e. the condition $f(X) = 0$ must be satisfied), then the same difficulty faced with (1.1) will be faced here again.

Similar to the previous constraint, this type is represented by the symbol g. If the design function contains m inequality constraints, then they can be represented as $[g_1, g_2, \dots, g_m]$, or $g_q : q = 1, 2, \dots, m$, or simply by using the vector notation $[G]$.

In mathematics, the sign that represents equality constraints is limited to (= "equal to"). The sign ($\neq$ "not equal to") means that the solution obtained for f should not equal to a predefined value. However, this sign does not give any additional information. For example, it does not show whether the solution is greater or less than the predetermined value. Therefore, four possible signs could be used to represent inequality constraints, which are classified into two main groups:

- **Strict inequalities**
  - $g_q(X) > \alpha$: $g_q(X)$ is **greater than α**
  - $g_q(X) < \beta$: $g_q(X)$ is **less than β**
- **Not-strict inequalities**
  - $g_q(X) \geqslant \alpha$: $g_q(X)$ is **greater than or equal to α** (also known as: **not less than α** or **at least α**)
  - $g_q(X) \leqslant \beta$: $g_q(X)$ is **less than or equal to β** (also known as: **not greater than β** or **at most β**)

For the strict types, it is difficult to determine the endpoints (minimum and maximal points) of $g_q(X)$ to satisfy (>) and (<), respectively. The reason is that the boundary of α and β cannot be reached (i.e. open intervals). For example, if ($\alpha = 3.0$) then the condition ($g_q(X) > \alpha$) is satisfied by infinite solutions $\{g_q(X) = 3 + 10^{-c} : c = 1, 2, \dots, \infty\}$. It shows that $g_q(X) \downarrow$ as $c \uparrow$, but it cannot reach the minimal point. Based on this, the non-strict types are used instead to provide endpoints, which are easy to be defined and coded in any numerical programming language.

To have the standard inequality constraint format, only the sign "$\leqslant$" must be used with zero on the right-hand side. Thus, (1.2) has to be reformulated as follows:

$$f(x_1, x_2, x_3) = 4.5 - \left[(a - x_1)^2 + (b - x_2)^2 + (c - x_3)^2\right] \leqslant 0 \tag{1.13}$$

Thus, for n-dimensional problems given with m inequality constraints, they can be mathematically expressed as follows:

$$\min_{x} f(x_1, x_2, \dots, x_n)$$

Subjected to:

$$\begin{aligned} &g_1(x_1, x_2, \dots, x_n) \leqslant 0 \\ &g_2(x_1, x_2, \dots, x_n) \leqslant 0 \\ &g_m(x_1, x_2, \dots, x_n) \leqslant 0 \end{aligned} \tag{1.14}$$

1.1.5.3 Side Constraints

This type of constraints has many other names, like **domain**, **solution space**, **search space**, **variable bounds**, **choice set**, **feasible region**, **bound constraints**, etc. (Deb, 2010; Rao, 2009; NEOS, 2012). From its name, this type of constraints is associated with the design features where each element of $[X]$ has two bounds called the lower and upper limits or bounds. Due to the same mathematical representation difficulties of the inequality constraints, the closed intervals are used here to define the side constraints in the design function as $x_i \in [x_i^{min}, x_i^{max}] : i = 1, 2, \ldots, n$, or as $x^{min} \leqslant x_i \leqslant x^{max} : i = 1, 2, \ldots, n$. Using the vector notation, it can be expressed as $X \in [X^{min}, X^{max}]$ or as $X^{min} \leqslant X \leqslant X^{max}$.

As a rule of thumb, the optimization algorithm performs better as the span between the lower and upper bounds decreases. The reason behind this phenomenon is that the algorithm needs less effort to search for the optimal solution within a very narrow domain. It is like making a zoom-in into a small spot of a bigger search space, and thus this cropped domain becomes very rich in good solutions.

Although the side constraints are classified as one type of design constraints,[10], a problem containing only side constraints is considered as an unconstrained problem. The reason behind this is that, by nature, the logical problem has to be designed with side constraints so that the algorithm can search for the optimality within a specific search space. Suppose that the given problem is very complex and cannot be depicted or solved analytically and its search space is open (i.e. $X^{-\infty} \leqslant X \leqslant X^{\infty}$), then no one can determine the location of the optima with this infinite domain. Thus, for n-dimensional optimization problems, they can be expressed as follows:

$$\min_{x} f(x_1, x_2, \ldots, x_n)$$

Subjected to:

$$x_i^{min} \leqslant x_i \leqslant x_i^{max} \quad i = 1, 2, \ldots, n \tag{1.15}$$

1.1.6 General Principles

Before describing the final standard format of optimization problems when all the design constraints are assigned, it is important to cover the following general principles:

1.1.6.1 Feasible Space vs. Search Space

From the preceding design constraints, it has been seen that the obtained optimum solution becomes useable only if it is feasible. The space of the design variables could be open to infinity, or could be bounded between two limits. Moreover, even with these side constraints, the feasibility also depends on some equality and inequality constraints that have to be satisfied as well. These classifications create three different layers on the entire space of any constrained optimization problem:

- **Infinite Space:** lower and upper limits are open to infinity, $X \in [X^{-\infty}, X^{\infty}]$
- **Search Space:** lower and upper limits are bounded, $X \in [X^{min}, X^{max}]$
- **Feasible Space:** lower and upper limits are bounded and the functional constraints are satisfied, $X \in [X^{-\infty}, X^{\infty}]$, $[h(X)]_l = 0$, and $[g(X)]_m \leqslant 0$

The difference between these three layers is graphically described in Figure 1.4. Thus, the feasible space must be inside the search space; i.e. part of the search space. Otherwise, the solution is considered infeasible. Also, as the number of equality and inequality constraints increases, the feasible space is shrunk more and more, and thus the optimum solution becomes very hard to be found.

10 Except for some references, such as (NEOS, 2012) where unconstrained optimization problems are called bound-constrained optimization problems.

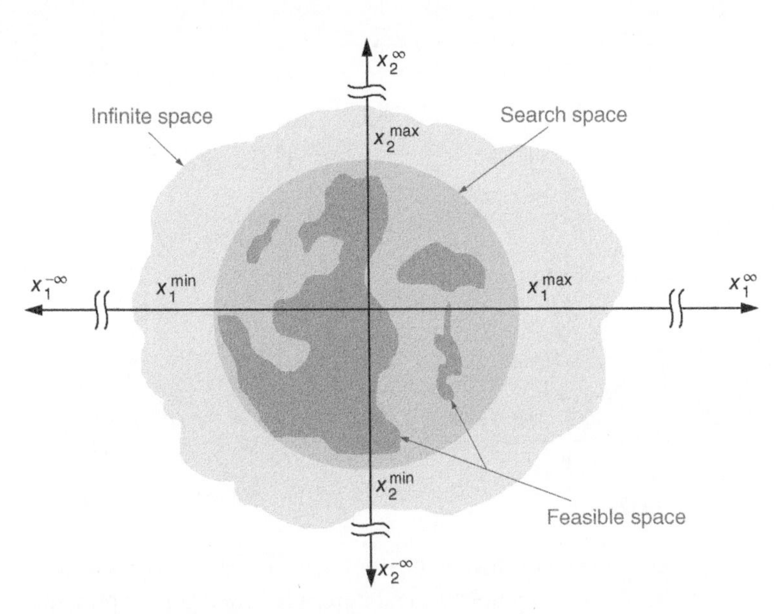

Figure 1.4 Infinite space vs. search space vs. feasible space.

1.1.6.2 Global Optimum vs. Local Optimum

It has been known that the definition of the optimum solution is either minimum or maximum solution. It could appear as a single optimum or multi-optimum solutions, as seen before in Figure 1.3. If multi-optimum solutions are not identical, then the most optimum solution of these points is called a **global optimum solution**, while the others are called **local optimum solutions**. Optimization problems could have groups of multi-global and multi-local optimum solutions. These local optima are considered as **traps** where the efficient optimization algorithm has the ability to escape from these traps and settle on or close to the global optimum solution(s) quickly and smartly.

Example 1.3 Consider the following arbitrary function:

$$f(x) = 30 - x + \sin(x) + \frac{5}{20x} + e^{0.3x} \quad ; \qquad 0 \leqslant x \leqslant 10$$

Design a MATLAB code to graphically represent this function when x moves from 0 to 10 with steps of $\Delta x = 0.01$. Then, use the command `fsolve` to find the global minimum and its solution. The best point on the plot can be used as the initial point for the minimization process.

Solution

The following is the code shown in `GlobalOptimum.m`:

```
1   clc
2   clear
3   x = 0:0.001:10;
4   f = sin(x) + sin(4*x);
5   plot(x, f, 'b', 'LineWidth', 2)
6   grid on; grid minor
7   xlabel('x'); ylabel('f(x)')
8   set(gca, 'FontSize', 13)
```

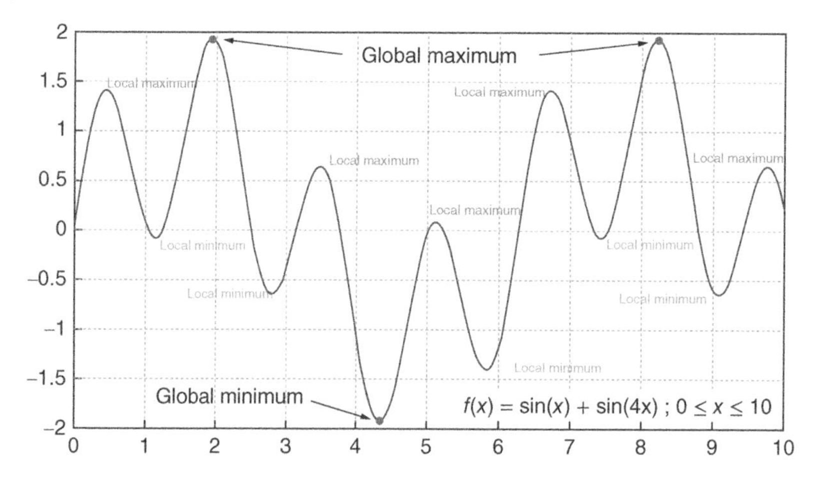

Figure 1.5 Global optimum vs. local optimum.

Running this simple code will generate the plot shown in Figure 1.5, which can also be accessed via `GlobalOptimum.fig`. As per the plot, the function has a mixture of local and global optimum solutions. Again, if the objective is to minimize f, then there will be only one global optimum solution where the two global maximum points are considered as the worst solutions, and vice versa if the objective is to maximize f.

If the global optimum solution is analytically[11] predefined, then this solution is denoted by an asterisk. This symbol is assigned to both the dependent and independent variables; as $f_{\min}(X^*)$ and X^*, respectively.

1.1.6.3 Types of Problem

Based on the availability of objective(s) and/or constraint(s), the design problem could be one of the four possible types, as summarized in Table 1.1. For more details, please refer to (Leondes, 1998; Eiben and Smith, 2003).

1.1.7 Standard Format

By taking into account all the preceding issues, any design function can be transformed into the following standard mathematical model (Venkataraman, 2009):

$$\min_{x} f(x_1, x_2, \dots, x_n)$$

Table 1.1 Types of problem.

	Objective function	
Constraints	**Yes**	**No**
Yes	Constrained optimization problem	Constraint satisfaction problem
No	Free optimization problem	No problem

11 Sometimes, the global optima can be numerically found with almost, or even exact, zero error.

$$
\begin{aligned}
\text{Subjected to: } & h_1(x_1, x_2, \dots, x_n) = 0 \\
& h_2(x_1, x_2, \dots, x_n) = 0 \\
& h_l(x_1, x_2, \dots, x_n) = 0 \\
& g_1(x_1, x_2, \dots, x_n) \leqslant 0 \\
& g_2(x_1, x_2, \dots, x_n) \leqslant 0 \\
& g_m(x_1, x_2, \dots, x_n) \leqslant 0 \\
& x_1^{\min} \leqslant x_1 \leqslant x_1^{\max} \\
& x_2^{\min} \leqslant x_2 \leqslant x_2^{\max} \\
& x_n^{\min} \leqslant x_n \leqslant x_n^{\max} \\
\text{where: } & f_{\min}(X^*) = a, \quad x_i^* = \{s_1, s_2, \dots, s_n\}
\end{aligned}
\tag{1.16}
$$

Also, it can be expressed as:

$$
\begin{aligned}
& \min_x f(x_1, x_2, \dots, x_n) \\
& \text{Subjected to: } h_p(x_1, x_2, \dots, x_n) = 0, \quad p = 1, 2, \dots, l \\
& \qquad g_q(x_1, x_2, \dots, x_n) \leqslant 0, \quad q = 1, 2, \dots, m \\
& \qquad x_i^{\min} \leqslant x_i \leqslant x_i^{\max}, \quad i = 1, 2, \dots, n \\
& \text{where: } f_{\min}(X^*) = a, \quad x_i^* = s_i
\end{aligned}
\tag{1.17}
$$

Or, if the vector notation form is employed, then it can be expressed as:

$$
\begin{aligned}
& \min_x f(X), \; [X]_n \\
& \text{Subjected to: } [h(X)]_p = 0, \quad p = 1, 2, \dots, l \\
& \qquad [g(X)]_q \leqslant 0, \quad q = 1, 2, \dots, m \\
& \qquad X^{\min} \leqslant X \leqslant X^{\max} \\
& \text{where: } f_{\min}(X^*) = a, \quad X^* = S
\end{aligned}
\tag{1.18}
$$

1.1.8 Constraint-Handling Techniques

The side constraints can be easily satisfied by controlling the independent variables to be within the decided bounds. The following equation can be coded in any numerical programming language to generate random values of each ith design variable:

$$
x_i = x_i^{\min} + \text{rand}\left(x_i^{\max} - x_i^{\min}\right)
\tag{1.19}
$$

where rand is a function that generates uniformly distributed random numbers in the interval between 0 and 1.

It is clear that the variable x_i is bounded between $x_i^{\min}$ and $x_i^{\max}$, which are respectively reached when rand = 0 and 1. In MATLAB, if we want to generate a matrix with random numbers, then we can use the following command:

```
1    xi = xi_min + (xi_max - xi_min) * rand(rows, cols)
```

For having discrete variables, we can round the continuous random value obtained by (1.19) as follows:

$$\begin{aligned} x_i &= \lfloor x_i^{\min} + \text{rand}\left(x_i^{\max} - x_i^{\min}\right)\rfloor \\ &= \text{round}\left[x_i^{\min} + \text{rand}\left(x_i^{\max} - x_i^{\min}\right)\right] \end{aligned} \tag{1.20}$$

which can be implemented in MATLAB as follows:

```
1       xi = round(xi_min + (xi_max - xi_min) * rand(rows, cols))
```

The other option for generating discrete random values is to use `randi` as follows:

```
1       xi = randi([xi_min xi_max], rows, cols)
```

Example 1.4 Suppose that you have three variables; x_1, x_2, and x_3. The first one is a continuous variable, which is bounded between −1.3 and 2.7. The second and third variables are discrete where $x_2 \in [-7, 1]$ and $x_3 \in [-1, 7]$. Design a MATLAB code to generate a 4×6 matrix of random values for each variable. Use `round` for x_2 and `randi` for x_3.

Solution

The program given in `RandomMatrices.m` and shown in the following text is used to generate matrices with continuous and discrete random values:

```
1       clc
2       clear
3       format short
4       x1 = -1.3 + (2.7 - -1.3) * rand(4, 6)
5       x2 = round(-7 + (1 - -7) * rand(4, 6))
6       x3 = randi([-1 7], 4, 6)
```

If we run this simple program, the following result will be displayed on the command window:

```
x1 =
   -0.4690    2.0772   -0.3893    0.4208    0.4555    1.0796
   -0.0950   -0.5209    0.4428   -0.5607   -0.8555   -0.2512
    0.5837   -0.3963   -0.0556    2.3195   -0.2677    1.1114
   -0.3780   -0.6172    2.3935    2.6190    0.3349    1.5449
x2 =
    -5    -4    -1    -3     1    -3
    -6    -3    -7    -2    -3    -2
    -5    -6     0    -5    -3    -2
    -4    -5    -1    -3    -5    -4
x3 =
     2     7     2    -1     5     5
     7     6     5     4     7     0
    -1    -1     0     3     7    -1
     6     1     5     6     2     5
```

Unfortunately, this direct solution approach does not work with functional constraints (i.e. equality and inequality). Rather, they require more complicated constraint-handling techniques.

Selecting the correct type is a very important step because the algorithm speed and accuracy can be markedly affected by inefficient techniques (Eiben and Smith, 2003). Figure 1.6 summarizes the most popular constraint-handling techniques (Deb, 2010; Venkataraman, 2009; Simon, 2013; Rao, 2009; Eiben and Smith, 2003; Yeniay, 2005; Kajee-Bagdadi, 2007; Yu and Gen, 2010).

The **penalty functions** are often used because the other approaches are hard to be modeled or/and need derivatives (Eiben and Smith, 2003). Besides, the **exterior penalty function (EPF)** is preferred in constrained EAs. The reason is that the **interior penalty function (IPF)** requires feasible individuals which in turn complicate the solution. In this section, a brief overview of some EPFs is given. Also, the **classical random search method**, which is classified as one of the **direct search methods**, is covered.

By referring to (1.16)–(1.18), EPFs can easily transform them into unconstrained optimization problems by employing either the **additive** or the **multiplicative** approach. These two approaches are respectively described as follows (Yeniay, 2005):

$$\min_X \phi(X), \text{ where } \phi(X) = \begin{cases} f(X), & \text{if } X \in \mathcal{F} \\ f(X) + P(X), & \text{if } X \notin \mathcal{F} \end{cases} \tag{1.21}$$

$$\min_X \phi(X), \text{ where } \phi(X) = \begin{cases} f(X), & \text{if } X \in \mathcal{F} \\ f(X) \times P(X), & \text{if } X \notin \mathcal{F} \end{cases} \tag{1.22}$$

where $P(X)$ is called the **penalty term**, which is equal to zero for feasible individuals ($X \in \mathcal{F}$) and be a positive value in case there is a violation of any constraint ($X \notin \mathcal{F}$). Thus, for minimization mode, the **penalized cost function** $\phi(X)$ becomes higher than its actual value $f(X)$. This $P(X)$ can be provided in different forms based on the type of penalty function employed. The most common form is:

$$\begin{aligned} P(X) &= \sum_{q=1}^{m} r_q\, \hat{g}_q(X) + \sum_{p=1}^{l} d_p\, \hat{h}_p(X) \\ \text{where: } \hat{g}_q(X) &= \left[\max\left(0, g_q(X)\right)\right]^{\beta} \\ \hat{h}_p(X) &= \left|h_p(X)\right|^{\gamma} \end{aligned} \tag{1.23}$$

where r_q and d_p are called the **penalty multipliers**. The coefficients β and γ are user-defined positive constants, which are commonly set equal to either 1 or 2 (Simon, 2013; Rao, 2009; Eiben and Smith, 2003).

Because $[H]_l = 0$ cannot be easily satisfied, so an acceptable tolerance ($\pm\varepsilon$) is adopted instead of crisp zero. Thus, the pth equality constrain is satisfied if:

$$-\varepsilon \leqslant h_p(X) \leqslant \varepsilon, \quad p = 1, 2, \dots, l \tag{1.24}$$

This equation can be split into two inequality constraints, which can be expressed using the standard format as follows:

$$\begin{aligned} h_p(X) - \varepsilon &\leqslant 0 \\ -h_p(X) - \varepsilon &\leqslant 0 \end{aligned} \tag{1.25}$$

Therefore, by using (1.25), (1.23) can be modified to be:

$$\begin{aligned} P(X) &= \sum_{q=1}^{m+l} r_q\, \hat{g}_q(X)\ ; \quad \text{where} \\ \hat{g}_q(X) &= \begin{cases} \left[\max\left(0, g_q(X)\right)\right]^{\beta}, & \text{for } q \in [1, m] \\ \left[\max\left(0, \left|h_q(X)\right| - \varepsilon\right)\right]^{\beta}, & \text{for } q \in [m+1, m+l] \end{cases} \end{aligned} \tag{1.26}$$

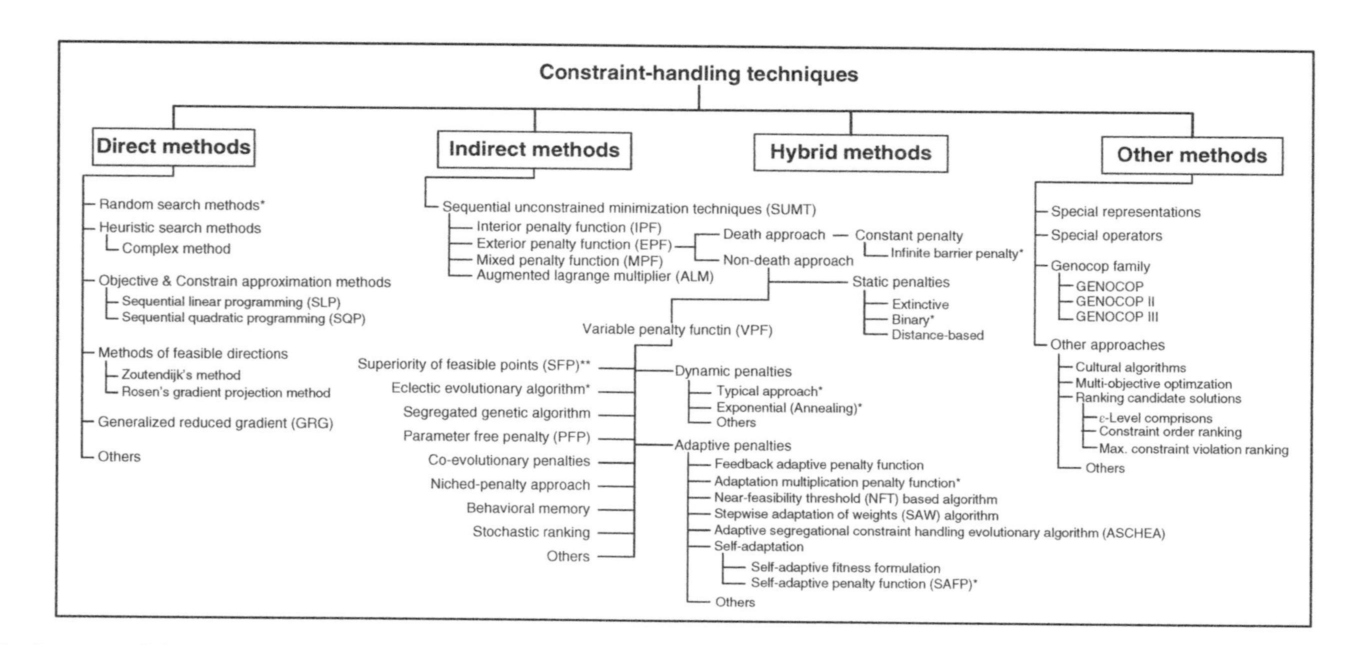

Figure 1.6 Summary of the most popular constraint-handling techniques.

Before describing the popular EPFs, it is important to mention that the penalized cost function "$\phi(X)$" does not apply to the random search method. This classical direct search method depends only on an internal *while*-loop to continue generating random elements of the design vector $[X]$ until satisfying all the design constraints. These ten constraint-handling techniques are briefly described in the following lines:

1.1.8.1 Random Search Method

This constraint-handling technique is very simple, and it can be easily programmed by following the steps of the pseudocode in Algorithm 1.

Algorithm 1 Rebuild infeasible individuals directly via the random search algorithm (RSA.)

Require: all the constraint values of each individual as a vector C
1: **for** $j \leftarrow 1$ *to* N **do** {where N = population size}
2: **while** any element of C_j is violated **do** {there are N vectors of C}
3: Randomly generate new design vector X
4: Determine new C_j
5: **end while**
6: **end for**

Unfortunately, this technique is not efficient and consumes high CPU time as the problem complexity increases (Rao, 2009). This complexity could be faced in different locations of the problem. For example, it could be the dimension of the problem, its type (convex or non-convex, explicit or implicit, unimodal or multimodal, etc.), type of design variables (continuous, discrete, mixed-integer, binary, etc.), number of functional constraints, type of functional constraints (equality or inequality), lower and upper bounds of design variables, etc.

1.1.8.2 Constant Penalty Function

In this type, the penalty term $P(X)$ is set to a very high value if there is any violation of any constraint. During the function evaluation process, this intensive penalty function rejects all the infeasible individuals. This is why it is called the **death penalty** approach (Rao, 2009; Simon, 2013).

The main drawback of this simple type is seen when the infeasible solutions are very close to the border of the feasible space. These individuals have some good information that could guide the optimization algorithm to reach the feasible space. However, this death EPF approach erases all these useful data. Thus, individuals having a few or many violations will be treated by the same rejection action. This means that all the infeasible individuals will completely disappear in the next generation (Eiben and Smith, 2003).

The most popular sub-type is called the **infinite barrier penalty** (Deb, 2010; Yeniay, 2005). It comes in an additive form where $P(X)$ given in (1.26) is calculated as follows:

$$P(X) = \mathcal{R} \sum_{q=1}^{m+l} \hat{g}_q(X) \tag{1.27}$$

where the penalty multiplier "r_q" is replaced with a very large constant number "$\mathcal{R}$"; it is usually set equal to 10^{20}. From this equation, the penalty term will reach almost infinity (i.e. $P(X) \approx +\infty$) if any violation is detected. Some of death penalty functions are defined by only $\mathcal{R}$ term since $\mathcal{R} \gg \sum_{q=1}^{m+l} \hat{g}_q(X)$. Thus, $P(X)$ is expressed as a high constant value, which is independent of the violation magnitudes.

1.1.8.3 Binary Static Penalty Function

It is an additive penalty. The distance function $\hat{g}_q(X)$ given in (1.26) can be expressed as a simple binary value, which equals 1 if the qth constraint is violated. Otherwise, it equals zero (Eiben and Smith, 2003). If r_q is taken as a constant value for all the functional constraints, then the final expression of this penalty function can be formulated as follows:

$$P(X) = r\sum_{q=1}^{l+m} \hat{g}_q(X), \quad \hat{g}_q(X) = \begin{cases} 1, & \text{if } \hat{g}_q(X) \text{ is violated} \\ 0, & \text{if } \hat{g}_q(X) \text{ is satisfied} \end{cases} \tag{1.28}$$

Now, the dimension of the vector r becomes one,[12] which is easy to be tuned. In contrast, the other static penalty functions, listed in Figure 1.6, are hard to be tuned. Besides, the binary static penalty can compromise between the speed and accuracy, which are very important winning factors in many engineering applications.

1.1.8.4 Superiority of Feasible Points (SFPs) – Type I

In this method, the penalized cost function given in (1.21) is expressed as follows:

$$\min_X \ \phi'(X) = \phi(X) + \theta(X)$$

$$\text{where: } \theta(X) = \begin{cases} 0, & \text{if } \mathcal{F} = \emptyset \text{ or } X \in \mathcal{F} \\ \alpha, & \text{if } \mathcal{F} \neq \emptyset \text{ and } X \notin \mathcal{F} \end{cases} \tag{1.29}$$

The value of α is the largest feasible individual:

$$\alpha = \max\ [f(Y)] \ : \ Y \in \mathcal{F} \tag{1.30}$$

This is done to ensure that the bad feasible individual is better, or at least, equal to the best infeasible individual (Yeniay, 2005). The penalty multiplier r_q given in (1.26) can be set as a fixed number (Michalewicz, 1995). In this book, r_q is set equal to 10.

1.1.8.5 Superiority of Feasible Points (SFP) – Type II

SFP-I does not cover the range when $f(X) < 0$, while SFP-II does. The second type has a similar expression, except for the value of α. Here, it is modified to be equal to the difference between the largest feasible and the smallest infeasible individuals as follows (Simon, 2013; Kajee-Bagdadi, 2007):

$$\alpha = \max \left[0, \max_{Y \in \mathcal{F}} f(Y) - \min_{Y \notin \mathcal{F}} \phi(Y)\right] \tag{1.31}$$

1.1.8.6 Eclectic Evolutionary Algorithm

The bad thing about the extinctive static penalty function shown in Figure 1.6 is that the solution quality is very sensitive to the values of the vector "r_q" (Yeniay, 2005). The eclectic **evolutionary algorithm** (EA) works in a similar principle of SFP types where the bad feasible individuals are considered to be better than the best infeasible individuals, but in a different way (Simon, 2013). Instead of using how much the constraints are violated, it uses the number of violated constraints as a basis to penalize the cost function as follows:

$$\phi(X) = \begin{cases} f(X), & \text{if } X \in \mathcal{F} \\ K\left[1 - \dfrac{v(X)}{(m+l)}\right], & \text{if } X \notin \mathcal{F} \end{cases} \tag{1.32}$$

12 i.e. it becomes a scalar value.

where K is a very large positive constant (it is taken as $K = 1 \times 10^9$ (Yeniay, 2005)), $v(X)$ is the number of satisfied or non-violated constraints, and $(m + l)$ is the total number of functional constraints.

1.1.8.7 Typical Dynamic Penalty Function

If the soft counter, that counts the number of iterations or generations inside the optimization algorithm, is used to simulate the time "t," then the amount of the penalized value is proportional to the number of iterations. This action provides two things:

- In the beginning, the penalization process will forgive the infeasible individuals by penalizing them with small values. Thus, the optimization algorithm is given a chance to collect some useful information about the search space being explored.
- As the number of iterations increases, the penalization level increases. Thus, after collecting enough information about the problem, the penalty function will start forcing the infeasible individuals to go inside the feasible search space.

The penalized cost function is defined as (Rao, 2009; Simon, 2013; Eiben and Smith, 2003):

$$\phi(X) = f(X) + (c.t)^{\alpha} P(X) \tag{1.33}$$

where c and α are constants. For example, they can be taken as 0.5 and 2, respectively (Simon, 2013). $P(X)$ is similar to that of (1.26), but with $r = 1$ and $\beta = 2$.

1.1.8.8 Exponential Dynamic Penalty Function

Instead of using the additive form, as in (1.33), this type of EPFs comes in a multiplicative form. The same assumptions are applied here, except that the penalty value grows in an exponential rate rather than a linear rate as in the typical dynamic approach. The new penalization process is defined as follows (Rao, 2009; Simon, 2013; Eiben and Smith, 2003):

$$\phi(X) = f(X) \times e^{\frac{P(X)}{\tau}} \tag{1.34}$$

where $\tau = \frac{1}{\sqrt{t}}$, which approaches zero as the number of iterations or generations approaches infinity (i.e. $t \to \infty \Rightarrow \tau \to 0 \Rightarrow \phi(X) \to \infty$).

This type of dynamic penalties is valid just for minimization where $f(X) \geqslant 0 \; \forall \; X$. Otherwise, the normalized version, as described in Simon (2013), has to be used instead.

1.1.8.9 Adaptive Multiplication Penalty Function

From its name, it comes in a multiplicative form. The cost function is penalized as follows (Yu and Gen, 2010):

$$\phi(X) = f(X) \times \left\{ 1 + \frac{1}{(m+l)} \sum_{q=1}^{m+l} \left[\frac{\hat{g}_q(X)}{\hat{g}_q^{\max}(X)} \right]^{\alpha} \right\} \tag{1.35}$$

Actually, there are many adaptive approaches presented in the literature. The first one was proposed by Hadj-Alouane and Bean (1997). After investigating the various types of adaptive EPFs, the adaptive multiplication approach is selected because it works based on the typical feedback and **near feasibility threshold** (**NFT**-based) approaches (Yeniay, 2005). The original equation has a subtraction arithmetic operator, while (1.35) has an addition arithmetic operator. This adjustment is essential to make it functional in the minimization mode. The qth constraint $\hat{g}_q(X)$ can be calculated by using (1.26) with $\beta = 1$. The symbol $\hat{g}_q^{\max}(X)$ denotes the biggest functional constraint, which can be obtained as follows:

$$\hat{g}_q^{\max}(X) = \max\left[\varepsilon, \max\left(\hat{g}_q(X)\right)\right] \tag{1.36}$$

The epsilon ε in (1.36) is very important to avoid dividing by zero when all the individuals are feasible. For example, ε can be set equal to 10^{-20}. Thus, if there is no infeasible individual, then the summation term becomes zero instead of getting an error due to dividing zero by zero.

1.1.8.10 Self-Adaptive Penalty Function (SAPF)

The SAPF algorithm works based on the following two conditions (Simon, 2013):

- If the ratio of the feasible individuals to the entire **population size** N is low, then the penalized cost function "$\phi(X)$" has to be small for infeasible individuals having a few violations.
- If that ratio is high, then only infeasible individuals having low cost "$f(X)$" should be penalized with small penalty terms.

To identify the best individual in the current population, SAPF penalizes infeasible individuals by two terms, $d(X)$ and $P(X)$, as follows (Sivananaithaperumal et al., 2011):

$$\phi(X) = d(X) + P(X) \tag{1.37}$$

where $d(X)$ is called the distance value. SAPF needs special care for constructing (1.37). The steps can be summarized as follows (Simon, 2013; Sivananaithaperumal et al., 2011):

- Firstly, normalize $f(X)$ of each individual as:

$$N(X) = \frac{f(X) - f_{\min}}{f_{\max} - f_{\min}} \tag{1.38}$$

  where $N(X) \in [0, 1]\ \forall\ X$; the best individual occurs when $N(X) = 0$, and vice versa when $N(X) = 1$.
- Secondly, compute the normalized violation magnitude of the qth constraint as:

$$M(X) = \frac{1}{m+l}\sum_{q=1}^{m+l}\left(\frac{\hat{g}_q(X)}{\hat{g}_q^{\max}(X)}\right) \tag{1.39}$$

  where $\hat{g}_q(X)$ and $\hat{g}_q^{\max}(X)$ can be obtained by using the expressions presented in the preceding penalty functions. Thus, $M(X) = 0\ \forall\ X \in \mathcal{F}$ and $M(\overline{X}) > 0\ \forall\ \overline{X} \notin \mathcal{F}$.
- Thirdly, for each individual, compute the distance $d(X)$ as:

$$d(X) = \begin{cases} M(X), & \text{if } \mathcal{F} = \emptyset \\ \sqrt{N^2(X) + M^2(X)}, & \text{if } \mathcal{F} \neq \emptyset \end{cases} \tag{1.40}$$

  where ($\mathcal{F} = \emptyset$) means that all the population individuals are infeasible.
- Fourthly, the penalty term in (1.37) can be calculated as:

$$P(X) = (1 - r)Y_1(X) + rY_2(X)$$

$$\text{where: } Y_1(X) = \begin{cases} 0, & \text{if } \mathcal{F} = \emptyset \\ M(X), & \text{if } \mathcal{F} \neq \emptyset \end{cases}$$

$$Y_2(X) = \begin{cases} 0, & \text{if } X \in \mathcal{F} \\ N(X), & \text{if } X \notin \mathcal{F} \end{cases}$$

$$r = \frac{\textit{number of feasible individuals}}{\textit{population size}} \tag{1.41}$$

As can be clearly seen from all these equations, (1.37)–(1.41), one of the main disadvantages of SAPF is that it consumes more CPU time compared with other penalty functions.

All these constraints handling techniques will be practically implemented in Chapter 7 to optimize the settings of protective relays. It has to be remembered that each one of these techniques has its own strengths and weaknesses.

1.1.9 Performance Criteria Used to Evaluate Algorithms

To evaluate the performance of any optimization algorithm, it is necessary to define the performance criteria based on which the algorithm performance can be assessed and compared with other algorithms. The first stage is to run or execute the algorithm to solve some standard **benchmark functions.**[13] These functions are classified as: **unimodal** (*having one optimum solution*) and **multimodal** (*having multiple optimum solutions*). Also, they can be classified as: **unconstrained/constrained, static/dynamic, convex/non-convex, smooth/non-smooth, non-noisy/noisy, non-shifted/shifted, non- rotated/rotated, single-objective/multi-objective**, etc. The other properties that might be considered are: **continuity, separability, differentiability, scalability**, etc.

The next step is to let the optimization algorithm to run multiple times and collecting the fitness obtained from each random run. Then, the **best, worst, mean**, and **standard deviation**[14] of the final solutions are extracted. By these fundamental records, someone can start comparing with other results reported in the literature. It is important to ensure that the initialization stage has the same parameters used in other studies (i.e. population size N, number of generations G, etc.) to get a fair comparison. For the minimization mode, the best of the best and the worst of the best-obtained solutions per certain number of simulation runs or trials can be respectively computed as follows:

$$f_{\text{best\_ever}} = \min\left(f_{\text{best},1}, f_{\text{best},2}, \cdots, f_{\text{best},i}, \cdots, f_{\text{best},T_r}\right) \tag{1.42}$$

$$f_{\text{worst\_ever}} = \max\left(f_{\text{best},1}, f_{\text{best},2}, \cdots, f_{\text{best},i}, \cdots, f_{\text{best},T_r}\right) \tag{1.43}$$

where T_r is the total number of trials and $f_{\text{best},i}$ represents the best solution obtained in the ith trial. Also, the mean and standard deviation can be respectively calculated as follows (Johnson, 2000):

$$f_{\text{mean}} = \frac{\sum_{i=1}^{T_r} f_{\text{best},i}}{T_r} \tag{1.44}$$

$$f_{\text{std\_dev}} = \sqrt{\frac{\sum_{i=1}^{T_r} \left(f_{\text{best},i} - f_{\text{mean}}\right)^2}{T_r - 1}} \tag{1.45}$$

For standard benchmark functions, the global optimal solution (f^*) is known and given. Based on this, the absolute error (Err_{abs}) between the estimated global optimal solution $f_{\text{best},i}$ and the exact solution f^* can be calculated as follows:

$$\text{Err}_{\text{abs},i} = \left|f_{\text{best},i} - f^*\right| \tag{1.46}$$

Thus, (1.42)–(1.45) can be, respectively, replaced by:

$$\text{Smallest error}: \quad \text{Err}_{\text{smallest\_ever}} = \min\left(\text{Err}_{\text{abs},1}, \text{Err}_{\text{abs},2}, \cdots, \text{Err}_{\text{abs},T_r}\right) \tag{1.47}$$

$$\text{Largest error}: \quad \text{Err}_{\text{largest\_ever}} = \max\left(\text{Err}_{\text{abs},1}, \text{Err}_{\text{abs},2}, \cdots, \text{Err}_{\text{abs},T_r}\right) \tag{1.48}$$

---

13 Also known as **test functions.**

14 Sometimes, the **median** metric is also computed.

$$\text{Mean error}: \quad \text{Err}_{\text{mean}} = \frac{\sum_{i=1}^{T_r} \text{Err}_{\text{abs},i}}{T_r} \tag{1.49}$$

$$\text{Standard deviation}: \quad \text{Err}_{\text{std\_dev}} = \sqrt{\frac{\sum_{i=1}^{T_r} \left(\text{Err}_{\text{abs},i} - \text{Err}_{\text{mean}}\right)^2}{T_r - 1}} \tag{1.50}$$

Example 1.5 Suppose that you have designed a multi-dimensional optimization algorithm that can find the near-global optimal settings of a group of protective relays. After running the algorithm for 10 independent runs, you got the following optimal solutions in MATLAB:

```
The best of Trial # 1 is 1.6597
The best of Trial # 2 is 1.7671
The best of Trial # 3 is 1.7433
The best of Trial # 4 is 1.849
The best of Trial # 5 is 1.7632
The best of Trial # 6 is 1.949
The best of Trial # 7 is 1.7063
The best of Trial # 8 is 1.8431
The best of Trial # 9 is 1.8036
The best of Trial # 10 is 1.8129
```

The task here is to add some extra codes at the end of the program to show the worst, best, median, mean, and standard deviation of these ten optimal solutions.

Solution

The following program is given in `PerformanceEvaluation.m`:

```
1   clc
2   clear
3   Optimal = [1.6597, 1.7671, 1.7433, 1.849, 1.7632, 1.949, 1.7063, ...
        1.8431, 1.8036, 1.8129];
4   Worst = max(Optimal)
5   Best = min(Optimal)
6   Median = median(Optimal)
7   Mean = mean(Optimal)
8   StdDev = std(Optimal)
```

If we run this simple program, the following result will be generated in the command window:

```
Worst =
     1.9490
Best =
     1.6597
Median =
     1.7853
Mean =
     1.7897
StdDev =
     0.0814
```

In addition to the aforementioned criteria, the algorithm computational speed[15] can also be used as one performance criterion. Advanced performance evaluations can be done by conducting some statistical and sensitivity tests; as reported in Al-Roomi and El-Hawary (2016b). Also, some additional performance evaluations are shown in Bartz-Beielstein (2006) with an extensive description and new ideas.

1.1.10 Types of Optimization Techniques

In general, the optimization techniques are classified into two main categories, called classical (or traditional) and modern (or non-traditional) techniques. It is hard to collect them all in one tree-diagram. Instead, the most popular types are briefly summarized in Figure 1.7 (Momoh, 2001a; Rao, 2009; Simon, 2013; Venkataraman, 2009; Deb, 2010; Fletcher, 1987; Forst and Hoffmann, 2010; Chong and Żak, 2001; Nocedal and Wright, 2006; Lee and El-Sharkawi, 2008).

1.2 Classical Optimization Algorithms

Traditional optimization algorithms are the most known methods. They need no special knowledge from other fields of science, like biology and its branches. These algorithms are straight-forward, which follow systematic mathematical steps, like finding some derivatives, constructing matrices, tracing the error deviation between two iterations, etc.

Some advantages of these optimization techniques are summarized as follows (Momoh, 2001a; Fletcher, 1987; Forst and Hoffmann, 2010; Chong and Żak, 2001; Nocedal and Wright, 2006; Lee and El-Sharkawi, 2008):

- They are very fast optimization algorithms. If the initial guess is good, then these techniques become very useful to be embedded in systems that need fast decisions, such as power system protection. The high computational speed performance comes from their simple structures and also because they are **single-point algorithms**. Thus, dealing with just one individual per each iteration will definitely accelerate the computational speed and save part of the memory.[16]
- These techniques are very old and thus they are well-established and available everywhere in thousands of books, which make them easy to be reached. Add to that, there are many ready-made codes with different programming formats and languages such as **Fortran**, **MATLAB**, **JAVA**, **C/C++**, **Python**, **Julia**, etc.
- They provide one unique solution every time they are executed with the same initialization parameters.
- They have a solid mathematical foundation and principles.
- These **gradient-based methods** can be used as **fine-tuning** sub-algorithms in hybrid global optimization techniques, as will be covered later.

---

15 Also known in the literature as the processing speed (CPU time).

16 This term (i.e. the computational speed) may not be correct in some conditions. Suppose, there is a very *high*-dimensional problem ($n = 10^4 \rightarrow 10^6$) and it needs to be optimized. Solving it by classical methods may take tens of hours or even multiple days. On the opposite side, modern optimization algorithms could find acceptable near optimal solutions within just a few iterations.

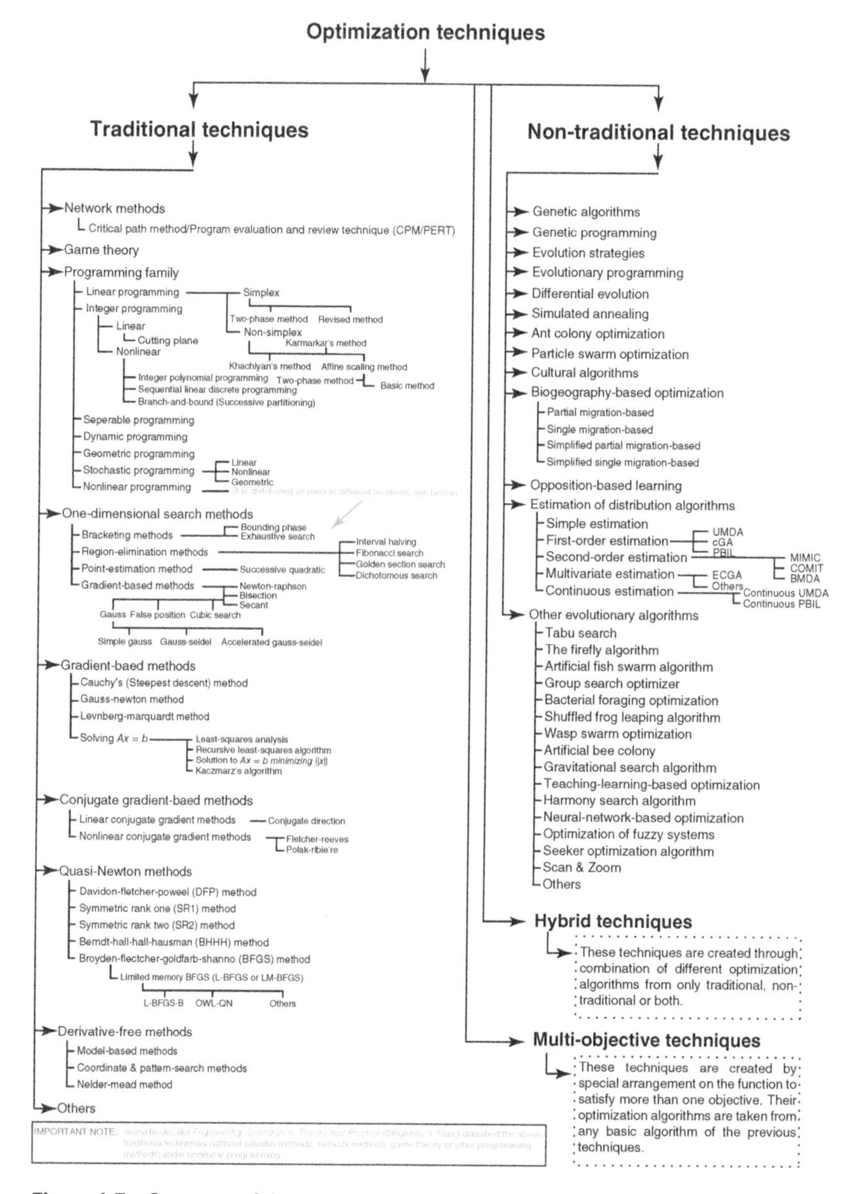

Figure 1.7 Summary of the most popular optimization techniques.

Unfortunately, there are many limitations associated with this category of optimization algorithms. Some of these weaknesses are summarized as follows:

- Some of them are very complicated algorithms, hard to be implemented, and require lots of advanced mathematical expressions.
- Some of them are restricted to *one*-dimensional problems.
- Some of them have matrices and/or require derivatives.
- Most of them are prone to easily trap into local optima, especially if the initial guess is not good.
- This field is very old and thus it is hard to get a creditable contribution

In this book, some classical optimization methods are used[17]. However, they are not individually applied. Instead, they are hybridized with **meta-heuristic optimization algorithms** as fine-tuners. It has to be mentioned that the Python's and MATLAB's built-in libraries are used for these classical optimization algorithms.

1.2.1 Linear Programming

Linear optimization or **linear programming** (**LP**) is one of **operations research** (**OR**) tools used to find the best or optimal solution of linear **mathematical models**. In real-world applications, the nonlinearity enters in different parts of the problem, such as in its objective function as well as equality and inequality constraints. Thus, LP is considered as a special type of **mathematical programming** that requires the following conditions (Marriott and Stuckey, 1998; Forst and Hoffmann, 2010; Rao, 2009; Venkataraman, 2009):

- Linear objective function
- Linear constraints
- Non-negative decision variables

1.2.1.1 Historical Time-Line

Based on the proverb "***necessity is the mother of invention***", the origin of LP can be traced back to the World War II[18] when the armies tried to find a proper way to deal with the military problems effectively and economically; especially when the army A wants to increase the losses of the enemy B with some limited and scarce resources. The history told us that the first one is the Russian mathematician Leonid Kantorovich who in 1939 proposed a method to solve linear problems. However, the work was published in 1959 (Schrijver, 1998). Also, at the same time, the Dutch-American mathematician and economist Tjalling Charles Koopmans independently proposed a method to solve linear economic problem (Sierksma, 2002). Three years later (i.e. in 1941), the American mathematician and physicist Frank Lauren Hitchcock successfully formulated a method that can solve linear transportation problems (Schrijver, 1998). After around five years (i.e. between 1946 and 1947), George Bernard Dantzig independently generalized the formulation of linear programming as a suitable tool to solve planning problems in US Air Force. His first paper was titled "***Programming in a Linear Structure***" (Dantzig, 1948). The term "linear programming" is coined by Koopmans in 1948, and one year later Dantzig published the simplex method.

Nowadays, LP is a very popular linear problem solver for many applications in mathematics, science, engineering, and business. LP is considered as an entry-level to understand more complicated programming methods. As can be clearly seen in Figure 1.7, LP is just one type of **programming family**, and it is available in two main categories called the **simplex** and **non-simplex methods**. Simplex LP problems can be solved by two ways called **tabular** and **algebraic** forms. The latter is an extension of the conventional algebraic method, which is used to overcome some weaknesses such as requiring many fixed steps and non-smart enough to jump from infeasible sets. Also, LP problems can be solved by searching within the plot boundaries of the constraints, which is known as the **graphical method**.

---

17 Please note that the Newton–Raphson algorithm and its modified versions, which are classified as classical optimization algorithms, are used in many studies of electric power systems engineering.

18 Some authors said that the origin is dated back to the beginning of the nineteenth century when Fourier suggested a method to solve linear inequalities problem in 1827, and that method is known as ***Fourier-Motzkin elimination*** (**FME**) (Sierksma, 2002).

1.2.1.2 Mathematical Formulation of LP Problems

By referring to the preceding three conditions of LP and the standard formulations given in Section 1.1.7, the mathematical model can be expressed in many forms. For example, it can be as follows:

$$\begin{aligned}
\text{Maximize:} \quad & c_1x_1 + c_2x_2 + \cdots + c_nx_n \\
\text{Subjected to:} \quad & a_{11}x_1 + a_{12}x_2 + \cdots + a_{1n}x_n \leqslant b_1 \\
& a_{21}x_1 + a_{22}x_2 + \cdots + a_{2n}x_n \leqslant b_2 \\
& \vdots \\
& a_{m1}x_1 + a_{m2}x_2 + \cdots + a_{mn}x_n \leqslant b_n \\
\text{and:} \quad & x_1 \geqslant 0\ ;\ x_2 \geqslant 0\ ;\ \cdots\ ; x_n \geqslant 0
\end{aligned} \tag{1.51}$$

Or, it can be formulated in vector notation as follows:

$$\begin{aligned}
\text{Maximize:} \quad & [C]\,[X] \\
\text{Subjected to:} \quad & [A]\,[X] \leqslant [b] \\
\text{and:} \quad & [X] \geqslant [0]
\end{aligned} \tag{1.52}$$

where the vectors of n variables and m constraints are defined as follows:

$$\begin{aligned}
[C]_{1\times n} &= \langle c_1, c_2, \ldots, c_n \rangle \\
[X]_{n\times 1} &= \langle x_1, x_2, \ldots, x_n \rangle^T \\
[A]_{m\times n} &= [a_{ij}] \\
[b]_{m\times 1} &= \langle b_1, b_2, \ldots, b_m \rangle^T
\end{aligned}$$

1.2.1.3 Linear Programming Solvers

Because LP is used since a long time ago, so it is not surprising that many free and commercial software and programming languages have special packages and/or libraries to solve LP problems (LPPs) by **simplex**, **revised simplex**, **interior-points**, etc. For instance, MATLAB, Python, **Mathematica**, **Maple**, **GNU Octave**, **LINGO**, and **MS Excel**. To give a brief idea, let's solve the numerical problem given in Example 1.6.

Example 1.6 Using MATLAB, find the maximal solution to the following LP problem using all the algorithms available in the `linprog` package:

$$z(x_1, x_2) = x_1 + x_2$$

Subjected to:

$$\begin{aligned}
x_1 + x_2 &\leqslant 12 \\
-2x_1 + x_2 &\leqslant \frac{9}{2} \\
x_1 - 3x_2 &\leqslant \frac{10}{3} \\
-x_1 - x_2 &\leqslant -3 \\
0 \leqslant x_1 &\leqslant 10 \\
0 \leqslant x_2 &\leqslant 15
\end{aligned}$$

Solution

The problem can be mathematically expressed as follows:

$$\max_{x_1,x_2} f^T X \text{ such that } \begin{cases} A \cdot X \leqslant b \\ X^{\min} \leqslant X \leqslant X^{\max} \end{cases}$$

where A, b, $X^{\min}$, and $X^{\max}$ are expressed in vector notation as follows:

$$A = \begin{bmatrix} 1 & 1 \\ -2 & 1 \\ 1 & -3 \\ -1 & -1 \end{bmatrix}, \quad b = \begin{bmatrix} 12 \\ \frac{9}{2} \\ \frac{10}{3} \\ -3 \end{bmatrix}, \quad X^{\min} = \begin{bmatrix} 0 \\ 0 \end{bmatrix}, \quad X^{\max} = \begin{bmatrix} 10 \\ 15 \end{bmatrix}$$

The following MATLAB code is given in `LP_2DimProb.m`:

```
1    clc
2    clear
3    algorithm = {'dual-simplex', 'interior-point-legacy', 'interior-point'};
4    % Linear inequality constraints:
5    A = [1 1; -2 1; 1 -3; -1 -1];
6    b = [12 9/2 10/3 -3]';
7    % Linear {Mathematical\index{s}{mathematical model} constraints:
8    Aeq = [];
9    beq = [];
10   % Lower and Upper Bounds:
11   lb = [0 0];
12   ub = [10 15];
13   % Objective function "x(1) + x(2)":
14   f = [-1 -1];
15   for i = 1:3
16       disp(['-------------------------| ', algorithm{i}, ' ...
             |-------------------------'])
17       % Solve the linear program
18       options = optimoptions('linprog', 'Algorithm', algorithm{i}, ...
             'display', 'iter');
19       tic;
20       X = linprog(f, A, b, Aeq, beq, lb, ub, [], options)
21       toc;
22       % Maximal point
23       Obj = X(1) + X(2)
24   end
```

If we run this simple program, the following result will be generated in the command window:

```
-------------------------| dual-simplex |-------------------------
LP preprocessing removed 1 inequalities, 0 equalities,
0 variables, and 2 non-zero elements.
  Iter      Time            Fval  Primal Infeas    Dual Infeas
     0     0.001   -2.500000e+01   2.478279e+01   0.000000e+00
     2     0.001   -1.200000e+01   0.000000e+00   0.000000e+00
Optimal solution found.
X =
    2.5000
    9.5000
Elapsed time is 0.015232 seconds.
Obj =
```

```
     12.0000
--------------------| interior-point-legacy |-------------------
  Residuals:    Primal      Dual       Upper     Duality      Total
                Infeas     Infeas      Bounds      Gap          Rel
                A*x-b    A'*y+z-w-f  {x}+s-ub  x'*z+s'*w     Error
  ---------------------------------------------------------------
  Iter    0:   3.21e+02 5.00e+00 2.65e+02 2.20e+03 2.36e+01
  Iter    1:   6.97e+00 1.40e-15 5.76e+00 1.71e+02 8.73e-01
  Iter    2:   1.33e-15 1.13e-15 0.00e+00 2.88e+01 7.06e-01
  Iter    3:   2.39e-15 2.54e-16 0.00e+00 2.85e-03 2.37e-04
  Iter    4:   4.55e-15 1.89e-16 0.00e+00 1.42e-07 1.19e-08
  Iter    5:   1.05e-14 5.68e-17 0.00e+00 1.47e-14 1.04e-15
Optimization terminated.
X =
    4.9178
    7.0822
Elapsed time is 0.013267 seconds.
Obj =
    12
----------------------| interior-point |----------------------
LP preprocessing removed 1 inequalities, 0 equalities,
0 variables, and 2 non-zero elements.
 Iter          Fval   Primal Infeas    Dual Infeas  Complementarity
    0   -4.310189e+00   9.769963e-15      1.038030e+01        5.190151e+00
    1   -5.249546e+00   1.509903e-14      1.347566e+00        3.929496e-01
    2   -1.199662e+01   2.842171e-14      6.737830e-04        1.441714e-01
    3   -1.199978e+01   8.881784e-15      5.117355e-08        2.237315e-04
    4   -1.200000e+01   1.776357e-14      1.110223e-16        1.863603e-15
Minimum found that satisfies the constraints.
Optimization completed because the objective function is non-decreasing
in feasible directions, to within the selected value of the function
tolerance, and constraints are satisfied to within the selected value of
the constraint tolerance.
X =
    6.2772
    5.7228
Elapsed time is 0.019607 seconds.
Obj =
    12
```

As can be seen from the results obtained by different LP algorithms, the fitness is always 12, but the decision variables could change from one solver to another based on the technique used (whether it is a dual-simplex, interior-point, etc.), the lowest acceptable error ε, etc. For this particular linear optimization problem, it is found that the `interior-point-legacy` algorithm is the fastest one. The `dual-simplex` algorithm consumed 14.81% and the `interior-point` algorithm consumed 47.79% more CPU time. In general, LP is very efficient if the optimization problem is expressed in a linear form. In the next chapters, we will see how to implement LP to optimally coordinate protective relays. Also, we will see how to linearize this engineering problem in order to be able to use LP in **multi-start mode**.

1.2.2 Global-Local Optimization Strategy

Let's return back to (1.5) and take $f(x)$ as an objective that needs to be optimized in **nonlinear programming (NLP)**. The optimal solution will not change if the same algorithm and its settings are not changed. This is true if t remains constant. Otherwise, the variation in t will cause β to change, which eventually forces the optima of f to move away from its initial optimal location. The same phenomenon can be seen in LP, **integer LP (ILP)**, **integer NLP (INLP)**, **mixed-integer LP (MILP)**, and **mixed-integer NLP (MINLP)**, etc.[19] Although the focus here is on LP, the same concept can be applied to any other classical algorithm.

1.2.2.1 Multi-Start Linear Programming

One way to tackle this issue is to initiate LP in multiple runs and then finding the best of the best. That is, an external `for`-loop is embedded in the algorithm, and then a random generating unit is placed before LP to provide new parameters in each run. Thus, the search space can be globally explored by that stochastic unit. The best random parameters are chosen to optimize the problem locally via LP. This mechanism is depicted in Figure 1.8.

Example 1.7 Consider Example 1.6. Assume that the objective function was linearized by omitting the following sinusoidal term when $\delta = 1$ radian:

$$z(x_1, x_2) = \left(1.1884x_1 + 1.1884x_2\right) \cdot \sin(\delta)$$

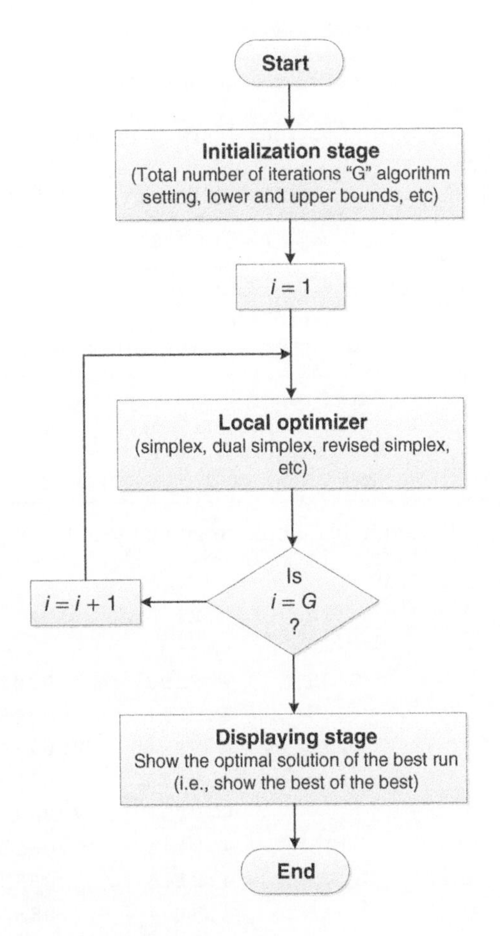

Figure 1.8 Mechanism of multi-start LP.

19 ILP and INLP are sub-categories of a wider title called **combinatorial optimization** where classical, meta-heuristic, and hybrid optimization algorithms could be seen.

Modify the preceding program to find the maxima at $\delta = \{0.25, 0.5, \dots, 6.25\}$. Based on the speed performance, only the `interior-point-legacy` algorithm should be used.

Solution

The problem can be mathematically expressed as follows:

The following MATLAB code is given in `MultiStart_LP_2DimProb.m`:

```
clc
clear
% variation in the radian angle parameter
Δ = 0.25:0.25:6.25;
% Linear inequality constraints:
A = [1 1; -2 1; 1 -3; -1 -1];
b = [12 9/2 10/3 -3]';
% Linear **check **{Mathematical constraints}:
Aeq = [];
beq = [];
% Lower and Upper Bounds:
lb = [0 0];
ub = [10 15];
for i = 1:length(Δ)
    % Objective function "[x(1) + x(2)] * sin(Δ)":
    f = [-1.1884 -1.1884]*sin(Δ(i));
    % Solve the linear program
    options = optimoptions('linprog', 'Algorithm', ...
        'interior-point-legacy', 'display', 'off');
    X(i,:) = linprog(f, A, b, Aeq, beq, lb, ub, [], options);
    % Maximal point
    Obj(i) = (1.1884*X(i, 1) + 1.1884*X(i, 2))*sin(Δ(i));
end
% Showing all the best solutions
Tab = table(Δ', X(:,1), X(:,2), Obj');
Tab.Properties.VariableNames = {'Δ' 'x1' 'x2' 'Obj'}
% Showing the best of the best
[zmax, ind] = max(Obj);
Global_Xopt = X(ind,:)
Best_Fitness = zmax
Best_Δ = Δ(ind)
```

If we run this simple program, the following result will be generated in the command window:

```
Tab =
    delta      x1        x2         Obj
    -----    ------    ------    --------
    0.25     4.9298    7.0702      3.5282
     0.5     4.9252    7.0748       6.837
    0.75     4.9211    7.0789      9.7207
       1     4.9178    7.0822          12
    1.25     4.9155    7.0845      13.533
     1.5     4.9145    7.0855      14.225
    1.75     4.9148    7.0852      14.032
       2     4.9164    7.0836      12.967
    2.25     4.9191    7.0809      11.096
     2.5     4.9228    7.0772      8.5347
```

```
    2.75      4.9271      7.0729      5.4428
       3      4.9319      7.0681      2.0125
    3.25       1.262       1.738    -0.38574
     3.5      1.1951      1.8049     -1.2506
    3.75      1.1676      1.8324     -2.0377
       4       1.154       1.846     -2.6982
    4.25        1.19        1.81     -3.1908
     4.5      1.3379      1.6621     -3.4851
    4.75      1.2876      1.7124     -3.5627
       5      1.3887      1.6113     -3.4188
    5.25      1.1663      1.8337     -3.0623
     5.5      1.1567      1.8433     -2.5154
    5.75      1.1722      1.8278     -1.8121
       6      1.2139      1.7861    -0.99617
    6.25      1.2672      1.7328    -0.11829
Global_Xopt =
    4.9145      7.0855
Best_Fitness =
   14.2251
Best_delta =
    1.5000
```

1.2.2.2 Hybridizing LP with Meta-Heuristic Optimization Algorithms as a Fine-Tuning Unit

For simple optimization problems, the multi-start strategy is a good choice to go with. However, this approach is inefficient to cover the entire domain searching for the optimal spot. For instance, the following 2-dimensional problem is known as Trefethen's function:[20]

$$\begin{aligned} f(X) = e^{\sin\left(50x_1\right)} &+ \sin\left(60e^{x_2}\right) + \sin\left(70\sin\left(x_1\right)\right) \\ &+ \sin\left(\sin\left(80x_2\right)\right) - \sin\left(10\left(x_1 + x_2\right)\right) + \frac{1}{4}\left(x_1^2 + x_2^2\right) \end{aligned} \tag{1.53}$$

Of course, it is a nonlinear function, and thus we cannot optimize it using LP. However, we want to use it as a clear example to illustrate the problem faced with all (LP, NLP, ILP, INLP, MILP, and MINLP). The 3D plot of this unconstrained benchmark function, in the domain of $[\ -1.0 \quad -0.5\ ]^T \leqslant [\ x_1 \quad x_2\ ]^T \leqslant [\ 1.0 \quad 0.5\ ]^T$, is shown in Figure 1.9. As can be clearly seen, there are lots of peaks and valleys. Thus, for both minimization and maximization modes, the optimal value depends on the initial point or/and pre-defined parameters; similar to δ in the preceding example. Because these local points are too many, so determining the optimal one through the multi-start strategy is a very difficult task and it may be semi or completely impossible.

The other reason is that the multi-start strategy mainly depends on a random process. Thus, with many pre-defined parameters, the mission to get the optimal set of these parameters is so hard. To clarify this point, suppose that you have a linearized problem that should be optimized by LP and this linearization process requires you to pick a set of 100 parameters,[21] let's say $A = \left[a_1, a_2, \ldots, a_{100}\right]$, before initiating LP. If the optimal set of these parameters is $A^* = \left\{a_1^*, a_2^*, \ldots, a_{100}^*\right\} = \{0.11, 0.22, \ldots, 11\}$, what is the chance to reach that optimal set by

20 One of the benchmark functions used in unconstrained optimization (Al-Roomi, 2015a).

21 This is the case with linearized cost functions of optimal relay coordination problems.

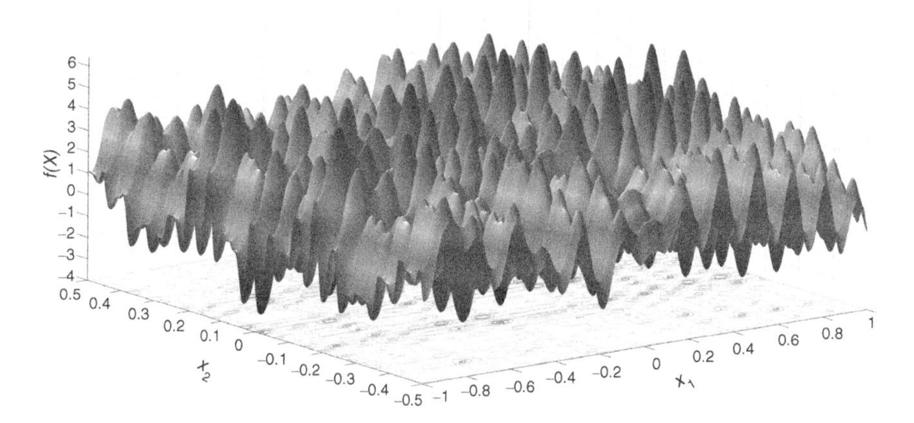

Figure 1.9 3D plot of Trefethen's function given in (1.53).

only depending on a pure random process? Of course, it is almost impossible. It requires some sort of intelligence to settle on that optimal set quickly without expensive computation. This is the main reason for hybridizing LP with modern meta-heuristic optimization algorithms. We want to compromise between the exploration level and the exploitation level. We want to merge the probabilistic and gradient process without forgetting the stochastic process provided by the mutation stage.[22] This hybrid optimization algorithm is illustrated in Figure 1.10. In this book, we will hybridize LP with only one meta-heuristic optimization algorithm. However, the same concept can be applied to any other meta-heuristic optimization algorithm.

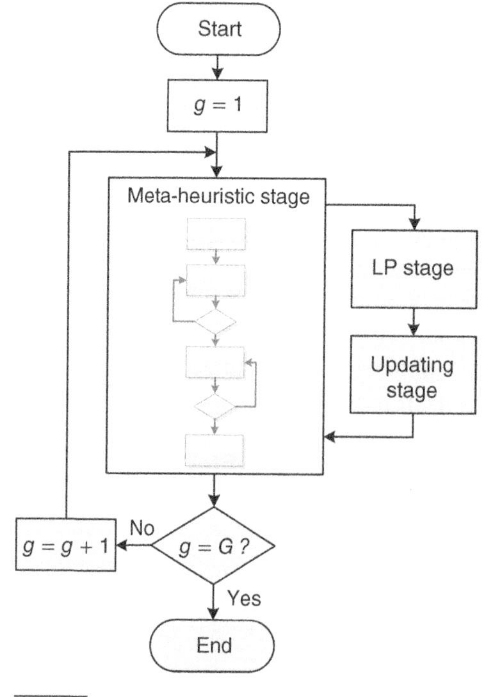

Figure 1.10 Main mechanism of hybrid meta-heuristic/LP optimization algorithms.

22 Or its equivalents in some non-evolutionary-based meta-heuristic optimization algorithms.

1.3 Meta-Heuristic Algorithms

Meta-heuristic optimization algorithms are also called **modern optimization algorithms**, **non-classical optimization algorithms**, and **non-traditional optimization algorithms**. They are classified into **probabilistic** and **stochastic algorithms** (Deb, 2010; Venkataraman, 2009; Eiben and Smith, 2003; Luke, 2009; Rao, 2009; Simon, 2013; Floreano and Mattiussi, 2008). The disadvantages of classical optimization algorithms motivate many researchers to think about other innovative approaches that can solve all the headache problems of the classical optimization algorithms. To think about approaches that can converge accurately and quickly to the area where the global optima are located without using any derivatives, matrices, or even initial points. Many mathematical theories, principles, and foundations taken from different disciplines have been utilized to design the mechanism of these unfamiliar algorithms. For example:[23]

- Based on the branches of physics science: the **simulated annealing** (**SA**) algorithm from the physics of matter, and the **gravitational search algorithm** (**GSA**) from the law of gravity.
- Based on the branches of biology science: the **genetic algorithm** (**GA**) from the genetic science, and the **biogeography based-optimization** (**BBO**) from the biogeography science.

With these inspirations, many researchers have successfully applied these models to solve highly complicated problems by just plugging them into the design function and then pressing the run button.

These modern optimization techniques can also be classified according to their number of generated individuals or candidate solutions per each iteration or generation. Algorithms having only one individual per each iteration (like traditional techniques) are called **single-point** or **single-solution algorithms**. Such algorithms are the **tabu search** (**TS**) and SA algorithms. Whereas, the term **population-based** (**PB**) **algorithm** is used to identify the algorithms that generate multiple individuals per each generation. Such algorithms are the **particle swarm optimization** (**PSO**) and the **ant colony optimization** (**ACO**) algorithms. Of course, the preceding non-traditional single-solution algorithms also can escape from trapping into local optima, and they are considered as **global optimizers**.

Also, it is essential to differentiate between **heuristic**, **meta-heuristic** and **hyper-heuristic** terms, because many references just mentioned them as "heuristic" methods without any clear definition. It has to be known that the heuristic-based techniques are problem-dependent algorithms that can learn from the given information about the design function and then adapting with that to provide good results. Such techniques could fail due to the chance to trap into local optima. Now, suppose that no useful information is given about the optimization problem; like the proper path that can guide the algorithm to reach the optimum solution or what the optima looks like. Then the techniques that can reach the space, where the global optima are located without knowing how and from where to start, are called meta-heuristic-based techniques.[24] These algorithms benefit from the data of the previously obtained solutions to determine the location of the best solution within the search space. Therefore, they might not guarantee to settle exactly on the global optimum solution. Instead, they might provide some approximate and **near-global** solutions.

---

23 It has to be said that there are different sub-categories and sub-sub-categories of these algorithms. For instance, some of them are called **evolutionary algorithms** (**EAs**), which are part of a more broad sub-category called the **nature-inspired algorithms**. Also, each EA has many versions. For example, some versions of GA are binary GA, real GA, stud GA (sGA), micro GA (μGA), etc.

24 Some references called them **black-box optimization algorithms**.

Hyper-heuristic is the most advanced technique compared with the preceding two techniques. It depends on the heuristic or meta-heuristic technique to create its own search space. Therefore, the word "hyper-heuristic" can be translated as a "heuristic search for heuristic" (Luke, 2009; Talbi, 2009; Gendreau and Potvin, 2010).

Some advantages of modern optimization algorithms are (Deb, 2010; Eiben and Smith, 2003; Venkataraman, 2009; Luke, 2009; Rao, 2009; Simon, 2013):

- No need any more to find derivatives or constructing matrices.
- Very robust and can converge to the space of the global optima.
- They are relatively easier to be understood.
- The designers need less time and limited libraries to create their programs.
- The designed programs can be used as general purpose optimizers for any plugged-in design function.
- This branch is new, and thus the door is widely open to conduct many types of research in this field.

On the other hand, some of the main disadvantages of modern optimization algorithms are:

- The population-based techniques are time-consuming. The processing time will increase significantly as the population size increases.
- The final solution is tuned after completing many generations because the algorithms are **probabilistic-/stochastic-based methods**.
- Still, the number of available references and codes is smaller than that of the traditional techniques, especially for the most recent invented techniques.
- To understand their principles and how they work, the researcher needs to study some special topics in physics, biology, or/and other branches of science. This may become hard for those people who have just pure background in the field of mathematics, engineering, computer science, or economics.

In Sections 1.3.1 and 1.3.2, some detailed information about the meta-heuristic optimization algorithms, used in this book, is given.

1.3.1 Biogeography-Based Optimization

The mechanism of this new population-based evolutionary algorithm is inspired by an old scientific study conducted in biogeography[25] by the ecologists Robert H. MacArthur and Edward O. Wilson in the period between 1960–1967 (MacArthur and Wilson, 1963, 1967). This study is known as "***The Theory of Island Biogeography***". The theory proposes that the dynamic equilibrium between immigrated and extinct species controls the endemic species on isolated islands.[26]

The **immigration rate** (λ) and the **emigration rate** (μ) can be set in many ways.[27] To simplify the mathematical process, MacArthur and Wilson used a simplified linear migration model

25 Biogeography is a branch of biological science. It heavily relies on theories and data taken from ecology, population biology, systematics, evolutionary biology, and earth science (Lomolino et al., 2009). Biogeography seeks to describe, analyze, and explain the geographic patterns and changes in the distribution of ecosystems and fossil species of plants (**flora**) and animals (**fauna**) through geological space and time (Jones, 1980; Cox and Moore, 1993).
26 In island biogeography, the word "***island***" could be an aquatic island, desert oasis, lakes or ponds, mountain-tops (sky-islands), caves, individual plants, microcosms or even patches of terrestrial ecosystems (Nierenberg, 1995; Myers and Giller, 1990; Alroomi et al., 2013b).
27 The emigration and immigration rates can be modeled as exponential, logistic, linear, etc. (Lomolino et al., 2009; Cody, 2006; MacArthur and Connell, 1966). Also, the maximum emigration and immigration rates can be unequal (i.e. $I \neq E$; where $I = \lambda_{max}$ and $E = \mu_{max}$) (MacArthur and Wilson, 1967; MacArthur and Connell, 1966). Moreover,

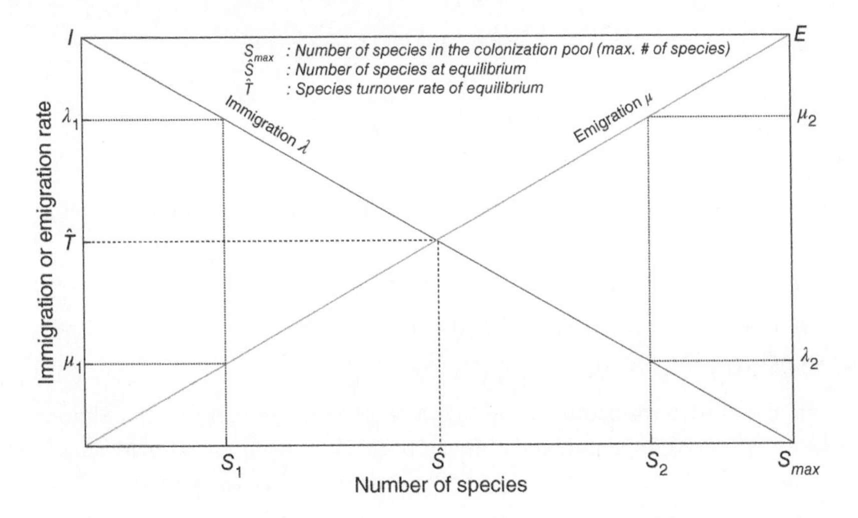

Figure 1.11 Simplified equilibrium model of biota in a single island.

with equal maximum immigration and emigration rates (i.e. $I = \lambda_{\max} = E = \mu_{\max}$) as shown in Figure 1.11. The symbol $\hat{T}$ denotes the species turnover rate, which happens when the species density settles on the equilibrium state $\hat{S}$. The symbol $S_{\max}$ denotes the maximum number of endemic species on that island (MacArthur and Wilson, 1963, 1967; MacArthur and Connell, 1966). Thus, $\lambda_{\max}$ or I happens when there is no available species on the ith island, and $\mu_{\max}$ or E happens when all the nests are occupied by the emigrated species from the mainland and/or other islands (MacArthur, 1972).

Simon (2008a) presented the first version of BBO. The population size (N) is simplified to be equal to $S_{\max}$. Therefore, λ_i and μ_i of the basic migration model, depicted in Figure 1.11, can be determined as follows:

$$\mu_i = \left(\frac{E}{N}\right) i \tag{1.54}$$

$$\lambda_i = 1 - \mu_i = I\left(1 - \frac{i}{N}\right) \tag{1.55}$$

Suppose that at time (t) the island contains i species with probability ($Pr_i(t)$), then the variation of the probability from t to $t + \Delta t$ can be described as follows (MacArthur and Wilson, 1963, 1967):

$$Pr_i(t + \Delta t) = Pr_i(t)(1 - \lambda_i \Delta t - \mu_i \Delta t) + Pr_{i-1}(t)\lambda_{i-1}\Delta t + Pr_{i+1}(t)\mu_{i+1}\Delta t \tag{1.56}$$

Considering (1.56), to have i species at time ($t + \Delta t$), one of the following three conditions should be satisfied (MacArthur and Wilson, 1963, 1967):

1. i species at time t, and no migrated species during the interval Δt;
2. $(i - 1)$ species at time t, and one species immigrated;
3. $(i + 1)$ species at time t, and one species emigrated.

---

the equilibrium location ($\hat{S}$) can be shifted to the left-side or to the right-side based on the type of rate functions, the area of island and/or the distance or isolation between the recipient island and the source island or mainland (Lomolino et al., 2009; MacArthur and Wilson, 1967; Losos and Ricklefs, 2010; Al-Roomi and El-Hawary, 2016c).

From calculus, it is known that the ratio $\left(\frac{\Delta Pr_i}{\Delta t}\right)$ approaches $\dot{P}r_i(t)$ as $\Delta t \to 0$:

$$\begin{aligned}\dot{P}r_i(t) &\cong \lim_{\Delta t \to 0} \frac{Pr_i(t+\Delta t) - Pr_i(t)}{\Delta t} \\ &\cong -(\lambda_i + \mu_i)Pr_i(t) + \lambda_{i-1}Pr_{i-1}(t) + \mu_{i+1}Pr_{i+1}(t)\end{aligned} \tag{1.57}$$

By considering the preceding three conditions, (1.57) can be re-expressed with the following three cases:

$$\dot{P}r_i(t) = \begin{cases} -(\lambda_i + \mu_i)Pr_i(t) + \mu_{i+1}Pr_{i+1}(t), & \text{if } i = 0 \\ -(\lambda_i + \mu_i)Pr_i(t) + \lambda_{i-1}Pr_{i-1}(t) + \mu_{i+1}Pr_{i+1}(t), & \text{if } 1 \leqslant i \leqslant N-1 \\ -(\lambda_i + \mu_i)Pr_i(t) + \lambda_{i-1}Pr_{i-1}(t)), & \text{if } i = N \end{cases} \tag{1.58}$$

The value of $\dot{P}r_i(t)$ can also be determined by using the matrix technique presented in Simon (2008a), which is successfully proved in Igelnik and Simon (2011). Thus, using the known values of $Pr_i(t)$ and $\dot{P}r_i(t)$, the value of $Pr_i(t + \Delta t)$ given in (1.56) can be approximated as follows:

$$Pr_i(t+\Delta t) \cong Pr_i(t) + \dot{P}r_i(t)\Delta t \tag{1.59}$$

Equation (1.59) is the final form that should be used in the BBO program to calculate $Pr(t + \Delta t)$. To find $Pr(t)$, Simon (2008a) used two methods; one is iterative and the other is based on eigenvector.

Example 1.8 Suppose that the number of islands is $N = 7$. Calculate $Pr_i(t)$. (**Hint:** iterate (1.58) with z iterations; where $z = N - 1$.)

Solution

First, find λ_i and μ_i:

$$\lambda_i = \begin{bmatrix}\lambda_0 & \lambda_1 & \lambda_2 & \lambda_3 & \lambda_4 & \lambda_5 & \lambda_6\end{bmatrix}^T$$
$$\mu_i = \begin{bmatrix}\mu_0 & \mu_1 & \mu_2 & \mu_3 & \mu_4 & \mu_5 & \mu_6\end{bmatrix}^T$$

By using (1.54)–(1.55) with $E = I = 1$ and $S_{\max} = z = N - 1 = 6$:

$\mu_0 = \left(\frac{1}{6}\right) \times 0 = 0.0000$	$\lambda_0 = 1 \times \left(1 - \frac{0}{6}\right) = 1.0000$
$\mu_1 = \left(\frac{1}{6}\right) \times 1 = 0.1667$	$\lambda_1 = 1 \times \left(1 - \frac{1}{6}\right) = 0.8333$
$\mu_2 = \left(\frac{1}{6}\right) \times 2 = 0.3333$	$\lambda_2 = 1 \times \left(1 - \frac{2}{6}\right) = 0.6667$
$\mu_3 = \left(\frac{1}{6}\right) \times 3 = 0.5000$	$\lambda_3 = 1 \times \left(1 - \frac{3}{6}\right) = 0.5000$
$\mu_4 = \left(\frac{1}{6}\right) \times 4 = 0.6667$	$\lambda_4 = 1 \times \left(1 - \frac{4}{6}\right) = 0.3333$
$\mu_5 = \left(\frac{1}{6}\right) \times 5 = 0.8333$	$\lambda_5 = 1 \times \left(1 - \frac{5}{6}\right) = 0.1667$
$\mu_6 = \left(\frac{1}{6}\right) \times 6 = 1.0000$	$\lambda_6 = 1 \times \left(1 - \frac{6}{6}\right) = 0.0000$

Now, select initial point:

$$Pr_i^{(0)}(t) = \begin{matrix} [\,\frac{1}{N} & \frac{1}{N} & \frac{1}{N} & \frac{1}{N} & \frac{1}{N} & \frac{1}{N} & \frac{1}{N}\,]^T \\ \downarrow & \downarrow & \downarrow & \downarrow & \downarrow & \downarrow & \downarrow \\ S=0 & S=1 & S=2 & S=3 & S=4 & S=5 & S=6 \end{matrix}$$

$$\therefore \; Pr_i^0(t) = \begin{bmatrix} Pr_0^0 & Pr_1^0 & Pr_2^0 & Pr_3^0 & Pr_4^0 & Pr_5^0 & Pr_6^0 \end{bmatrix}^T$$
$$= \begin{bmatrix} \frac{1}{7} & \frac{1}{7} & \frac{1}{7} & \frac{1}{7} & \frac{1}{7} & \frac{1}{7} & \frac{1}{7} \end{bmatrix}^T$$
$$= \begin{bmatrix} 0.1429 & 0.1429 & 0.1429 & 0.1429 & 0.1429 & 0.1429 & 0.1429 \end{bmatrix}^T$$

By using (1.58): (**Remember:** for simplified migration model; $\lambda_i + \mu_i = 1$ "always")

$$\text{at } S = 0 : \dot{Pr}_0(t) = -(\lambda_0 + \mu_0)Pr_0 + \cancel{\lambda_{-1}Pr_{-1}} + \mu_1 Pr_1$$
$$= -1.0000(0.1429) + 0.1667(0.1429)$$
$$= -0.1191$$

$$\text{at } S = 1 : \dot{Pr}_1(t) = -(\lambda_1 + \mu_1)Pr_1 + \lambda_0 Pr_0 + \mu_2 Pr_2$$
$$= -1.0000(0.1429) + 1.0000(0.1429) + 0.3333(0.1429)$$
$$= \mathbf{0.0476}$$

$$\text{at } S = 2 : \dot{Pr}_2(t) = -(\lambda_2 + \mu_2)Pr_2 + \lambda_1 Pr_1 + \mu_3 Pr_3$$
$$= -1.0000(0.1429) + 0.8333(0.1429) + 0.5000(0.1429)$$
$$= \mathbf{0.0476}$$

$$\text{at } S = 3 : \dot{Pr}_3(t) = -(\lambda_3 + \mu_3)Pr_3 + \lambda_2 Pr_2 + \mu_4 Pr_4$$
$$= -1.0000(0.1429) + 0.6667(0.1429) + 0.6667(0.1429)$$
$$= \mathbf{0.0476}$$

$$\text{at } S = 4 : \dot{Pr}_4(t) = -(\lambda_4 + \mu_4)Pr_4 + \lambda_3 Pr_3 + \mu_5 Pr_5$$
$$= -1.0000(0.1429) + 0.5000(0.1429) + 0.8333(0.1429)$$
$$= \mathbf{0.0476}$$

$$\text{at } S = 5 : \dot{Pr}_5(t) = -(\lambda_5 + \mu_5)Pr_5 + \lambda_4 Pr_4 + \mu_6 Pr_6$$
$$= -1.0000(0.1429) + 0.3333(0.1429) + 1.0000(0.1429)$$
$$= \mathbf{0.0476}$$

$$\text{at } S = 6 : \dot{Pr}_6(t) = -(\lambda_6 + \mu_6)Pr_4 + \lambda_5 Pr_5 + \cancel{\mu_7 Pr_7}$$
$$= -1.0000(0.1429) + 0.1667(0.1429)$$
$$= -0.1191$$

By using (1.59) and setting $\Delta t = 1$, $Pr_i^{(1)}(t)$ can be calculated as:

$$Pr_i^{(1)}(t) = Pr_i^{(0)}(t) + \dot{Pr}_i^{(0)}(t)\Delta t$$

$$\therefore \; Pr_i^{(1)}(t) = \begin{bmatrix} 0.1429 \\ 0.1429 \\ 0.1429 \\ 0.1429 \\ 0.1429 \\ 0.1429 \\ 0.1429 \end{bmatrix} + \begin{bmatrix} -0.1191 \\ 0.0476 \\ 0.0476 \\ 0.0476 \\ 0.0476 \\ 0.0476 \\ -0.1191 \end{bmatrix} \times 1.00 = \begin{bmatrix} 0.0238 \\ 0.1905 \\ 0.1905 \\ 0.1905 \\ 0.1905 \\ 0.1905 \\ 0.0238 \end{bmatrix}$$

If we repeat these steps until reaching the 7th iteration, i.e. $Pr_i^{(7)}(t)$, the following result will be obtained:

$$\begin{aligned} Pr_i^7(t) &= Pr_i(t+\Delta t) \\ &= \begin{bmatrix} 0.0134 & 0.1073 & 0.2009 & 0.3569 & 0.2009 & 0.1073 & 0.0134 \end{bmatrix}^T \end{aligned}$$

From Example 1.8, it is important to ensure that the sum of $Pr_i^7(t)$ is equal to one because this probability vector is determined by a numerical method. In the preceding example, it shows one that is ok. However, if it is not equal to one, then the following corrected probability vector has to be used:

$$Pr_{i,\text{corrected}}(t) = \frac{Pr_i(t)}{\sum_{i=0}^{z} Pr_i(t)} \tag{1.60}$$

The second approach, that was also proposed by Simon (2008a), is done through a direct way using an eigenvector technique. To know it, suppose that $S_{\max}$ equals the number of islands[28] (N) minus one (i.e. $z = S_{\max} = N - 1$). The value of $\dot{Pr}_i(t)$ in (1.58) can be determined as follows:

$$\dot{Pr} = APr, \quad \text{where: } Pr = \left[Pr_0, Pr_1, \dots, Pr_z\right]^T$$

$$A = \begin{bmatrix} -(\lambda_0+\mu_0) & \mu_1 & 0 & \cdots & 0 \\ \lambda_0 & -(\lambda_1+\mu_1) & \mu_2 & \ddots & \vdots \\ \vdots & \ddots & \ddots & \ddots & \vdots \\ \vdots & \ddots & \lambda_{z-2} & -(\lambda_{z-1}+\mu_{z-1}) & \mu_z \\ 0 & \cdots & 0 & \lambda_{z-1} & -(\lambda_z+\mu_z) \end{bmatrix} \tag{1.61}$$

For the simplified linear migration model[29] where $E = I$, the previous matrix can be defined as:

$$A = EA' = E\begin{bmatrix} -1 & \frac{1}{z} & 0 & \cdots & 0 \\ \frac{z}{z} & -1 & \frac{2}{z} & \ddots & \vdots \\ \vdots & \ddots & \ddots & \ddots & \vdots \\ \vdots & \ddots & \frac{2}{z} & -1 & \frac{z}{z} \\ 0 & \cdots & 0 & \frac{1}{z} & -1 \end{bmatrix} \tag{1.62}$$

It can be observed that the zero eigenvalue of A' happens with the following eigenvector:

$$\begin{aligned} &v = [v_1, v_2, \dots, v_{z+1}]^T \\ &v_w = \begin{cases} \dfrac{z!}{(z+1-w)!(w-1)!}, & \text{for } w = 1, 2, \dots, w' \\ v_{z+2-w}, & \text{for } w = w'+1, w'+2, \dots, z+1 \end{cases} \\ &\text{where: } w' = \left\lceil \frac{z+1}{2} \right\rceil \end{aligned} \tag{1.63}$$

Also, the $(z + 1)$ eigenvalues of the biogeography A' are computed as:

$$e = \left\{0, \frac{-2}{z}, \frac{-4}{z}, \dots, -2\right\} \tag{1.64}$$

28 Just for simplification.
29 It is a special case (MacArthur and Wilson, 1967).

Equation (1.64) was proven later in Igelnik and Simon (2011). This matrix technique is easier than the iterative technique because it can give a direct way to find the steady-state probability ($P(\infty)$) by applying the following theorem (Simon, 2008a):

Theorem 1.1: *The steady-state value for the probability of the number of each species is given by:*

$$Pr_i(\infty) = \frac{v}{\sum_{w=1}^{z+1} v_w} \tag{1.65}$$

Although the second method is easier and $Pr_i(\infty)$ can be directly computed without any iterations, this approach is not preferred in many numerical programming languages, because they set $N = \infty$ when $N > 170$. This infinity issue can be resolved if an additional sub-algorithm is used. However, dealing with long product operations requires extra CPU time (Alroomi et al., 2013a). Based on this, the iterative method is more flexible and more convenient. Thus, it is adopted in this book.

Example 1.9 Repeat Example 1.8 by using the steady-state probability. (**Hint:** first use (1.63), then use (1.65).)

Solution

If:

$$z = N - 1 = 6 \quad \Rightarrow \quad \text{means} \;\; S_1 = 0 \rightarrow S_{\max} = 6$$

Thus:

$$Pr_i = \begin{bmatrix} Pr_0 & Pr_1 & Pr_2 & Pr_3 & Pr_4 & Pr_5 & Pr_6 \end{bmatrix}^T$$

Then, the smallest integer w' is:

$$w' = \text{ceil}\left[\frac{(z+1)}{2}\right] = \text{ceil}\left(\frac{6+1}{2}\right) = \text{ceil}\,(3.5) = 4$$

Therefore:

$$v = \begin{bmatrix} v_1 & \cdots & v_{z+1} \end{bmatrix}^T = \begin{bmatrix} v_1 & \cdots & v_7 \end{bmatrix}^T$$

$$v_i = \begin{cases} \frac{z!}{(z+1-w)!(w-1)!}, & \text{if } (w = 1, \dots, w') \\ v_{z+2-w}, & \text{if } (w = w'+1, \dots, z+1) \end{cases}$$

$$= \begin{cases} \frac{6!}{(7-w)!(w-1)!}, & \text{if } (w = 1, 2, 3, 4) \\ v_{8-w}, & \text{if } (w = 5, 6, 7) \end{cases}$$

$$v_1 = \frac{6!}{(7-1)!\,(1-1)!} = \frac{6!}{6!0!} = 1$$

$$v_2 = \frac{6!}{(7-2)!\,(2-1)!} = \frac{6!}{5!1!} = 6$$

$$v_3 = \frac{6!}{(7-3)!\,(3-1)!} = \frac{6!}{4!2!} = 15$$

$$v_4 = \frac{6!}{(7-4)!\,(4-1)!} = \frac{6!}{3!3!} = 20$$

$$v_5 = v_{8-5} = v_3 = 15$$

$$v_6 = v_{8-6} = v_2 = 6$$

$$v_7 = v_{8-7} = v_1 = 1$$
$$\therefore v = \begin{bmatrix} 1 & 6 & 15 & 20 & 15 & 6 & 1 \end{bmatrix}^T$$

Now, we can apply ***Theorem 1.1***, i.e. (1.65), to calculate $Pr_i(\infty)$ as follows:

$$\because Pr_i(\infty) = \frac{v}{\sum_{w=1}^{z+1} v_w} = \frac{v}{\sum_{w=1}^{7} v_w} = \frac{\begin{bmatrix} 1 & 6 & 15 & 20 & 15 & 6 & 1 \end{bmatrix}^T}{(1+6+15+20+15+6+1)}$$

$$\therefore Pr_i(\infty) = \begin{bmatrix} \frac{1}{64} & \frac{6}{64} & \frac{15}{64} & \frac{20}{64} & \frac{15}{64} & \frac{6}{64} & \frac{1}{64} \end{bmatrix}^T$$

$$= \underset{S=0}{[0.0156} \quad \underset{S=1}{0.0938} \quad \underset{S=2}{0.2344} \quad \underset{S=3}{0.3125} \quad \underset{S=4}{0.2344} \quad \underset{S=5}{0.0938} \quad \underset{S=6}{0.0156]^T}$$

If the vector v is obtained, then $Pr_i(\infty)$ can be solved easily by Simon (2013):

$$Pr_i(\infty) = 2^{-z} v = \begin{bmatrix} 0.0156 & 0.0938 & 0.2344 & 0.3125 & 0.2344 & 0.0938 & 0.0156 \end{bmatrix}^T$$

It can be clearly seen that there are some slight differences between this answer and that of Example 1.8. The reason is that the previous probability is obtained after 7 iterations, while this answer is for steady-state condition (i.e. $t = \infty$).

In BBO, the objective function can be optimized if each island is considered as one individual and the independent variables of each individual are dealt as features. The solutions can be enhanced if these features are distributed between the source and recipient islands. The source island could become a recipient island for other better islands (Simon, 2008a). That is, the richness of species on an island is decided through a probabilistic process. If many good **biotic** and **abiotic** features[30] are available on an island, then it will be good land for immigrants. Each feature is called a **suitability index variable** (*SIV*), which represents one independent variable of such a problem in BBO. The **island suitability index** (*ISI*)[31] is the dependent variable, which varies with any change in any element of the vector *SIV*. Because BBO is a population-based algorithm, so optimizing n-dimensional problem with N-individuals can be mathematically represented as follows:

$$ISI_i = f_i(SIV_{i,1}, SIV_{i,2}, \ldots, SIV_{i,n}), \quad i = 1, 2, \ldots, N \tag{1.66}$$

Once the initialization stage is completed, the BBO algorithm should pass some sub-algorithms:

1.3.1.1 Migration Stage

The main idea of this stage is to share the good features of rich islands to modify poor islands. Because the selection is done through a probabilistic process, so the ith island is likely to be selected as a source of modification if ISI_i is high, and vice versa for the jth recipient island. From Figure 1.11, low λ_i and high μ_i indications mean a large number of endemic species are available on the ith island. Thus, the solution ISI_i is high. As an example, point S_x is located before $\hat{S}$, so λ_x is high and μ_x is low, and thus ISI_x is considered as a poor solution. On the opposite side, point S_y is located after $\hat{S}$, so λ_y is low and μ_y is high, and thus ISI_y is considered as a good solution. Based on this, μ_i and λ_i are used as metrics to know the solution quality of each island.

30 Biotic factors: predation, competition, interactions, etc. Abiotic factors: wind, water, sunlight, temperature, pressure, soil, etc. (MacDonald, 2003).

31 It is also called the **habitat suitability index** (*HSI*).

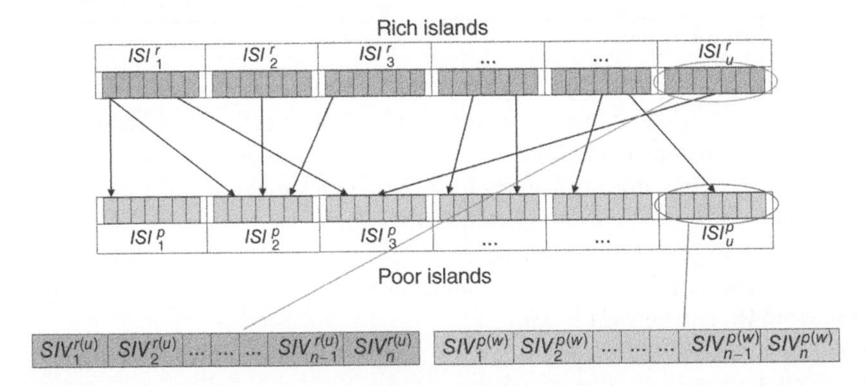

Figure 1.12 Migration process among different islands in BBO.

Through the migration process, the islands with low *ISI* could be improved per each new generation, and at the same time, the solution quality of the best islands are kept away from any corruption.

The original BBO algorithm comes with four migration forms, as described in Simon (2011), Alroomi et al. (2013b), and called partial, simplified partial, single, and simplified single migration based (PMB, SPMB, SMB, and SSMB) models. The first published BBO paper used the PMB model (Simon, 2008a), which is graphically described in Figure 1.12. As can be clearly seen from this figure, these rich and poor islands act as sources and recipients of those migrated n *SIV*. Each *SIV* s of a poor island is updated by *SIV* σ that is probabilistically selected from one rich island. For the SPMB model, the n *SIV* of poor islands are updated from the first best island(s), which in turn increases the probability to trap into local optimum solutions. The migration processes of the SMB and SSMB models are, respectively, similar to those of the PMB and SPMB models with one main difference: only one randomly selected *SIV* s of each poor island is modified. The last two models are faster, but with low convergence rates. The book will consider all the essential modifications presented in Alroomi et al. (2013a) as a basis for the proposed MpBBO-SQP algorithm. Thus, the original BBO algorithm (before being hybridized with SA and SQP) can save around 32.32% of its total CPU time and with better performance than that of the PMB-BBO model. The migration process used in this book is described by Algorithm 2.

1.3.1.2 Mutation Stage

As with many nature-inspired algorithms, this stage is very essential to increase the exploration level. The mutation process of the BBO algorithm can be defined as random natural events that affect the availability of the biotic and abiotic features on an isolated island, which in turn reflected on the total endemic species on that island. These events could be positive (like shipwrecks and wind pollination) that increase the species density, or they could be negative (like volcanoes, diseases, and earthquakes).

In BBO, the species count probability *Pr* is used exclusively to find the mutation rate (Simon, 2008a). Thus, many choices are available to researchers to select their preferable mutation rate, such as Gaussian, Cauchy and Lèvy mutation operators reported in Gong et al. (2010b). The original mutation rate, which is also used in this book, is described as follows (Simon, 2008a):

$$m_i = m_{\max}\left(1 - \frac{Pr_i}{Pr_{\max}}\right) \tag{1.67}$$

where $Pr_{\max}$ is the largest element of the vector Pr, and $m_{\max}$ is a user-defined maximum allowable value that m_i can reach.

Algorithm 2 Partial migration pseudocode.

Require: Let ISI_i denote the ith population member and contains n features
Require: Define emigration rate μ_i and immigration rate λ_i for each member
1: **for** $i \leftarrow 1$ *to* N **do** {where N = number of islands or individuals, see (1.66)}
2: **for** $s \leftarrow 1$ *to* n **do** {where n = number of features "SIV" or design variables}
3: Use λ_i to probabilistically select the immigrating island ISI_i
4: **for** $j \leftarrow 1$ *to* N **do** {Break once ISI_j is selected}
5: Use μ_j to probabilistically decide whether to emigrate to ISI_i
6: **if** ISI_j is selected **then** {where $ISI_i \neq ISI_j$}
7: Randomly select an SIV σ from ISI_j
8: Replace a random SIV s in ISI_i with SIV σ
9: **end if**
10: **end for**
11: next SIV
12: **end for**
13: next island
14: **end for**

As can be seen from (1.67), the mutation rate is inversely proportional to the probability rate (i.e. $m_i \propto^{-1} Pr_i$). This equation forces m_i to be equal to $m_{\max}$ at ($Pr_i = 0$), and equal to 0 at the largest element of Pr. It can be graphically represented as shown in Figure 1.13.

The mutation rate will flip the bell-shape graph of the probability rate. The main objective of using m_i rather than Pr_i is to have better control over the islands targeted for the mutation stage. That is, the islands located at or near the equilibrium point $\hat{S}$ will be preserved, while the other islands sorted on both sides will have a higher chance to be mutated and hence could be improved. The mutation process is described by Algorithm 3.

The mutation can be applied to all N islands. However, Simon (2008b) suggested that only the worst individuals have to be mutated. The range of these mutated individuals can be defined in

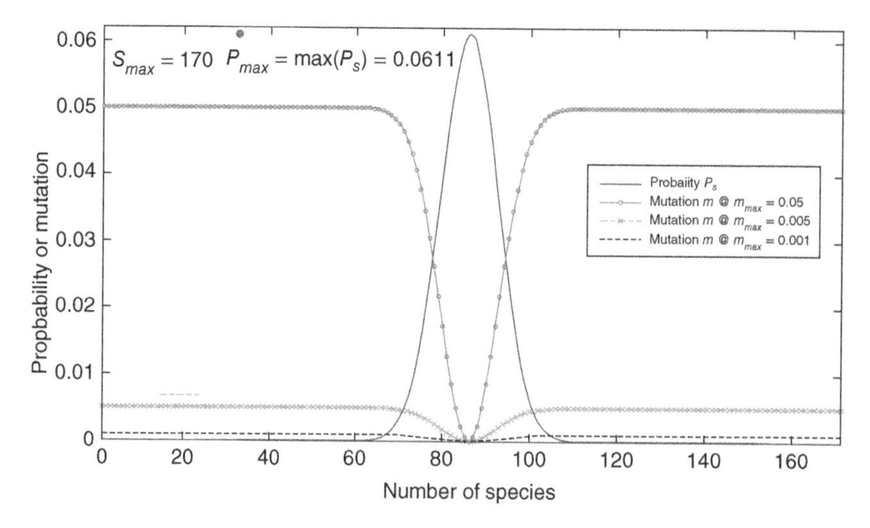

Figure 1.13 Comparison between Pr and m at different $m_{\max}$.

Algorithm 3 Original Mutation pseudocode.

1: **for** $i \leftarrow 1$ *to* N **do** {where N = number of islands or individuals, see (1.66)}
2: Calculate probability Pr_i based on λ_i and μ_i {by iterative or eigenvector method}
3: Calculate mutation rate m_i {using (1.67)}
4: **if** rand $< m_i$ and $i \geqslant R_m$ **then** {R_m is a user-defined mutation range}
5: Replace n *SIV* vector of ISI_i with a randomly generated n *SIV* vector
6: **end if**
7: **end for**

Algorithm 3 as follows:

$$R_m = \text{round}\left[\left(\frac{N}{d_m}\right) : (N)\right] \tag{1.68}$$

where d_m is a user-defined value and N is the total number of islands; see (1.66). If $d_m = 2$, then the worst half individuals are to be mutated. The percentage of the total mutated individuals is proportional to d_m.

Example 1.10 Consider the steady-state probability obtained in Example 1.9. Find the mutation rate $m(t)$ if $m_{\max} = 0.05$.

Solution

From Example 1.9, the probability rate is:

$$Pr_i(\infty) = \begin{bmatrix}0.0156 & 0.0938 & 0.2344 & 0.3125 & 0.2344 & 0.0938 & 0.0156\end{bmatrix}^T$$
$$Pr_{\max} = \max\left[Pr_i(t)\right] = 0.3125$$

Now, by using (1.67), $m(t)$ can be calculated as follows:

$$m(t) = 0.05\left(1 - \frac{\begin{bmatrix}0.0156 & 0.0938 & 0.2344 & 0.3125 & 0.2344 & 0.0938 & 0.0156\end{bmatrix}^T}{0.3125}\right)$$
$$= \begin{matrix}[0.0475 & 0.0350 & 0.0125 & 0.0000 & 0.0125 & 0.0350 & 0.0475]^T \\ \downarrow & \downarrow & \downarrow & \downarrow & \downarrow & \downarrow & \downarrow \\ S=0 & S=1 & S=2 & S=3 & S=4 & S=5 & S=6\end{matrix}$$

1.3.1.3 Clear Duplication Stage

If this optional stage is used in BBO, then the diversity of the problem features could increase. The reason behind this is that the emigrated *SIV* σ will take the same value and place in other island(s), so these duplicated features may have an insignificant impact on their *ISI*. For *one*-dimensional problems, duplicated *SIV* will give duplicated islands. In this situation, the exploration level will decrease and the algorithm may quickly settle on a non-global optimum solution.[32] The main purpose of this stage is to check all n *SIV* of all N *ISI* whether they are duplicated or not. If any duplicated feature is detected, then it is replaced by a new randomly generated feature. This process is described by Algorithm 4 (Simon, 2008b).

It is important to know that this sub-algorithm must be partially deactivated for discrete features of **mixed-integer optimization problems** and completely deactivated for **combinatorial optimization problems** if the set has a few discrete features. Based on the step-size resolution and

32 The blended BBO, given in Ma and Simon (2011), is immune to this duplication phenomenon.

Algorithm 4 Clear duplication pseudocode.

Require: Check all n SIV on all N ISI
1: **while** there is a duplicated SIV **do**
2:    **for** $i \leftarrow 1$ *to* N **do** {where N = number of islands or individuals, see (1.66)}
3:       **if** any duplicated SIV s is detected **then**
4:          Replace the duplicated SIV s in ISI_i with a randomly generated SIV σ
5:       **end if**
6:    **end for**
7: **end while**

the side constraints of discrete variables, ignoring this vital step could lead to trapping into infinite loops.

Example 1.11 Design a MATLAB code to scan the entered vector. The program should be able to detect duplicated elements and replace them with new random values.

Solution
First, we need to find the following vector (refer to (1.66) for n and N):

$$IP = \begin{bmatrix} SIV_{j,1} & SIV_{j,2} & \cdots & SIV_{j,N} \end{bmatrix}, \; IP_{\min} = SIV_j^{\min}, \; IP_{\max} = SIV_j^{\max}$$

where $\{j = 1, 2, \ldots, n\}$.

For each jth feature, the program given in `ClearDuplication.m` should be executed:

```
1   clc
2   clear
3   while length(IP)≠length(unique(IP))
4       A=unique(IP);
5       [B,bin]=histc(IP,A);
6       c=0;
7       for i=1:length(B)
8           if (B(i)>1)
9               c=1+c;
10              Rep(c)=A(i);
11          end
12      end
13      for i=1:c
14          d=0;
15          for j=1:length(IP)
16              if IP(j)==Rep(i)
17                  d=1+d;
18                  if d>1
19                      IP(j)=IPmin+(IPmax-IPmin)*rand;
20                  end
21              end
22          end
23      end
24  end
```

1.3.1.4 Elitism Stage

Suppose that the good individuals obtained in the last generation are ruined by the previous BBO stages (i.e. migration, mutation, and/or clear duplication). Then, those good solutions will be lost forever if this optional stage is not activated in the BBO program. This stage can provide a rollback option to rescue the last state of the corrupted best islands, or **elites**, and then recycle them back into the population of the next generation (Simon et al., 2009).

1.3.1.5 The Overall BBO Algorithm

The classical BBO algorithm can be described by the following steps:

1. Initialize the BBO parameters (N, I, E, m_{max}, etc.).
2. Find species count probabilities (Pr_i) and mutation rate m_i based on the calculated emigration rate μ_i and immigration rate λ_i by (1.54)–(1.55).
3. Generate N random islands where each island represents one solution to a given n-dimensional problem with non-duplicated *SIV*s.
4. Sort the solutions (i.e. N *ISI*), so that the first best solution ISI_1 should be mapped with the highest number of species and the lowest immigration rate λ_i (or the highest emigration rate μ_i), and continue the descending order until reaching the worst solution (i.e. ISI_N).
5. Save the elite solutions for the next generation; it is an optional stage (Simon et al., 2009).
6. Probabilistically, select the source islands based on μ_i and the islands that need to be modified (i.e. "the recipient islands") based on λ_i, and then do the migration process. After that, update all N *ISI* before ending this step.
7. Do mutation process for the islands based on their probabilities calculated in step (2). After that, clear any duplicated *SIV* once the mutation process is completed. Then, update all N *ISI*.
8. Return to step (4) for the next iteration. This loop can be terminated either after reaching an acceptable tolerance or completing a desired number of generations.

The overall mechanism of the BBO algorithm is depicted by the flowchart shown in Figure 1.14. The software shown in Figure 1.15 is developed to create many possible BBO structures flexibly without knowing any programming skills. It gives users the ability to hybridize BBO with many other sub-algorithms and many options.

1.3.2 Differential Evolution

The DE algorithm is known as one of the most popular and simplest population-based evolutionary algorithms. It was presented by R. Storn and K.V. Price in 1995 Storn and Price (1995) and Storn (1996). This algorithm can be programmed easily and quickly without facing that much of challenges, and thus it gets high attention from many researchers.

Because DE is a population-based algorithm, N candidate solutions are generated at the initialization stage. Each individual contains n independent variables. Suppose that the parameters N and n are respectively used to represent the population size and the dimension of each individual. In DE, each new individual is generated by selecting three different individuals from the same population.

Referring to the literature, someone may note that there are many versions of DE. The simplest DE algorithm is called classic DE or DE/rand/1/bin.[33] The mechanism of the classic DE algorithm is described by Algorithm 5. More details about the other DE versions can be found in Simon (2013).

33 The other variations are: DE/rand/1/either-or, DE/rand/1/L, DE/rand/2/bin, DE/rand/2/L, DE/best/1/bin, DE/best/1/L, DE/best/2/bin, DE/best/2/L, DE/target-to-best/1/bin, DE/target/1/bin, DE/target/1/L, DE/target/2/bin, and DE/target/2/L.

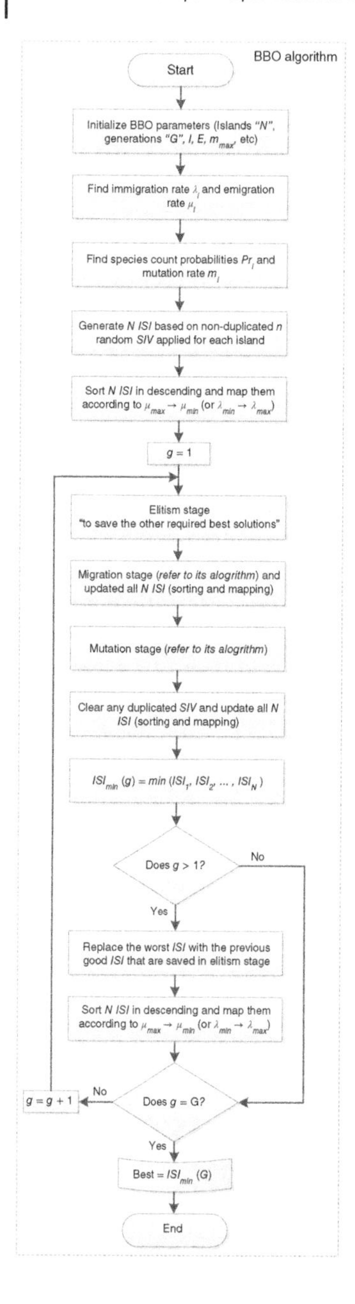

Figure 1.14 Flowchart of the BBO algorithm.

1.4 Hybrid Optimization Algorithms

In general, the main objective of these highly advanced non-traditional techniques is to accumulate the strengths and eliminate or minimize the weaknesses of individual optimization algorithms. The hybridization could be done by using at least two optimization algorithms. These algorithms can be taken from classical and/or meta-heuristic categories. In addition, the designers have a large space to maneuver and they can apply their hybridizations and/or modifications in many locations of the new algorithm. Therefore, the overall algorithm will integrate all of these improvements.

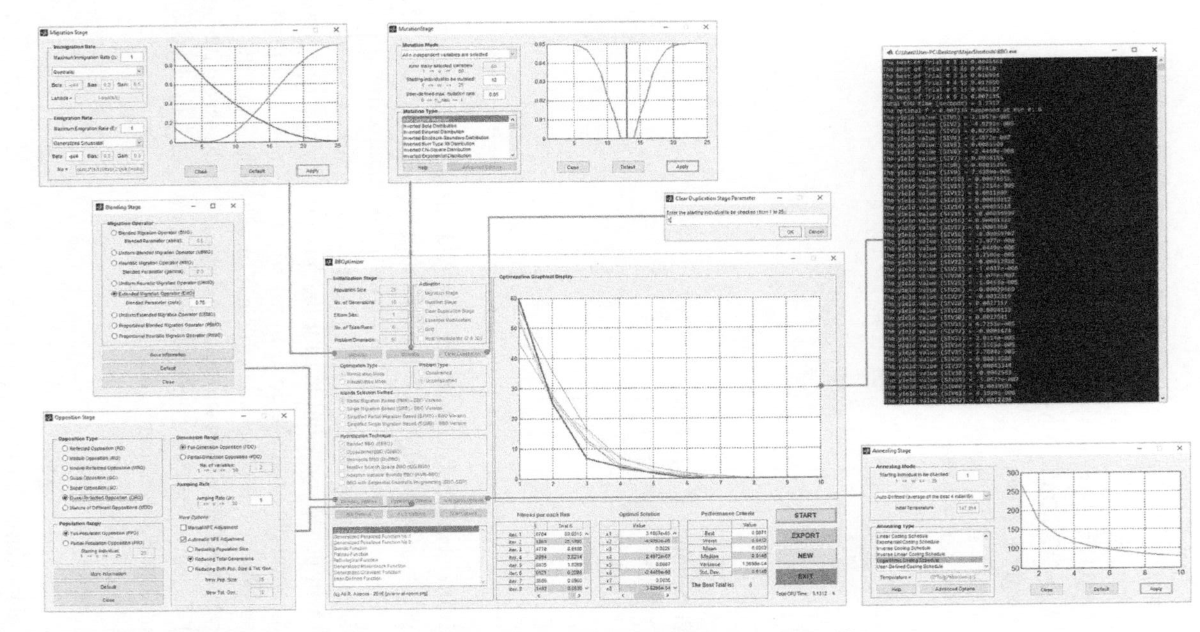

Figure 1.15 Our integrated biogeography-based optimization algorithm software; developed in MATLAB.

Algorithm 5 Classic differential evolution pseudocode.

Require: F = step-size parameter $\in [0.4, 0.9]$
Require: C_r = crossover rate $\in [0.1, 1]$
Require: Initialize a population of candidate solutions $\{X_i\}$ for $i \in [1, N]$
1: **while** not(termination criterion) **do**
2: **for** each individual X_i, $i \in [1, N]$ **do**
3: $r_1 \leftarrow$ random integer $\in [1, N]$ $r_1 \neq i$
4: $r_2 \leftarrow$ random integer $\in [1, N]$ $r_2 \notin \{i, r_1\}$
5: $r_3 \leftarrow$ random integer $\in [1, N]$ $r_3 \notin \{i, r_1, r_2\}$
6: $V_i \leftarrow X_{r_1} + F\left(X_{r_2} - X_{r_3}\right)$ (mutant vector)
7: $\zeta_r \leftarrow$ random integer $\in [1, n]$
8: **for** each dimension $j \in [1, n]$ **do**
9: $r_{c_j} \leftarrow$ random integer $\in [0, 1]$
10: **if** $(r_{c_j} < C_r)$ or $(j = \zeta_r)$ **then**
11: $U_{i,j} \leftarrow V_{i,j}$
12: **else**
13: $U_{i,j} \leftarrow X_{i,j}$
14: **end if**
15: **end for**
16: **end for**
17: **for** each population index $i \in [1, N]$ **do**
18: **if** $f(U_i) < f(X_i)$ **then**
19: $X_i \leftarrow U_i$
20: **end if**
21: **end for**
22: **end while**

Minimizing the disadvantages associated with each individual optimization algorithm does not mean that the overall hybrid optimization algorithm will have zero disadvantages. Each additional sub-algorithm added to the new optimization algorithm will consume an extra CPU time and, at the same time, will complicate the programming code. Thus, the new optimization algorithm resulted from this hybridization will be slower[34] and hard to be modified by other programmers. Moreover, because the hybridization phase can be done anywhere on the main algorithm, so there is no standard format to follow. Further, if there is insufficient information about the proposed hybrid optimization algorithm, then this algorithm will stay lonely and no body, except the programmer(s), knows its mechanism. The goal of the following Sections 1.4.1 and 1.4.2 is to reveal the mechanism of hybrid optimization algorithms used in this book.

1.4.1 BBO-LP

Unlike gradient-based algorithms, EAs are probabilistic-based single/multi-point search techniques, the fact that makes them very slow convergence and time consuming algorithms. Thus, many researchers prefer to use only LP and NLP techniques or, at least, hybridizing them with EAs (Noghabi et al., 2009; Bedekar and Bhide, 2011a).

34 Unless modifying/bypassing some stages of the main algorithm or/and reducing the simulation parameters.

To accelerate the convergence rate and accuracy, a hybrid BBO-LP algorithm is designed. LP is selected instead of NLP for speed and simplicity. However, the LP algorithm can be incorporated with EAs only if the objective function and its design constraints are expressed in a linear form or if they are linearized as will be seen later in Chapter 4. The framework of this hybrid optimization algorithm is illustrated by the flowchart shown in Figure 1.16.

The mechanism of this hybrid optimization algorithm can be briefly described in the following steps:

- First, BBO is executed normally, and the fitness per generation is selected. It could be one or more solutions based on the elitism parameter and user preference.

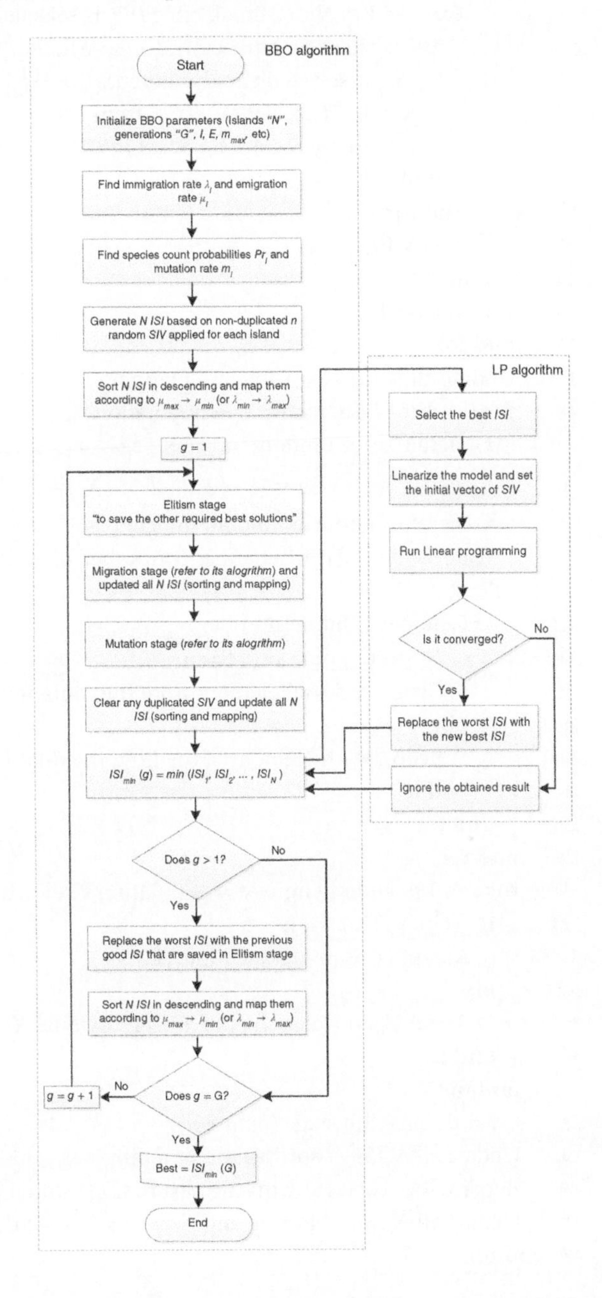

Figure 1.16 Flowchart of the BBO-LP algorithm.

Algorithm 6 Pseudocode of the proposed hybrid BBO/DE algorithm.

Require: Initialization stage: step-size parameter F, crossover rate C_r, problem dimension n, population size N, number of generations G, etc.

```
1: for g ← 1 to G do {where G = number of generations}
2:   Elitism stage (optional)
3:   for p ← 1 to N do {where N = number of islands or individuals}
4:     for s ← 1 to n do {where n = number of features "SIV" or design variables}
5:       Use λ_p to probabilistically select the immigrating island ISI_p
6:       for q ← 1 to N do {Break once ISI_q is selected}
7:         Use μ_q to probabilistically decide whether to emigrate to ISI_p
8:         if ISI_q is selected then {where ISI_p ≠ ISI_q}
9:           Randomly select an SIV σ from ISI_q
10:          Randomly replace an SIV s in ISI_p with SIV σ
11:        end if
12:      end for
13:      next SIV
14:    end for
15:    next island
16:  end for
17:  Update all N ISI → sorting and mapping
18:  for i ← 1 to N do {where N = population size}
19:    Generate three random integers {r_1, r_2, r_3} ∈ [1, N], where r_1 ≠ r_2 ≠ r_3 ≠ i
20:    Create mutant vector V_i = X_{r_1} + F × (X_{r_2} − X_{r_3})
21:    Generate a uniform random integer ζ ∈ [1, n]
22:    for j ← 1 to n do {where n = number of independent variables, or dimension of the
       problem}
23:      Generate random number r_{c,j} ∈ [0, 1]
24:      if (r_{c,j} < C_r) or (j = ζ) then
25:        Generate a trial element from the mutant vector U_{i,j} = V_{i,j}
26:      else
27:        Keep the jth element of the ith individual U_{i,j} = X_{i,j}
28:      end if
29:    end for
30:  end for
31:  for i ← 1 to N do {where N = population size}
32:    if f(U_i) ⩾ f(X_i) then
33:      Accept U_i as an updated individual
34:    else
35:      Reject U_i and keep the previous individual X_i
36:    end if
37:  end for
38:  Clear duplication stage (optional)
39:  Update all N ISI → sorting and mapping
40:  Replace the worst ISI with the past best ISI stored in the elitism stage (optional)
41:  Update all N ISI → sorting and mapping (optional)
42: end for
```

- Then, the independent variables of the preceding fitness (i.e. n *SIV*) are set as initial point to the linearized problem in the LP sub-algorithm.
- After that, LP is executed for each elite solution to fine-tune the independent variables and exploit the solutions as much as possible.
- Finally, the worst solutions in the population are replaced with the fine-tuned solutions obtained by the LP stage.[35]
- Repeat all the preceding steps again in the next generation.

This hybrid algorithm acts like a **multi-start** LP algorithm, but with the power of EAs to explore the entire search space quickly and effectively. The speed can be saved here as a result of being using just a few iterations and a small population size compared with conventional EAs. Of course, if BBO-LP is initiated with the same parameters of BBO, then the former algorithm will be slower.

1.4.2 BBO/DE

As a global optimizer, BBO has a good exploitation level, but it lacks exploration level (Pattnaik et al., 2010; Gong et al., 2010a, b; Lohokare et al., 2013). In contrast, DE has a good exploration level, and it can reach the space where the global optimal solution is located in Gong et al. (2010a). Therefore, merging the strength of DE (i.e. the good exploration level) with the strength of BBO (i.e. the good exploitation level) together in one superior optimization algorithm has been suggested by many researchers (Al-Roomi and El-Hawary, 2016a). In this book, a new fully discretized hybrid optimizer built based on the BBO and DE algorithms is proposed. Here, the BBO mutation stage given in Algorithm 3 is completely replaced with the mutant process of the classic DE algorithm given in Algorithm 5. To match DE with BBO, DE should have N candidate solutions and n independent variables, so each new individual injected into the existing population of BBO is generated by DE from three randomly selected and induplicated individuals. The mechanism of this new hybrid BBO/DE optimizer is depicted in Algorithm 6. The DE algorithm can be described through lines 18 to 37 of Algorithm 6. This fully discretized hybrid optimization algorithm will be used later to solve some relay coordination problems.

Problems

Written Exercises

W1.1 Consider the following equation:

$$f(x, y, z) = \sin(x) \cdot \tanh(y) - \frac{a \cdot x^{\delta(u)}}{\ln(z)} \cdot \frac{\csc(z^c)}{\log_b(\xi(v) \cdot y)}$$

Find:

1. Independent variable(s)
2. Dependent variable(s)
3. Constants
4. Parameters
5. External unobserved variables

35 This is a very important step to guarantee that the best solutions, obtained by BBO before initiating LP, are not ruined.

W1.2 Consider the plot shown in Figure 1.17, which represents the following function:

$$y(z) = \text{sinc}(x) \cdot \sin\left(x^2\right)$$

Find the locations of the best and worst solutions if:

1. optimization ≡ maximization
2. optimization ≡ minimization

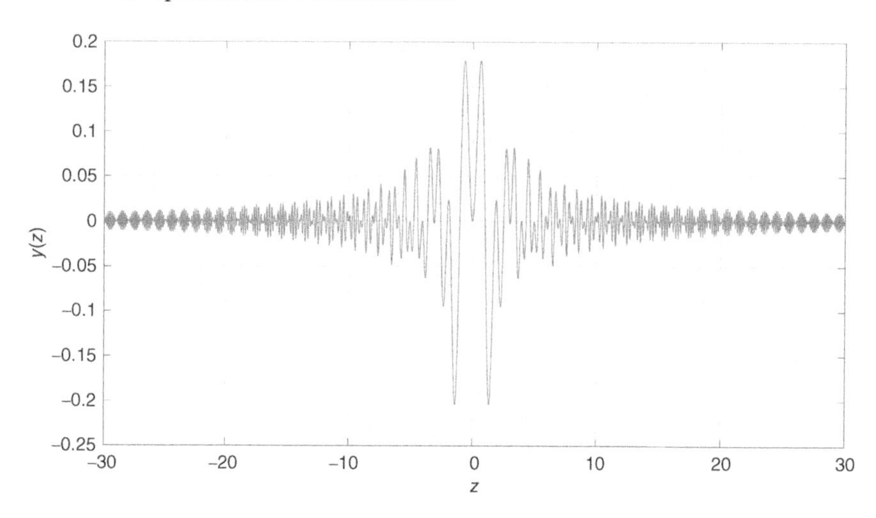

Figure 1.17 Plot of the function used in Exercise W1.2.

W1.3 Suppose that a data scientist designed a *two*-dimensional meta-heuristic optimization algorithm to find the minima of

$$f(X) = x_1 \cdot \sin\left(x_2\right) + x_2 \cdot \cos\left(x_1\right)$$

The program was coded five years ago and without any comments. If the department head asks to modify the code to find the maxima, what is your recommendation to the data scientist to accomplish that task quickly and correctly?

W1.4 Suppose that the function that needs to be minimized has a value of $f(X) = 1.035$ and its constraint-handler has a penalty term of $P(X) = 20$. Calculate the penalized cost function using the additive and multiplicative approaches.

W1.5 If the objective function of Exercise W1.4 is to maximize $f(X)$ feasibly, how do you calculate the penalized cost function with and without using the duality principle?

W1.6 Suppose that you are designing an adaptive optimizer where the objective is to find optimal solutions within a very short time. If the optimization problem is highly non-linear, non-convex, and contains many design constraints, then can you use the random search methods to handle these constraints? Why?

W1.7 Why does the literature contain many types of penalty functions?

W1.8 Although the death penalty function is so easy to implement, many engineers and data scientists do not prefer it! What is the reason behind this contradiction? Explain.

W1.9 Suppose a combinatorial optimization algorithm is used to minimize a problem with discrete variables. If the following optimal solutions are obtained after 7 independent runs, calculate the best, worst, median, mean, mode, variance, and standard deviation: $1, 5, 1, 3, 2, 1, 4$.

W1.10 State twenty meta-heuristic optimization algorithms.

W1.11 State two non-population-based meta-heuristic optimization algorithms

W1.12 Suppose that you want to apply BBO to an n-dimensional combinatorial optimization problem. If each one of these n discrete variables is bounded between ten elements as follows:

$$SIV_i \in [\text{value 1}, \text{value 2}, \ldots, \text{value 10}]$$

Can we activate the clear-duplication stage: If BBO is initiated with 10 islands? If BBO is initiated with 20 islands? Why?

W1.13 Figure 1.18 shows the 3D plot of Matyas' benchmark function. This minimization problem can be mathematically expressed as follows (Al-Roomi, 2015a):

$$f(X) = 0.26\left(x_1^2 + x_2^2\right) - 0.48x_1x_2$$

$$-10 \leq x_i \leq 10,\ i = 1, 2,\ f_{\min}(X^*) = 0,\ x_i^* = 0$$

Suppose that the total number of islands is $N = 7$ and the species probability is calculated using the steady-state technique. Minimize this benchmark function by BBO using only one generation.

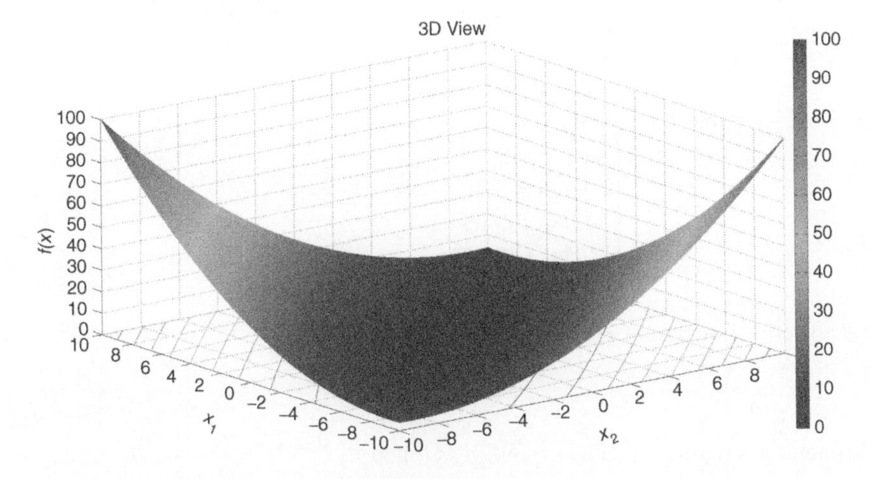

Figure 1.18 Matyas' benchmark function.

Computer Exercises

C1.1 Design a MATLAB code to plot the following function and then find its minimum value:

$$f(x) = [\text{sinc}(x) + \text{excsc}(x)] \cdot \cos(x)$$

where $x \in \left[\frac{\pi}{9}, \frac{9\pi}{10}\right]$ with a step-size resolution of $\Delta x = 0.001$. (**Hint:** You can use the command `fmincon`.)

C1.2 Design a MATLAB code to plot the following function and then find its minimum value:

$$f(x) = \text{versin}(x) + \text{covercos}(x)$$

where $x \in [-\pi, \pi]$ with a step-size resolution of $\Delta x = 0.001$. (**Hint:** You can use the command `fsolve`, `fminsearch`, or `fmincon`.)

C1.3 Design a MATLAB code to plot the following function and then find its minimum value:

$$f(x) = \text{hacoversin}(x) + \text{havercos}^{-1}(x)$$

where $x \in [-3, 3]$ with a step-size resolution of $\Delta x = 0.001$. (**Hint:** You can use the command `fsolve`, `fminsearch`, or `fmincon`.)

C1.4 Design a MATLAB code to plot the following function and then find its minimum value:

$$f(x) = \Re\left\{\sec(x) + \sec^{-1}(x) + \text{sech}(x) + \text{sech}^{-1}(x) + \text{exsec}(x)\right\}$$

where $x \in [-1.5, 1.5]$ with a step-size resolution of $\Delta x = 0.001$. (**Hint:** You can use the command `fsolve`, `fminsearch`, or `fmincon`.)

C1.5 Modify the MATLAB program given in Example 1.6 to solve the following linear maximization problem using `dual-simplex`:

$$\max_{x,y}\left(x + \frac{y}{3}\right) \text{ subject to } \begin{cases} x + y \leqslant 2 \\ x + y/4 \leqslant 1 \\ x - y \leqslant 2 \\ x/4 + y \geqslant -1 \\ x + y \geqslant 1 \\ -x + y \leqslant 2 \\ x + y/4 = 1/2 \\ -1.0 \leqslant x \leqslant 1.5 \\ -0.5 \leqslant y \leqslant 1.25 \end{cases}$$

C1.6 If the actual objective function of Exercise C1.5 is:

$$\max_{x,y}\left(x + \frac{y}{\xi(\theta)}\right)$$

and the parameter ξ is mathematically expressed as follows:

$$\xi(\theta) = \frac{1}{1000\ \text{hacoversin}(\theta)}$$

where $\theta \in [0, 2\pi]$ and $\Delta\theta = 0.01$.
Modify the MATLAB program given in Example 1.7 to find the best θ.

C1.7 Generate the 3D plot shown in Figure 1.9. (**Hint:** The effect is done by modifying the lighting and/or material used in the plot. To have the same effect shown in Figure 1.9, you can click on `Insert` ▶ `Light`.)

C1.8 Design a MATLAB code to find $Pr(t)$ by using (1.58) and then plot it for 200 islands. You should get a plot similar to the one shown in Figure 1.19. (**Hint:** Refer to Example 1.8.)

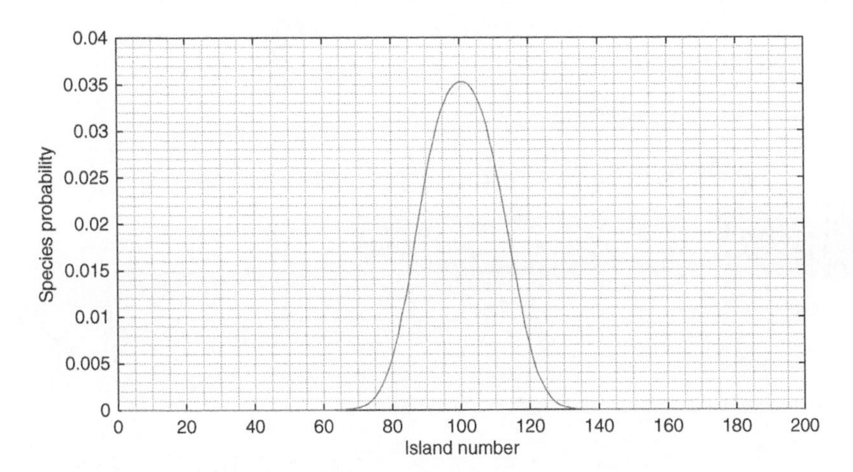

Figure 1.19 Probability of species among 200 islands.

C1.9 Design a MATLAB code to find $Pr_i(\infty)$ for 170 islands by using (1.65). (**Hint:** Refer to Example 1.9.)

C1.10 Suppose the input features of BBO are:

$$IP = [12.3, 11, 10.3, 5.7, 12.3, 8.98, 10.3]$$

where $IP_{\min} = 5$ and $IP_{\max} = 13$.
Modify the MATLAB program given in Example 1.11 to clear the duplicated elements.

C1.11 Figure 1.20 shows the 3D plot of Branin's benchmark function no.1. This minimization problem can be mathematically expressed as follows (Al-Roomi, 2015a):

$$f(X) = \left(x_2 - \frac{5.1x_1^2}{4\pi^2} + \frac{5x_1}{\pi} - 6\right)^2 + 10\left(1 - \frac{1}{8\pi}\right)\cos(x_1) + 10$$

$$-5 \le x_1 \le 10; \quad 0 < x_2 \le 15$$

$$f_{\min}(X^*) = 0.39\,788\,735\,772\,973\,816, \; x_i^* \approx (-\pi, 12.275), (\pi, 2.275), (9.42\,478, 2.475)$$

Suppose that the BBO algorithm is initiated with a population size of $N = 50$, 1000 generations, 3 elite solutions, $m_{\max} = 0.1$, and the species probability is calculated using the steady-state technique. Code BBO in MATLAB and then minimize this benchmark function.

C1.12 Figure 1.21 shows the 3D plot of Rosenbrock's benchmark function. This minimization problem can be mathematically expressed as follows (Al-Roomi, 2015a):

$$f(X) = \sum_{i=1}^{n-1} \left[100\left(x_{i+1} - x_i^2\right)^2 + \left(x_i - 1\right)^2\right]$$

$$-30 \le x_i \le 30, \quad i = 1, 2, \quad f_{\min}(X^*) = 0, \; x_i^* = 1$$

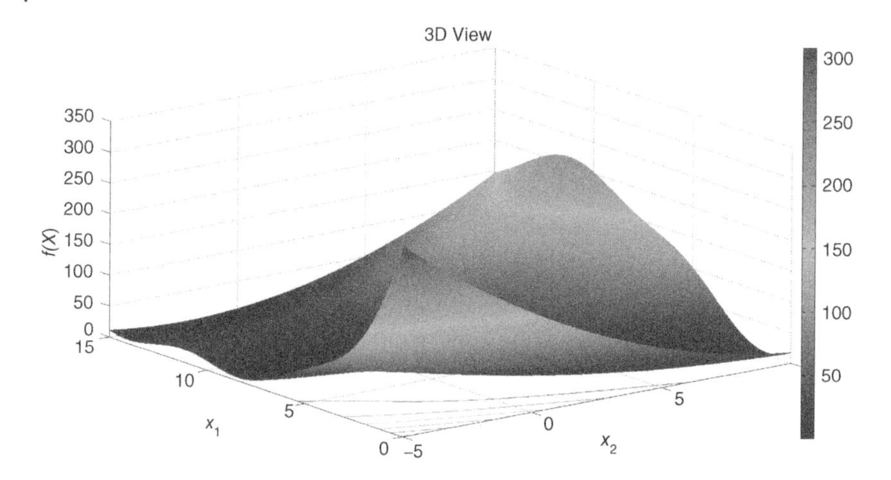

Figure 1.20 Branin's benchmark function no.1. Source: Al-Roomi (2015a).

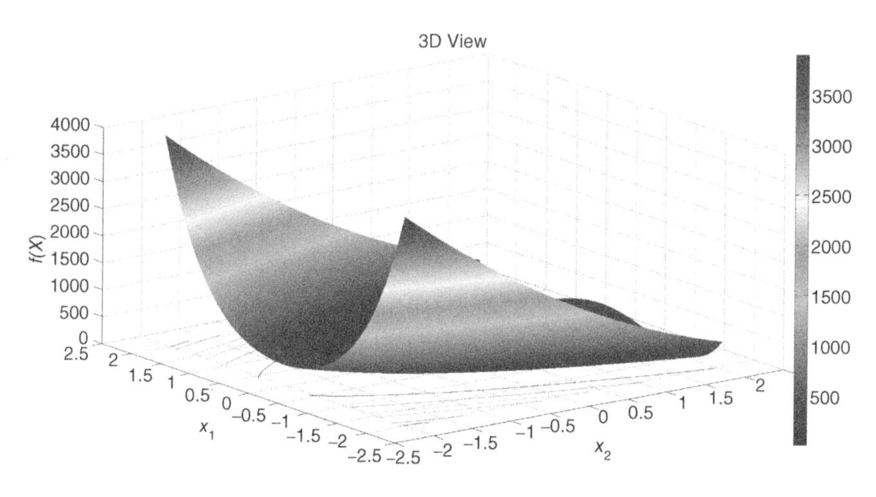

Figure 1.21 Generalized Rosenbrock's valley function. Source: Al-Roomi (2015a).

Suppose that the classic DE algorithm is initiated with a population size of $N = 50$, 100 generations, 3 elite solutions, crossover rate of $Cr = 0.5$, and the step-size parameter of $F = 0.5$. Code DE in MATLAB and then minimize this benchmark function.

2

Fundamentals of Power System Protection

Aim

The goal of this chapter is to give the reader why power system protection is so important. It classifies open- and short-circuit faults, shows different protection zones, explains the operational philosophy of primary and backup relays, lists the design criteria that should be considered during designing protection schemes, introduces overcurrent relays with their types and sub-types, and finally covers some case studies. Overall, this chapter clarifies the main concepts of protection and why do we need to coordinate protective devices.

Any electric power system consists of three principal parts:

1. Power generation
2. Power transmission
3. Power distribution

The connection between these parts can be graphically described in Figure 2.1. In the past, the structure of these parts was very simple; and even for power outages and load shedding, it was not a big issue because the dependency on the electricity was very low at that time.

With modern civilization, a huge amount of electricity is consumed to power appliances, street lights, factories, transportation and telecommunication systems, etc. Thus, the dependency becomes very high and no one can imagine how human life will be without electrical energy. To meet these needs, the new systems are designed and managed in a way that the energy can be economically and reliably delivered to the utilization points (ALSTOM, 2002). Thus, the modern systems are very advanced and at the same time are very complicated to operate and control without any failure. As a protection engineer, these failures are called "**faults**."

If any fault exists for a long time without clearance, then lots of bad scenarios could happen. Such scenarios are breaking of cable insulations due to heat generated by high current, damaging devices and equipment,[1] shortage in power delivered to customers, etc.

2.1 Faults Classification

In electric power systems, the word "*fault*" means any abnormal flow of electric current. There are many causes to create faults, such as lightning strikes, ice, snowstorms, wind, flying objects,

1 Or at least, their prospective lifespan will reduce.

Optimal Coordination of Power Protective Devices with Illustrative Examples, First Edition. Ali R. Al-Roomi.

Companion website: www.wiley.com/go/al-roomi/optimalcoordination

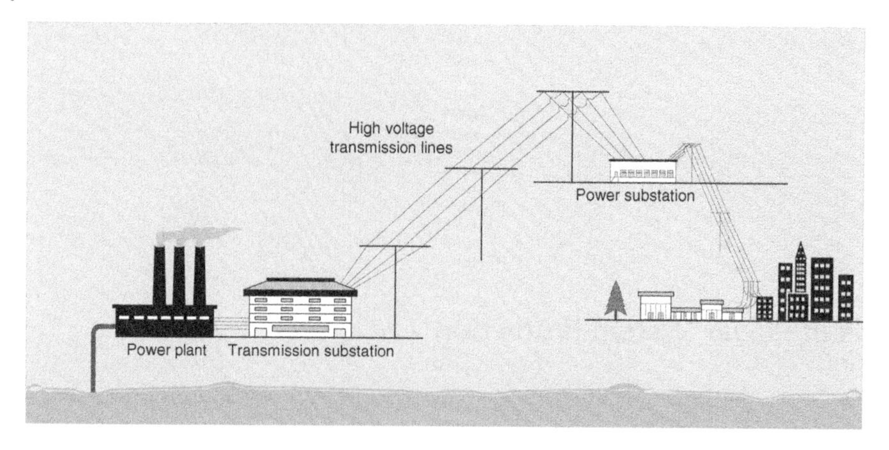

Figure 2.1 General representation of complete power system cycle.

physical contact by animals, falling trees, human errors, insulation aging, etc. The fault, as a major subject, could be categorized under two types:

1. **Open-Circuit Fault** (or **Series Fault**): it happens when the current path is interrupted.
2. **Short-Circuit Fault** (or **Shunt Fault**): it happens when the current flow is bypassed.

In most situations, the series fault does not lead to catastrophes except in some special cases. This fault type could happen if one line or more is broken without touching the other lines or grounded parts. Otherwise, it could lead to a shunt fault (Paithankar and Bhide, 2003).

The shunt fault is more dangerous, and it can hit any part of the power system. If it is not cleared as fast and selective as possible, then this fault type could lead to cascading trips that would eventually lead to a system trip "**blackout**." This is illustrated in Figure 2.2. The faults can be classified into:

1. **Passive Fault:** it happens if the values of some variables are higher than the designed maximum limits. some examples are *overloading*, *over voltage*, *power swing*, *under frequency*, etc.
2. **Active Fault:** it is a real fault and it needs to be cleared quickly.

Faults can also be classified into two types based on the fault duration:

1. **Transient Fault** (or **Temporary Fault**): this fault disappears by itself after a short time. The main procedure to clear it is by switching-off the faulty part and then re-energized again. This fault type can decrease the lifespan of conductors and/or equipment. But, usually, it does not damage them. In many cases, the transient fault happens in overhead transmission/sub-transmission lines as a result of several causes, such as momentary tree contacts, animal contacts, contact between conductors due to wind or snowfall, insulator flashover due to lightning strikes, etc. In practice, an automatic recloser is used to return the temporary faulty cable into service after waiting a specific time.
2. **Persistent Fault** (or **Permanent Fault**): the more dangerous scenario is that when it happens with a persistent fault because that fault will remain even after opening and re-closing the circuit. If this fault type happens, a quick trip action is urgently needed. Some examples of such fault on overhead cables are big flying objects, human errors, aging insulators, etc. For underground cables, mechanical damage is the most common cause.

For three-phase (3ϕ) systems, there are five types of shunt fault as shown in Figure 2.3. These possibilities could be increased to seven in the applications of underground mining where the sixth

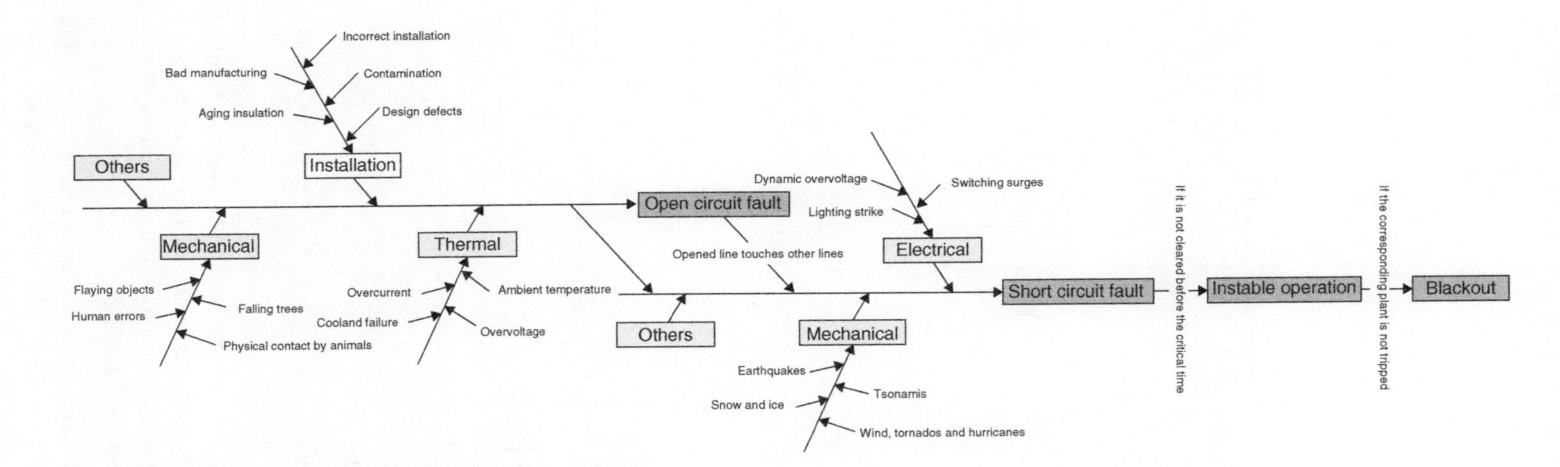

Figure 2.2 Cause and effect diagram of different fault scenarios.

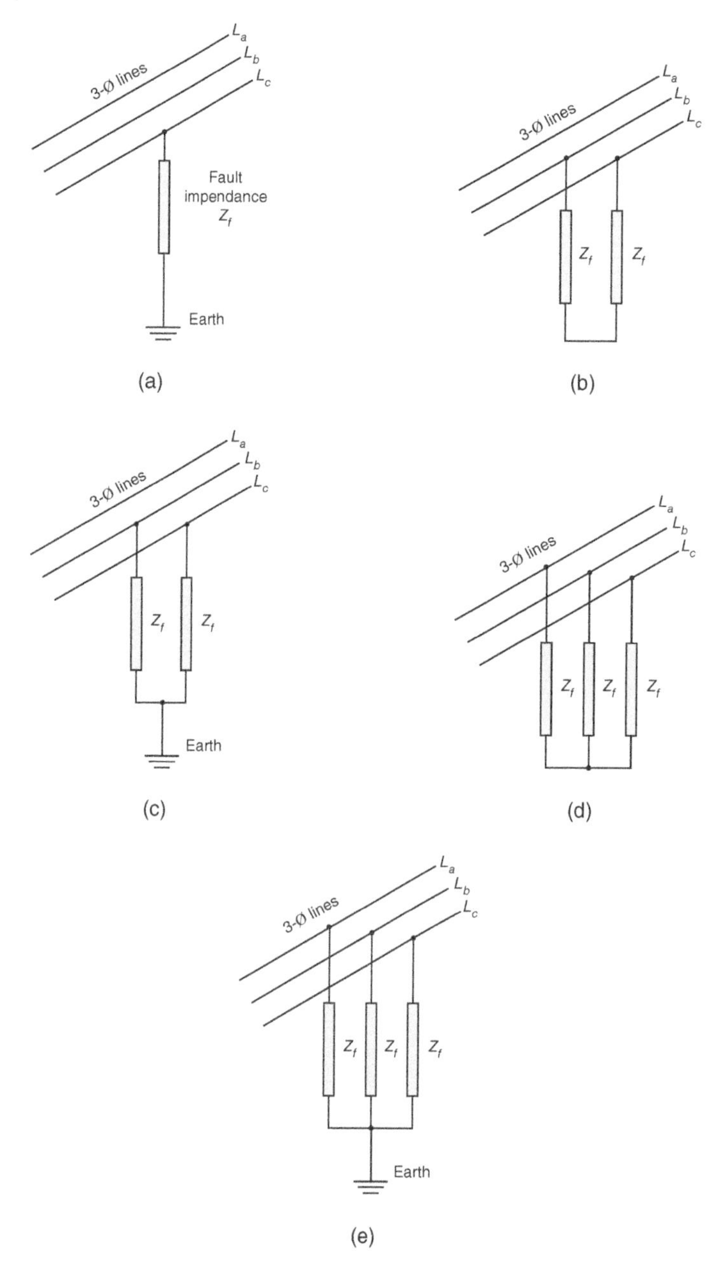

Figure 2.3 The possibilities of active short-circuit fault. (a) L-G fault, (b) L-L fault, (c) L-L-G fault, (d) L-L-L fault, and (e) L-L-L-G fault.

and seventh possibilities are *phase-to-pilot* fault and *pilot-to-ground* fault (Hewitson et al., 2004). In practice, *phase-to-ground* fault is the most probable type whereas *three-phase* fault is the most severe type as listed in Table 2.1 (Blackburn and Domin, 2006; Bandyopadhyay, 2006). These active fault classifications are grouped into two groups:

1. **Symmetrical Faults** (or **Balanced Faults**)**:** there are only two possibilities of balanced faults which are the last two types (i.e. 3ϕ faults) of Figure 2.3. They are called symmetrical because each phase has equal sinusoidal waves (i.e. *voltage* and *current*) on the axis and are shifted by

Table 2.1 The probability of faults occurrence corresponding to their types.

Fault type	Probability of occurrence	Severity level
Single phase-to-ground	70–80%	Lowest
Phase-to-phase	10–8%	
Phase-to-phase-to-ground	17–10%	
Three-phase	3–2%	Highest

Table 2.2 The probability of fault occurrence on different power system elements.

Power system element	Probability of occurrence (%)
Overhead lines	50
Underground cables	9
Transformers	10
Generators	7
Switchgears	12
CT, PT, etc.	12

120° between each other. Thus, the self-impedances of these three lines are equal; and at the same time, if there are any mutual impedances between these lines, then their values will be equal too.

2. **Asymmetrical Faults** (or **Unbalanced Faults**): the remaining of the five fault possibilities are classified as unbalanced faults. For these types, the phases are not affected equally where the magnitudes and angles of their voltages, currents, and impedances will not be the same anymore.

Table 2.2 shows the probability of fault occurrence on different power system elements (Paithankar and Bhide, 2003). It shows that 50% of faults occur in overhead lines.

2.2 Protection System

As a sense of engineer, eliminating all the causes or sources of these faults is an impossible mission due to many factors, like uncontrolled/unpredicted abnormal events, and high cost of having a highly secured and reliable system, especially if some of these faults are not predefined or rarely happen. Instead, the practical and relatively economic approach is to eliminate the effects of these causes (i.e. the faults themselves) by fast detection and clearing actions. Thus, any fault can be prevented from jeopardizing system integrity and its consequences can be mitigated effectively and thus the system can be preserved from any unstable operating condition (Padiyar, 2008; Grainger and Stevenson, 1994; Glover et al., 2012). The common devices used in power system protection are:

1. Non-electrical relays (such as: bimetallic, Buchholz, and pressure relief relays);
2. Electrical relays (such as: distance, overcurrent, differential, over/under voltage, over/under frequency, reverse power, and over flux relays);

3. Reclosers;
4. Sectionalizers; and
5. Fuses.

The fault can be easily detected if the system voltages and/or currents are known. From these two fundamental signals, all the required information can be extracted. Thus, these two fundamental signals need to be measured quickly and accurately. To accomplish this task, a **current transformer** (**CT**) and a **potential transformer** (**PT**[2]) are used here, as shown in Figure 2.4. It has to be said that there are two stages of step-down current and voltage quantities. The first one, which has a huge stepping action, is done through these main transformers, while the second stage is done through what are called **auxiliary transformers**. These small transformers are placed inside protective relays. Some issues have to be taken into account regarding the received signals, such as the effects of decaying DC components, harmonic, and CT saturation. The protective relays can be classified according to their available technology:

- **Hardware-Based Protective Relays:**
  - Electromechanical (or Electromagnetic) Relays
  - Solid-State (or Static) Relays
  - Digital Relays
- **Numerical** (or **Software-Based**) **Protective Relays:**
  - Microprocessor-Based Relays
  - DSP-Based Relays

From the aforementioned classification, the electromechanical relays (such as: moving coil, attracted armature, induction, and motor operated devices) are the first generation OCRs, having appeared early in the last century (1900s). They are limited by the need for periodic maintenance (PM) and calibration because of their mechanical moving parts. They have a few discrete values of *PS* and *TMS*, which means that the feasible search space is very confined and thus it is hard to be feasibly optimized. Also, they have slower response due to their **over-shoot time**[3] caused by inertia. Moreover, their simple technology make them blind to each other, and the designer is forced to use three to four relays to protect all the three phases plus the ground line as illustrated in Figure 2.5. There are some advantages, however, to using these devices, such as their stability and insensitivity to the network conditions; and they are still in service because of their lifespan. Moreover, because of their longevity, there are many skilled experts to coordinate this kind of relays (Mason, 1956; McCleer, 1987; Paithankar and Bhide, 2003; Bandyopadhyay, 2006).

The second-generation OCRs were created to utilize analog electronic technology to mimic the first generation. These relays are called solid-state or static relays. The main technical problem faced with these relays is that the ambient temperature can affect their stability. Also, precise

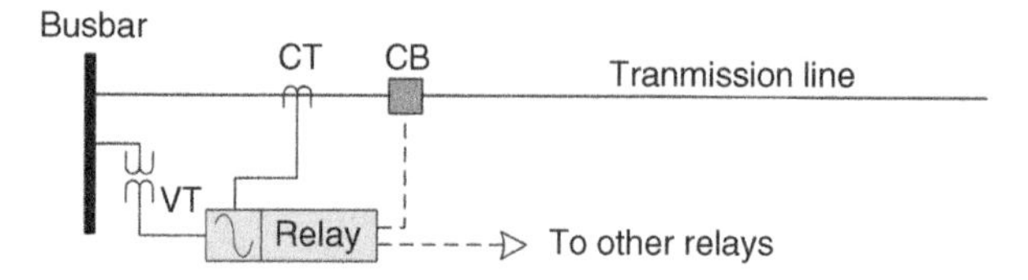

Figure 2.4 Illustrated 1ϕ protection system.

---

2 It is also known as a **voltage transformer** (**VT**).
3 It is also called the **over-travel delay** and the **coasting time** (Al-Roomi, 2014).

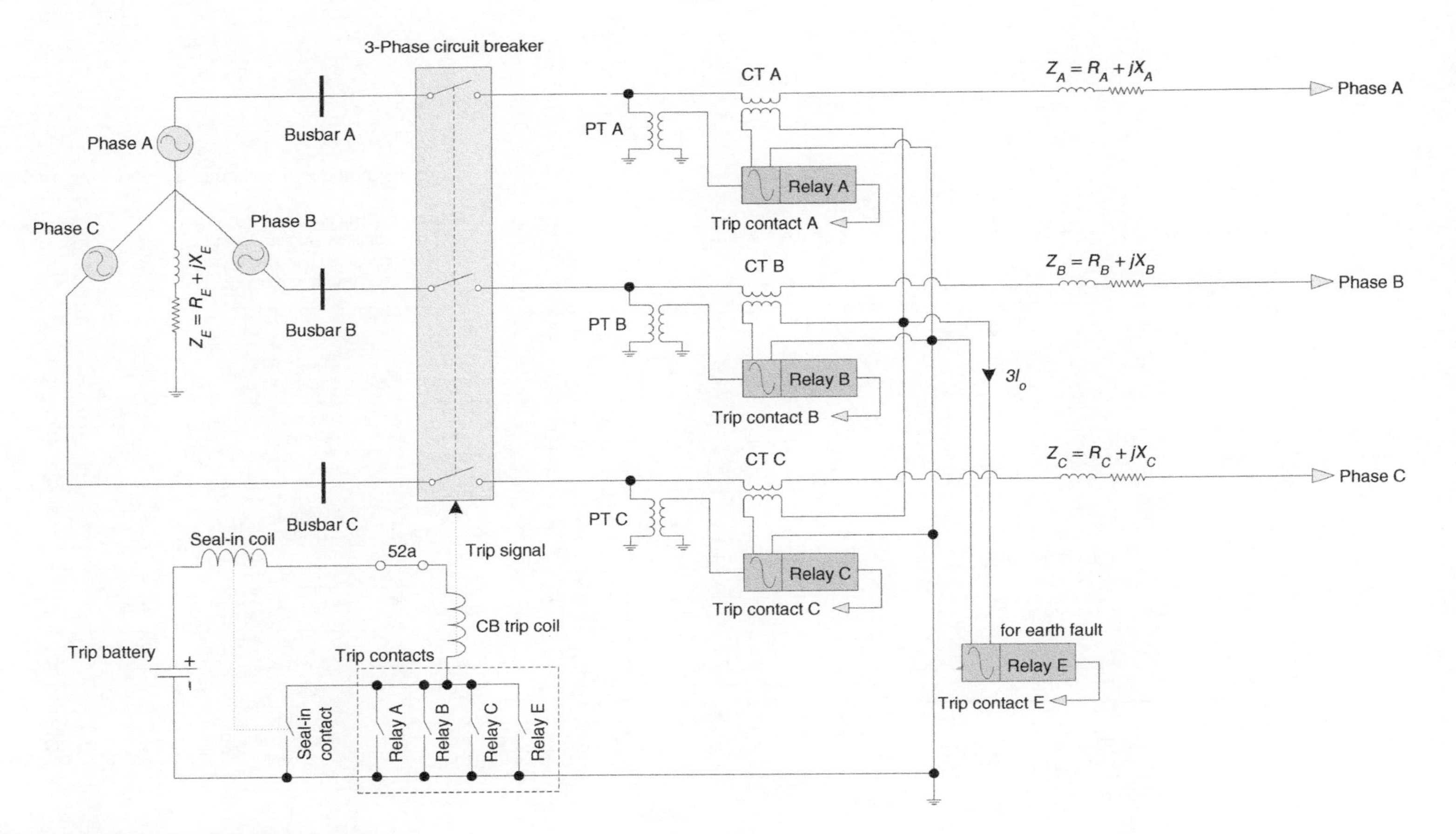

Figure 2.5 General 3ϕ protection system.

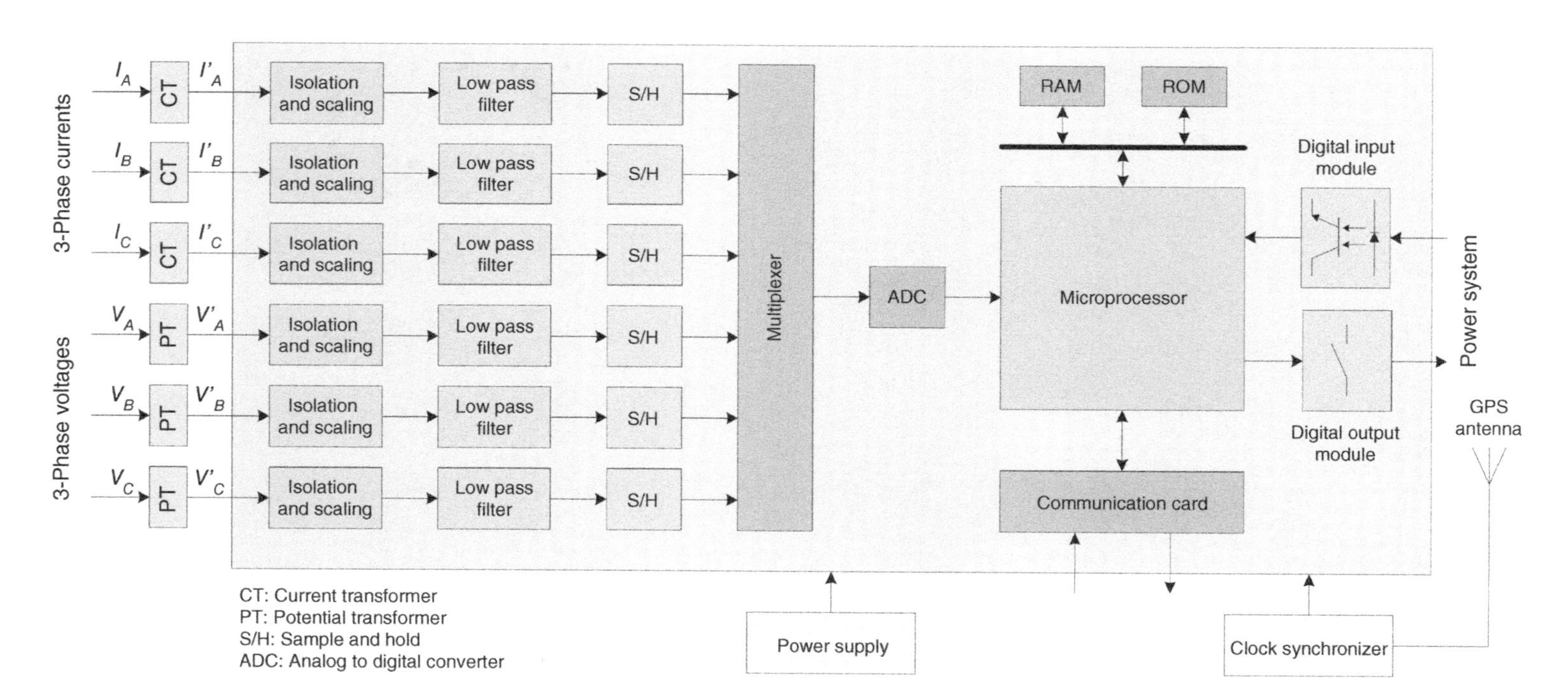

Figure 2.6 General hardware of a numerical relay.

passive components (such as resistors, capacitors, and inductors) are required to reduce the total error (Gers and Holmes, 2004).

In less than one decade, the third generation digital relays appeared. Many manufacturers (such as *SIEMENS*, *AREVA*, *ALSTOM*, *GE*, *ABB*, and *SEL*) successfully designed new techniques to allow two-way communications between these relays through some standard protocols (Elmore, 1994; ALSTOM, 2002; Gers and Holmes, 2004).

Although all the inherent weaknesses of the electromechanical and solid-state relays can be permanently resolved by the third generation relays, they are still hardware-based relays. Thus, to have programmable relays, the fourth generation numerical relays were invented. Because the relay manufacturers found that the numerical relays share the same hardware (analog and digital inputs modules, auxiliary CTs and PTs, low pass filters, multiplexer, analog to digital converter, microprocessor, RAM and ROM, digital output module, communication card, power supply, etc.), so some of these relays can work as **general purpose relays.**[4] These innovative relays are called numerical relays; and because of their capabilities, some researchers considered them as **intelligent electronic devices** (**IEDs**). They are manufactured based on microcontrollers, microprocessors, or even digital signal processors (DSPs) for high computational applications. All the relay settings, instructions, and operations can be updated, modified, or even upgraded through some special software provided by their original equipment manufacturers (OEMs) (Wood et al., 1989; Johns and Salman, 1995; Gers and Holmes, 2004; Rebizant et al., 2011). The operational philosophy of these state-of-the-art relays can be changed by uploading new algorithms. The internal structure of these relays is shown in Figure 2.6.

2.3 Zones of Protection

In order to make protection designs adequate, power systems are divided into multiple zones where each zone can be individually protected against its corresponding faults. If the faults are located inside the zone, then they are called **in-zone faults**,[5] otherwise they are considered as **out-zone faults**.[6] This approach can also minimize the total number of isolated components. Figure 2.7 describes the concept of protection zones through a simple system. Note that the overlaps between the zones are important to ensure that all the spots are protected (Elmore, 1994; Gers and Holmes, 2004).

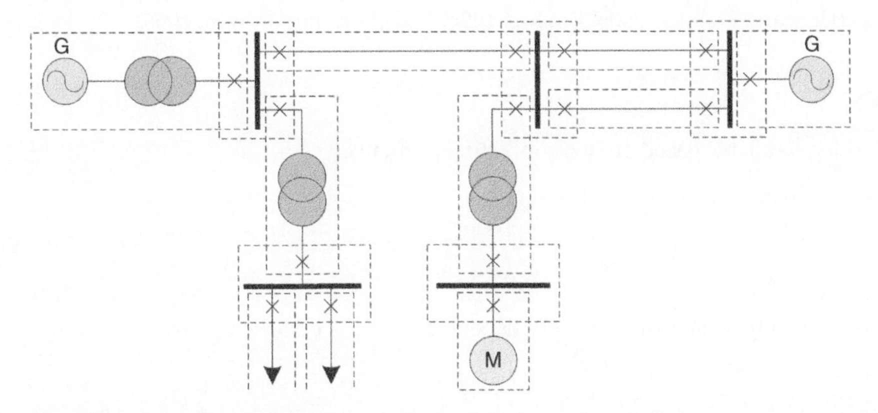

Figure 2.7 Zones of protection.

4 They are also known as **multi-function relays**.
5 Also called **internal faults**.
6 Also called **external faults**.

Practically, these zones are protected by using combinations of different protective relays. Except for the general purpose protective relays, there are many types of protective relays that can be classified according to their working principles:

- Distance relays
- Overcurrent relays
- Differential relays
- Over/under voltage relays
- Over/under frequency relays
- Reverse power relays
- Over flux relays
- etc.

For example, power transformers are protected by a group of relays, such as differential, overcurrent, overflux, etc.

2.4 Primary and Backup Protection

Good protection system designs can be created if each zone has a number of primary and backup relays. The first set of relays act as a first wall of protection for in-zone faults, and the second set of relays act as a second wall of protection for out-zone faults. The backup relays could be installed locally at the same location of the primary relays so that the same internal faults can be seen. They are known as **local backup relays**. Also, they can be installed in other zones to see the preceding faults as external faults. They are known as **remote backup relays**. These backup relays are very important because the probability of having failures on any one of the primary protective relays always exists. This means that the second line of defense has to be ready all the time. Some examples will be taken later to show how to select the corresponding backup relays for each primary relay.

2.5 Performance and Design Criteria

Some general criteria have to be fulfilled to accomplish the assigned protection task. These criteria can be classified as (Gers and Holmes, 2004; Elmore, 1994; Blackburn and Domin, 2006):

2.5.1 Reliability

The system is considered reliable based on two conflicting elements:

1. Dependability
2. Security

2.5.1.1 Dependability

The protection system can be designed to be very dependable if it operates correctly for in-zone faults.

2.5.1.2 Security

The protection system can be designed to be very secure if it does not operate incorrectly for out-zone faults.

2.5.2 Sensitivity

The protection system should detect any fault under any level of severity. The good design should take into account both the lowest and highest faults to set the protective relays correctly.

2.5.3 Speed

Isolating the fault as quick as possible is very important issue, and it is one of the reasons why the existing advanced technologies are employed in the protective relaying filed. But, sometimes, an intentional delay is required to differentiate between the primary and backup protection and to avoid misoperation due to load fluctuations.

2.5.4 Selectivity

If the correct set of relays is selected to clear the fault, then this action will enhance the system stability and increase the total amount of power delivered to the customers and end-users. The word "selectivity" can be shortly defined as ***"the maximum continuity of service with minimum system disconnection (Blackburn and Domin**, 2006)."*

2.5.5 Performance versus Economics

There is no doubt that the design of any protection system is better:

- as the relays become faster with low inherent errors, and
- as each zone is equipped with multiple types of relays.

However, the total cost of such a design will be very high. Therefore, a good design can be achieved if it balances performance and money.

2.5.6 Adequateness

One of the reasons for having highly expensive numerical relays is when the relay designs have multifunctions features. Instead of embedding all the features inside one relay, it is better to create a variety of models for different assigned protection tasks.

2.5.7 Simplicity

If the protection system has a simple design, then it will provide easy operation, maintenance, troubleshooting, and flexible future upgrading process. The designed protection scheme can be accomplished in several ways with different complexities and options. Therefore, the simplest design does not always mean the cheapest design.

2.6 Overcurrent Protective Devices

Electric protective relays are widely used in power systems. However, there are other devices that can also be used to protect some specific electric components. Such these devices are Buchholz relays, pressure relief relays, fuses, and bimetallic relays. The first two protective devices have limited applications and they are extensively described in Ram and Vishwakarma (1995). The last two protective devices are explained in the following text.

2.6.1 Fuses

Fuses can be considered as the oldest protective devices that are still used nowadays (Paithankar and Bhide, 2003). This is because of their simple construction and working principle. To clarify this point, let's see the major elements of protective relays shown in Figure 2.4. Any short-circuit fault occurs on that transmission line will be detected by measuring the two fundamental variables, i.e. current and voltage, respectively via main CT and PT and auxiliary CT and PT. Then a filtration stage is embedded to deal with decaying DC components, harmonic, and CT saturation. After that, a special computing algorithm is coded to deal with the incoming signals and detect any possible fault. Finally, a trip signal is sent from that protective relay to its corresponding circuit breaker to interrupt the flow of the short-circuit current.

Instead, all the preceding essential stages[7] are combined into a one small and compact device called a fuse. The most distinct properties of this protective device are its speed and price compared with all the known types of protective relays. Also, fuses do not need to use any CTs, PTs, or even circuit breakers (CBs) to accomplish their jobs (Al-Roomi and El-Hawary, 2018b).

The working principle of fuses is very simple. The thermal energy of a conducting material is proportional with the flowing capacity and duration of electrons. By deforming a part of that material in a way so that the energy absorption and melting processes are concentrated in a weak link,[8] that link will melt and vaporize once the magnitude and duration of the flowing current exceeds the threshold level. This phenomenon occurs during shunt faults where the fault current steeply increases due to low impedances of the new formed short paths. Figure 2.8 shows four possible shapes of the previously described weak link (McCleer, 1987).

The main disadvantages of these unique protective devices are their uncontrollable operating time and the irreversible interrupting action. Thus, if the fuse weak link is melted, then the whole fuse unit needs to be replaced with a new one. Some good theoretical and practical descriptions about fuses can be found in McCleer (1987), Das (2012), and Hegde et al. (2015).

As a summary, fuses are considered as very fast non-directional and non-adjustable overcurrent protective devices. There are many types of fuses available in the market, which can be classified based on: (i) operating time, (ii) voltage and insulation level, (iii) type of system, (iv) maximum fault current level, and (v) load current. Different classes are used in the market to describe the application(s) of each type, such as: RK1, RK5, CC, T, K, G, J, L, and R (Gers and Holmes, 2004; Das, 2012). There are three major characteristics called cut-off characteristic, time-current characteristic, and I^2T characteristic (Bakshi and Bakshi, 2010). Because the operating principle and mechanism of fuses depend on heat transfer, the actual characteristic curves could be different than the standard ones based on the current ambient temperature (SIEMENS, n.d.).

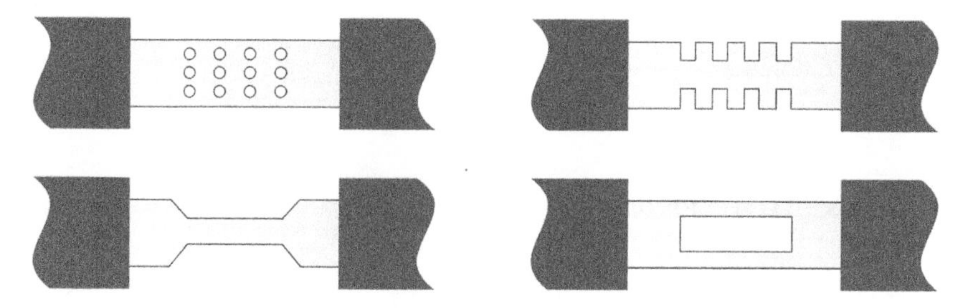

Figure 2.8 Different weak link regions of fuses internal structure. Source: McCleer (1987).

7 i.e. sensing, filtration, comparison, detection, and interruption stages.
8 i.e. its area is less than that of the end regions.

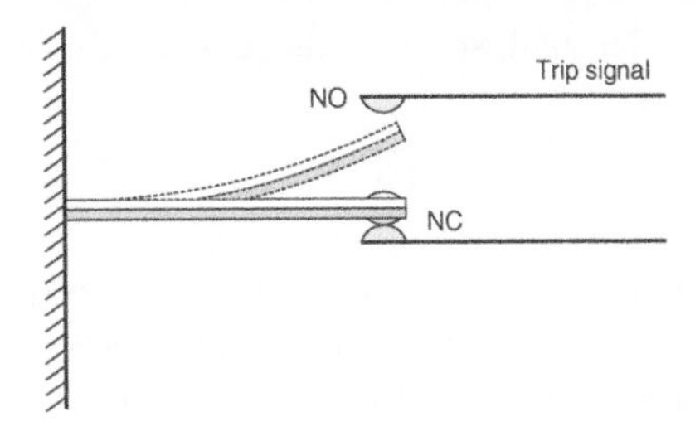

Figure 2.9 Bimetallic relay.

2.6.2 Bimetallic Relays

These devices work on a very basic physical phenomenon where the body expansion increases proportionally with the applied temperature. If two different materials with different expansion rates are combined together, the overall expansion will be in a circular shape instead of linear as shown in Figure 2.9. If some contacts are attached to the right-end of these two strips, a trip signal can be generated across that normally-open (NO) contact once the thermal energy exceeds its threshold limit and duration time.

The inherent problem associated with this type of protective devices is its slow response, and thus bimetallic relays are mainly used to protect motors against long overloading condition (Paithankar and Bhide, 2003). This main drawback restricts them to be used in protecting generation and transmission parts because they need a quick isolation process. Based on this, bimetallic relays cannot be considered as proper candidates to replace fuses.

The operating philosophy of other overcurrent protective devices, such as reclosers and sectionalizers, can be found in Blackburn and Domin (2006), Gers and Holmes (2004), ALSTOM (2002), Hewitson et al. (2004), Ram and Vishwakarma (1995), and Das (2012), which is out of scope of the present book.

2.6.3 Overcurrent Protective Relays

As understood from Sections 2.6.1 and 2.6.2, overcurrent protection can be achieved by the use of fuses or relays (Hodgkiss, 1995). OCRs are more suitable for large and complex networks because the coordination between different devices becomes very hard. However, because fuses are cheap and available with different time responses, they are still used in combination with protective OCRs in some large networks.

In comparison with other expensive relays, OCRs can compromise between different design criteria, and this is why they are popular and widely used in power system protection (Elmore, 1994; Urdaneta et al., 1997).

Non-directional OCRs have only currents as inputs from power networks. These input currents are provided by **current transformers** (CTs). The main setting of OCRs is called **plug setting** (**PS**); which is also known as **current setting multiplier** (**CSM**) in some references. The other setting is called **time multiplier setting** (**TMS**); which is also known as **time setting multiplier** (**TSM**) in some references. It has to be remembered that there are two standards here; European and North American. Thus, in some references, the plug setting is also known as **pickup setting** (I_p), **pickup current setting** (**PCS**), and **current tap setting** (**CTS**). For the time multiplier setting, the other names are **time dial setting** (**TDS**) and **time lever setting** (**TLS**). The other important note is that the range of *TMS* is different than that of *TDS*. More information will be covered later in Chapter 3 when both standards are incorporated to find the best relay settings.

The fault severity can be seen by OCRs via the following equation:

$$PSM = \frac{I_{\text{relay}}}{PS} \tag{2.1}$$

where PSM denotes the plug-setting multiplier,[9] I_{relay} is the fault current seen by the relay; referred to the secondary side of the CT. If $PSM < 1$, it means that no fault exists in the system.

Based on the relation between fault current and relay operating time, which is called the **time-current characteristic curve** (TCCC), OCRs can be categorized into three main types (Gers and Holmes, 2004):

2.6.4 Instantaneous OCR (IOCR)

This type of relays, which is also known as **Definite Current OCR** (DCOCR), has only PS. The TCCC is shown in Figure 2.10. It shows that IOCR operates when the current exceeds the predefined I_{ins}. Thus, it can be mathematically expressed as follows:

$$T = \begin{cases} T_{\text{ins}}, & \text{if } I_{\text{relay}} \geqslant I_{\text{ins}} \\ \infty, & \text{otherwise} \end{cases} \tag{2.2}$$

where I_{ins} is the predetermined value that the fault current has to reach, and T_{ins} is the inherent time delay that IOCR cannot pass.

This type of relays cannot be applied when it has to differentiate between two faulty points if the source impedance is much larger than the impedance between these two points; i.e. $Z_S \gg Z_L$. For example, if there are two faults, one at the busbar A and the other at the busbar B of Figure 2.11, then the operating time of the nearest fault will be almost the same as that of the farthest fault.

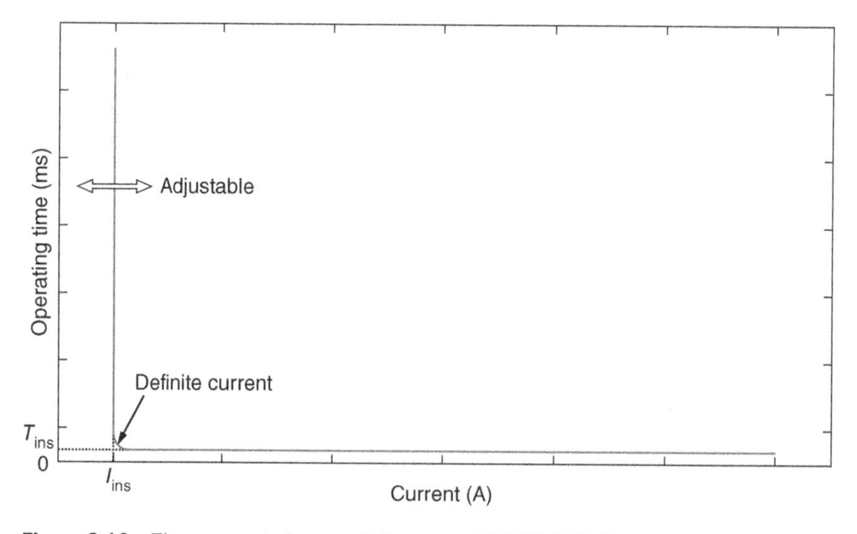

Figure 2.10 Time-current characteristic curve of ITOCR/DCOCR.

9 For the North American standard, this variable is known as the **multiples of pickup current** (**M**) in some references where $M = \frac{I_{\text{relay}}}{I_p}$ (Urdaneta et al., 1988; Anderson, 1998; IEEE, 1999).

2.6.5 Definite Time OCR (DTOCR)

In addition to *PS*, this type has a **time delay setting** (T_{set}). The relay operates only if the fault current exceeds the predefined value and the fault duration persisted for a time longer than the pre-specified delay. Thus, the fault at different locations can be distinguished easily as shown in Figure 2.12.

By referring to Figure 2.11, the main disadvantage of this type is when the fault is located near the source. For this situation, higher fault currents are expected. Instead of taking immediate/fast action to clear that highly dangerous fault, the relay will wait until reaching T_{set}. From that figure, the following equation will decide if DTOCR should operate or not:

$$T = \begin{cases} T_{set}, & \text{if } I_{relay} \geqslant I_{set} \\ \infty, & \text{otherwise} \end{cases} \tag{2.3}$$

where I_{set} is the predetermined value that the fault current has to reach, and T_{set} is the predetermined time delay that DTOCR has to wait for.

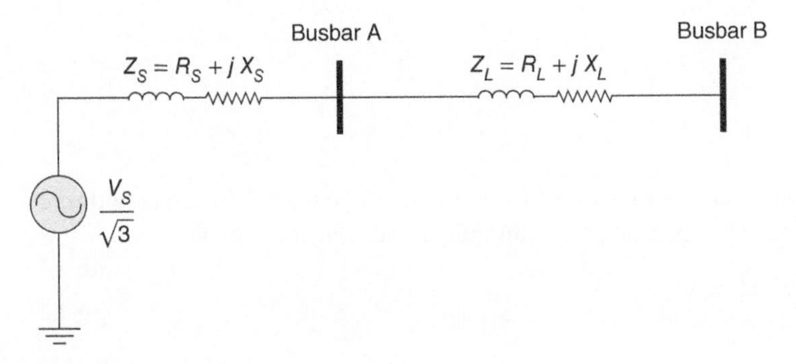

Figure 2.11 Illustration of two-bus radial network.

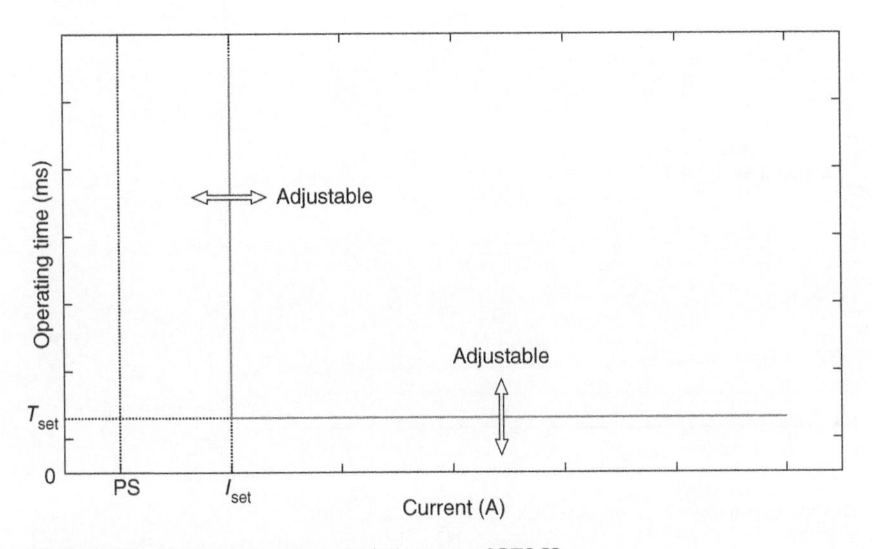

Figure 2.12 Time-current characteristic curve of DTOCR.

2.6.6 Inverse Time OCR (ITOCR)

This type of relays has both *PS* and *TMS*. It can solve the preceding problems of DTOCR because its operating time is inversely proportional to the short-circuit current value:

$$T \propto \frac{1}{PSM} \tag{2.4}$$

This can be depicted in Figure 2.13.

Someone may ask about the mathematical model of (2.4)! Actually, many models have been suggested to mimic the actual operating time of this kind of OCRs. Many of these models will be studied later in the Chapter 3. These models are so important for engineers to simulate the behavior of protective relays in computers, so good and optimal settings can be obtained. In the past,[10] the primitive way to calculate the operating time of ITOCRs is to use some standard log sheets provided by OEMs.

Example 2.1 The log sheet of the CO8 time inverse characteristic curve[11] is shown in Figure 2.14. Calculate T when:

1. $PSM = 2$ and $TDS = 1/2$
2. $PSM = 2$ and $TDS = 1$
3. $PSM = 4$ and $TDS = 7$

Solution

Let's draw a vertical line at each plug setting, and then horizontal line at each intersect with the corresponding time dial setting. The operating time can then be calculated as follows:

- For $PSM = 2$ and $TDS = 1/2$: $T \approx 1$ s
- For $PSM = 2$ and $TDS = 1$: $T \approx 2.13$ s
- For $PSM = 4$ and $TDS = 7$: $T \approx 4$ s

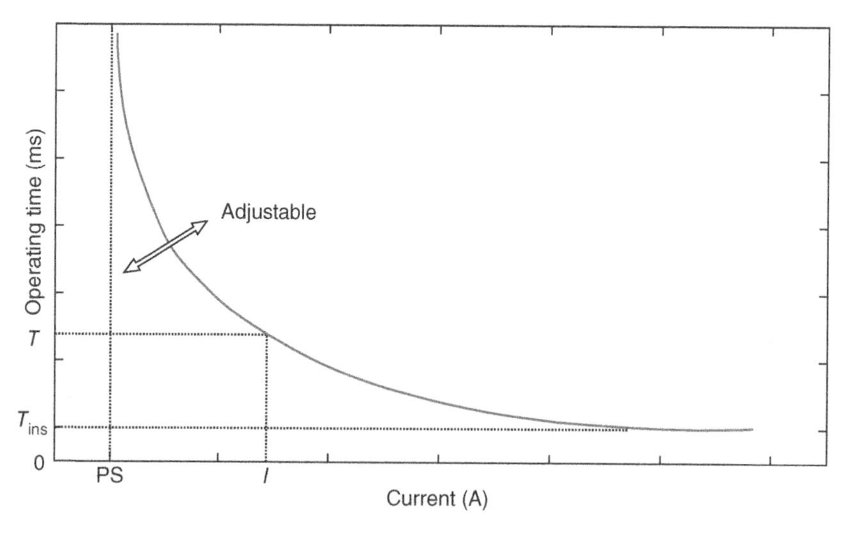

Figure 2.13 Time-current characteristic curve of ITOCR.

10 It is still used by technicians to quickly calculate the operating time of OCRs so they can determine if these devices are functioning well or need calibration, repair, or replacement.
11 The complete list of the CO series relays will be covered later in Chapter 3.

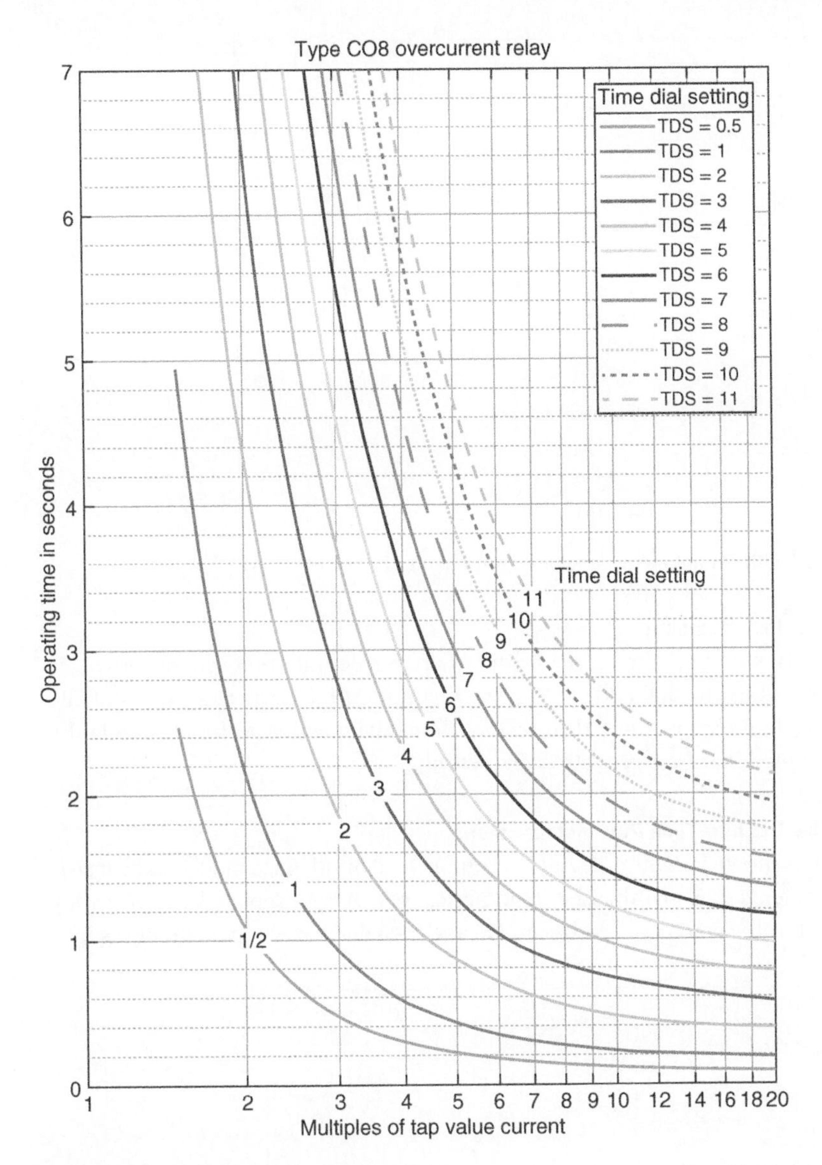

Figure 2.14 Type CO8 time-current curves

2.6.7 Mixed Characteristic Curves

We have seen the three principle curves of OCRs. Each one of them has its pros and cons. The question here is: *Can we hybridize between these three curves so that the advantages can be merged and the disadvantages can be minimized?* Yes, we can do that because these elements are built as separate units. We have four possible hybrid configurations, which are listed as follows:

2.6.7.1 Definite-Time Plus Instantaneous

This combination of definite-current and definite-time elements is shown in Figure 2.15. It works as a definite-time OCR for a specific range of short-circuit currents located between I_{set} and I_{ins}. Once I_{relay} exceeds the maximum allowable limit (I_{ins}), the relay will take immediate action to clear the fault.

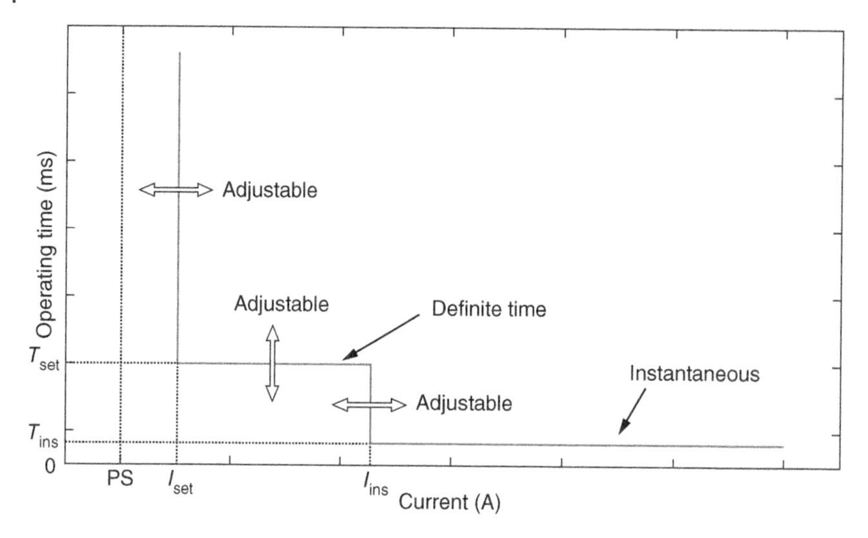

Figure 2.15 OCR equipped with instantaneous and definite-time elements.

2.6.7.2 Inverse-Time Plus Instantaneous

The same principle is shown in Figure 2.16, except that the definite-time characteristic is replaced with the inverse-time characteristic. Thus, the operating time decreases exponentially until reaching I_{ins} where the relay is allowed to take an immediate action to isolate the faulty component. It has to be noted that I_{ins} is a controllable setting.

2.6.7.3 Inverse-Time Plus Definite-Time Plus Instantaneous

This characteristic curve is shown in Figure 2.17. It is a complicated multi-stage model that consists of the three principle elements. It is mainly used in numerical relays.[12] As depicted, the operating time of OCR starts at high value when I_{relay} is close to PS, and then decreases exponentially as I_{relay}

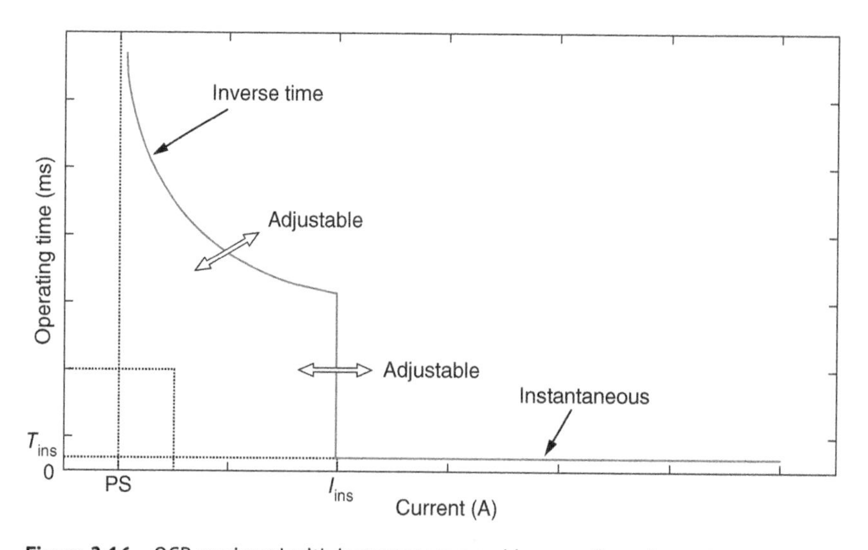

Figure 2.16 OCR equipped with instantaneous and inverse-time elements.

12 It will be discussed in Chapter 3; specifically, in programmable/customized TCCCs.

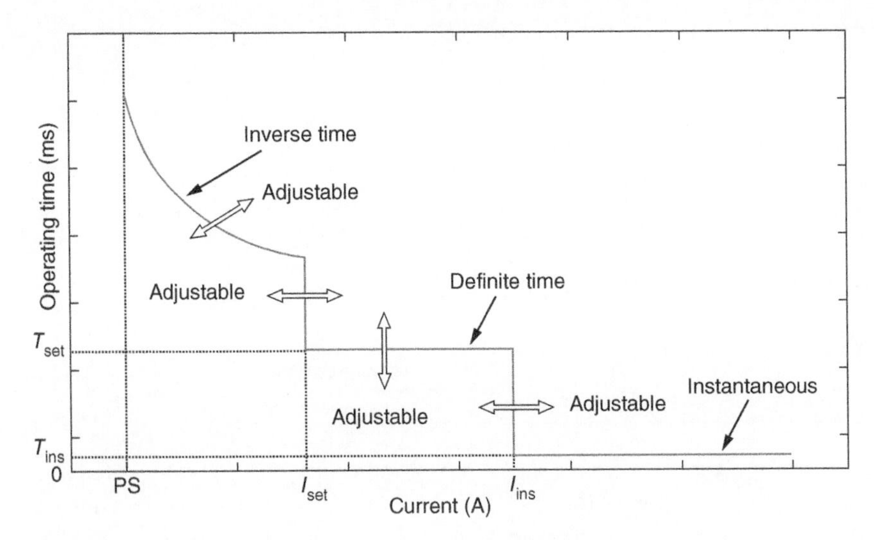

Figure 2.17 OCR equipped with instantaneous, definite-time, and inverse-time elements.

increases. Once I_{relay} passes I_{set}, the operating time will equal I_{set}. Thus, T will remain constant until passing the next threshold (I_{ins}) where the relay is allowed to operate instantaneously.

Someone may think about re-arranging these elements as follows:

Definite-Time ⇒ Inverse-Time ⇒ Instantaneous

This kind of special relays can be programmed in numerical relays that support customized TCCCs. However, this arrangement violates the goal of using the inverse-time characteristic. This will be further explained in Chapter 4 when the inverse-time grading is covered.

2.6.7.4 Inverse-Time Plus Definite-Time

This combination is illustrated in Figure 2.18. Thus, T decreases exponentially as I_{relay} increases. Compared with Figure 2.16, the relay here will not take an immediate action when I_{relay} reaches the threshold (i.e. I_{set}). Instead, the operating time will be constant and equal to T_{set}.

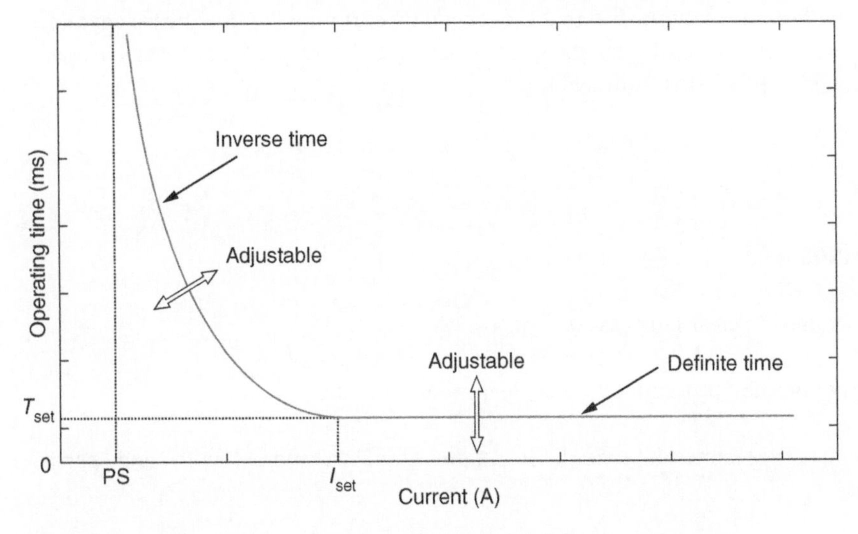

Figure 2.18 OCR equipped with instantaneous and definite-time elements.

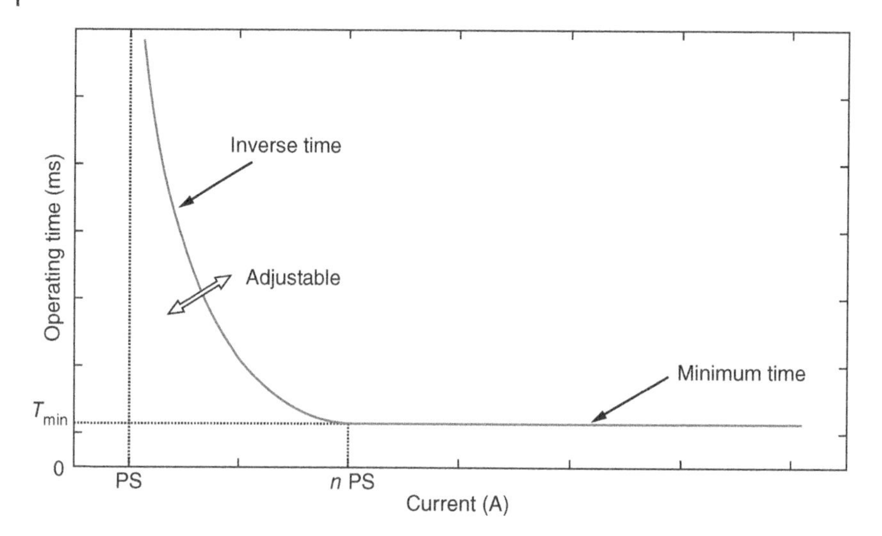

Figure 2.19 Characteristic of inverse definite minimum time over current relay (IDMT OCR).

Again, although we can swap between the two elements to have (definite-time plus inverse-time), this arrangement is not recommended and makes the usage of inverse-time characteristic meaningless.

2.6.7.5 Inverse Definite Minimum Time (IDMT)

It can be seen as a realized version of the inverse-time curve shown in Figure 2.13 or as a special version of the mixed curve shown in Figure 2.18. This curve, which is very popular and known as the IDMT curve, is shown in Figure 2.19. We can look to it as a mixed curve of (inverse-time plus definite-time) where I_{set} equals a specific multiples of PS and the operating time reaches an uncontrollable value called T^{min}.

The reason behind getting this special curve is that the inverse-time curve shown in Figure 2.13 represents an ideal relay which cannot be achieved. As the fault current on the primary side of CT increases, the stepped-down current on the secondary side of CT increases too. This will continue until the transformer becomes saturated so that the secondary current will not increase anymore. This phenomenon will occur at n multiples of PS. After that limit, the relay will not see any increase in I_{relay} and thus T will not pass that saturated limit.

Problems

Written Exercises

W2.1 Why do we need to clear faults as soon as possible?

W2.2 What is the common procedure used to clear transient faults?

W2.3 What are the other names of "series fault," "shunt fault," "temporary fault," and "persistent fault"?

W2.4 For temporary and persistent faults, what is the most common type faced in overhead lines? Is it the same for underground cables?

W2.5 Both fuses and bimetallic relays operate based on excessive heat generated by current. Can we replace high-speed fuses with bimetallic relays? Explain.

W2.6 State three types of overcurrent protective devices?

W2.7 What are the four technologies used in manufacturing overcurrent relays?

W2.8 General purpose or multi-function relays can act as any type of protective relays. Why most of the numerical relays offered in the market are designed to perform a specific task, such as overcurrent, distance, differential, etc.?

W2.9 An overcurrent relay is used to protect a line. The current transformer has a ratio of 80. If PSM is twice PS, determine the fault current seen by the relay if PS equals 3.5 A. What is the actual fault current before stepping it down?

W2.10 What is the main drawback of DCOCR?

W2.11 What is the main drawback of DTOCR?

W2.12 Using the log sheet of the CO8 time inverse characteristic curve, which is shown in Figure 2.14, calculate the following:
1. TDS =? if $T = 0.43$ s and $PSM = 5$
2. T =? if $PSM = 7$ and $TDS = 5$
3. PSM =? if $T = 4.3$ s and $TDS = 2$

Computer Exercises

C2.1 Develop a MATLAB program to solve Exercise W2.9 automatically.

C2.2 Modify the code designed in Exercise C2.1 to show the results when all the ANSI standard CT-ratios given in Appendix A are considered.

C2.3 Remodify the code to show the results and then plot them in one graph.

3

Mathematical Modeling of Inverse-Time Overcurrent Relay Characteristics

Aim

This chapter is dedicated to clarify the confusion behind the models used to calculate the operating time of inverse-time overcurrent relays. The literature contains many equations where some of them are based on polynomial equations, while the others are based on exponential equations. Further, some of the equations belong to the European standard, while the others belong to the North American standard. Moreover, the big companies have their own standards, such as the models used with IAC and CO series. Add to all these issues, the setting used to adjust the trip contacts has two different names with different ranges. Thus, the reader will definitely face lots of challenges and might end up with programming a wrong code and thus leads to getting incorrect results.

3.1 Computer Representation of Inverse-Time Overcurrent Relay Characteristics

In Chapter 2, one method has been presented to calculate the operating time of ITOCRs, which is by tracing the intersection between the multiples of pickup current and the time dial setting of curves plotted in log sheets. This primitive calculation is impractical if protection engineers want to simulate these relays in computers. Three popular approaches are presented in the literature:

3.1.1 Direct Data Storage

This is very simple approach. It stores the actual relay settings and its behavior as digits in computer memory. Thus, it will act like a lookup table for predefined input data, so it is a very fast method to retrieve a value from memory. If the input data is not stored in the computer, then the closest points can be utilized to estimate the output. To give a brief idea about this approach, let's take the following example:

Example 3.1 Table 3.1 shows some actual settings of the log sheet shown in Figure 2.14. Estimate the following unknowns using both the classical linear interpolation and the logarithmic-based non-linear interpolation:

1. T_1 [Compare the results if $T_2^{\text{actual}} = 0.1053$ s]
2. M_5 [Compare the results if $M_5^{\text{actual}} = 2.5$]
3. TDS_8 [Compare the results if $TDS_8^{\text{actual}} = 3.45$]

Optimal Coordination of Power Protective Devices with Illustrative Examples, First Edition. Ali R. Al-Roomi.

Companion website: www.wiley.com/go/al-roomi/optimalcoordination

Table 3.1 Some actual data of Co-8 relay stored in computer memory.

No.	Multiples of tap current	Time dial setting	Operating time
1	12	1/2	0.1108 s
2	14	1/2	T_2 =?
3	16	1/2	0.1017 s
4	2	1	2.1633 s
5	M_5 =?	1	1.3133 s
6	3	1	0.9238 s
7	8	2	0.5489 s
8	8	TDS_8 =?	0.9468 s
9	8	4	1.0978 s

Solution

- **Estimating T_2:** For the first three rows, we can see that TDS is fixed. Therefore, the relationship is between M and T:
  - **Classical Linear Interpolation**:

$$12 \rightarrow 0.1108\,\text{s}$$
$$14 \rightarrow T_2$$
$$16 \rightarrow 0.1017\,\text{s}$$

$$\because \frac{T_2 - 0.1017}{0.1108 - 0.1017} = \frac{14 - 16}{12 - 16} = \frac{-2}{-4} = \frac{1}{2}$$

$$\therefore T_2 = \frac{1}{2}(0.1108 - 0.1017) + 0.1017$$
$$= 0.1062\,\text{s}$$

The absolute error is:

$$|T_2^{\text{actual}} - T_2| = |0.1053 - 0.1062| = 9 \times 10^{-4}\text{s}$$

  - **Logarithmic-Based Non-linear Interpolation**:

$$\because \frac{T_2 - 0.1017}{0.1108 - 0.1017} = \frac{\log(14) - \log(16)}{\log(12) - \log(16)}$$

$$\therefore T_2 = 0.1017 + \frac{\log(14) - \log(16)}{\log(12) - \log(16)} \cdot (0.1108 - 0.1017)$$
$$= 0.1059\text{s}$$

The absolute error is:

$$|T_2^{\text{actual}} - T_2| = |0.1053 - 0.1059| = 6 \times 10^{-4}\text{s}$$

- **Estimating M_5:** For the second three rows, we can see that *TDS* is fixed. Therefore, the relationship is between *M* and *T*:
  - **Classical Linear Interpolation**:

$$\begin{aligned} 2 &\rightarrow 2.1633\,\text{s} \\ M_5 &\rightarrow 1.3133\,\text{s} \\ 3 &\rightarrow 0.9238\,\text{s} \end{aligned}$$

$$\because \frac{M_5 - 2}{3 - 2} = \frac{1.3133 - 2.1633}{0.9238 - 2.1633}$$

$$\begin{aligned} \therefore M_5 &= 2 + \frac{1.3133 - 2.1633}{0.9238 - 2.1633} \cdot (3 - 2) \\ &= 2.6858 \end{aligned}$$

The absolute error is:

$$|M_5^{\text{actual}} - M_2| = |2.5 - 2.6858| = 0.1858$$

  - **Logarithmic-Based Non-linear Interpolation**:

$$\because \frac{M_5 - 2}{3 - 2} = \frac{\log(1.3133) - \log(2.1633)}{\log(0.9238) - \log(2.1633)}$$

$$\begin{aligned} \therefore M_5 &= 2 + \frac{\log(1.3133) - \log(2.1633)}{\log(0.9238) - \log(2.1633)} \cdot (3 - 2) \\ &= 2.5865\,\text{s} \end{aligned}$$

The absolute error is:

$$|M_5^{\text{actual}} - M_5| = |2.5 - 2.5865| = 0.0865$$

- **Estimating TDS_8:** For the third three rows, we can see that *M* is fixed. Therefore, the relationship is between *TDS* and *T*:
  - **Classical Linear Interpolation**:

$$\begin{aligned} 2 &\rightarrow 0.5489\,\text{s} \\ TDS_8 &\rightarrow 0.9468\,\text{s} \\ 4 &\rightarrow 1.0978\,\text{s} \end{aligned}$$

$$\because \frac{TDS_8 - 2}{4 - 2} = \frac{0.9468 - 0.5489}{1.0978 - 0.5489}$$

$$\begin{aligned} \therefore TDS_8 &= 2 + \frac{0.9468 - 0.5489}{1.0978 - 0.5489} \cdot (4 - 2) \\ &= 3.4498 \end{aligned}$$

The absolute error is:

$$|TDS_8^{\text{actual}} - TDS_8| = |3.45 - 3.4498| = 1.9129 \times 10^{-4}$$

- **Logarithmic-Based Non-linear Interpolation**:

$$\because \frac{TDS_8 - 2}{4 - 2} = \frac{\log(0.9468) - \log(0.5489)}{\log(1.0978) - \log(0.5489)}$$

$$\therefore TDS_8 = 2 + \frac{\log(0.9468) - \log(0.5489)}{\log(1.0978) - \log(0.5489)} \cdot (4 - 2)$$
$$= 3.5730$$

The absolute error is:

$$|TDS_8^{\text{actual}} - TDS_8| = |3.45 - 3.5730| = 0.1230 \times 10^{-4}$$

We can observe the following points:

1. For estimating T, both interpolation techniques provide good accuracy.
2. For estimating M, the logarithmic-based non-linear interpolation performs better.
3. For estimating TDS, the classical linear interpolation performs better.

In general, this approach is applicable to protective devices with limited settings and/or have fixed time-current characteristics (IEC, 1989; ABB, 1992). Thus, it cannot be used with solid-state, digital "hardware-based," and numerical "software-based" relays.

3.1.2 Curve Fitting Formulas

This is the preferred approach, which can be used easily and smoothly without concerning about the data availability, data size, and interpolation accuracy. Also, it is a user-friendly approach for engineers and programmers, and it can represent many types of ITOCRs just by changing the values of the coefficients associated with the model. Furthermore, it can be represented by one or a few lines, which means that it can be flexibly presented in reports and other document forms.

By referring to (2.4), to model the operating time of this type, different equations have been presented. Actually, settling on one standard formula was a very challenging task and it was deeply discussed by the IEEE and IEC committees. Some of these formulas are listed in the following text (Urdaneta et al., 1988; Zocholl et al., 1989; IEC, 1989; Anderson, 1998; IEEE, 1999):

3.1.2.1 Polynomial Equations

These equations can be easily obtained by fitting the real relay data through using linear regression (LR). Least squares method can be used to obtain the coefficients of the polynomial model quickly without referring to any iterative techniques (Kutner et al., 2004). These polynomial equations have been suggested by Several investigators and researchers. Albrecht et al. (1964) suggested the following form:

$$T = \left[\sum_{j=1}^{m} \sum_{i=1}^{n} \alpha_{ji} (TDS)^j \left(\frac{I}{I_p} \right)^i \right]^k \tag{3.1}$$

where I is the fault current, I_p is the pickup setting, and $\{\alpha_{ji}, m, n, \text{and } k\}$ are constants.

Remember that, I_p is also called CTS where the multiples of the tap settings is calculated as $M = I/I_p$. In IEC, we use plug setting (PS) and PSM. We try to list these notations in different locations because all these notations are still used in many references. Thus, it is important to understand their meanings.

The logarithmic-based polynomial equation suggested by the EEI project is presented in the following text (Radke, 1963; Zocholl et al., 1989):

$$\log(T - \mathrm{DC}) = a_0 + a_1 \left[\log\left(\frac{I}{I_p}\right)\right] + a_2 \left[\log\left(\frac{I}{I_p}\right)\right]^2 + a_3 \left[\log\left(\frac{I}{I_p}\right)\right]^3 + a_4 \left[\log\left(\frac{I}{I_p}\right)\right]^4 \tag{3.2}$$

where DC and a's are constants.

Unfortunately, the polynomial equations shown in (3.1) and (3.2) are not accurate. The main drawbacks associated with them are:

1. The time-current characteristic is not asymptotic to the pickup current.
2. The relay operating time does not decrease monotonically as the current increases.

Sachdev and Fleming (1978) and Singh et al. (1980) developed four polynomial equations:

$$\log T = \frac{a_{-1}}{\log\left(\frac{I}{I_p}\right)} + a_0 + a_1 \left[\log\left(\frac{I}{I_p}\right)\right] + a_2 \left[\log\left(\frac{I}{I_p}\right)\right]^2 + \cdots \tag{3.3}$$

$$\log T = a_0 + \frac{a_1}{\left[\log\left(\frac{I}{I_p}\right)\right]} + \frac{a_2}{\left[\log\left(\frac{I}{I_p}\right)\right]^2} + \frac{a_3}{\left[\log\left(\frac{I}{I_p}\right)\right]^3} + \cdots \tag{3.4}$$

$$T = a_0 + \frac{a_1}{\left(\frac{I}{I_p} - 1\right)} + \frac{a_2}{\left(\frac{I}{I_p} - 1\right)^2} + \frac{a_3}{\left(\frac{I}{I_p} - 1\right)^3} + \cdots \tag{3.5}$$

$$T = a_0 + \frac{a_1}{\left(\frac{I}{I_p}\right) - 1} + \frac{a_2}{\left(\frac{I}{I_p}\right)^2 - 1} + \frac{a_3}{\left(\frac{I}{I_p}\right)^3 - 1} + \cdots \tag{3.6}$$

As can be seen, (3.3) and (3.4) are logarithmic-based polynomial equations where the logarithm is placed on both sides of the equations. In Zocholl et al. (1989), the IEEE-PES committee discussed the following three possible composite models:

1. Developing a polynomial model to act as a function of the relay current, such as (3.5) or (3.6), at a predefined time dial setting.
2. Developing a polynomial model to estimate the relay operating time by different time dial settings for a selected relay current.
3. Combining the two polynomial models to have a multivariate function of I, I_p, and *TDS*.

For the first IEEE-PES committee's model, if (3.5) is selected, then the following polynomial equation can be formulated as a predefined *TDS*:

$$\frac{T}{TDS} = a_0 + \frac{a_1}{\left(\frac{I}{I_p} - 1\right)} + \frac{a_2}{\left(\frac{I}{I_p} - 1\right)^2} + \frac{a_3}{\left(\frac{I}{I_p} - 1\right)^3} + \cdots \tag{3.7}$$

This polynomial model is implemented in Razavi et al. (2008) and Mohammadi et al. (2010) for the IEC normal inverse TCCC. By replacing *TDS* with *TMS* and taking $M = I/I_p = I/PS$, (3.7) can be expressed as follows:

$$\frac{T}{TMS} = a_0 + \frac{a_1}{(M-1)} + \frac{a_2}{(M-1)^2} + \frac{a_3}{(M-1)^3} + \frac{a_4}{(M-1)^4}$$
$$= 1.987\,72 + \frac{8.579\,22}{(M-1)} - \frac{0.461\,29}{(M-1)^2} + \frac{0.036\,446\,5}{(M-1)^3} - \frac{0.000\,319\,901}{(M-1)^4} \tag{3.8}$$

Abyaneh et al. (2003) did a transformation between TDS and TMS. The model given in (3.5) is expressed for the normal inverse overcurrent type of CDG11 manufactured by G.E.C. as follows:

$$T = 4.2437 - \frac{18.8563}{(M-1)} + \frac{92.6691}{(M-1)^2} - \frac{149.1507}{(M-1)^3} + \frac{75.6537}{(M-1)^4} \tag{3.9}$$

For the second IEEE-PES committee's model, the following mathematical expression is used:

$$T = b_0 + b_1 TDS + b_2 TDS^2 + b_3 TDS^3 \tag{3.10}$$

For the normal inverse overcurrent type of CDG11, the forth order (i.e. $b_4 TDS^4$) is considered. The coefficients of this model are (Abyaneh et al., 2003):
$b_0 = 0.0192$, $b_1 = 1.8107$, $b_2 = 1.1860$, $b_3 = 2.1121$, and $b_4 = 1.1160$.

For the third IEEE-PES committee's model, T can be expressed using the following general form (Urdaneta et al., 1988):

$$T = f(TDS, I_p, I) \tag{3.11}$$

Because we know that $M = I/I_p$, so (3.11) can be re-expressed as follows:

$$T = f(TDS, M) \tag{3.12}$$

The last equation is decomposed into two sub-equations (Urdaneta et al., 1988):

$$T = u(TDS) \cdot v(M) \tag{3.13}$$

where u is the polynomial model given in (3.10) and v is the polynomial model given in (3.5).

Substituting (3.10) and (3.5) into (3.13) and taking $M = I/I_p$ yields:

$$T = \left(b_0 + b_1 TDS + b_2 TDS^2 + b_3 TDS^3\right)\left[a_0 + \frac{a_1}{(M-1)} + \frac{a_2}{(M-1)^2} + \frac{a_3}{(M-1)^3} + \frac{a_4}{(M-1)^4}\right] \tag{3.14}$$

For this empirical model, a Westinghouse CO[1] -9 type relay was used[2] by Sachdev and Fleming (1978). The values of these polynomial coefficients are (Elrafie and Irving, 1993):
$a_0 = 0.929\,647\,8$, $a_1 = 6.792\,136$, $a_2 = 14.032\,59$, $a_3 = -8.430\,325$, $a_4 = 2.679\,891$, $b_0 = 1.860\,068 \times 10^{-2}$, $b_1 = 5.607\,550\,2 \times 10^{-2}$, $b_2 = 3.012\,818\,7 \times 10^{-3}$, and $b_3 = 1.234\,002\,4 \times 10^{-8}$.

Damborg et al. (1984) and Ramaswami et al. (1984) expanded (3.14) and then cancelled some terms in order to find the best-fit polynomial model:

$$\begin{aligned}
T &= u(TDS) \cdot v(M) \\
&= \left(b_0 + b_1 TDS + b_2 TDS^2 + b_3 TDS^3\right)\left[a_0 + \frac{a_1}{(M-1)} + \frac{a_2}{(M-1)^2} + \frac{a_3}{(M-1)^3} + \frac{a_4}{(M-1)^4}\right] \\
&= a_0 b_0 + \cancel{\frac{a_1 b_0}{M-1}} + \cancel{\frac{a_2 b_0}{(M-1)^2}} + \cancel{\frac{a_3 b_0}{(M-1)^3}} + \cancel{\frac{a_4 b_0}{(M-1)^4}} \\
&\quad + a_0 b_1 TDS + \cancel{a_1 b_1 \frac{TDS}{M-1}} + a_2 b_1 \frac{TDS}{(M-1)^2} + a_3 b_1 \frac{TDS}{(M-1)^3} + \cancel{a_4 b_1 \frac{TDS}{(M-1)^4}} \\
&\quad + \cancel{a_0 b_2 TDS^2} + a_1 b_2 \frac{TDS^2}{M-1} + a_2 b_2 \frac{TDS^2}{(M-1)^2} + \cancel{a_3 b_2 \frac{TDS^2}{(M-1)^3}} + a_4 b_2 \frac{TDS^2}{(M-1)^4} \\
&\quad + a_0 b_3 TDS^3 + \cancel{a_1 b_3 \frac{TDS^3}{M-1}} + \cancel{a_2 b_3 \frac{TDS^3}{(M-1)^2}} + \cancel{a_3 b_3 \frac{TDS^3}{(M-1)^3}} + \cancel{a_4 b_3 \frac{TDS^3}{(M-1)^4}}
\end{aligned}$$

---

1 **COx** are brands of the Westinghouse's electromechanical OCRs, which were acquired by ABB since 1989. The acronym "CO" stands for "**Circuit Opening.**" The OCRs listed under this series are: **CO2, CO5, CO6, CO7, CO8, CO9**, and **CO11** (AREVA, 2011; Gers and Holmes, 2004; Al-Roomi, 2014).

2 This relay has a very inverse time-current characteristic curve (ABB, 1992; Anderson, 1998).

The reduced best-fit polynomial model is:

$$T = c_1 + c_2 TDS + c_3 \frac{TDS}{\left(\frac{I}{I_p} - 1\right)^2} + c_4 \frac{TDS^2}{\left(\frac{I}{I_p} - 1\right)} + c_5 \frac{TDS^2}{\left(\frac{I}{I_p} - 1\right)^2} + c_6 \frac{TDS}{\left(\frac{I}{I_p} - 1\right)^3} + c_7 \frac{TDS^2}{\left(\frac{I}{I_p} - 1\right)^4} \tag{3.15}$$

where $c_1 = a_0 b_0$, $c_2 = a_0 b_1$, $c_3 = a_2 b_1$, $c_4 = a_1 b_2$, $c_5 = a_2 b_2$, $c_6 = a_3 b_1$, and $c_7 = a_4 b_2$.

The values of these polynomial coefficients are:

$c_1 = 0.0344$, $c_2 = 0.0807$, $c_3 = 1.95$, $c_4 = 0.0577$, $c_5 = -0.0679$, $c_6 = -0.700$, and $c_7 = 0.0199$.

This 7 term model is valid for $TDS = 1$ to 7 and $M = 2$ to 40. The model is accurate within three cycles for small time delays or 5% for large time delays. For $M < 2$, the model has an error of 15%. The other less popular model derived by Damborg et al. consists of 15 terms and it is valid for $TDS = 1$ to 11 and $M = 1.5$ to 50. The mathematical expression of this 15 term polynomial model is:

$$T = c_0 + \frac{c_1}{\left(\frac{I}{I_p} - 1\right)} + c_2 TDS + \frac{c_3}{\left(\frac{I}{I_p} - 1\right)^2} + c_4 \frac{TDS}{\left(\frac{I}{I_p} - 1\right)} + c_5 TDS^2 + \frac{c_6}{\left(\frac{I}{I_p} - 1\right)^3} + c_7 \frac{TDS}{\left(\frac{I}{I_p} - 1\right)^2} + c_8 \frac{TDS^2}{\left(\frac{I}{I_p} - 1\right)} + c_9 \frac{TDS^2}{\left(\frac{I}{I_p} - 1\right)^4} + \frac{c_{10}}{\left(\frac{I}{I_p} - 1\right)^4} + c_{11} \frac{TDS}{\left(\frac{I}{I_p} - 1\right)^3} + c_{12} \frac{TDS^2}{\left(\frac{I}{I_p} - 1\right)^2} + c_{13} \frac{TDS^2}{\left(\frac{I}{I_p} - 1\right)^3} + c_{14} \frac{TDS}{\left(\frac{I}{I_p} - 1\right)^4} \tag{3.16}$$

where the numerical values of these coefficients for the same CO9 relay are:

$c_0 = 0.038\,50$, $c_1 = -0.940\,29$, $c_2 = 0.062\,79$, $c_3 = 4.006\,90$, $c_4 = 0.643\,30$, $c_5 = 0.001\,23$, $c_6 = -4.652\,90$, $c_7 = -0.057\,63$, $c_8 = -0.004\,70$, $c_9 = 0.032\,33$, $c_{10} = 1.386\,30$, $c_{11} = 1.155\,80$, $c_{12} = 0.084\,50$, $c_{13} = -0.106\,17$, and $c_{14} = -0.432\,49$.

Example 3.2 Consider the 7 term polynomial model given in (3.15). Write a MATLAB program to plot the operating time of CO9 relay at different values of M and TDS. Move M from 2 to 20 in a step-size of 0.1, and TDS from 1 to 7 in a step-size of 1. Use a log-scale on the x-axis.

Solution

The following is the code shown in `CO9_Damborg7TermEq.m`:

```
1   clc
2   clear
3   c = [0.0344, 0.0807, 1.95, 0.0577, -0.0679, -0.700, 0.0199]; % relay ...
        coefficients
4   sz = 0.1; % step-size resolution
5   TDS = 1:1:7; % time dial settings
6   M = 2:sz:20; % multiples of pickup current
7   for i=1:length(TDS)
8       for j=1:length(M)
9           T(j,i) = c(1) + c(2)*TDS(i) + c(3)*TDS(i)/((M(j)-1)^2) + ...
                c(4)*(TDS(i))^2/(M(j)-1) + c(5)*(TDS(i))^2/((M(j)-1)^2) + ...
                c(6)*TDS(i)/((M(j)-1)^3) + c(7)*(TDS(i))^2/((M(j)-1)^4);
10      end
11      glabel{i} = ['TDS = ', num2str(TDS(i))];
```

```
12        end
13        semilogx(M, T, 'LineWidth', 2); % plotting using log-scale on the x-axis
14        grid on; grid minor
15        set(gca, 'FontSize', 13)
16        set(gca, 'xtick', [1:1:10,12:2:20]); set(gca, 'ytick', 0:1:7);
17        xlim([1, M(end)]); ylim([0, 7])
18        title('Type CO9 Overcurrent Relay')
19        xlabel('Multiples of Tap Value Current')
20        ylabel('Operating Time in Seconds')
21        leg = legend(glabel);
22        title(leg, 'Time Dial Setting')
```

If we run this code, the plot shown in Figure 3.1 will be generated. This plot can also be accessed by opening `CO9_Damborg7TermEq.fig`.

Example 3.3 Let's take the data-set provided by M.S. Sachdev and coworkers Zocholl et al. (1989), which is tabulated in Table 3.2. Find the coefficients of (3.6) using the least error squares (LES) curve fitting method. For the accuracy, use five decimal places.

Solution

Substituting the values of M and T of the 1st row in (3.6) gives:

$$a_0 + \frac{a_1}{2.5 - 1} + \frac{a_2}{2.5^2 - 1} + \frac{a_3}{2.5^3 - 1} + \frac{a_4}{2.5^4 - 1} = 5.10$$

which yields

$$a_0 + 0.666\,67a_1 + 0.190\,48a_2 + 0.068\,38a_3 + 0.026\,27a_4 = 5.10$$

If we continue the procedure until reaching the 16th row, we will have 16 equations with 5 unknown coefficients. These equations can be expressed in vector notation as follows:

$$\begin{bmatrix} 1.000\,00 & 0.666\,67 & 0.190\,48 & 0.068\,38 & 0.026\,27 \\ 1.000\,00 & 0.500\,00 & 0.125\,00 & 0.038\,46 & 0.012\,50 \\ 1.000\,00 & 0.400\,00 & 0.088\,89 & 0.023\,88 & 0.006\,71 \\ 1.000\,00 & 0.333\,33 & 0.066\,67 & 0.015\,87 & 0.003\,92 \\ 1.000\,00 & 0.285\,71 & 0.051\,95 & 0.011\,10 & 0.002\,44 \\ 1.000\,00 & 0.250\,00 & 0.041\,67 & 0.008\,06 & 0.001\,60 \\ 1.000\,00 & 0.200\,00 & 0.028\,57 & 0.004\,65 & 0.000\,77 \\ 1.000\,00 & 0.166\,67 & 0.020\,83 & 0.002\,92 & 0.000\,42 \\ 1.000\,00 & 0.142\,86 & 0.015\,87 & 0.001\,96 & 0.000\,24 \\ 1.000\,00 & 0.125\,00 & 0.012\,50 & 0.001\,37 & 0.000\,15 \\ 1.000\,00 & 0.111\,11 & 0.010\,10 & 0.001\,00 & 0.000\,10 \\ 1.000\,00 & 0.090\,91 & 0.006\,99 & 0.000\,58 & 0.000\,05 \\ 1.000\,00 & 0.076\,92 & 0.005\,13 & 0.000\,36 & 0.000\,03 \\ 1.000\,00 & 0.066\,67 & 0.003\,92 & 0.000\,24 & 0.000\,02 \\ 1.000\,00 & 0.058\,82 & 0.003\,10 & 0.000\,17 & 0.000\,01 \\ 1.000\,00 & 0.052\,63 & 0.002\,51 & 0.000\,13 & 0.000\,01 \end{bmatrix} \cdot \begin{bmatrix} a_0 \\ a_1 \\ a_2 \\ a_3 \\ a_4 \end{bmatrix} = \begin{bmatrix} 5.100\,00 \\ 3.500\,00 \\ 2.600\,00 \\ 2.100\,00 \\ 1.800\,00 \\ 1.580\,00 \\ 1.330\,00 \\ 1.200\,00 \\ 1.100\,00 \\ 1.020\,00 \\ 0.950\,00 \\ 0.870\,00 \\ 0.810\,00 \\ 0.770\,00 \\ 0.730\,00 \\ 0.700\,00 \end{bmatrix}$$

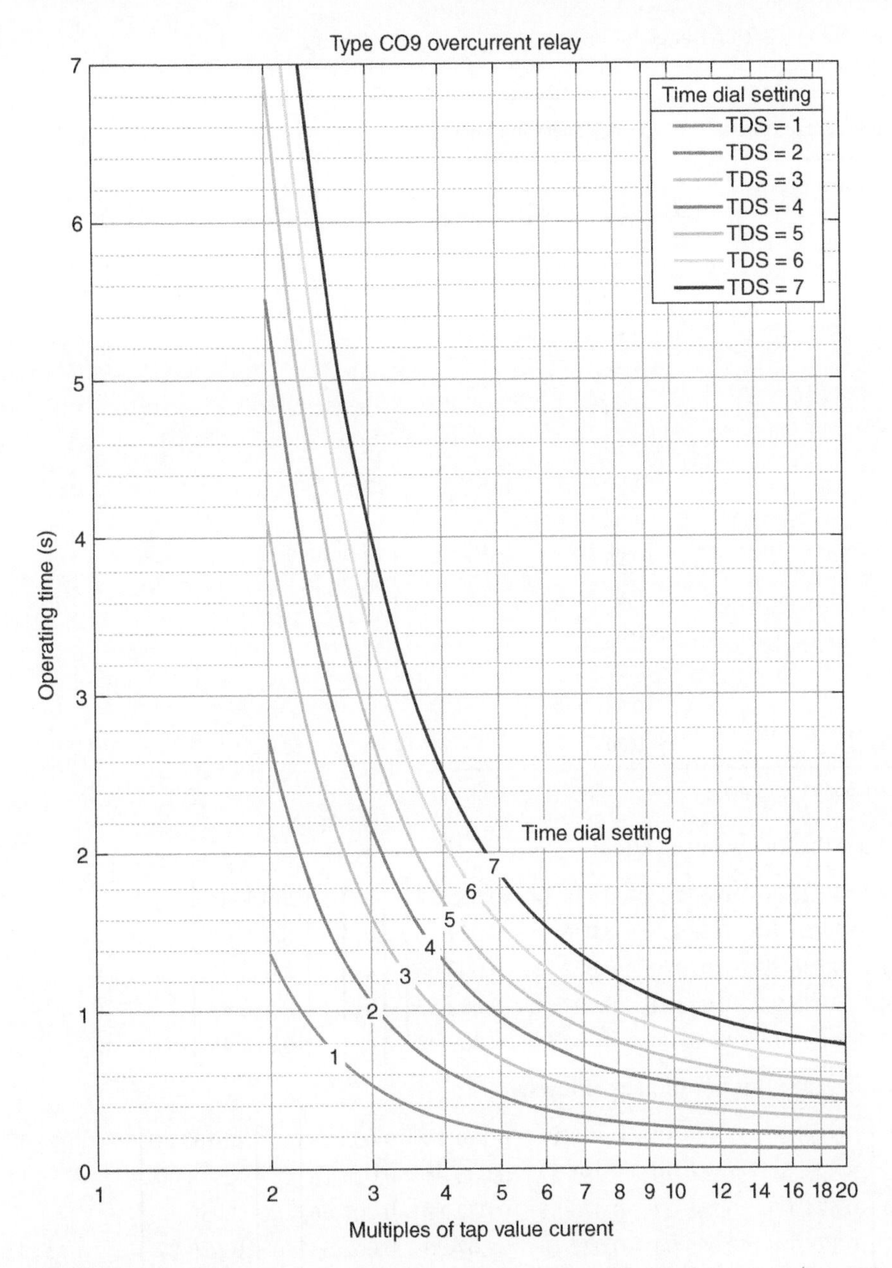

Figure 3.1 Type CO9 time-current curves plotted based on Damborg's 7 term polynomial model.

That is, we have

$$\underset{16\times\cancel{5}}{[\mathbf{A}]} \cdot \underset{\cancel{5}\times 1}{[\mathbf{x}]} = \underset{16\times 1}{[\mathbf{b}]}$$

As can be seen, both sides of the equation will be a 16×1 vector, while we have just five unknowns. Thus, we need to pre-multiply both sides by the transpose of [**A**]:

$$\underset{5\times\cancel{16}}{[\mathbf{A}]^{\mathrm{T}}} \cdot \underset{\cancel{16}\times\cancel{5}}{[\mathbf{A}]} \cdot \underset{\cancel{5}\times 1}{[\mathbf{x}]} = \underset{5\times\cancel{16}}{[\mathbf{A}]^{\mathrm{T}}} \cdot \underset{\cancel{16}\times 1}{[\mathbf{b}]}$$

Table 3.2 Data-set used by Sachdev et al. for inverse time-current characteristic.

No.	Multiplies of tap current	Relay operating time (s)
1	2.5	5.10
2	3.0	3.50
3	3.5	2.60
4	4.0	2.10
5	4.5	1.80
6	5.0	1.58
7	6.0	1.33
8	7.0	1.20
9	8.0	1.10
10	9.0	1.02
11	10.0	0.95
12	12.0	0.87
13	14.0	0.81
14	16.0	0.77
15	18.0	0.73
16	20.0	0.70

Source: Data from Zocholl et al. (1989).

Thus, we will have the following equation:

$$\begin{bmatrix} 16.000\,00 & 3.527\,30 & 0.674\,17 & 0.179\,14 & 0.055\,24 \\ 3.527\,30 & 1.250\,70 & 0.288\,27 & 0.086\,94 & 0.029\,15 \\ 0.674\,17 & 0.288\,27 & 0.070\,55 & 0.022\,18 & 0.007\,66 \\ 0.179\,14 & 0.086\,94 & 0.022\,18 & 0.007\,20 & 0.002\,55 \\ 0.055\,24 & 0.029\,15 & 0.007\,66 & 0.002\,55 & 0.000\,92 \end{bmatrix} \cdot \begin{bmatrix} a_0 \\ a_1 \\ a_2 \\ a_3 \\ a_4 \end{bmatrix} = \begin{bmatrix} 26.160\,00 \\ 8.928\,00 \\ 2.059\,46 \\ 0.626\,87 \\ 0.212\,48 \end{bmatrix}$$

Now, we can easily solve this equation as follows:

$$\therefore \begin{bmatrix} a_0 \\ a_1 \\ a_2 \\ a_3 \\ a_4 \end{bmatrix} = \begin{bmatrix} 16.000\,00 & 3.527\,30 & 0.674\,17 & 0.179\,14 & 0.055\,24 \\ 3.527\,30 & 1.250\,70 & 0.288\,27 & 0.086\,94 & 0.029\,15 \\ 0.674\,17 & 0.288\,27 & 0.070\,55 & 0.022\,18 & 0.007\,66 \\ 0.179\,14 & 0.086\,94 & 0.022\,18 & 0.007\,20 & 0.002\,55 \\ 0.055\,24 & 0.029\,15 & 0.007\,66 & 0.002\,55 & 0.000\,92 \end{bmatrix}^{-1} \cdot \begin{bmatrix} 26.160\,00 \\ 8.928\,00 \\ 2.059\,46 \\ 0.626\,87 \\ 0.212\,48 \end{bmatrix}$$

$$= \begin{bmatrix} 0.271\,14 \\ 11.042\,71 \\ -76.032\,67 \\ 254.346\,84 \\ -206.996\,65 \end{bmatrix}$$

$$\therefore T = 0.271\,14 + \frac{11.042\,71}{M-1} - \frac{76.032\,67}{M^2-1} + \frac{254.346\,84}{M^3-1} - \frac{206.996\,65}{M^4-1}$$

3.1.2.2 Exponential Equations

The same thing can be seen here! This topic was very active since 1960s. There is no single formula that can be used for both the IEEE and IEC standards. However, these exponential models are much flexible than the polynomial models[3] because they have a few number of coefficients. That is, we can easily change the mode of the characteristic curve from, for example, inverse to very inverse (or to extremely inverse) with a very small adjustment to the original model. Before reaching the final confirmed models, let's first see the earlier models. Warrington (1962) proposed the following exponential model:

$$T = C + \frac{K_{TDS}}{M^n - 1} \tag{3.17}$$

where C, K, and n are constants.[4]

After about three years, Hieber (1965) proposed the following more complicated equation:

$$T = C + \frac{K_{TDS}}{(M - h + wM^{-2M})^q} - b\left(\frac{M}{50}\right)^n \tag{3.18}$$

where b, h, q, and w are constants.

The accuracy of these two equations is between 5% and 10%. Based on the characteristic, (3.17) is suitable for the European relays. Also, this nonlinear equation has a few coefficients, which means that the optimal values can be easily obtained by adequate curve fitting tools. On the opposite side, (3.18) has many coefficients, so finding a good initial point to optimize these coefficients, during fitting the curve, requires an extra effort[5] (Zocholl et al., 1989).

In IEEE (1999), the derived time-current equation for electromechanical relays is:

$$T = \begin{cases} TDS\left(\frac{A}{M^2 - 1}\right), & \text{for } 0 < M < 1 \\ TDS\left(\frac{A}{M^p - 1} + B\right), & \text{for } M > 1 \end{cases} \tag{3.19}$$

where A is a constant and it equals:

$$A = \frac{K_d \theta}{\tau_s} \tag{3.20}$$

where K_d is the drag magnet damping factor, θ is the disk travel, and τ_s is the initial spring torque.

The part that emulates the dynamics of the induction disk electromechanical overcurrent relay is (IEEE, 1999):

$$T = \frac{A}{M^2 - 1} = \frac{A}{\left(\frac{I}{I_p}\right)^2 - 1} \tag{3.21}$$

This equation is very useful, which can be used to derive the reset characteristic.

---

3 We are not talking about the accuracy of exponential and polynomial models. The accuracy depends on many factors; including the models being compared.

4 The term C is used to account for the effect of friction and hysteresis of the magnetic circuit (Zocholl et al., 1989).

5 The primitive approach is to manually select different initial points, which is an exhausting approach. With recent state-of-the-art optimization techniques, this point can be tackled by many innovative ways. For example, we can design a multi-start algorithm to investigate many random initial points, or we can hybridize the curve fitting algorithm as an objective function inside a meta-heuristic optimization algorithm so that the latter one can act as a global optimizer and the former one can act as a local optimizer.

The Reset Characteristic For an electromechanical ITOCR, suppose that the induction disk has an initial displacement from its reset position when I is set to zero. This will lead the disk to move in an opposite direction toward the reset position. By referring to (3.20), the preceding phenomenon can be seen by setting $I = 0$:

$$T = \frac{A}{\left(\frac{0}{I_p}\right)^2 - 1} = -A = \frac{-K_d\theta}{\tau_s} \tag{3.22}$$

This value is the reset time required for the disk to reach the reset position and the negative sign indicates the backward rotation. Thus, the reset time T_r is calculated as follows:

$$|T_r| = \frac{K_d\theta}{\tau_s} \tag{3.23}$$

and the reset characteristic for $(0 < M < 1)$ is:

$$T = \frac{T_r}{M^2 - 1} \tag{3.24}$$

Table 3.3 shows the coefficients of some CO series relays; one of the most two popular series of ITOCRs in North America. These values have been determined by fitting the actual observations taken from real relays, which can be done by nonlinear regression (NLR) analysis.[6] Some references, like p. 63 of (Anderson, 1998), take these coefficients as the confirmed data to emulate real electromechanical protective relays. *Is it the end of the story?* No! The other most popular series of ITOCRs in North America is IAC[7] series. Table 3.4 shows the coefficients belong to some relays marketed under IAC series. As can be clearly seen, the best-fit curves of the CO series have different coefficients than those of the IAC series.

Table 3.3 Characteristic curve coefficients of type CO induction relay.

Constant	Moderately inverse	Very inverse	Extremely inverse
A	0.047	18.92	28.08
B	0.183	0.492	0.130
p	0.02	2.00	2.00
T_r	5.4	21.0	26.5

Table 3.4 Characteristic curve coefficients of type IAC induction relay.

Constant	Moderately inverse	Very inverse	Extremely inverse
A	0.056	20.29	20.33
B	0.045	0.489	0.081
p	0.02	2.00	2.00
T_r	4.3	22.3	22.7

6 For polynomial equations, we can use linear regression analysis because the coefficients (which are the unknowns) have linear relationship between each other (Kutner et al., 2004).
7 The acronym "IAC" stands for "**Inverse Alternate Current**."

Example 3.4 Using the coefficients tabulated in Tables 3.3 and 3.4, design a MATLAB code to plot the reset and operating time of very inverse induction relays of type CO and IAC series. Assume that M moves from 0.01 to 30 in steps of 0.1 and $TDS = 1$.

Solution
For this mission, (3.19) and (3.24) should be used. The following is the code shown in `ResetTimeCurves.m`:

```
1   clc
2   clear
3   % Model A: very inverse curve - type CO
4   A1 = 18.92;
5   B1 = 0.492;
6   p1 = 2.0;
7   Tr1 = 21.0;
8   % Model B: very inverse curve - type IAC
9   A2 = 20.29;
10  B2 = 0.489;
11  p2 = 2.0;
12  Tr2 = 22.3;
13  sz = 0.1; % step-size resolution
14  M = 0.01:sz:30; % multiples of pickup current
15  for i=1:length(M)
16      if M(i) < 1
17          T1(i) = abs(Tr1/(M(i)^2 - 1)); % reset time of the CO9 relay
18          T2(i) = abs(Tr2/(M(i)^2 - 1)); % reset time of the IAC 53 relay
19      else
20          T1(i) = A1/(M(i)^p1 - 1) + B1; % operating time of the CO9 relay
21          T2(i) = A2/(M(i)^p2 - 1) + B2; % operating time of the IAC 53 ...
                relay
22      end
23  end
24  loglog(M, T1, M, T2, 'LineWidth', 2); % plotting using log-scale on ...
        the x-axis
25  glabel = {'Type CO', 'Type IAC'};
26  grid on; grid minor
27  set(gca, 'FontSize', 13)
28  xlim([0.011, 100]); ylim([0.1, 10000])
29  title('Very Inverse Characteristic Curves of Type CO and IAC ...
        Overcurrent Relays')
30  xlabel('Multiples of Tap Value Current')
31  ylabel('Operating Time in Seconds')
32  leg = legend(glabel);
33  title(leg, 'Time Dial Setting')
```

If we run this code, the loglog plot shown in Figure 3.2 will be generated. This plot can also be accessed by opening `ResetTimeCurves.fig`.

Based on these small differences, and not to be biased toward CO series, the standard coefficients are taken based on the average of those tabulated in Tables 3.3 and 3.4, which are listed in Table 3.5. These are the IEEE C37.112-1996 standard (IEEE, 1999).

For the European standard, the operating time of protective relays can be calculated by (3.19), but with using different coefficients and dealing with $\{TMS, PS, PSM\}$ instead of $\{TDS, I_p, M\}$. Thus,

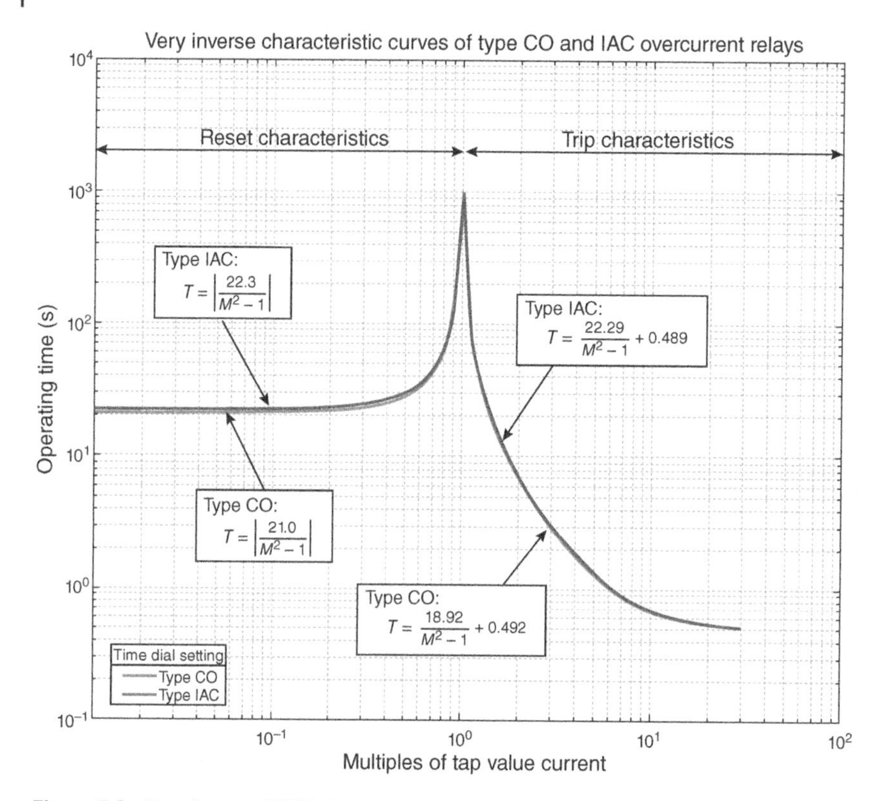

Figure 3.2 Very inverse TCCCs for two models of induction type ITOCRs.

Table 3.5 IEEE standard characteristic curve coefficients of induction relays manufactured in North America.

Constant	Moderately inverse	Very inverse	Extremely inverse
A	0.0515	19.610	28.200
B	0.1140	0.4910	0.1217
p	0.0200	2.0000	2.0000
T_r	4.8500	21.600	29.100

for the IEC-based relays, we should deal with this expression:

$$T = \frac{k\ TMS}{\left(\frac{I}{PS}\right)^{\alpha} - 1} = \frac{k\ TMS}{PSM^{\alpha} - 1} \tag{3.25}$$

where $TMS \in [0.1, 1.1]$ compared with $TDS \in [0.5, 11]$ (ALSTOM, 2002).

Table 3.6 shows the standard values of the coefficients k and α (ABB, 1992; Al-Roomi and El-Hawary, 2017a).

Now, the question that might be asked here is: *What is about the solid-state (or static) relays?* For these relays, we do not have any movement parts and thus we do not need to concern about the inertia that causes over-travel delay (Al-Roomi and El-Hawary, 2019a). Thus, to get the

Table 3.6 IEC standard characteristic curve coefficients of induction relays manufactured in Europe.

Type of curve	k	α
Standard Inverse	0.02	0.140
Very Inverse	1.00	13.50
Extremely Inverse	2.00	80.00
Ultra-Inverse	2.5	315.2
Long-Time Inverse	1.00	120.0

Source: Based on ABB (1992); Al-Roomi and El-Hawary (2017a).

best-fit general equation for solid-state relays, a small constant term is added to (3.19) when $M > 1$ (Anderson, 1998):

$$T = TDS\left(\frac{A}{M^p - \varsigma} + B\right) + \kappa \tag{3.26}$$

where ς and κ are constants.

For standard North American relays, $\varsigma = 1$ (unitless) and $\kappa = 10$ ms. Thus, the general formula given in (3.26) becomes:

$$T = TDS\left(\frac{A}{M^p - 1} + B\right) + 0.01 \tag{3.27}$$

Because these relays are more flexible and can be easily calibrated and adjusted (Verma, 1976), many characteristic curves can be defined. Table 3.7 shows the coefficients of (3.27) for many characteristic curves (Anderson, 1998).

The exponent p can be changed from one to two just by replacing an inexpensive module (Anderson, 1998). Also, the nonlinearity of the equation can be realized by linear components, which is presented as a hardware in Verma (1976).

3.1.2.3 Artificial Intelligence

We have seen that the reason for preferring polynomial models over exponential models is the ability to fit the curves by using LR analysis. NLR analysis can generate very accurate models, which is

Table 3.7 Characteristic curve coefficients for tripping time equation of solid-state relays given in (3.27).

Characteristic curve	A	B	p
Definite time	0.20	0.180	1
Moderately inverse time	0.55	0.180	1
Short time	0.20	0.015	1
Modified inverse time	1.35	0.055	1
Modified very inverse time	1.35	0.015	1
Inverse time	5.40	0.180	2
Very inverse time	5.40	0.110	2
Extremely inverse time	5.40	0.030	2

Source: Data from Anderson (1998).

the case here with exponential models. However, it is not an easy task. Now, imagine that all these barriers have vanished! This is one of the main advantages of using advanced machine learning (ML) tools.[8] For example, we can use support vector regression[9] (SVR) to approximate the actual reset and operating time of ITOCRs. The other possible option is to use artificial neural networks (ANNs). Compared with LR and NLR, ANNs are more powerful and efficient in many highly complicated problems. They can be used to solve all the technical problems, difficulties, and challenges faced with LR and NLR. The knowledge of these networks, which represents the response or output variable of LR and NLR, can be built without referring to any mathematical expression. Abyane et al. (1997) hybridized neural network with fuzzy logic (FL) to accurately emulate ITOCRs without using numerical analysis nor mathematical equations. The other example is (Wu et al., 2012), which used singular value decomposition (SVD) to model ITOCRs.

However, the main disadvantage of these methods is their working principle. They are hard to program and use plus the possibility to require special software to run them. For instance, let's look at ANNs. These virtual brain neural networks act as black-boxes. Actually, this is a double-edged sword. It can be seen as a source of strength because ANNs can estimate the system response directly without trying to express any weary mathematical equation. However, this approach makes the whole process secret and nobody can know what is going on inside these mysterious networks. Based on this, it limits many possible applications and reduces the explainability and interpretability. Also, some of the drawbacks associated with ANNs are concentrated in the selection of features, topology, number of hidden layers, and number of neurons assigned to each layer. Add to that, it is hard to know the best learning algorithm and its settings or hyperparameters. Moreover, the best set of activation functions used in the hidden layers and the output layer is an open question. Furthermore, a long CPU time is required to train ANNs with big data. These data should be normalized before feeding them to ANNs. Last but not least, ANNs do not guarantee to reach optimal results nor providing 100% reliability (Illingworth, 1989; Baughman and Liu, 1995; Al-Roomi and El-Hawary, 2020h).

3.1.3 Special Models

Despite all the models suggested earlier, the literature also contains a few other models that do not belong to IEC, IEEE, or CO/IAC series. Some of these special models are presented in the following text:

3.1.3.1 RI-Type Characteristic

This special characteristic is principally used to coordinate electromechanical relays. The mathematical expression of this characteristic is (ABB, 1998):

$$T = \frac{TMS}{0.339 - 0.236PSM^{-1}} \tag{3.28}$$

As can be clearly seen, we are dealing with $\{TMS, PSM\}$ instead of $\{TDS, M\}$. Thus, to make this equation compatible with (3.25), we can multiply both sides by $(\frac{-0.236}{-0.236})$ so that the expression given in (3.28) can be re-arranged as follows:

$$T = \frac{k\ TMS}{\left(\frac{I}{PS}\right)^{\alpha} - P} = \frac{k\ TMS}{PSM^{\alpha} - P} = \frac{-4.2373\ TMS}{PSM^{-1} - 1.436\,44} \tag{3.29}$$

where $k = -4.2373$ s, $\alpha = -1$, and $P = -1.43644$.

---

8 LR and NLR are classified as two types of classical ML tools.

9 A special type of support vector machine (SVM) (Al-Roomi and El-Hawary, 2020h).

3.1.3.2 RD-Type Characteristic

It is also known as the RXIDG-type characteristic (Schneider Electric, 2015). This special model is principally used in earth-fault protection for having a high degree of selectivity even at high resistance faults. The protection can operate in a selective way even if it is not directional (ABB, 1998). The mathematical expression of this model is:

$$T = 5.8 - 1.35 \ln \left(\frac{PSM}{\overline{TMS}} \right) \tag{3.30}$$

where $\overline{TMS}$ is TMS with a restricted range.[10]

3.1.3.3 FR Short Time Inverse

This European standard is perfectly compatible with the formula given in (3.25) (Al-Roomi and El-Hawary, 2017a):

$$T = \frac{k\ TMS}{\left(\frac{I}{PS}\right)^{\alpha} - P} = \frac{k\ TMS}{PSM^{\alpha} - P} = \frac{0.05\ TMS}{PSM^{0.04} - 1} \tag{3.31}$$

3.1.3.4 UK Rectifier Protection

This model merges between the IEC and IEEE models. It uses TDS instead of TMS with $\alpha \neq 1$ and $B = 0$ (Schneider Electric, 2015):

$$T = \frac{45\,900\ TDS}{M^{5.6} - 1} \tag{3.32}$$

3.1.3.5 BNP-Type Characteristic

Similarly, this model merges between the IEC and IEEE models. It uses TMS instead of TDS with $B \neq 0$ (Schneider Electric, 2015):

$$T = TMS \left(\frac{1000}{PSM^2 - 1} + 0.655 \right) \tag{3.33}$$

This European standard is perfectly compatible with the formula given in (3.25).

3.1.3.6 Standard CO Series Characteristics

The values tabulated in Table 3.3 were used to obtain the coefficients of the IEEE standard model; which are tabulated in Table 3.5. The standard values of the North American relays manufactured under the CO series are tabulated in Table 3.8. These coefficients should be used with (3.19).

Table 3.8 Standard characteristic curve coefficients of type CO relays.

Type of curve	A	B	p
US short time inverse (CO2)	0.023 94	0.016 94	0.02
US long time (CO5)	4.842	1.967	1.1
US definite minimum time (CO6)	0.3164	0.1934	1.4
US moderately inverse time (CO7)	0.0094	0.0366	0.02
US time inverse (CO8)	5.95	0.18	2
US very inverse time (CO9)	4.12	0.0958	2
US extreme inverse time (CO11)	5.57	0.028	2

10 For Schneider Electric's relays: $\overline{TMS} \in [0.3, 1]$ with a step-size resolution of $\Delta\overline{TMS} = 0.01$ (Schneider Electric, 2015).

Example 3.5 Write a MATLAB code to re-generate the semilogarithmic plot shown in Figure 2.14.

Solution

The code given in the following text can be found by opening `CO8.m`:

```
1   clc
2   clear
3   % relay coefficients
4   a = 2.00; % alpha
5   b = 5.95000; % beta
6   g = 0.18000; % gamma
7   sz = 0.1; % step-size resolution
8   % relay settings
9   TDS = [1/2, 1:11]; % time dial settings
10  M = [1.5:sz:20]; % multiples of pickup current
11  for i=1:length(TDS)
12      for j=1:length(M)
13          T(j,i) = TDS(i) * (b/(M(j) \couhata - 1) + g);
14      end
15      glabel{i} = ['TDS = ', num2str(TDS(i))];
16  end
17  semilogx(M, T, 'LineWidth', 2); % plotting using log-scale on the x-axis
18  grid on; grid minor
19  set(gca, 'FontSize', 13)
20  set(gca, 'xtick', [1:1:10,12:2:20]); set(gca, 'ytick', 0:1:7);
21  xlim([1, M(end)]); ylim([0, 7])
22  title('Type CO8 Overcurrent Relay')
23  xlabel('Multiples of Tap Value Current')
24  ylabel('Operating Time in Seconds')
25  leg = legend(glabel);
26  title(leg, 'Time Dial Setting')
```

If we run this code, we will get a plot similar to the one shown in Figure 2.14.

3.1.3.7 IAC and ANSI Special Equations

The same reason holds here. First, Table 3.4 does not represent the standard coefficients of the North American relays manufactured under the CO series. Second, the IEEE standard model given in (3.19) is not used here. Instead, the following expression is used:

$$
\begin{aligned}
T &= TDS\left[A + \frac{B}{\left(\frac{I}{I_p}\right) - C} + \frac{D}{\left(\frac{I}{I_p}\right)^2 - C} + \frac{E}{\left(\frac{I}{I_p}\right)^3 - C}\right] \\
&= TDS\left[A + \frac{B}{M - C} + \frac{D}{M^2 - C} + \frac{E}{M^3 - C}\right]
\end{aligned}
\tag{3.34}
$$

where $\{A, B, C, D, E\}$ are constants and tabulated in Table 3.9 (Basler Electric, 2000; ISA, 2011; Schneider Electric, 2017a, b; Labrador, 2018).

Example 3.6 Write a MATLAB code to calculate the trip time of IAC and ANSI characteristic curves modeled based on (3.34). Do that for $M = 0.5, 1, 2, 3, 4, 5, 10, 15, 20, 30$, and 40. Take $TDS = 3$.

Table 3.9 Standard characteristic curve coefficients of ANSI and type IAC relays.

Type of curve	*A*	*B*	*C*	*D*	*E*
Inverse (IAC 51)	0.2078	0.863	0.8	−0.4180	0.1947
Very tnverse (IAC 53)	0.09	0.7955	0.1	−1.2885	7.9586
Short inverse (IAC 55)	0.0428	0.0609	0.62	−0.0010	0.0221
Long inverse (IAC 66)	80	0	2	2	2
Extremely inverse (IAC 77)	0.004	0.6379	0.62	1.7872	0.2461
ANSI moderately inverse	0.1735	0.6791	0.8	−0.08	0.1271
ANSI normally inverse	0.0274	2.2614	0.3	−0.1899	9.1272
ANSI very inverse	0.0615	0.7989	0.34	−0.284	4.0505
ANSI extremely inverse	0.0399	0.2294	0.5	3.0094	0.7222

Source: Based on Basler Electric (2000); ISA (2011); Schneider Electric (2017a); Angel Labrador (2018).

The program should be able to present the results in a table where the relay types are placed in the row header and the multiples of pickup current are placed in the column header. If $M = 2$, which is the fastest model? What happen if $M > 2$? Why does IAC 66 show NaN when $M = 2$?

Solution

If we open `IAC_ANSI_Models.m`, the following code will be found:

```
1   clc
2   clear
3   % relay coefficients
4   A = [0.2078; 0.09; 0.0428; 80; 0.004; 0.1735; 0.0274; 0.0615; 0.0399];
5   B = [0.863; 0.7955; 0.0609; 0; 0.6379; 0.6791; 2.2614; 0.7989; 0.2294];
6   C = [0.8; 0.1; 0.62; 2; 0.62; 0.8; 0.3; 0.34; 0.5];
7   D = [-0.4180; -1.2885; -0.0010; 2; 1.7872; -0.08; -0.1899; -0.284; ...
        3.0094];
8   E = [0.1947; 7.9586; 0.0221; 2; 0.2461; 0.1271; 9.1272; 4.0505; 0.7222];
9   L = {'Inverse (IAC 51)'; 'Very Inverse (IAC 53)'; 'Short Inverse (IAC ...
        55)'; 'Long Inverse (IAC 66)'; 'Extremely Inverse (IAC 77)';...
10      'ANSI Moderately Inverse'; 'ANSI Normally Inverse'; 'ANSI Very ...
            Inverse'; 'ANSI Extremely Inverse'}; % labels
11  % relay settings
12  TDS = 3; % time dial setting
13  M = [2, 3, 4, 5, 10, 15, 20, 30, 40, 60]; % multiples of pickup current
14  for i=1:length(A)
15      for j=1:length(M)
16          T(i,j) = TDS * (A(i) + B(i)/(M(j) - C(i)) + D(i)/(M(j)^2 - ...
                C(i)) + E(i)/(M(j)^3 - C(i)));
17      end
18  end
19  Tab = table(L, T(:,1), T(:,2), T(:,3), T(:,4), T(:,5), T(:,6), ...
        T(:,7), T(:,8), T(:,9), T(:,10));
20  Tab.Properties.VariableNames = {'M' 'Two' 'Three' 'Four' 'Five' 'Ten' ...
        'Fifteen' 'Twenty' 'Thirty' 'Forty' 'Sixty'}
```

Running this code will generate the table shown in the following text:

M	Two	Three	Four	Five	Ten	Fifteen	Twenty	Thirty	Forty	Sixty
'Inverse (IAC 51)'	2.4702	1.6696	1.3592	1.1927	0.89276	0.8003	0.75518	0.71069	0.68867	0.66679
'Very Inverse (IAC 53)'	3.5572	1.5462	1.0125	0.79296	0.49625	0.42005	0.38324	0.3464	0.32777	0.30888
'Short Inverse (IAC 55)'	0.26889	0.20732	0.1833	0.17052	0.14791	0.14111	0.13783	0.13462	0.13304	0.13148
'Long Inverse (IAC 66)'	NaN	241.1	240.53	240.31	240.07	240.03	240.02	240.01	240	240
'Extremely Inverse (IAC 77)'	3.0851	1.4839	0.93844	0.67477	0.27071	0.16919	0.12426	0.083125	0.06396	0.045721
'ANSI Moderately Inverse'	2.1962	1.4318	1.1474	0.99872	0.73991	0.66301	0.62606	0.59002	0.57233	0.55485
'ANSI Normally Inverse'	7.475	3.5549	2.3093	1.7222	0.80328	0.54929	0.42857	0.31101	0.25316	0.19581
'ANSI Very Inverse'	2.9819	1.4429	0.97581	0.76174	0.43621	0.34779	0.30579	0.26481	0.24459	0.22449
'ANSI Extremely Inverse'	3.4469	1.5389	0.93291	0.65853	0.28505	0.20802	0.17786	0.15315	0.1428	0.13378

From the preceding table, we can observe the following:

- When $M = 2$, IAC 55 shows the lowest wait time, which means that it is the fastest model at $TDS = 3$ and $M = 2$.
- Also, we can see that the operating time of IAC 77 is extremely decreasing. It goes below that of IAC 55 when $M \geqslant 20$.

By referring to (3.34) and Table 3.9, we can see that $C = 2$, which leads to:

$$T_{\text{IAC66}} = 3\left[80 + \frac{0}{2-2} + \frac{2}{2^2-2} + \frac{2}{2^3-2}\right]$$

It is clear that the second term inside the square bracket will be $\frac{0}{0}$, which is the reason why T_{IAC66} gives NaN when $M = 2$. Thus, the value $(B = 0)$ means that the whole term should be deactivated. To solve this technical issue programmatically, we need to add an `if`-statement, so when $M = 2$ and $B = 0$ the second term is not counted. The following correction can be applied to the inner loop of the nested `for`-loop:

```
1    for i=1:length(A)
2        for j=1:length(M)
3            if M(j) == 2 && B(i) == 0
4                T(i,j) = TDS * (A(i) + D(i)/(M(j)^2 - C(i)) + ...
                     E(i)/(M(j)^3 - C(i)));
5            else
6                T(i,j) = TDS * (A(i) + B(i)/(M(j) - C(i)) + D(i)/(M(j)^2 ...
                     - C(i)) + E(i)/(M(j)^3 - C(i)));
7            end
8        end
9    end
```

If we run this corrected code, the operating time of IAC 66 is 244 s when $TDS = 3$ and $M = 2$.

3.1.4 User-Defined Curves

So far, we have seen how to approximate the actual operation of electromechanical/electromagnetic and solid-state/static inverse-time overcurrent relays. For programmable relays, the story becomes more interesting! Suppose that we can program the operation of protective relays to accept any time-current characteristic curve defined by end-users. This can be done by two approaches:

1. **Fixed Formula with User-Defined Parameters:** This can be achieved just by changing the parameters of the model used to calculate the operating time. For example, using the IEC standard model with user-defined k and α. This is the most popular one because it can be easily programmed in optimization algorithms to optimally coordinate protective relays without

changing the main structure of the objective function being minimized. This will be seen in Section 3.2.

2. **Customized Formulas:** This can also be achieved by modifying the characteristic curve to satisfy the user's requirements (ALSTOM, 2002; Schneider Electric, 2017a). We have seen this in the RD-type characteristic where a logarithmic function is used. Also, we have seen before that we can mix between TCCCs of DCOCRs, DTOCRs, and ITOCRs. To clarify it more, let's suppose that an engineer wants to customize a TCCC to be like the fuse[11] characteristic using the following expression:

$$T = \frac{100\ TDS}{M^2} \tag{3.35}$$

This curve is known as "I^2T curve." The operating time is inversely proportional to the square of M and proportional to TDS with a fixed gain of 100.

3.2 Dealing with All the Standard Characteristic Curves Together

As we have seen in Section 3.1, there are many equations that can be used to emulate inverse-time overcurrent relays. These equations create lots of confusions, especially for new researchers in this field. This is why we give more attention to this point because we will need this information later to optimally coordinate protective relays. For example, if someone selects the 15 term model given in (3.16), then he/she cannot compare the final results with others obtained by the 7 term model given in (3.15). Similarly, the results obtained by any one of these two polynomial equations, which are developed by Damborg et al., should not be compared with those obtained by Sachdev's polynomial equations. Otherwise, the performance comparison will be unfair and thus leads to incorrect conclusion. *Why do we say that?* Because, until the last decade, many researchers solved the ORC problem based on the polynomial equations (Elrafie and Irving, 1993). Nowadays, the exponential equations are preferred to emulate the operating time of ITOCRs. However, we cannot select any exponential model! To be more specific, two common standard models are used today. The first one is called the European standard model, which is based on the IEC[12] 60255-151:2009 and the British standard (BS142). The other one is called the North American standard model, which is based on the ANSI/IEEE C37.112-1996 standard. The other survived models are those given in (3.28)–(3.34) for the RD-type and IAC series characteristic curves.

3.2.1 Differentiating Between Time Dial Setting and Time Multiplier Setting

Except for (3.30) and (3.34), the other standard models can be explained by the following flexible general formula:

$$T_T = TDM \times \left[\frac{\beta}{\left(\frac{I}{I_p}\right)^{\alpha} - \xi} + \gamma \right] \tag{3.36}$$

where T_T is the trip time (or time to trip), TDM is either TDS (i.e. $TDM \in [0.5, 11]$) or TMS (i.e. $TDM \in [0.1, 1.1]$), and $\{\alpha, \beta, \gamma, \xi\}$ are constants.

Table 3.10 lists all the coefficients of the popular models used in overcurrent protection (ALSTOM, 2002; Schneider Electric, 2015; Al-Roomi, 2020; Hase et al., 2020). We can see that the

11 We are not comparing the operating speed of relays and fuses.
12 It is a successor to the IEC 255-3 standard (IEEE, 1999).

Table 3.10 Most popular standard coefficients for calculating the operating time of European and North American relays.

Type of curve[a]	Standard	TDM	α	β	γ	ξ
IEC Standard Inverse (SI)	IEC/A	*TMS*	0.02	0.14	0	1
IEC Very Inverse (VI)	IEC/B	*TMS*	1	13.5	0	1
IEC Extremely Inverse (EI)	IEC/C	*TMS*	2	80	0	1
IEC Ultra-Inverse (UI)	IEC	*TMS*	2.5	315.2	0	1
IEC Long Time Inverse (LTI)	IEC/UK	*TMS*	1	120	0	1
IEC Short Time Inverse (STI)	IEC/FR	*TMS*	0.04	0.05	0	1
IEEE Long Time Inverse	IEEE	*TDS*	0.02	0.086	0.185	1
IEEE Long Time Very Inverse	IEEE	*TDS*	2	28.55	0.712	1
IEEE Long Time Extremely Inverse	IEEE	*TDS*	2	64.07	0.25	1
IEEE Moderately Inverse	IEEE (IEC/D)	*TDS*	0.02	0.0515	0.114	1
IEEE Very Inverse	IEEE (IEC/E)	*TDS*	2	19.61	0.491	1
IEEE Extremely Inverse	IEEE (IEC/F)	*TDS*	2	28.2	0.1217	1
IEEE Short Time Inverse	IEEE	*TDS*	0.02	0.167 58	0.118 58	1
IEEE Short Time Extremely Inverse	IEEE	*TDS*	2	1.281	0.005	1
US Moderately Inverse (U1)	US	*TDS*	0.02	0.0104	0.2256	1
US Inverse[b](U2)	US	*TDS*	2	5.95	0.18	1
US Very Inverse (U3)	US	*TDS*	2	3.88	0.963	1
US Extremely Inverse (U4)	US	*TDS*	2	5.67 (ISA, 2011), 5.64 (SEL, 2013)	0.0352 (ISA, 2011), 0.024 34 (SEL, 2013)	1
US Short Time Inverse (U5)	US	*TDS*	0.02	0.003 42	0.002 62	1
CO short time inverse (CO2)	CO	*TDS*	0.02	0.023 94	0.016 94	1
CO long time (CO5)	CO	*TDS*	1.1	4.842	1.967	1
CO definite minimum time (CO6)	CO	*TDS*	1.4	0.3164	0.1934	1
CO moderately inverse time (CO7)	CO	*TDS*	0.02	0.0094	0.0366	1
CO time inverse (CO8)	CO	*TDS*	2	5.95	0.18	1
CO very inverse time (CO9)	CO	*TDS*	2	4.12	0.0958	1
CO extremely inverse time (CO11)	CO	*TDS*	2	5.57	0.028	1
UK Rectifier Protection	RECT	*TDS*	5.6	45 900	0	1
BNP (EDF)	EDF	*TMS*	2	1000	0.655	1
RI	RI	*TMS*	−1	−4.2373	0	1.436 44

a) According to IEC 602555-151 standard definition.
b) It is similar to CO8.

side constraint of the variable *TDM* is not constant. Thus, in optimization algorithms, we need to add an `if`-statement to adjust the lower and upper bounds based on the model used to emulate overcurrent relays.

The same thing is applied to the reset time. For this mission, the following flexible general formula can be used:

$$T_R = TDM \times \left[\frac{T_r}{1 - \left(\frac{I}{I_p}\right)^{\lambda}} \right] \tag{3.37}$$

where T_R is the reset time (or time to reset) and $\{T_r, \lambda\}$ are constants.

Table 3.11 lists the constant values of T_r and λ for some standard models (Schneider Electric, 2015; Hase et al., 2020).

We might see slight differences in these coefficients in some manufacturers' technical/instruction/reference manuals, especially with those published based on obsolete standards or based on an extended model. For example, the instruction manual shown in Basler Electric (2000) uses the extended model given in (3.26) for solid-state ITOCRs. Thus, the flexible general formula given in (3.36) is extended to accept the new constant term:

$$T_T = TDM \times \left[\frac{\beta}{\left(\frac{I}{I_p}\right)^{\alpha} - \xi} + \gamma \right] + \kappa \tag{3.38}$$

and the reset time is expressed as follows:

$$T_R = TDM \times \left[\frac{T_r}{1 - \left(\frac{I}{I_p}\right)^{2}} \right] \tag{3.39}$$

As can be clearly seen, the only difference between (3.37) and (3.39) is that $\lambda = 2$ "always." Table 3.12 shows the coefficients implemented to emulate ITOCRs manufactured under BS142 and the CO and IAC series using (3.38)–(3.39).

Example 3.7 Consider the customized characteristic curve given in (3.35). Compared with (3.36), what are the values of α, β, γ, and ξ?

Solution

First, let's equate both models:

$$\frac{100\cancel{TDM}}{M^2} = \cancel{TDM} \times \left[\frac{\beta}{\left(\frac{I}{I_p}\right)^{\alpha} - \xi} + \gamma \right]$$

which can be simplified to:

$$\frac{100}{M^2} = \frac{\beta}{M^{\alpha} - \xi} + \gamma$$

By comparing both sides, we can quickly reach to this answer:

$$\alpha = 2,\ \beta = 100,\ \gamma = 0,\ \xi = 0$$

Table 3.11 Most popular standard coefficients for calculating the reset time of European and North American relays.

Type of curve	Standard	RTDM	T_r	λ
IEC Standard Inverse (SI)	IEC/A	*RTMS*	8.2 (Schneider Electric, 2015), 13.5 (SEL, 2013; Hase et al., 2020)	6.45 (Schneider Electric, 2015), 2 (SEL, 2013; Hase et al., 2020)
IEC Very Inverse (VI)	IEC/B	*RTMS*	50.92 (Schneider Electric, 2015), 47.3 (SEL, 2013; Hase et al., 2020)	2.4 (Schneider Electric, 2015), 2 (SEL, 2013; Hase et al., 2020)
IEC Extremely Inverse (EI)	IEC/C	*RTMS*	44.1 (Schneider Electric, 2015), 80.0 (SEL, 2013; Hase et al., 2020)	3.03 (Schneider Electric, 2015), 2 (SEL, 2013; Hase et al., 2020)
IEC Long Time Inverse (LTI)	IEC	*RTMS*	40.62 (Schneider Electric, 2015), 120.0 (SEL, 2013; Hase et al., 2020)	0.4 (Schneider Electric, 2015), 2 (SEL, 2013; Hase et al., 2020)
IEC Short Time Inverse (STI)	IEC/FR	*RTMS*	4.85 (SEL, 2013; Hase et al., 2020)	2 (SEL, 2013; Hase et al., 2020)
IEEE Moderately Inverse	IEEE (IEC/D)	*RTDS*	4.85 (Schneider Electric, 2015)	2 (Schneider Electric, 2015)
IEEE Very Inverse	IEEE (IEC/E)	*RTDS*	21.6 (Schneider Electric, 2015)	2 (Schneider Electric, 2015)
IEEE Extremely Inverse	IEEE (IEC/F)	*RTDS*	29.1 (Schneider Electric, 2015)	2 (Schneider Electric, 2015)
US Moderately Inverse (U1)	US	*RTDS*	1.08 (SEL, 2013; Hase et al., 2020)	2 (SEL, 2013; Hase et al., 2020)
US Inverse (U2)	US	*RTDS*	5.95 (SEL, 2013; Hase et al., 2020)	2 (SEL, 2013; Hase et al., 2020)
US Very Inverse (U3)	US	*RTDS*	3.88 (SEL, 2013; Hase et al., 2020)	2 (SEL, 2013; Hase et al., 2020)
US Extremely Inverse (U4)	US	*RTDS*	5.67 (Hase et al., 2020), 5.64 (SEL, 2013)	2 (SEL, 2013; Hase et al., 2020)
US Short Time Inverse (U5)	US	*RTDS*	0.323 (SEL, 2013; Hase et al., 2020)	2 (SEL, 2013; Hase et al., 2020)
CO Short Time Inverse (CO2-P20)	CO	*RTDS*	0.323 (Schneider Electric, 2015)	2 (Schneider Electric, 2015)
CO Short Time Inverse (CO2-P40)	CO	*RTDS*	2.261 (Schneider Electric, 2015)	2 (Schneider Electric, 2015)

Example 3.8 Suppose that $TDS = TDM = 1$. Design a MATLAB code to plot all the relay models listed in Table 3.10 when M moves from 1.5 to 20. For the coefficients of the US Extremely Inverse (U4), use the one reported in ISA (2011). The x-axis should be on a logarithmic scale.

Solution

Because *TMS* and *TDS* are equal, so we can use (3.36) without worrying about the range of *TDM*. The following code, which can be browsed via `MultipleTCCCsPlotter.m`, can plot all the models:

Table 3.12 Standard characteristic curve coefficients of type CO and IAC and BS relays.

Type of curve	α	β	γ	ξ	κ	T_r
CO Short Time Inverse (CO2)	1.2969	0.2663	0.033 93	1.000	0.028	0.5000
CO Long Time (CO5)	1.0000	5.6143	2.185 92	1.000	0.028	15.750
CO Definite Minimum Time (CO6)	1.5625	0.4797	0.213 59	1.000	0.028	0.8750
CO Moderately Inverse Time (CO7)	0.5000	0.3022	0.128 40	1.000	0.028	1.7500
CO Time Inverse (CO8)	2.0938	8.9341	0.179 66	1.000	0.028	9.0000
CO Very Inverse Time (CO9)	2.0469	5.4678	0.108 14	1.000	0.028	5.5000
CO Extremely Inverse Time (CO11)	2.0938	7.7624	0.027 58	1.000	0.028	7.7500
IAC Inverse (IAC 51)	0.4375	0.2747	0.104 20	1.000	0.028	0.8868
IAC Very Inverse (IAC 53)	1.9531	4.4309	0.099 10	1.000	0.028	5.8231
IAC Short Inverse (IAC 55)	0.9844	0.0286	0.020 80	1.000	0.028	0.0940
IAC Long Inverse (IAC 66)	0.3125	2.3955	0.000 02	1.000	0.028	7.8001
IAC Extremely Inverse (IAC 77)	2.0469	4.9883	0.012 90	1.000	0.028	4.7742
BS Very Inverse (BS142-B)	1.0469	1.4636	0.000 00	1.000	0.028	3.2500
BS Extremely Inverse (BS142-C)	2.0469	8.2506	0.000 00	1.000	0.028	8.0000

```
1	clc
2	clear
3	% relay coefficients
4	a = [0.02, 1, 2, 2.5, 1, 0.04, 0.02, 2, 2, 0.02, 2, 2, 0.02, 2, 0.02, ...
	    2, 2, 2, 0.02, 0.02, 1.1, 1.4, 0.02, 2, 2, 2, 5.6, 2, -1]; % alpha
5	b = [0.14, 13.5, 80, 315.2, 120, 0.05, 0.086, 28.55, 64.07, 0.0515, ...
	    19.61, 28.2, 0.16758, 1.281, 0.0104, 5.95, 3.88, 5.67, 0.00342, ...
	    0.02394, 4.842, 0.3164, 0.0094, 5.95, 4.12, 5.57, 45900, 1000, ...
	    -4.2373]; % beta
6	g = [0, 0, 0, 0, 0, 0, 0.185, 0.712, 0.25, 0.114, 0.491, 0.1217, ...
	    0.11858, 0.005, 0.2256, 0.18, 0.963, 0.0352, 0.00262, 0.01694, ...
	    1.967, 0.1934, 0.0366, 0.18, 0.0958, 0.028, 0, 0.655, 0]; % gamma
7	x = ones(1,length(g)); x(end) = 1.43644; % xi
8	L = {'IEC Standard Inverse (SI)', 'IEC Very Inverse (VI)', 'IEC ...
	    Extremely Inverse (EI)', 'IEC Ultra-Inverse (UI)', 'IEC Long Time ...
	    Inverse (LTI)', 'IEC Short Time Inverse (STI)',...
9	    'IEEE Long Time Inverse', 'IEEE Long Time Very Inverse', 'IEEE ...
	        Long Time Extremely Inverse', 'IEEE Moderately Inverse', ...
	        'IEEE Very Inverse', 'IEEE Extremely Inverse',...
10	    'IEEE Short Time Inverse', 'IEEE Short Time Extremely Inverse', ...
	        'US Moderately Inverse (U1)', 'US Inverse (U2)', 'US Very ...
	        Inverse (U3)', 'US Extremely Inverse (U4)',...
11	    'US Short Time Inverse (U5)', 'CO Short Time Inverse (CO2)', 'CO ...
	        Long Time (CO5)', 'CO Definite Minimum Time (CO6)', 'CO ...
	        Moderately Inverse Time (CO7)', 'CO Time Inverse (CO8)',...
12	    'CO Very Inverse Time (CO9)', 'CO Extreme Inverse Time (CO11)', ...
	        'UK Rectifier Protection', 'BNP (EDF)', 'RI'}; % labels
13	sz = 0.1; % step-size resolution
14	% relay settings
15	TDM = 1; % time dial/multiplier setting
16	M = 1.5:sz:20; % multiples of pickup current
```

```
    for i=1:length(a)
        for j=1:length(M)
            T(j,i) = TDM * (b(i)/(M(j) \couhata(i) - x(i)) + g(i));
        end
        glabel{i} = [L{i}];
    end
    semilogx(M, T, 'LineWidth', 2); % plotting using log-scale on the x-axis
    grid on; grid minor
    set(gca, 'FontSize', 13)
    set(gca, 'xtick', [1:1:10,12:2:20]);
    set(gca, 'ytick', 0:2:20);
    xlim([1, M(end)]);
    ylim([0, 20])
    title('Time-Current Characteristic Curves of Many Standard Models')
    xlabel('Multiples of Tap Value Current')
    ylabel('Operating Time in Seconds')
    leg = legend(glabel);
    title(leg, 'Time Dial/Multiplier Setting')
```

If we run this code, the plots shown in Figure 3.3 will be generated. This figure can also be accessed by opening `MultipleTCCCPlots.fig`.

3.2.2 Dealing with Time Dial Setting and Time Multiplier Setting as One Variable

Although the preceding approach is easy to code, it creates lots of programmatic problems. For example, if we want to optimally coordinate a group of North American and European inverse-time overcurrent relays where each one of them has its own TCCC. These characteristic curves could be inverse, very inverse, extremely inverse, etc.

Because some relays are adjusted by *TDS* and others by *TMS*, so many stages of meta-heuristic optimization algorithms will not work properly! To clarify it more, let's take the following example:

Example 3.9 Suppose a mesh network has 100 ITOCRs with multiple TCCCs and these relays need to be optimally coordinated by DE. The relay settings of three random candidates are:

$$X_{r_1} = \begin{bmatrix} PS_{r_1} \\ TMS_{r_1} \end{bmatrix} = \begin{bmatrix} 0.8 \\ 0.1 \end{bmatrix}$$

$$X_{r_2} = \begin{bmatrix} TS_{r_2} \\ TDS_{r_2} \end{bmatrix} = \begin{bmatrix} 2 \\ 7.4 \end{bmatrix}$$

$$X_{r_3} = \begin{bmatrix} PS_{r_3} \\ TMS_{r_3} \end{bmatrix} = \begin{bmatrix} 1.5 \\ 0.23 \end{bmatrix}$$

Using the classic mutation strategy and a step-size of $F = 0.6$, find the mutant vector.

Solution

By referring to Algorithm 5, the classic mutation strategy (aka DE/rand/1/bin) is:

$$V_i \leftarrow X_{r_1} + F\left(X_{r_2} - X_{r_3}\right)$$

For *PS* and tap setting (*TS*), the calculation is smooth and straightforward. On the opposite side, $TMS \in \left[TMS^{\min}, TMS^{\max}\right]$ is not compatible with $TDS \in \left[TDS^{\min}, TDS^{\max}\right]$. Thus, the following

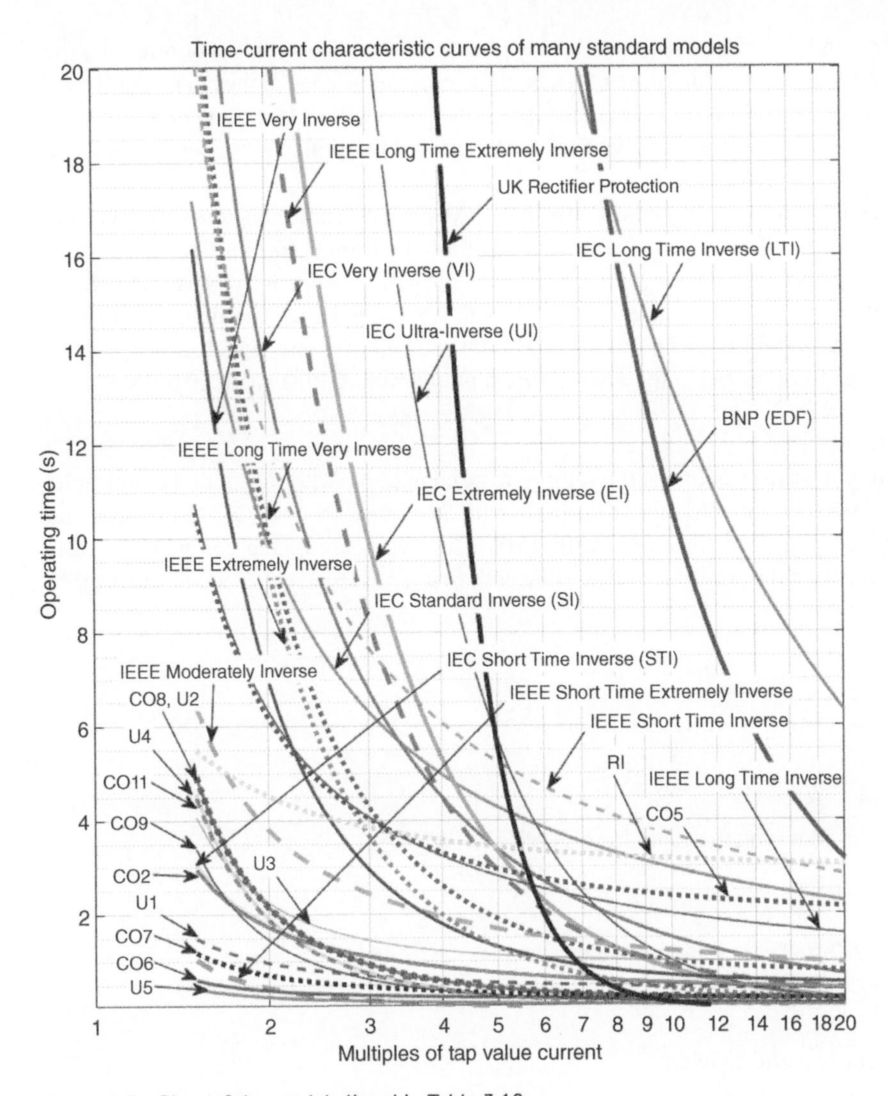

Figure 3.3 Plots of the models listed in Table 3.10.

problem will be faced if the ith relay is manufactured in Europe:

$$\begin{aligned} TMS_i &= TMS_{r_1} + F\left(TDS_{r_2} - TMS_{r_3}\right) \\ &= 0.1 + 0.6(7.4 - 0.23) \\ &= 4.402 \end{aligned}$$

As can be seen, $TMS_i > TMS^{\max}$. Yes, someone may say that we can clip the mutant vector to be within the allowable search space. If we use this option, then two things need to be taken into account. First, we are pushing many candidates to settle on the lower and upper bounds because of incompatibility between TDS and TMS. Second, the program does not know which bounds we are working on. Thus, the programmer could need to define many `if`-statements, which leads to a complicated optimization code.

The same thing can be seen in most meta-heuristic optimization algorithms. For example, the migration stage of BBO, the blending stage of GA, and the opposition stage of the opposition-based learning (OBL) algorithm. Thus, the approach presented in Section 3.2.1 is beneficial if the researchers want to study the relay behavior in their labs or if the protection engineers want to adjust the relay settings in their workshops.

The aim of this book is to optimally coordinate protective devices. Some chapters will have only ITOCRs. Therefore, we will heavily deal with optimization algorithms to find the best settings of relays with multiple TCCCs. Thus, we need to figure out how to effectively solve the preceding incompatibility issue. The following are two ways to solve this issue. The first one is based on ALSTOM/AREVA guidelines given in ALSTOM (2002). The other one, which is much more accurate, is borrowed from my old notes when I started programming BBO about 10 years ago.

3.2.2.1 Fixed Divisor

The objective here is to transform one of the settings to be compatible with the other. For example, we can multiply *TMS* by a suitable value to comply with *TDS*. Instead ALSTOM (2002) uses the opposite one; i.e. reducing *TDS* to comply with *TMS*. It is a very simple method where *TDS* is divided by a suitable fixed divisor. To find this value, let's first try to change the value from $TDS^{\min}$ to $TMS^{\min}$:

$$\frac{TDS^{\min}}{p} = TMS^{\min}$$

If $TDS^{\min} = 0.5$ and $TMS^{\min} = 0.1$, then p equals:

$$p = \frac{TDS^{\min}}{TMS^{\min}} = \frac{0.5}{0.1} = 5$$

The same thing for changing the value from $TDS^{\max}$ to $TMS^{\max}$:

$$\frac{TDS^{\max}}{q} = TMS^{\max}$$

If $TDS^{\max} = 9$ and $TMS^{\max} = 1$, then q equals:

$$q = \frac{TDS^{\max}}{TMS^{\max}} = \frac{9}{1} = 9$$

Taking the average of p and q yields:

$$\frac{p+q}{2} = \frac{5+9}{2} = 7$$

As we can see, the divisor is not constant if we move from the lower bound to the upper bound. Also, these bounds are not constant for both North American and European relays where each manufacturer has its own specifications. For example, the universal relays P3U10, P3U20, and P3U30 of Schneider Electric have $TDS \in [0.5, 20]$ (Schneider Electric, 2017b). The relay SEL-421 of Schweitzer Engineering Laboratories has $TDS \in [0.5, 15]$ (SEL, 2013). The CO relays of ABB have $TDS \in [0.1, 12.5]$ (ABB, 1992). Also, in many references, $TDS \in [0.5, 11]$ (El-Hawary, 1983; Anderson, 1998; Gers and Holmes, 2004). The same thing for *TMS*. For example, the series of MiCOM P140 of AREVA have $TMS \in [0.1, 1]$ (ALSTOM, 2002). In general, based on relay technology, $TMS^{\min}$ is in the range between 0.025 and 0.1 and $TMS^{\max}$ is in the range between 1.1 and 1.5 (Al-Roomi and El-Hawary, 2019a).

Therefore, the fixed divisor depends on many factors. For the series of MiCOM P140, (3.36) becomes:

$$T_T = \begin{cases} \frac{TDS}{7} \times \left[\frac{\beta}{\left(\frac{I}{I_p}\right)^{\alpha} - \xi} + \gamma \right], & \text{North American curves} \\ TMS \times \left[\frac{\beta}{\left(\frac{I}{I_p}\right)^{\alpha} - \xi} + \gamma \right], & \text{European curves} \end{cases} \tag{3.40}$$

This leads us to the following important points:

1. If we have TDS and we want to define a characteristic curve in a European relay:

$$T_T = \frac{TDS}{7} \times \left[\frac{\beta}{\left(\frac{I}{I_p}\right)^{\alpha} - \xi} + \gamma \right] \tag{3.41}$$

2. If we have TMS and we want to define a characteristic curve in a North American relay:

$$T_T = 7\ TMS \times \left[\frac{\beta}{\left(\frac{I}{I_p}\right)^{\alpha} - \xi} + \gamma \right] \tag{3.42}$$

Thus, we do not care about the lower and upper bounds of TDS because it is transformed to comply with TMS. We have to remember that, in optimization algorithms, if all the relays are manufactured in Europe, then we can use just TMS for all the supported characteristics, and vice versa for TDS if only North American relays are used.

Example 3.10 Suppose that $TMS = [0.1, 0.5, 1]$, $TDS = [0.5, 5, 9]$ and $M = 2 \rightarrow 20$ in steps of $\Delta M = 0.1$. Plot the IEC and the IEEE extremely inverse models for each time multiplier/dial setting. Then, re-plot the previous IEEE model with considering (3.41). All the plots should be within the same figure with a logarithmic scale on the x-axis.

Solution

The following code can be found in `IEEEwithTMS.m`:

```
1   clc
2   clear
3   % relay coefficients
4   a = [2, 2]; % alpha
5   b = [80, 28.2]; % beta
6   g = [0, 0.1217]; % gamma
7   % relay settings
8   TMS = [0.1, 0.5, 1]; % time multiplier setting
9   TDS = [0.5, 5, 9]; % original time dial setting
10  M = 2:0.1:20; % multiples of pickup current
11  j = 1;
```

```
12      for i=1:length(TMS)
13          T(:,j) = ITOCR(TMS(i), M, a(1), b(1), g(1)); % for TMS
14          T(:,j+1) = ITOCR(TDS(i), M, a(2), b(2), g(2)); % for original TDS
15          T(:,j+2) = ITOCR(TDS(i)/7, M, a(2), b(2), g(2)); % for attenuated TDS
16          L{j} = ['IEC Extremely Inverse | TMS = ', num2str(TMS(i))]; % ...
                append TMS label
17          L{j+1} = ['IEEE Extremely Inverse | TDS = ', num2str(TDS(i))]; % ...
                append original TDS label
18          L{j+2} = ['IEEE Extremely Inverse | TDS = ', num2str(TDS(i)), ...
                '/7']; % append attenuated  TDS label
19          j = j + 3; % update the counter
20      end
21      semilogx(M, T, 'LineWidth', 2); % plotting using log-scale on the x-axis
22      grid on; grid minor
23      set(gca, 'FontSize', 13)
24      set(gca, 'xtick', [1:1:10,12:2:20]);
25      set(gca, 'ytick', 0:2:30);
26      xlim([1, M(end)]);
27      ylim([0, 30])
28      title('TMS vs Original TDS vs Attenuated TDS ')
29      xlabel('Multiples of Tap Value Current')
30      ylabel('Operating Time in Seconds')
31      leg = legend(L);
32      title(leg, 'Time Dial/Multiplier Setting')
33      function T = ITOCR(TDM, M, A, B, G)
34          T = TDM * (B./(M. \couhatA - 1) + G);
35      end
```

Executing this program will generate the plots shownin Figure 3.4, which can also be seen by opening `IEEEwithTMS.fig`.

3.2.2.2 Linear Interpolation

From the previous method, the conversion does not match with the actual bounds of TMS. In the previous example, we have taken $TDS \in [0.5, 9]$ and $TMS \in [0.1, 1]$. Because $p \neq q \neq 7$, so $TDS^{min}/7 = 0.0714 < TMS^{min}$ and $TDS^{max}/7 = 1.2857 > TMS^{max}$. Also, $(TDS^{max} - TDS^{min})/7 = 1.2143 > (TMS^{max} - TMS^{min}) = 0.9$. Furthermore, this fixed divisor does not reflect the same value on the range of TMS. For example, if TDS is located at the middle (i.e. $TDS = \frac{TDS^{max}-TDS^{min}}{2} = \frac{9-0.5}{2} = 4.25$, the transformed value is supposed to be reflected at the middle of the range of TMS (i.e. it should equals $\frac{TMS^{max}-TMS^{min}}{2} = \frac{1-0.1}{2} = 0.45$). Instead, $\frac{TDS}{7} = \frac{4.25}{7} = 0.6071$, which is not at the center of TMS. This will become worse if relays from different models, technologies, or manufacturers are used. Thus, the divisor 7 is just an approximation for a specific type(s) of ITOCRs.

To solve this technical issue, we can use linear interpolation to make a projection between the two standard variables. This is valid for both directions, from TDS to TMS and vice versa. The procedure is depicted in Figure 3.5.

We can see that $TMS^{min} \neq TDS^{min}$ and $TMS^{max} \neq TDS^{max}$. If we know TDS_p, then *what is the equivalent value on the TMS scale?*

First, we define the reflection factors as follows:

$$RF_{p\to q} = \frac{\Delta TMS}{\Delta TDS} \tag{3.43}$$

$$RF_{q\to p} = \frac{\Delta TDS}{\Delta TMS} = \frac{1}{RF_{p\to q}} \tag{3.44}$$

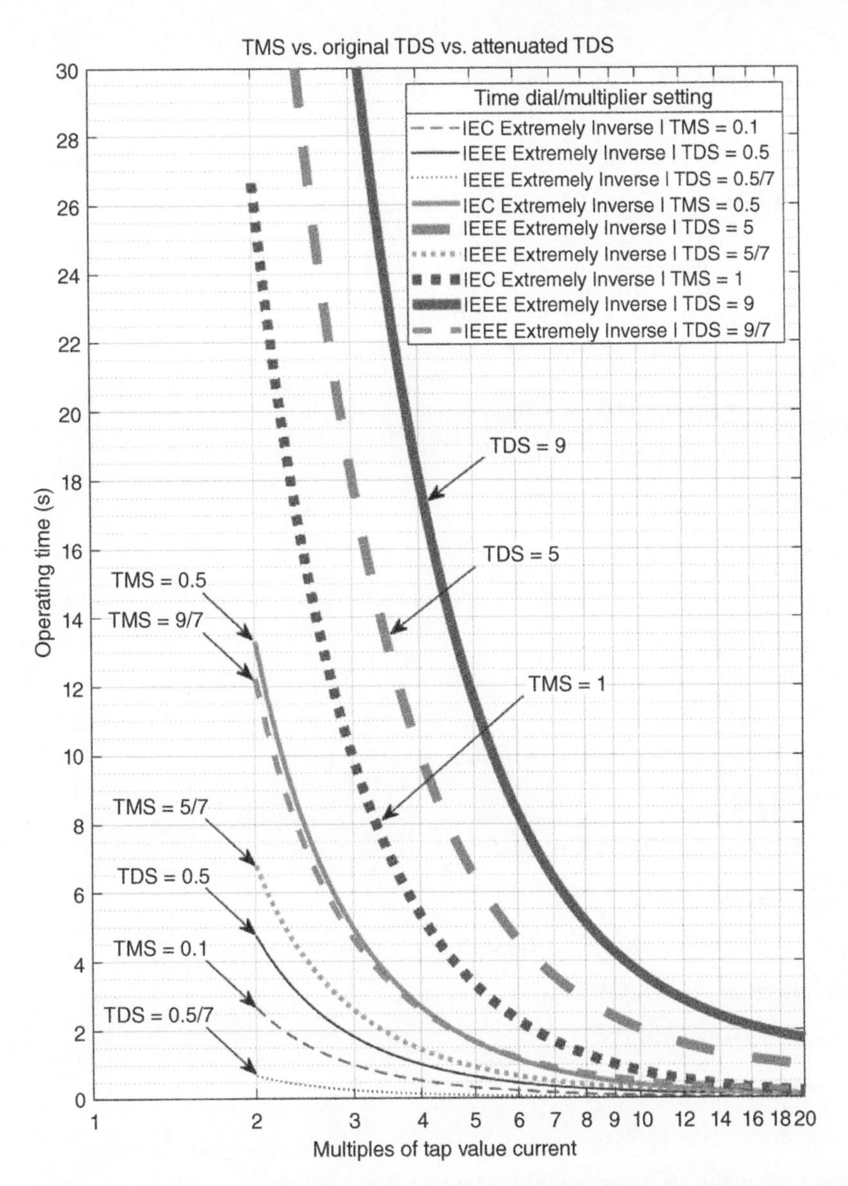

Figure 3.4 Modeling the IEEE extremely inverse model in a European relay using (3.41).

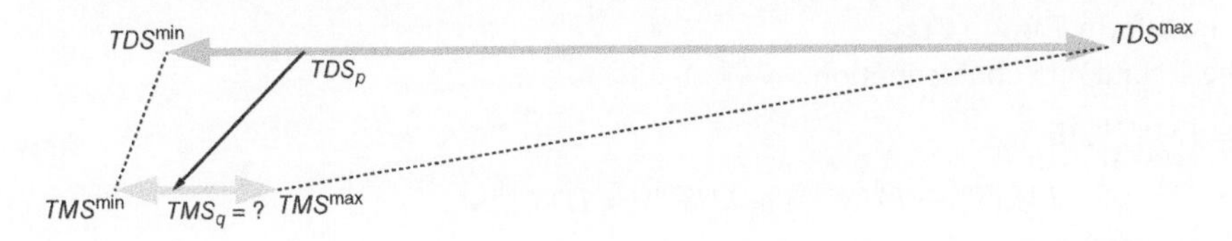

Figure 3.5 Projecting *TDS* to the search space of *TMS* via linear interpolation.

where

$$\Delta TMS = TMS^{\max} - TMS^{\min} \tag{3.45}$$

$$\Delta TDS = TDS^{\max} - TDS^{\min} \tag{3.46}$$

Then, we link *TMS* with *TDS* through the following linear relationship:

$$\frac{TDS_p - TDS^{\min}}{TDS^{\max} - TDS^{\min}} = \frac{TMS_q - TMS^{\min}}{TMS^{\max} - TMS^{\min}} \tag{3.47}$$

Now, we can express TMS_q as a function of TDS_p as follows:

$$\begin{aligned} TMS_q &= f\left(TDS_p\right) \\ &= \left(TMS^{\max} - TMS^{\min}\right)\left[\frac{TDS_p - TDS^{\min}}{TDS^{\max} - TDS^{\min}}\right] + TMS^{\min} \\ &= \frac{\Delta TMS}{\Delta TDS}\left(TDS_p - TDS^{\min}\right) + TMS^{\min} \end{aligned} \tag{3.48}$$

Therefore, the first reflection equation is:

$$TMS_q = RF_{p\to q}\left(TDS_p - TDS^{\min}\right) + TMS^{\min} \tag{3.49}$$

which is used to transform *TDS* to *TMS*.

From the first reflection equation:

If $TDS_p = TDS^{\min}$, then;

$$TMS_q = RF_{p\to q} \times \left(\cancel{TDS^{\min} - TDS^{\min}}\right) + TMS^{\min} = TMS^{\min} \checkmark$$

If $TDS_p = TDS^{\max}$, then;

$$\begin{aligned} TMS_q &= RF_{p\to q} \times \left(TDS^{\max} - TDS^{\min}\right) + TMS^{\min} \\ &= \frac{\Delta TMS}{\cancel{\Delta TDS}} \times \cancel{\Delta TDS} + TMS^{\min} \\ &= TMS^{\max} - \cancel{TMS^{\min}} + \cancel{TMS^{\min}} = TMS^{\max} \checkmark \end{aligned}$$

which means that our reflection procedure is correct:

$$TDS^{\min} \longrightarrow TMS^{\min}$$

$$TDS^{\max} \longrightarrow TMS^{\max}$$

$$TDS_p \longrightarrow TMS_q$$

Similarly, by analogy, we can derive the second reflection equation as follows:

$$TDS_p = RF_{q\to p} \times \left(TMS_q - TMS^{\min}\right) + TDS^{\min} \tag{3.50}$$

which is used to transform *TMS* to *TDS*.

Again, from the second reflection equation:

If $TMS_q = TMS^{\min}$, then;

$$TDS_p = RF_{q\to p} \times \left(\cancel{TMS^{\min} - TMS^{\min}}\right) + TDS^{\min} = TDS^{\min} \checkmark$$

If $TMS_q = TMS^{\max}$, then;

$$\begin{aligned} TDS_p &= RF_{q\to p} \times \left(TMS^{\max} - TMS^{\min}\right) + TDS^{\min} \\ &= \frac{\Delta TDS}{\cancel{\Delta TMS}} \times \cancel{\Delta TMS} + TDS^{\min} \\ &= TDS^{\max} - \cancel{TDS^{\min}} + \cancel{TDS^{\min}} = TDS^{\max} \checkmark \end{aligned}$$

Therefore, the reflection process will work in the same principle for the reverse transformation (i.e. from TMS_q to TDS_p):

$$TMS^{\min} \longrightarrow TDS^{\min}$$

$$TMS^{\max} \longrightarrow TDS^{\max}$$

$$TMS_q \longrightarrow TDS_p$$

Once these two reflection equations are derived, we can now swap between *TDS* and *TMS* smoothly. Thus, in optimization algorithms, we can just use the `if`-statement during evaluating the fitness. That is, instead of using two ranges for *TDS* and *TMS*, we are dealing now with just one range. Therefore, the problem addressed in Example 3.9 can be solved. It is valid for DE, BBO, GA, OBL, etc.

Example 3.11 Suppose that we have a North American numerical relay which accepts customized/programmable TCCCs.

1. Design a model to mimic the IEC very inverse TCCC based on *TDS*.
2. Calculate the trip time when $TDS = 6$ and $PSM = 8$.

Take $TMS^{\min} = 0.1$, $TDS^{\min} = 0.5$, $\Delta TMS = 1$, and $\Delta TDS = 10$.

Solution

Because it is a North American relay, so we are dealing with *TDS*. On the opposite side, we want to model one of the IEC standard TCCCs, which means that we have to deal with *TMS*. Therefore, we need to use the first reflection equation given in (3.49) to transform the internal setting of the relay from *TDS* to *TMS* before being fed to the model.

Based on (3.36) and the standard coefficients tabulated in Table 3.10, the IEC very inverse model can be mathematically expressed as follows:

$$T_T = \frac{13.5\ TMS}{\frac{I}{I_p} - 1}$$

By applying (3.49), we can use *TDS* instead of *TMS* as follows:

$$T_T = \left(\frac{13.5}{PSM - 1}\right) \cdot \left[RF_{p\rightarrow q}\left(TDS - TDS^{\min}\right) + TMS^{\min}\right]$$

To calculate the relay operating time at the given condition, first we use (3.43) to find $RF_{p\rightarrow q}$:

$$RF_{p\rightarrow q} = \frac{\Delta TMS}{\Delta TDS} = \frac{1}{10} = 0.1$$

Therefore, the trip time is:

$$T_T = \left(\frac{13.5}{6 - 1}\right) \cdot [0.1\,(6 - 0.5) + 0.1] = 1.755\text{s}$$

3.2.3 General Guidelines Before Conducting Researches and Studies

The preceding Sections 3.2.1 and 3.2.2 show us how to differentiate between *TDS* and *TMS* and how to deal with them together. However, as we have said at the beginning of this section, it is still unfair to conduct a comparative study even if we select the same TCCC model and covert *TMS* to/from *TDS* correctly. Many optimal coordination studies are conducted based on only one unified variable

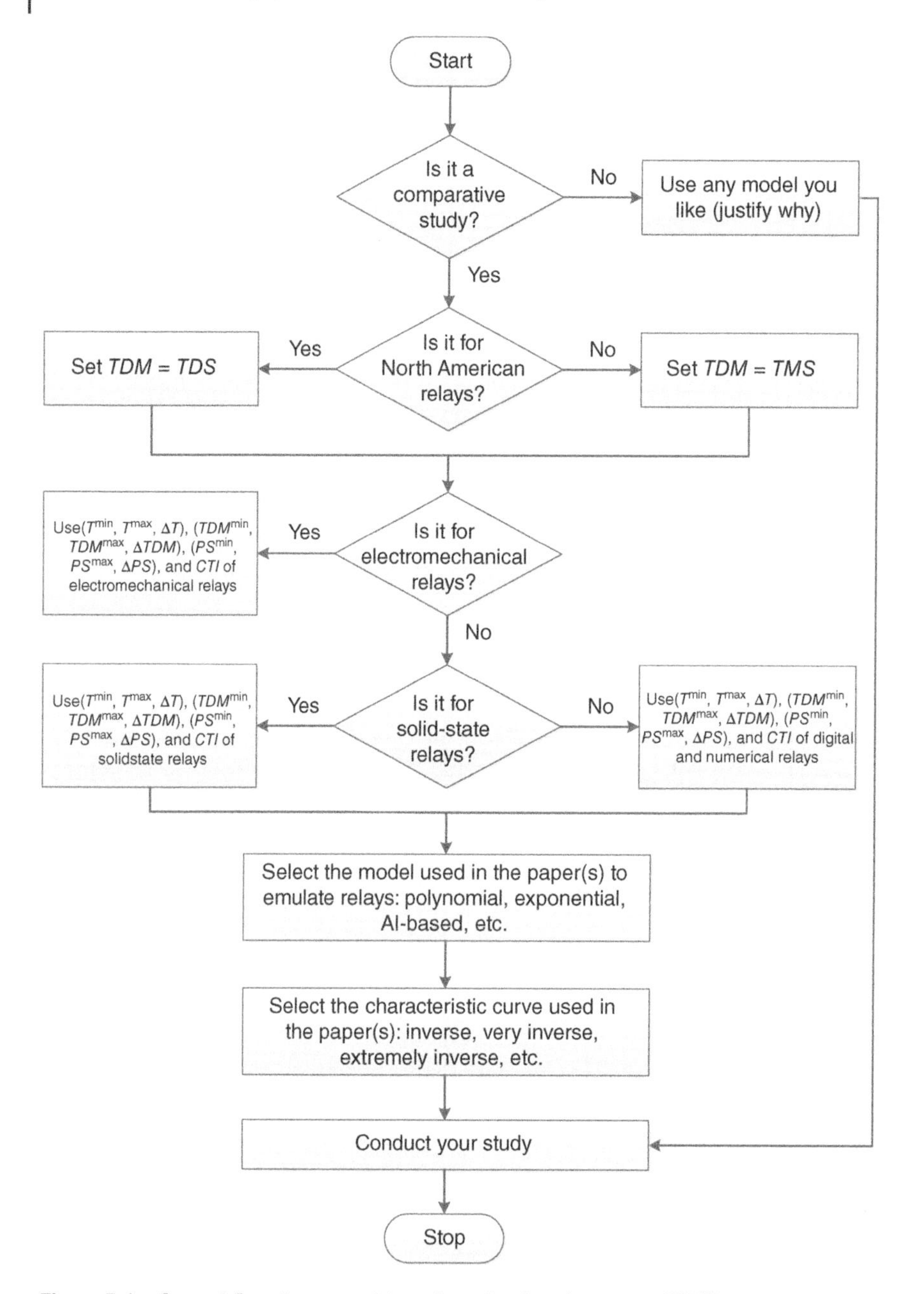

Figure 3.6 General flowchart to guide us for selecting the proper ITOCRs.

(i.e. *TMS* or *TDS*) and one set of model coefficients (for example, modeling only the IEC standard inverse or the IEEE very inverse). Further, because we have electromechanical, solid-state, digital, and numerical relays, so the variable bounds and their step-size resolutions could be identical if all the relays are the same and could be different from one relay to another. The same thing is applied to *CTI* (Al-Roomi and El-Hawary, 2019a).

To address all the aforementioned points, the general flowchart shown in Figure 3.6 can be used as a guideline before conducting our researches and studies. As can be seen, if the goal of conducting your study is not to compare the results with others reported in the literature, then you are free to choose any relay technology, model, side constraints, etc. However, you need to justify your selection. For example, you cannot use electromechanical or solid-state relays as the preferable choice for the next generation smart grids where only state-of-the-art numerical relays and IEDs are used.

Problems

Written Exercises

W3.1 Repeat Example 3.3 to find the coefficients of (3.5). For the accuracy, use four decimal places.

W3.2 Repeat Example 3.3 to find the coefficients of (3.4). For the accuracy, use four decimal places. (**Hint:** You will need to de-transform log T to have just T.)

W3.3 Repeat Example 3.3 to find the coefficients of (3.3). For the accuracy, use four decimal places. (**Hint:** You will need to de-transform log T to have just T.)

W3.4 Suppose that the model suggested to estimate the operating time of ITOCR is:

$$T = \zeta_0 + \frac{\sigma W_1}{\zeta_1} + \frac{\sigma W_2}{\zeta_2} + \frac{\sigma W_3}{\zeta_3} + \frac{\sigma W_4}{\zeta_4}$$

Using the results given in Example 3.3, find ζ's coefficients if $\sigma = 1$ and $W_i = \left(M^i - 1\right)^{-1}$ where $i = 1, \dots, 4$. Do not use any curve fitting to find these coefficients.

W3.5 Repeat Exercise W3.4 with $\sigma = 1.5$. Again, solve this problem without using any curve fitting.

W3.6 Consider the polynomial model given in (3.6). To correctly mimic the actual operating time of ITOCRs, the model should exhibit an asymptotic behavior at $M = 1$ and monotonic decrease in the operating time as M increases. Prove that when $M \to 1$, $T \to +\infty$ only if:

$$\sum_{j=1}^{n} \frac{a_j}{j} > 0$$

If not, $T \to -\infty$

W3.7 Using the expression given in the last written exercise, analyze the model given in Example 3.3 to check whether T approaches positive or negative infinity when $M = 1$.

W3.8 Consider the polynomial model given in (3.5). Find the conditions that satisfy $T \to +\infty$ or $-\infty$ when $M \to 1$.

W3.9 Consider the polynomial model given in (3.4). Find the conditions that satisfy $T \to +\infty$ or 0 when $M \to 1$.

W3.10 Repeat Exercise W3.9 for the polynomial model given in (3.3).

W3.11 Calculate the operating time of the RI-type characteristic when $TMS = 0.3$ and $PSM = 18$.

W3.12 Repeat Exercise W3.11 for the RD-type characteristic.

W3.13 Repeat Exercise W3.11 for the BNP-type characteristic.

W3.14 Repeat Exercise W3.11 for the UK model used in rectifier protection. Take $M = PSM$, $TDS \in [0.5, 12]$ and $TMS \in [0.1, 1.1]$. (**Hint:** Use the linear interpolation method to transform TMS to TDS.)

W3.15 Fix the problem shown in Example 3.9 by transforming TDS to TMS via the fixed divisor method. Does the new TMS_i lie between ($TMS^{\text{min}} = 0.1$) and ($TMS^{\text{max}} = 1.1$)? Repeat it for $F = 0.4$ and $TDS_{r_2} = 1.4$. Does it still lie between TMS^{min} and TMS^{max}? Why?

W3.16 Repeat Exercise W3.15 by using the linear interpolation method. Take $TDS \in [0.1, 12.5]$. Is there any difference in the answer? Why?

Computer Exercises

C3.1 Resolve Example 3.3 in MATLAB using the same steps. Build the matrices manually.

C3.2 Repeat Exercise W3.4 with $\sigma = 1.125$ and $W_i = \left(M^{i/\sigma} - 1\right)^{-1}$. (**Hint:** Utilize the MATLAB code developed in Exercise C3.1 to find the new coefficients.)

C3.3 Write a MATLAB code to compare the operating time estimated by the 7 and 15 term models, suggested by Damborg et al., and then display the results in a table. Use the following vectors:

$$\begin{bmatrix} M \\ TDS \end{bmatrix} = \begin{bmatrix} 1.5 & 2 & 5 & 10 & 15 & 20 & 25 & 30 & 35 & 40 & 45 & 50 \\ 0.5 & 1 & 2 & 3 & 4 & 5 & 6 & 7 & 8 & 9 & 10 & 11 \end{bmatrix}$$

C3.4 Modify the MATLAB code given in `C09_Damborg7TermEq.m` to graphically compare the behavior of the 7 and 15 term models in one figure.

C3.5 Write a MATLAB code to re-solve Example 3.3 by using its built-in regression toolbox.

C3.6 Using the MATLAB code developed in Exercise C3.5, fit the following model:

$$T = d_0 + \frac{d_1}{(\log_2 M)} + + \frac{d_2}{(\log_2 M)^2} + \frac{d_3}{(\log_2 M)^3} + \frac{d_4}{(\log_2 M)^4}$$

Use the data-set given in Example 3.3.

C3.7 Repeat Exercise C3.6 with $\log_{10}$.

C3.8 Repeat Example 3.4 with the IEEE moderately inverse coefficients tabulated in Table 3.5.

C3.9 Repeat Example 3.4 with the IEEE extremely inverse coefficients tabulated in Table 3.5.

C3.10 Modify the code given in Example 3.4 to plot the three characteristic curves of both CO and IAC series relays as well as the IEEE standard curves. Each type (i.e. moderately inverse, very inverse, and extremely inverse) should be displayed in a horizontally-aligned subplot. Each subplot should combine both CO and IAC.

C3.11 Modify the code given in Example 3.8 to plot the solid-state models tabulated in Table 3.7.

C3.12 Modify the code given in Example 3.5 to generate the semilogarithmic plot of the IEC standard inverse model. Take $TMS = [0.1, 0.2, 0.3, \ldots, 1.1]$.

C3.13 Write a MATLAB code to plot the models listed in Table 3.9. (**Hint:** Merge the codes given in Examples 3.6 and 3.8.)

C3.14 Modify the code given in Example 3.8 to add a customized TCCC similar to the I^2T curve given in (3.35).

C3.15 Using the linear interpolation method, modify the code given in Example 3.8 to plot all the models for a programmable European relay. Take $TDS \in [0.5, 11]$ and $TMS \in [0.025, 1.5]$.

C3.16 Repeat Exercise C3.15 for a programmable North American relay. Take $TDS \in [0.1, 12.5]$ and $TMS \in [0.1, 1.1]$.

4

Upper Limit of Relay Operating Time

Aim

This chapter aims to give some guidelines that should be considered during setting the upper limit of relay operating times. Although this side constraint can be proven intuitively, the specific numerical value can be obtained by satisfying two important criteria. Firstly, the trip time does not pass the thermal threshold of power system components. Secondly, the trip time should not be longer than the critical time required to preserve system stability. The first criterion is described from the point of view of thermal protection, while the second one is briefly described by studying the transient stability of a simple lossless single-machine infinite bus(SMIB) system. That is, the maximum isolation time should be small enough to protect electric components from excessive heat as well as preserving system stability.

In Chapter 3, we have seen that the variable bounds of *TDS*/*TMS* and I_p/*PS* and their step-size resolutions depend on the model and relay technology. This is also the case for accounting for the error or time delay[1] that should be given to primary relays to take their action before initiating backup protective devices. Thus, for relay operating time (T), it is obvious that its lower bound ($T^{\min}$) also depends on the relay model and technology. What is not covered yet is the upper limit of this functional constraint. That is, *do we need to define* $T^{\max}$*?* If yes, then *how to define this time limit?* This chapter gives some guidelines to set $T^{\max}$ based on two criteria; thermal limit and operation stability of power system components. The reader can skip this chapter if the side constraints of T are given and known. However, the material provided here can enhance our understanding and empower us to solve more complicated coordination problems.

4.1 Do We Need to Define $T^{\max}$?

We can prove that this limit is required either intuitively or mathematically. Let's leave the mathematical answer to the second question. The intuitive answer can be reached by looking at the cause and effect diagram shown in Figure 2.2. The logic here says the following statements:

1. We need protection system to clear faults.
2. We have to clear faults to avoid excessive heat.
3. We must avoid excessive heat to prevent damage to power components.

Thus, this quick answer tells us that there is a maximum time to take proper actions before break down. If we want to dive into details, then the next question should be answered as well.

1 Refer to (5.3).

Optimal Coordination of Power Protective Devices with Illustrative Examples, First Edition. Ali R. Al-Roomi.

Companion website: www.wiley.com/go/al-roomi/optimalcoordination

4.2 How to Define T^{max}?

If there is a fault, the isolation process should be taken as fast as possible. However, because we want to satisfy the selectivity criterion between primary/backup (P/B) relay pairs, so some times, we need to give some delays. But, we have to ensure that these delays should not exceed the maximum allowable chances given to protective devices. Also, we have to differentiate between the selectivity criterion and the time threshold given to P/B relays to isolate the fault before facing big troubles, such as insulation failure, instable operation, or partial/full damage to equipment. There are two mathematical ways to prove that the upper limit of T is required:

4.2.1 Thermal Equations

The first way, which is briefly mentioned in the previous answer, is to prevent any damage to power components due to excessive heat because we know that the total power loss is proportional to the square of current flowing in the medium. For 3ϕ lines, we have:

$$P_L = 3|I|^2R \tag{4.1}$$

where P_L is the active power loss due to heat.

During short-circuit faults, $|I_f| \gg |I|$, so the faulty components are exposed to excessive heat. Because each physical component has a thermal limit, so an urgent isolation process is required to take that faulty component out of service.

Depending on the operating time of ITOCRs is not enough because the equation does not provide enough information regarding the thermal limits of power system components. That is, the relay operating speed depends on PS and TMS (for the European ITOCRs) or CTS/I_p and TDS (for the North American ITOCRs) and these settings can be changed by protection engineers.[2] For this, we need to refer to thermal overload protection. There are different equations used to calculate the operating time of thermal relays based on the power component they are installed for; whether overhead lines and cables, motors, or transformers. These models are listed in the following text for some ABB Relion series relays (Smith, 2015; Smith and Jain, 2017).

4.2.1.1 Thermal Overload Protection for 3ϕ Overhead Lines and Cables

For feeder thermal overload protection[3] functions, the operating time of load and pre-load current can be obtained by solving a first order differential equation. From the derivation given in Appendix E:

$$\frac{d\theta}{dT} = \frac{1}{\tau}\left(\frac{I^2R}{hA} - \theta\right) \tag{4.2}$$

where

θ temperature (above ambient)
T time
τ time constant
I highest phase current (measured in RMS)
R resistance

---

2 To be more specific, the correct setting of PS or I_p should be taken based on overload factor, which takes into account the thermal limit. This will be covered later in Chapter 6; specifically during explaining the side constraints of PS/I_p.

3 The ANSI device number is 49F.

h heat transfer coefficient[4]
A surface area of conductor

The solution to (4.2) is:

$$\theta = \theta_0 + \left(\frac{I^2R}{hA} - \theta_0\right)\left(1 - e^{-T/\tau}\right) \tag{4.3}$$

where

$$\theta_0 = \frac{I_0^2 R}{hA} \tag{4.4}$$

$$\theta_{\text{final}} = \frac{I^2 R}{hA} \tag{4.5}$$

$$\frac{R}{hA} = \frac{\theta_n}{I_n^2} \tag{4.6}$$

and

θ_0 initial temperature (above ambient)
θ_f final temperature (above ambient)
I_n nominal current (reference current)
θ_n nominal temperature at reference current (above ambient)

Solving (4.3) for time:

$$T = \tau \times \ln\left[\frac{I^2 - I_0^2}{I^2 - \left(I_n\sqrt{\frac{\theta}{\theta_n}}\right)^2}\right] \tag{4.7}$$

where

$$I_0 = I_{\text{pre}} \text{ (pre-load current)} \tag{4.8}$$

$$I_n = I_{\text{ref}} \text{ (reference current)} \tag{4.9}$$

$$\theta = \theta_{\max} - \theta_{\text{env}} \text{ (trip/operating values)} \tag{4.10}$$

$$\theta_n = \theta_{\text{rise}} \text{ (temperature rise)} \tag{4.11}$$

Substituting (4.8)–(4.11) into (4.7) yields:

$$T = \tau \times \ln\left[\frac{I^2 - I_{\text{pre}}^2}{I^2 - \left(I_{\text{ref}}\sqrt{\frac{\theta_{\max} - \theta_{\text{env}}}{\theta_{\text{rise}}}}\right)^2}\right] \tag{4.12}$$

The trip current (I_T) can be calculated as follows:

$$I_T = I_n\sqrt{\frac{\theta}{\theta_n}} \tag{4.13}$$

4 The unit of this quantity in heat-transfer discipline is W/m^2 k, which means that the temperature unit should be in Kelvin "absolute value." However, ABB Relion series thermal relays use Celsius (Smith, 2015; Smith and Jain, 2017; ABB, 2018) in time and temperature calculations. Thus, we will follow this in our calculations.

which can be substituted into (4.7) to have another expression:

$$T = \tau \times \ln \left[\frac{I^2 - I_0^2}{I^2 - I_T^2} \right] \tag{4.14}$$

Also, we have to mention that the formula given in (4.3) is valid for the initial temperature. Because the protected element will gain an extra heat, so the new calculated temperature should take into account the increment in the temperature. Thus, we have to replace θ with θ_i and θ_0 with θ_{i-1} to have the following modified formula:

$$\theta_i = \theta_{i-1} + \left(\frac{I^2 R}{hA} - \theta_0 \right) \left(1 - e^{-\Delta T/\tau}\right) \tag{4.15}$$

which is called the thermal counter (ABB, 2018).

The final temperature rise can be calculated by the following temperature estimator:

$$\theta_f = \left(\frac{I}{I_{\text{ref}}} \right)^2 \cdot \theta_{\text{ref}} \tag{4.16}$$

where θ_{ref} is the reference temperature.

This temperature can be used to calculate the present time to trip:

$$T_T = -\tau \cdot \ln \left(\frac{\theta_f - \theta_T}{\theta_f - \theta_i} \right) \tag{4.17}$$

which is activated when θ_f exceeds the set trip temperature (θ_T).

Also, the time required to lockout release is calculated via the following equation[5]:

$$T_R = -\tau \cdot \ln \left(\frac{\theta_f - \theta_R}{\theta_f - \theta_i} \right) \tag{4.18}$$

where θ_R is the lockout release temperature.

Example 4.1 Write a MATLAB code to plot the cable temperature as a function of time when T starts from 0 to 8 minutes. Use the following data:

$$\tau_1 = 30\,\text{s},\ \tau_2 = 90\,\text{s},\ \theta_0 = 40\,^\circ\text{C, and}\ \theta_f = 105\,^\circ\text{C}$$

Solution

The code given in `ThermalOverloadPlotter1.m` is:

```
1   clc
2   clear
3   Tau1 = 30; % time constant no.1
4   Tau2 = 90; % time constant no.2
5   theta_initial = 40; % initial temperature
6   theta_final = 105; % final temperature
7   dT = 1; % difference in time
8   T = 0:dT:480; % time period
9   theta1old = theta_initial;
10  theta2old = theta_initial;
11  for i = 1:length(T)
12      theta1(i) = theta1old + (theta_final - theta1old)*(1 - ...
            exp(-dT/Tau1));
```

5 That is, the calculation of the cooling time to a set value (ABB, 2018).

```
13          theta2(i) = theta2old + (theta_final - theta2old)*(1 - ...
                exp(-dT/Tau2));
14          theta1old = theta1(i); % update the cable temperature
15          theta2old = theta2(i); % update the cable temperature
16      end
17      plot(T, theta1, 'k', T, theta2, '--r', 'LineWidth', 2)
18      grid on; grid minor
19      legend({'\tau = 300 sec', '\tau = 5000 sec'}, 'Location', 'southeast')
20      set(gca, 'FontSize', 13)
21      xlim([0, T(end)]); ylim([theta_initial-5, theta_final+5])
22      title('Cable Temperature vs Time')
23      xlabel('Time in Seconds')
24      ylabel('Cable Temperature in ^{\circ}C')
```

If we run the code given earlier, the plot shown in Figure 4.1 can be generated. This plot can also be accessed by opening `ThermalOverloadPlotter1.fig`.

Example 4.2 Suppose a Relion thermal overload relay is fed with the following values:

$$\tau = 250\,\text{s},\ I_{\text{pre}} = 0.8\,\text{pu},\ I_{\text{ref}} = 1\,\text{pu},\ I = 1.75\,\text{pu},\ \theta_{\text{env}} = 40\,^{\circ}\text{C, and } \theta_{\text{max}} = 110\,^{\circ}\text{C}$$

Calculate the trip time when:

1. $\theta_{\text{rise}} = 65\,^{\circ}\text{C}$
2. $\theta_{\text{rise}} = 85\,^{\circ}\text{C}$

Solution

Using (4.12), the trip time is:

1. For $\theta_{\text{rise}} = 65\,^{\circ}\text{C}$:

$$T = 250 \times \ln\left[\frac{1.75^2 - 0.8^2}{1.75^2 - \left(1 \times \sqrt{\frac{110-40}{65}}\right)^2}\right] = 49.7226\,\text{s}$$

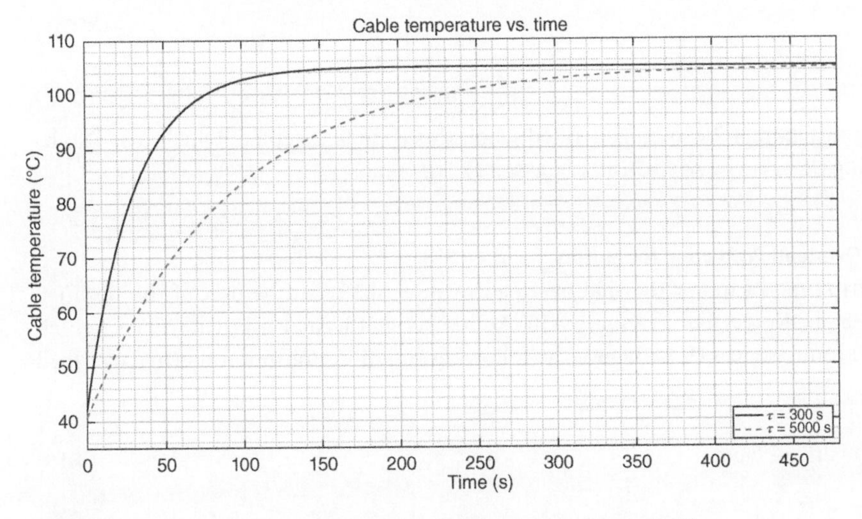

Figure 4.1 Rise temperature of the thermal relay given in Example 4.1.

2. For $\theta_{rise} = 85\,°\text{C}$:

$$T = 250 \times \ln\left[\frac{1.75^2 - 0.8^2}{1.75^2 - \left(1 \times \sqrt{\frac{110-40}{85}}\right)^2}\right] = 19.6960\,\text{s}$$

Example 4.3 Write a MATLAB code to plot the time–temperature curve for $\theta_{rise} \in [40,100]\,°\text{C}$. Use the following data:

$$\tau_1 = 300\,\text{s},\ \tau_2 = 1500\,\text{s},\ I_{pre} = 0.75\,\text{pu},\ I_{ref} = 1\,\text{pu},\ I = 1.5\,\text{pu},\ \theta_{env} = 30\,°\text{C},\text{ and } \theta_{max} = 115\,°\text{C}$$

Solution

The code given in `ThermalOverloadPlotter2.m` is:

```
1   clc
2   clear
3   Tau1 = 300;
4   Tau2 = 1500;
5   I = 1.5;
6   Ip = 0.75;
7   Iref = 1;
8   theta_max = 115;
9   theta_env = 30;
10  theta_rise = 40:0.25:100;
11  for i = 1:length(theta_rise)
12      Num = I^2 - Ip^2;
13      Den = I^2 - (Iref*sqrt((theta_max-theta_env)/theta_rise(i)))^2;
14      T1(i) = Tau1*log(Num/Den);
15      T2(i) = Tau2*log(Num/Den);
16  end
17  plot(theta_rise, T1, theta_rise, T2, 'LineWidth', 2)
18  grid on; grid minor
19  legend({'\tau = 300 sec', '\tau = 1500 sec'})
20  set(gca, 'FontSize', 13)
21  title('Thermal Overload Protection for Overhead Lines')
22  xlabel('Rise Temperature in ^{\circ}C')
23  ylabel('Time in Seconds')
```

The plot shown in Figure 4.2 can be generated when this code is executed, which can also be accessed by opening `ThermalOverloadPlotter2.fig`.

4.2.1.2 Thermal Overload Protection for Motors

To calculate the trip time of motor thermal overload relays,[6] (4.12) can be used if the negative sequence current is neglected. For example, ABB Relion series thermal overload relays, which are intelligent electronic devices (IEDs), have the following equations for motor thermal overload protection:

$$\sqrt{\frac{\theta}{\theta_n}} = k\ (\text{overload factor}) \tag{4.19}$$

6 The ANSI device number is 49M.

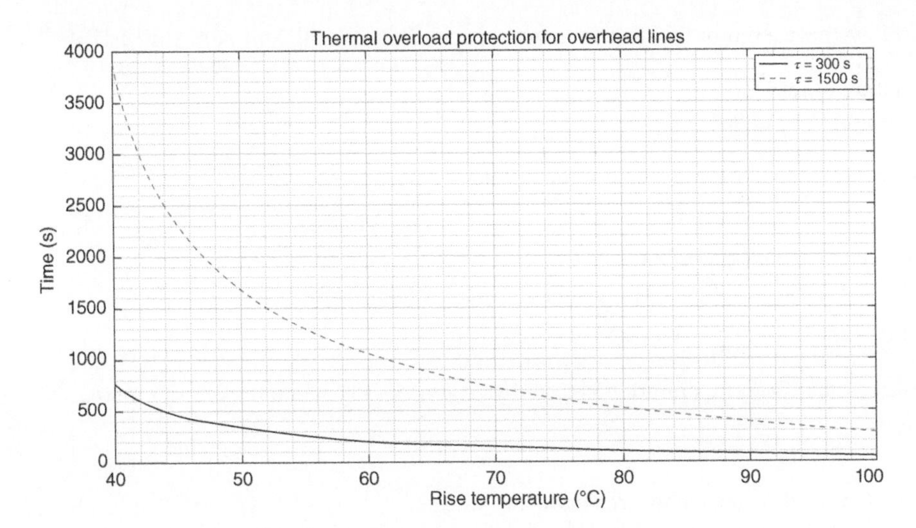

Figure 4.2 Operating time of the thermal relay given in Example 4.3.

$$I_n = I_{\text{rated}} \text{ (motor rated current)} \tag{4.20}$$

$$\tau = \tau_{\text{start}} \text{ (motor starting time constant)} \tag{4.21}$$

$$\tau = \tau_{\text{normal}} \text{ (motor running time constant)} \tag{4.22}$$

Based on these two time constants, there are two operating conditions:

1. starting condition; and
2. running condition.

The general time equation can be obtained by substituting (4.8), (4.19), and (4.20) into (4.7):

$$T = \tau \times \ln \left[\frac{I^2 - I_{\text{pre}}^2 \times p}{I^2 - (kI_{\text{rated}})^2} \right] \tag{4.23}$$

where p is weighting factor to scale the pre-load current, which is taken as 100% for motor protection.

We need to substitute (4.21) into (4.23) for $\tau = \tau_{\text{start}}$ just during the starting period because $I > 2.5 \times I_{\text{rated}}$. Otherwise, the time constant of (4.23) should equal τ_{normal} because $I \in (0.12 \times I_{\text{rated}}, 2.5 \times I_{\text{rated}})$ during the normal operation (Smith, 2015; Smith and Jain, 2017).

Let's return back to (4.1). The temperature is proportional to the square of current. This principle can also be used to calculate the overload factor as follows:

$$I_T = k \cdot I_{\text{rated}} \tag{4.24}$$

Linking the trip current with the rated current:

$$\frac{\theta_T - \theta_{\text{env}}}{\theta_{\text{rated}} - \theta_{\text{env}}} = \left(\frac{I_T}{I_{\text{rated}}} \right)^2 \tag{4.25}$$

Taking the square root of both sides and then multiplying them by I_{rated}, the overload factor can be calculated by using either temperature or current as follows:

$$k = \sqrt{\frac{\theta_T - \theta_{\text{env}}}{\theta_{\text{rated}} - \theta_{\text{env}}}} = \frac{I_T}{I_{\text{rated}}} \tag{4.26}$$

Example 4.4 If we have a motor with the following specification (Smith and Jain, 2017):

- Class F insulation of 150 °C
- Class B temperature rise of 90 °C

Calculate the overload factor if the ambient temperature is 35 °C.

Solution

Using (4.26), we can calculate k as follows:

$$k = \sqrt{\frac{155 - 35}{125 - 35}} = \sqrt{\frac{120}{90}} = \sqrt{1.3333} = 1.1547$$

It can be observed that the temperature increases by around 33.33% when the current increases by around 15.5%.

4.2.1.3 Thermal Overload Protection for Transformers

For transformer thermal overload relays,[7] two time constants are required to describe the warming. The first one is a short time constant (τ_1), and it is used for transformer windings. The other one is a long time constant (τ_2), and it is used for transformer oil. However, some manufacturers program IEDs with models having only one equivalent time constant (ABB, 2018). The temperature rise in transformer can be calculated as follows (Smith, 2015):

$$\theta = \Delta\theta + \theta_{\text{env}} \tag{4.27}$$

where $\Delta\theta$ is the temperature calculated by the following thermal counter (Smith and Jain, 2017; ABB, 2018):

$$\Delta\theta = \left[p \cdot \left(\frac{I}{I_{\text{ref}}}\right)^2 \cdot \theta_{\text{ref}}\right] \cdot \left(1 - e^{\frac{-\Delta T}{\tau_1}}\right) + \left[(1-p) \cdot \left(\frac{I}{I_{\text{ref}}}\right)^2 \cdot \theta_{\text{ref}}\right] \cdot \left(1 - e^{\frac{-\Delta T}{\tau_2}}\right) \tag{4.28}$$

Substituting (4.16) into (4.28) yields:

$$\Delta\theta = p \cdot \theta_f \cdot \left(1 - e^{\frac{-\Delta T}{\tau_1}}\right) + (1-p) \cdot \theta_f \cdot \left(1 - e^{\frac{-\Delta T}{\tau_2}}\right) \tag{4.29}$$

To simplify the model, we can set the weighting factor ($p = 0$) so that the short time constant is vanished and the single equivalent time constant will equal the long time constant:

$$\Delta\theta = \left[\left(\frac{I}{I_{\text{ref}}}\right)^2 \cdot \theta_{\text{ref}}\right] \cdot \left(1 - e^{\frac{-\Delta T}{\tau}}\right) \tag{4.30}$$

or

$$\Delta\theta = \theta_f \times \left(1 - e^{\frac{-\Delta T}{\tau}}\right) \tag{4.31}$$

where

$$\tau = \tau_2 \text{ (long time constant)} \tag{4.32}$$

$$\theta_n = \theta_{\text{rise}} \text{ (temperature rise)} \tag{4.33}$$

7 The ANSI device number is 49T.

In Relion series relays, θ is calculated as follows (Smith, 2015):

$$\theta = \frac{\theta_T}{100\%} \times \theta_{\text{max}} - \theta_{\text{env}} \text{ (operate value)} \tag{4.34}$$

Substituting (4.8), and (4.32)–(4.34) in (4.7) yields:

$$T = \tau_2 \times \ln\left[\frac{I^2 - I_{\text{pre}}^2}{I^2 - \left(I_{\text{ref}}\sqrt{\frac{\frac{\theta_T}{100\%}\times\theta_{\text{max}} - \theta_{\text{env}}}{\theta_{\text{rise}}}}\right)^2}\right] \tag{4.35}$$

Example 4.5 Write a MATLAB code to plot the time–temperature curve for protecting a transformer against excessive heat. Use the numerical data reported in Smith and Jain (2017), which is listed as follows:

$\tau_1 = 306$ s, $\tau_2 = 4920$ s, $I_{\text{ref}} = 1$ pu, $I = 1.5$ pu, $\theta_{\text{env}} = 40\,°\text{C}$, $\theta_{\text{ref}} = 65\,°\text{C}$, $\theta_{\text{max}} = 105\,°\text{C}$, $\Delta T = 0.05$ s, and $p = 0.4$

Calculate the trip time from the curve via the data cursor tool available in the figure.

Solution

The code given in `ThermalOverloadTransformer.m` is:

```
1  clc
2  clear
3  Tau1 = 306; % short time constant
4  Tau2 = 4920; % long time constant
5  theta_env = 40; % ambient temperature
6  theta_max = 105; % maximum temperature
7  theta_ref = 65; % reference temperature
8  I = 1.5; % highest phase current
9  Iref = 1; % reference current
10 theta_final = (I/Iref)^2*theta_ref; % final temperature (rise from ...
       ambient)
11 dT = 0.05; % difference in time
12 T = 0:dT:1000; % time period
13 p = 0.4; % weighting factor
14 theta1old = 0;
15 theta2old = 0;
16 for i = 1:length(T)
17     theta1(i) = theta1old + (theta_final - theta1old)*(1 - ...
           exp(-dT/Tau1));
18     theta2(i) = theta2old + (theta_final - theta2old)*(1 - ...
           exp(-dT/Tau2));
19     theta(i) = theta_env + p*theta1(i) + (1-p)*theta2(i);
20     theta1old = theta1(i); % update the cable temperature
21     theta2old = theta2(i); % update the cable temperature
22     if theta(i) ≥ theta_max
23         break
24     end
25 end
26 plot(T(1:length(theta)), theta, 'r', 'LineWidth', 2)
27 grid on; grid minor
28 set(gca, 'FontSize', 13)
```

```
29        xlim([0, 800]); ylim([0, 120])
30        title('Thermal Rise vs Time')
31        xlabel('Time in Seconds')
32        ylabel('Rise Temperature in ^{\circ}C')
```

If we run this code, the plot shown in Figure 4.3 can be generated. It can also be reached by opening `ThermalOverloadTransformer.fig`. The trip time calculated from the data cursor is 723.5 s.

From these examples, we have seen that the actual temperature is calculated using thermal overload protection. Based on this, the thermal image of the protected component is replicated. Thus, the thermal relay can initiate the trip signal when the calculated temperature exceeds the maximum temperature threshold defined in the relay. Compared with conventional non-directional OCRs, thermal overload relays have a thermal memory so that they continuously calculate the actual temperature of power components. That is, OCRs will fully reset after overload condition subsides, which might lead to insulation failures, while thermal relays do no. What we want to say is that each protective device should have a maximum trip time threshold. This limit should not be exceeded to avoid bad consequences on the entire power system.

4.2.2 Stability Analysis

The other point that should be taken into account, for setting $T^{\max}$, is system stability. It is one of the most important studies in electric power systems engineering, especially for generating units. Figure 4.4 shows a list of devices and functions used to protect ABB gas turbine – model GT13DM.

Studying power system stability needs a complete book. The reader can refer to (Kundur, 1994; Kimbark, 1995; Grigsby and Grigsby, 2007; Padiyar, 2008). In the next few pages, we will cover just the part needed to adjust $T^{\max}$ from the stability perspective. For this mission, we will focus on studying the importance of fault protection in transient system stability. A simple **single-machine infinite bus**[8] **(SMIB)** system is used. Because it is an optional topic, so we will cover it as a case study. The SMIB problem is analyzed by using the **equal-area criterion** method. The goal is to show that the system could lose its entire stability if the fault is not cleared before the **critical time (T_{cr})**. Because the fault severity increases as the fault location decreases, so the value of T_{cr} should be inversely proportional to the location of that fault. Let's take the following example:

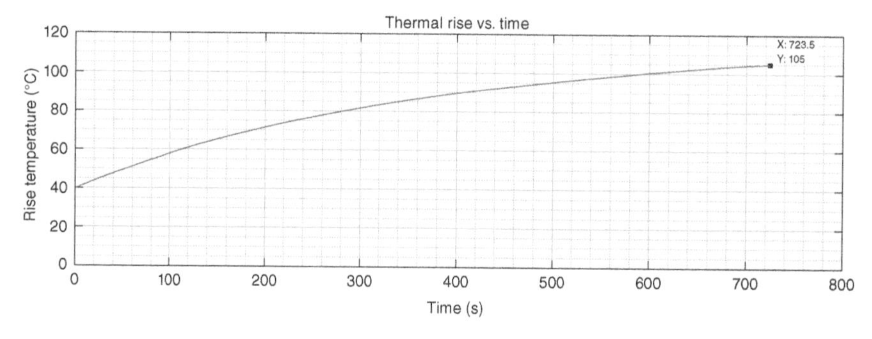

Figure 4.3 Illustration of temperature rise to trip for transformer thermal overload protection simulated in Example 4.5.

8 It is also known as a **one-machine infinite bus (OMIB)** in some references.

ANSI	Function	Input (DI) or direct function (F)	System	Line circuit breaker	Generator load breaker	Field circuit breaker	GT automatic program OFF command	Unit board circuit breaker	Remark
87G	Generator differential	F / F	A / B	•		•	•	•	b
24G.1	V/Hz protection	F / F	A / B						Alarm
24G.2	V/Hz protection	F / F	A / B		•	•			a, b
59.1	Over voltage 1	F / F	A / B		•	•	•		a
59.2	Over voltage 2	F / F	A / B		•	•	•		a
46.1	Negative sequence 1	F / F	A / B						b, Alarm
46.2	Negative sequence 2	F / F	A / B	•	•	•	•	•	a
59NB.1	Starting earth fault	F / F	A / B						Alarm, blocked if GLB is closed
59NB.2	Bus duct earth fault	F / F	A / B	•		•	•	•	Blocked if GLB is open
51V	Voltage restraint O/C	F / F	A / B	•		•	•	•	Logic between 51 and 27
21	Mininum impedance	F / F	A / B	•		•	•	•	b or VTS (60)
59GN	95% stator earth fault	F / F	A / B	•		•	•	•	
32.1	Reverse power 1 (high set.)	F / F	A / B		•	•			a, blocked if GLB is open
32.2	Reverse power 2 (low set.) for normal shut-down	F / F	A / B		•				a, blocked if GLB is open
40G.1	Loss of excitation	F / F	A / B	•		•	•	•	b or VTS (60)
40G.2	Loss of excitation with time integrator	F / F	A / B	•		•	•	•	b or VTS (60)
81.1	Under frequency 1	F / F	A / B						Alarm, blocked if GLB is open or U < 6
81.2	Under frequency 2	F / F	A / B	•					Blocked if GLB is open or U < 60V
81.3	Under frequency 3	F / F	A / B	•	•	•	•	•	a, blocked if U < 60V
81.O	Over frequency	F / F	A / B						Alarm
49G.1	Overload (stator winding temperature)	F / F	A / B						Alarm
49G.2	Overload (stator winding temperature)	F / F	A / B		•				a
BF	Generator load breaker failure	F / F	A / B	•		•	•	•	c
50/27 (41/51)	Inadvertent GLB closing	F / F	A / B	•		•	•	•	
27	Under voltage protection	F / F	A / B						Alarm
50E/51E	Excitation transformer over current protection	DI / DI	A / B		•	•	•		
87T	Transformer differntial	F	B	•		•	•	•	
24T	V/Hz protection	F	B	•					
63T	Buchholz relay 2	DI	B	•		•	•	•	
50T	Over current inst. time	F / F	A / B	•		•	•	•	
49T	Winding temperature	DI	B	•		•	•	•	
63TP	Pressure relief relay	DI	B	•		•	•	•	
26T	Oil temperature relay	DI	B	•		•	•	•	
87REF	Restricted earth fault	F	A	•		•	•	•	
51T + 67T	Directional, over current def. time protection	F	A	•		•	•	•	
87UT	Unit transformer differential	F	A	•		•	•	•	
63UT	Buchholz relay 2	DI	A	•		•	•	•	
26UT	Oil temperature relay	DI	A	•		•	•	•	
50UT	Over current inst. time	F	A	•		•	•	•	
51UT	Over current def. time	F	A	•		•	•	•	
50UB	Unit board over current inst. time	F	B					•	
51UB	Unit board over current def. time	F	B					•	
64F.1	Rotor E/F stage 1	F	A						Alarm
64F.2	Rotor E/F stage 2	DI / DI	A / B		•	•	•		a
60	Voltage balance/supervison	DI / DI	A / B						Alarm
NA	Voltage transformer	DI	A						Alarm (VT52, MCB)
NA	Voltage transformer	DI	B						Alarm (VT53, MCB)
NA	Start disconnector	DI / DI	A / B						Closed (interlocking)
NA	Generator emergency trip	DI / DI	A / B		•	•	•		Push buttons
NA	Generator load breaker	DI / DI	A / B						Closed (interlocking)
40E	Excitation failure	DI / DI	A / B		•	•	•		From excitation
NA	High voltage failure	DI / DI	A / B	•	•	•		•	From HVSWYD
NA	GT surge protection	DI / DI	A / B	•					From DCS

Remarks:
a - The Generator Load Breaker Tripping is Blocked if Over current function I > 2 is present.
b - The protection function is blocked if the Start Disconnector is closed.
c - Breaker failure function will be active if the Load breaker fails to open and also if the failure current exceed the capacity of the load breaker.

Figure 4.4 List of devices and functions used to protect ABB gas turbine - model GT13DM.

Example 4.6 Consider the SMIB problem shown in Figure 4.5. By referring to Appendix C, do the transient system stability analysis for both fault locations using the equal-area criterion method.

Solution

From Appendix C:

$$X_e = \frac{X_{e_1} X_{e_2}}{X_{e_1} + X_{e_2}} = 0.3 \text{ pu}, \quad X_l = X_{tr} + X_e = 0.45 \text{ pu}, \quad \text{and}$$

$$X_{tot} = X'_d + X_l = 0.75 \text{ pu}$$

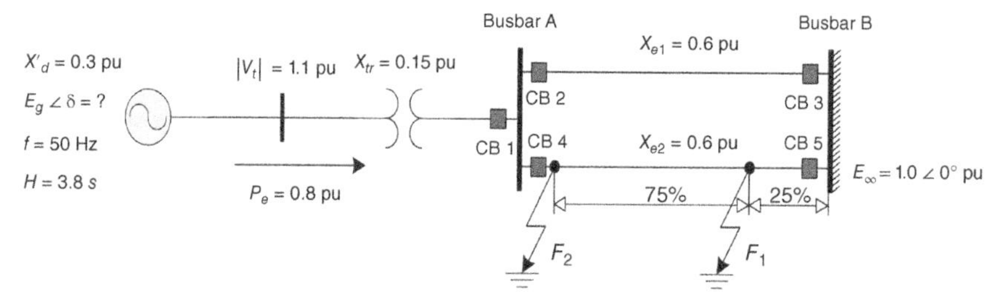

Figure 4.5 Single-machine infinite bus (SMIB).

$$\because\ P_e = \frac{|V_t|\,|E_\infty|}{X_l}\sin\theta$$

$$\therefore\ P_e = \frac{1.1(1.0)}{0.45}\sin\theta$$

$$\therefore\ \theta = 19.1033°$$

$$\therefore\ V_t = |V_t|\angle\theta = 1.1\angle 19.1033°\ \text{pu}$$

$$\because\ I = \frac{|V_t|\angle\theta - |E_\infty|\angle 0}{jX_l}$$

$$\therefore\ E_g = V_t + jX_d'I = 1.2230\angle 29.3798°\ \text{pu}$$

$$\therefore\ |E_g| = 1.2230\ \text{pu}\ \text{ and }\ \delta_o = 29.3798° = 0.512\,774\ \text{rad}$$

- **During Pre-Fault:**

$$\because P_e = \frac{|E_g|\,|E_\infty|}{X_{\text{tot}}}\sin\delta_o = P_{1_{\max}}\sin\delta_o \Rightarrow\ \therefore P_{1_{\max}} = \frac{P_e}{\sin\delta_o} = 1.630\,67\ \text{pu}$$

By re-calling the swing equation from (Padiyar, 2008; Grainger and Stevenson, 1994):

$$\frac{2H}{\omega_s}\frac{d^2\delta}{dt^2} = P_m - P_e = P_m - P_{1_{\max}}\sin\delta$$

where

P_m: shaft power input to the synchronous machine, MW
P_e: electrical power crossing the air gap, MW
$P_{\max}$: maximum electrical power can be produced, MW
ω_s: synchronous speed of the machine, rad/s
δ: angular displacement of the rotor, rad or degree
H: inertia constant of the machine, MJ/MVA

During the steady-state condition, the mechanical power (P_m) is equal to the electrical power (P_e):

$$P_m = P_e = P_{1_{\max}}\sin\delta_o = 0.8\ \text{pu}$$

$$\frac{d^2\delta}{dt^2} = \frac{\omega_s}{2H}P_m - \frac{\omega_s}{2H}P_{1_{\max}}\sin\delta = 10.5263\pi - 21.4562\pi\sin\delta$$

Simulink ODE[9] Solver needs differential equations, so let's take $\delta = x_1$ and $\frac{dx_1}{dt} = x_2$:

$$\because \frac{d\delta}{dt} = \omega - \omega_s$$

$$\therefore \frac{dx_1}{dt} = x_2$$

$$\therefore \frac{dx_2}{dt} = \frac{d\omega}{dt} = \frac{d^2\delta}{dt^2} = 10.5263\pi - 21.4562\pi \sin\delta$$

- **During Fault:**
  * If the fault occurs at F_1 (i.e. 75% away from the busbar A):

    During F_1, the circuit shown in Figure 4.5 is converted to the one shown in Figure 4.6. This circuit can be solved by using the Thevenin equivalent circuit of SMIB as covered in Glover et al. (2012) or by using the $Y - \Delta$ transformation technique as covered in Saadat (1999). Another easier approach can be done by constructing the network admittance matrix as follow (Grainger and Stevenson, 1994):

$$\mathbf{Y}_{\mathrm{BUS}} = \begin{bmatrix} -j2.222\,22 & 0 & j2.222\,22 \\ 0 & -j8.333\,34 & j1.666\,67 \\ j2.222\,22 & j1.666\,67 & -j6.111\,11 \end{bmatrix}$$

To represent it as a two-port network, node three, i.e. busbar A, should be eliminated. Thus, the admittance matrix can be reduced as follows:

$$\mathbf{Y}_{\mathrm{BUS_{red}}} = \begin{bmatrix} -j2.2222 - \dfrac{j2.2222(j2.2222)}{-j6.1111} & 0 - \dfrac{j1.6667(j2.2222)}{-j6.1111} \\ 0 - \dfrac{j2.2222(j1.6667)}{-j6.1111} & -j8.3334 - \dfrac{j1.6667(j1.6667)}{-j6.1111} \end{bmatrix}$$

$$= \begin{bmatrix} 1.414\,14\angle - 90^\circ & 0.606\,061\angle 90^\circ \\ 0.606\,061\angle 90^\circ & 7.878\,79\angle - 90^\circ \end{bmatrix}$$

By this [2 × 2] matrix, the problem is converted to two-port network as shown in Figure 4.7, and P_e can be easily computed by using the following general expression (Grainger and Stevenson, 1994; Saadat, 1999):

$$P_i = \sum_{j=1}^{\text{No. of buses}} |E_i|\,|E_j|\,|Y_{\mathrm{BUS_{red}}}(i,j)| \cos\left(\delta_i - \delta_j - \theta_{i,j}\right)$$

$$\therefore P_1 = P_e = |E_g E_\infty Y_{1,1}| \cos\left(\delta - \delta - \theta_{1,1}\right) + |E_g E_\infty Y_{1,2}| \cos\left(0 - \delta - \theta_{1,2}\right)$$

$$\therefore P_e = 0.741\,213 \sin\delta \Rightarrow \therefore P_{2_{\max}} = 0.741\,213 \text{ pu}$$

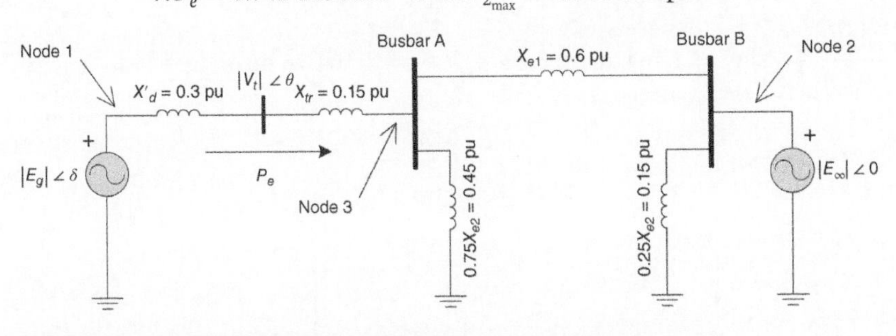

Figure 4.6 Single-line diagram of the SMIB problem when F_1 occurs.

9 ODE stands for ordinary differential equation.

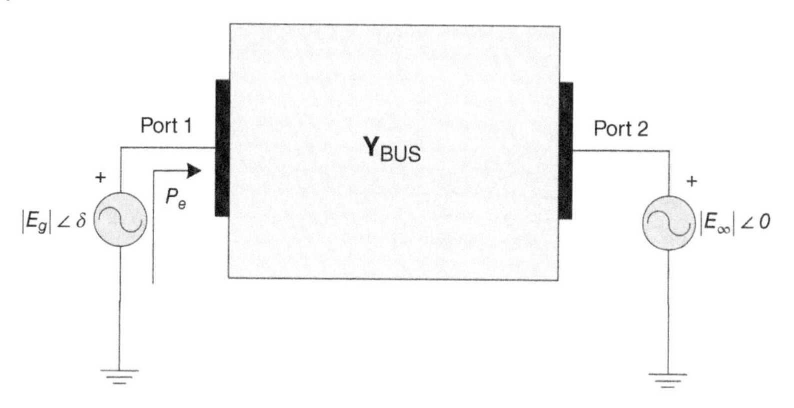

Figure 4.7 Expressing the SMIB network as a two-port network.

The swing equation becomes:

$$\because \frac{d\delta}{dt} = \omega - \omega_s \text{ and } \frac{2H}{\omega_s}\frac{d^2\delta}{dt^2} = P_m - P_e$$

$$\therefore \frac{dx_1}{dt} = x_2 = \omega - 100\pi \text{ and } \frac{d^2x_1}{dt^2} = \frac{dx_2}{dt} = 10.5263\pi - 9.7528\pi \sin\delta$$

* If the fault happens at F_2 (i.e. 0% away from the busbar A):

Now, let's consider F_2, which occurs exactly at busbar A, then Figure 4.5 is converted to the one shown in Figure 4.8. This circuit can be solved easily as follows:

$$\because P_e = \frac{|E_g|\,|E_x|}{X'_d + X_{tr}} \sin\delta = 0$$

$$\therefore P_{2_{max}} = 0$$

Thus, the swing equation becomes:

$$\because \frac{d\delta}{dt} = \omega - \omega_s \text{ and } \frac{2H}{\omega_s}\frac{d^2\delta}{dt^2} = P_m$$

$$\therefore \frac{dx_1}{dt} = x_2 = \omega - 100\pi \text{ and } \frac{d^2x_1}{dt^2} = \frac{dx_2}{dt} = 10.5263\pi$$

- **During Post-Fault (for both cases, F_1 and F_2):**

  If the trip signal is sent from the two-end protective relays to CB_4 and CB_5, respectively, then the faulty line can be taken out of service. Thus, the network shown in Figure 4.5 will change to the one shown in Figure 4.9. Now, let's apply the preceding steps:

$$\because P_e = \frac{|E_g|\,|E_\infty|}{X'_d + X_{tr} + X_{e_1}} \sin\delta = 1.164\,76 \sin\delta \Rightarrow \therefore P_{3_{max}} = 1.164\,76 \text{ pu}$$

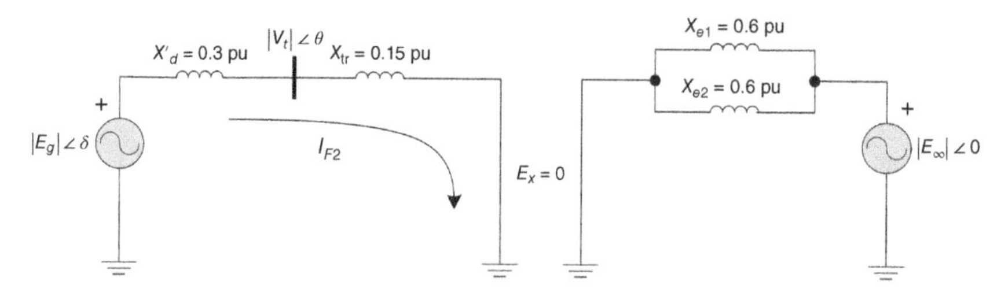

Figure 4.8 Single-line diagram of the SMIB problem when F_2 occurs.

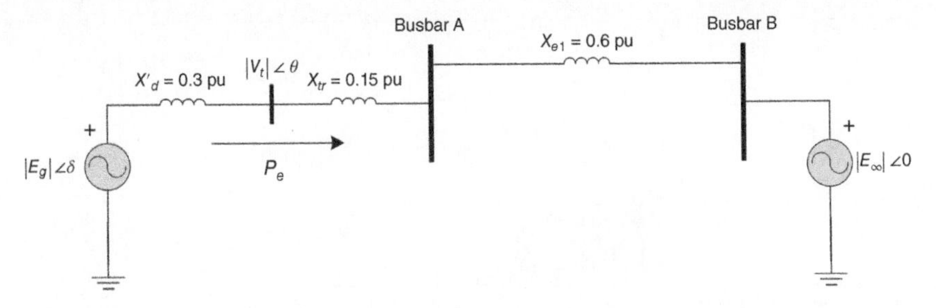

Figure 4.9 The network of SMIB during post-fault condition.

The swing equation becomes:

$$\frac{dx_1}{dt} = x_2 = \omega - 100\pi \text{ and } \frac{d^2x_1}{dt^2} = \frac{dx_2}{dt} = 10.5263\pi - 15.3258\pi \sin\delta$$

The main results extracted from all these steps are:

- **For a fault at F_1:**

$$P_{\max} = \begin{cases} P_{1_{\max}} = 1.630\,67 \text{ pu} & \text{during pre-fault} \\ P_{2_{\max}} = 0.741\,213 \text{ pu} & \text{during fault} \\ P_{3_{\max}} = 1.164\,76 \text{ pu} & \text{during post-fault} \end{cases}$$

- **For a fault at F_2:**

$$P_{\max} = \begin{cases} P_{1_{\max}} = 1.630\,67 \text{ pu} & \text{during pre-fault} \\ P_{2_{\max}} = 0 \text{ pu} & \text{during fault} \\ P_{3_{\max}} = 1.164\,76 \text{ pu} & \text{during post-fault} \end{cases}$$

The system stability can be preserved if the fault is cleared by taking the faulty line out of service. This can be done by opening the circuit breakers of both ends (i.e. CB_4 and CB_5) within an allowable time, which is less than or equal to t_{cr}. Thus, it is important to know t_{cr}, which can be determined by finding the **critical angle (δ_{cr})** when the equal-area criterion problem shown in Figure 4.10 is solved as follows:

- **Equal-Area Criterion for F_1:** By using the integration technique, the first area (A_1) can be calculated as follows:

$$A_1 = \int_{\delta_o}^{\delta_{cr}} \left(P_m - P_{2_{\max}} \sin\delta\right) d\delta = \left[P_m\delta + P_{2_{\max}} \cos\delta\right]_{\delta_o}^{\delta_{cr}}$$
$$= P_m\delta_{cr} + P_{2_{\max}} \cos\delta_{cr} - P_m\delta_o - P_{2_{\max}} \cos\delta_o$$

Similarly, the second area (A_2) can be calculated as follows:

$$A_2 = \int_{\delta_{cr}}^{\delta_{\max}} \left(P_{3_{\max}} \sin\delta - P_m\right) d\delta$$
$$= -P_{3_{\max}} \cos\delta_{\max} - P_m\delta_{\max} + P_{3_{\max}} \cos\delta_{cr} + P_m\delta_{cr}$$

From Figure 4.10:

$$\delta_{\max} = \pi - \delta_1, \text{ where } \delta_1 = \sin^{-1}\left(\frac{P_m}{P_{3_{\max}}}\right)$$

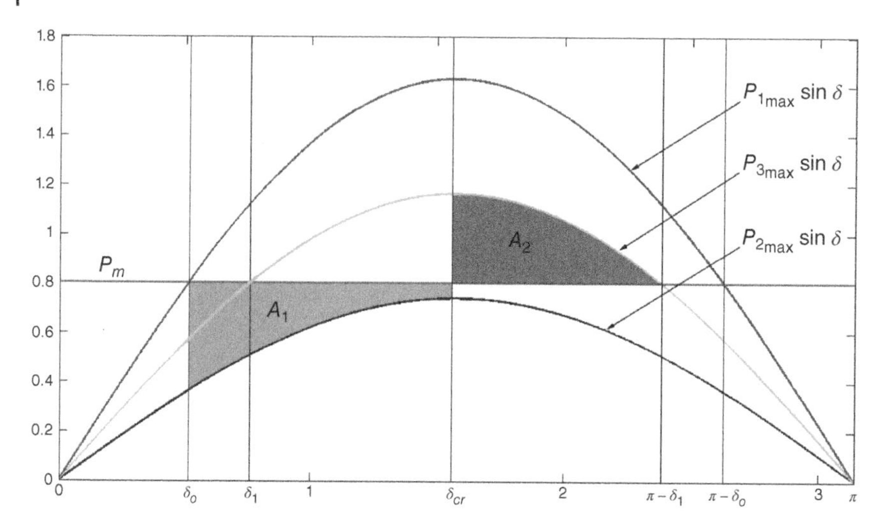

Figure 4.10 SMIB with equal-area criterion for F_1.

For $A_1 = A_2$:

$$P_m\delta_{\text{cr}} + \left(P_{3_{\max}} - P_{2_{\max}}\right)\cos\delta_{\text{cr}} = P_m\delta_{\text{cr}} + P_{3_{\max}}\cos\delta_{\max} + P_m\left(\delta_{\max} - \delta_o\right) - P_{2_{\max}}\cos\delta_o$$

$$\therefore\ \delta_{\text{cr}} = \cos^{-1}\left[\frac{P_{3_{\max}}\cos\delta_{\max} + P_m\left(\delta_{\max} - \delta_o\right) - P_{2_{\max}}\cos\delta_o}{\left(P_{3_{\max}} - P_{2_{\max}}\right)}\right]$$
$$= 89.3370^\circ = 1.559\ 22\ \text{rad}$$

To find $\delta(t)$, the second integration is applied to $\frac{d^2\delta}{dt^2}$, so:

$$\frac{d\delta}{dt} = \int \frac{d^2\delta}{dt^2}\,dt = \int \frac{\omega_s}{2H}\left(P_m - P_e\right)\,dt = \frac{\omega_s}{2H}\left(P_m - P_e\right)t + C$$

$$\delta(t) = \int \frac{d\delta}{dt}\,dt = \frac{\omega_s}{4H}\left(P_m - P_e\right)t^2 + \delta_o$$

At $t = t_{\text{cr}} \rightarrow \delta(t) = \delta_{\text{cr}}$, thus:

$$t_{\text{cr}} = \sqrt{\frac{\delta_{\text{cr}} - \delta_o}{\frac{\omega_s}{4H}\left(P_m - Pe\right)}}$$

It is clear that t_{cr} cannot be determined analytically, because of the nonlinearity of the swing equation (Saadat, 1999). Therefore, it has to be solved numerically. By simulating the system in MATLAB,[10] it has been found that $t_{\text{cr}} \approx 0.394\ 35$ s. Therefore, the protection system has a chance of 19.7175 cycles to isolate the fault before losing the system stability.

10 As will be seen later.

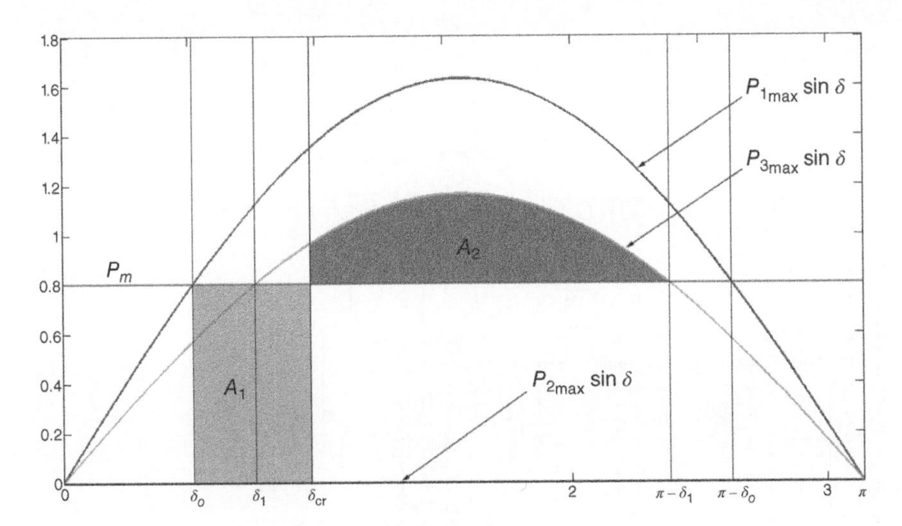

Figure 4.11 SMIB with equal-area criterion for F_2.

- **Equal-Area Criterion for F_2:** This fault will give $P_{2_{max}} = 0$, see Figure 4.11.
  By analogy:

$$\therefore \delta_{cr} = \cos^{-1}\left[\frac{P_{3_{max}}\cos\delta_{max} + P_m\left(\delta_{max} - \delta_o\right)}{P_{3_{max}}}\right] = 56.0321° = 0.977\,945 \text{ rad}$$

$$t_{cr} = \sqrt{\frac{\delta_{cr} - \delta_o}{\frac{\omega_s}{4H}P_m}} = 0.167\,729 \text{ s} = 167.729 \text{ ms}$$

Therefore, as the fault approaches the source, its severity will increase, and at the same time the chance available to the protection system to isolate that fault decreases. For this fault, the system stability can be preserved if the isolation process does not exceed 6.836 45 cycles.

Example 4.7 Using MATLAB and Simulink, simulate the results obtained in the previous example.

Solution

The effect of fault location on the stability of the SMIB system can be simulated by using Simulink's differential equation editor (DEE) as shown in Figure 4.12. The maximum real power (P_{max}) before, during, and after the fault occurrence are injected to DEE through the Signal Builder block. Also, the clearing time can be set by adjusting the width of each P_{max} in the previous block. All the project files are placed in a folder called `SMIB`.

If $t_{clear} = 0.10$ s, then the stability can be preserved for both faults:

$$t_{clear} < t_{cr,F_2} < t_{cr,F_1} \Rightarrow 100.000 \text{ ms} < 167.729 \text{ ms} < 394.350 \text{ ms}$$

If $t_{clear} = 0.17$ s, then the stability can be lost if F_2 occurs:

$$t_{cr,F_2} < t_{clear} < t_{cr,F_1} \Rightarrow 167.729 \text{ ms} < 170.000 \text{ ms} < 394.350 \text{ ms}$$

As the fault location is very far from the source, the system stability can be restored even with longer time. This can be observed if $t_{clear} = 0.39$ s because still $t_{clear} < t_{cr,F_1}$. Figure 4.13 shows all these conditions for both fault locations.

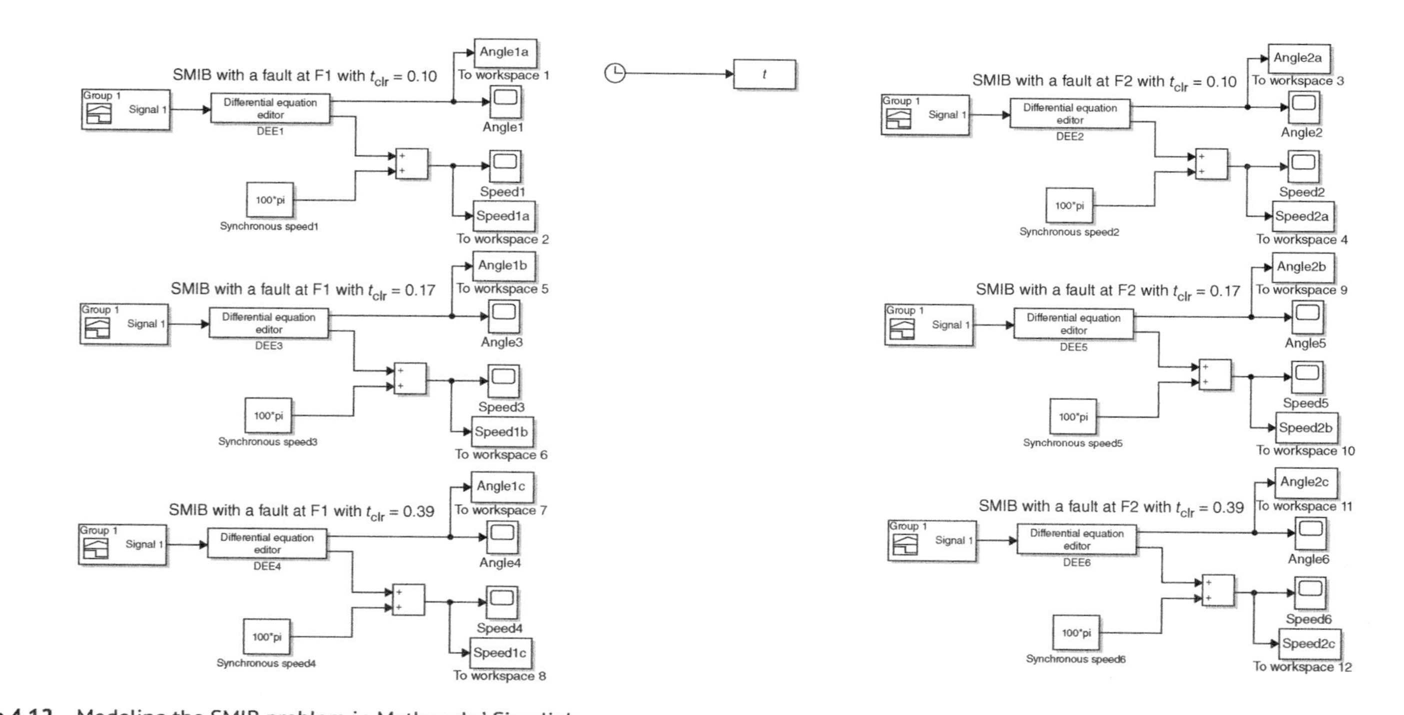

Figure 4.12 Modeling the SMIB problem in Mathworks' Simulink.

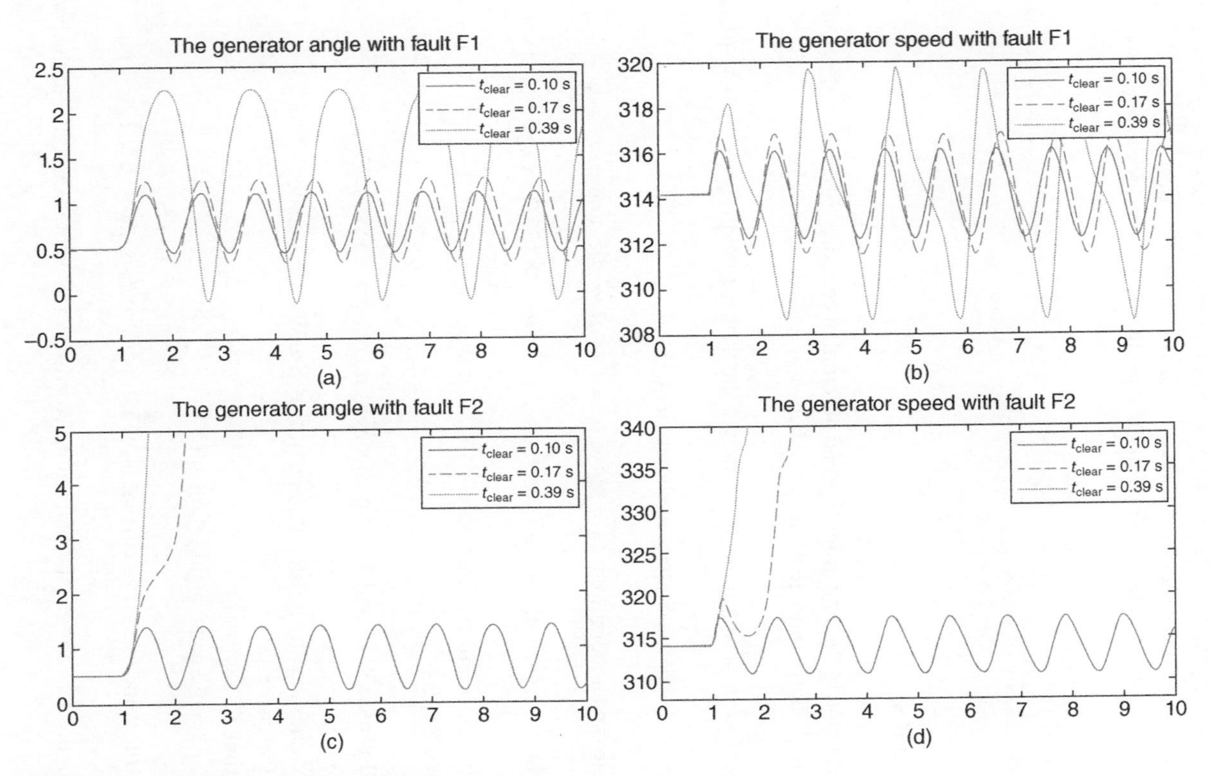

Figure 4.13 Simulink's results of SMIB with two different fault locations.

It can be clearly seen that the isolation speed of any protection system against any abnormal condition in the network is a very important issue for protection engineers. The isolation time can be minimized by reducing the time delay of sensing elements, the calculation speed of algorithms,[11] and the operating time of protective devices and/or the tripping speed of circuit breakers.

This book will mainly focus on one part of these factors, which is the minimization of the total operating time when the protective relays are optimally coordinated. The other factors depend on the technology used, and thus we don't have any control over them. However, Chapters 9, 10, 11, 12, 13, 16, and 17 are dedicated to solving different ORC problems when relay technology and TCCC are subject to change.

Problems

Written Exercises

W4.1 The derivation of the first order thermal differential equation given in (4.2) is available in Appendix E. Solve this equation and match your answer with (4.3).

W4.2 Consider (4.3) as a solution to (4.2). Derive the three equations given in (4.4)–(4.6).

W4.3 From (4.3), derive the expression shown in (4.7).

W4.4 Calculate the temperature if $\theta_0 = 35\,°\text{C}$, $\theta_f = 120°$, $T = 30$ s, and $\tau = 400$ s.

W4.5 Using the data given in Exercise W4.4, what is τ if $\theta = 50\,°\text{C}$? Check your answer.

W4.6 Calculate the nominal temperature if $I_0 = 0.8$ pu, $I_n = 1$ pu, $I = 1.5$ pu, $T = 8$ min, $\tau = 400$ s, and $\theta = 50\,°\text{C}$.

W4.7 Repeat Exercise W4.6 with $I = 2$ pu. Does θ_n increase or decrease? Why?

W4.8 Based on the data and result obtained in Exercise W4.7, calculate:
1. The trip current.
2. The time.

(**Hint** : T should equal eight minutes.)

W4.9 What is the value of the trip current if the ratio of θ to θ_0 is 200%? Use $I_n = 1$ pu.

W4.10 Based on the data and result obtained in Exercise W4.9, what is the value of the overload factor?

W4.11 Find T if $\tau = 150$ s, $I = 2.5$ pu, and $I_0 = 0.85$ pu. Use I_T calculated in Exercise W4.9. Why is it much faster than that obtained in Exercise W4.8?

11 For numerical relays and IEDs.

W4.12 A motor is protected by a thermal overload relay. The overload factor is 150% and the weighting factor is 100%. If the values of I, I_{pre}, and I_{rated} are respectively equal to 2.85, 0.87, and 1 per unit, calculate T if $\tau_{\text{start}} = 800$ s and $\tau_{\text{normal}} = 320$ s. (**Hint:** Use only one time constant in your calculation.)

W4.13 Suppose a thermal overload relay is used to protect a transformer. The time constant used for windings is 350 s and for oil is 5000 s. The surrounding temperature is 30 °C and the transformer nominal temperature is 45 °C measured at 1 per unit of current. If the maximum phase current is 1.25 per unit with 100 °C as a maximum temperature:

1. Calculate θ_f
2. Calculate $\Delta\theta$ when:
   - $p = 0.4$
   - $p = 0.0$
3. Calculate θ when $\theta_T = 95\%$
4. Calculate T when $I_{\text{pre}} = 0.8$ pu

Take $\Delta T = 1$ minute.

W4.14 Consider the SMIB system shown in Figure 4.14. Assume that the internal voltage magnitude (E_g) is 1.2 pu and the infinite bus voltage magnitude is 1 pu. The active component of the load[12] on the circuit is 3 pu when a 3ϕ short-circuit fault occurs at the middle of transmission line 3.

1. Find the initial power angle (δ).
2. Will the system remain stable under a sustained fault?

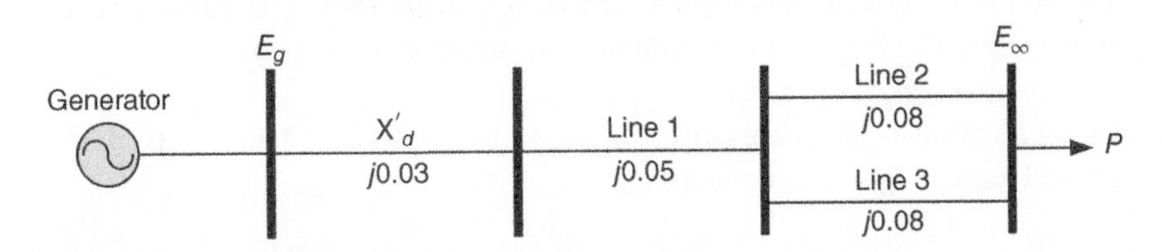

Figure 4.14 The SMIB system used in Exercise W4.14.

W4.15 Repeat Exercise W4.14 with $E_g = 1.44$ pu, then determine the maximum angle of oscillation under that sustained 3ϕ short-circuit fault.

W4.16 Repeat Exercise W4.14 with $E_g = 1.5$ pu and $P_{e_0} = 3.8$ pu.

W4.17 Repeat Exercise W4.14 with $E_g = 1.5$ pu and $P_{e_0} = 4.5$ pu.

W4.18 Consider the SMIB system shown in Figure 4.15. Assume that the internal voltage magnitude (E_g) is 1.5 pu and the infinite bus voltage magnitude is 1 pu. The active component of the load on the circuit is 2.4 pu when a 3ϕ short-circuit fault occurs at the middle of transmission line 3.

---

12 That is, the electrical power (P_{e_0}) equals the mechanical power (P_{M_0}).

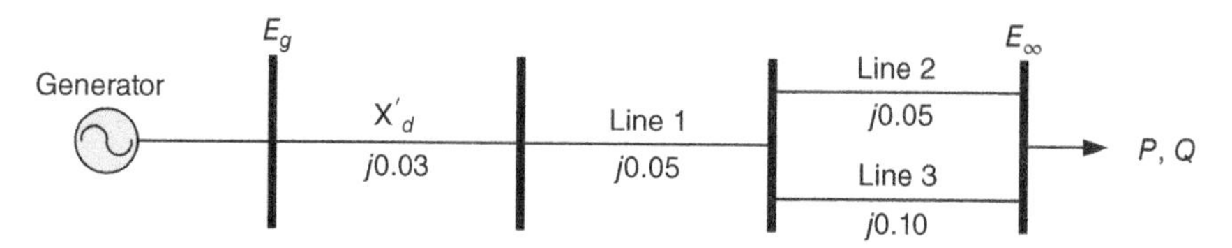

Figure 4.15 The SMIB system used in Exercise W4.18.

1. Find the initial power angle (δ) and the reactive component of the load.
2. Show analytically that the system will remain stable under a sustained fault.
3. Determine the maximum angle of oscillation under the preceding sustained 3ϕ short-circuit fault.

Computer Exercises

C4.1 Modify the code given in Example 4.2 to plot T as a function of I at two different time constants. The new data is:

$$\tau_1 = 250 \text{ s}, \tau_2 = 3600 \text{ s}, \theta_{\text{env}} = 40\,°\text{C}, \theta_{\text{rise}} = 65\,°\text{C}, \theta_{\text{max}} = 105\,°\text{C}, I_{\text{ref}} = 1 \text{ pu},$$
$I_{\text{pre}} = 0.8$ pu, $I \in [1, 2.5]$ pu, and $\Delta I = 0.01$ pu. Use a log-scale on the y-axis.

C4.2 Modify the code given in Example 4.2 to plot T as a function of two variables. The first variable is θ_{rise} and the second one is I where their domains are $\theta_{\text{rise}} \in [40,100]$ and $I \in [1.5, 3]$. The resolution is 1000 steps from the lower to the upper bound of each variable. Take the remaining information as follows: $\tau = 1500$ s, $\theta_{\text{env}} = 30\,°\text{C}$, $\theta_{\text{max}} = 105\,°\text{C}$, $I_{\text{ref}} = 1$ pu, and $I_{\text{pre}} = 0.775$ pu. Use the `meshc` command to generate your 3D plot.

C4.3 Repeat Exercise C4.2 with two time constants; $\tau = 306$ s and $\tau = 4920$ s. The two 3D plots should be generated with a log-scale on the z-axis.

C4.4 Write a MATLAB code to plot the trip current (I_T) as a function of the trip temperature (θ_T). Take $I_{\text{rated}} = 1$ pu, $\theta_{\text{env}} = 37\,°\text{C}$, $\theta_{\text{rated}} = 110\,°\text{C}$, $\theta_T \in [135, 225]\,°\text{C}$, and $\Delta\theta = 0.01\,°\text{C}$.

C4.5 Write a MATLAB code to plot the trip current (I_T) as a function of the rated temperature (θ_{rated}). Take $I_{\text{rated}} = 1$ pu, $\theta_{\text{env}} = 40\,°\text{C}$, $\theta_T = 170\,°\text{C}$, $\theta_{\text{rated}} \in [90, 110]\,°\text{C}$, and $\Delta\theta = 0.01\,°\text{C}$.

C4.6 Write a MATLAB code to re-generate the plots shown in Figures 4.10 and 4.11.

C4.7 Assume that the active power (P_e) gradually increases from 0.8 to 2.5 pu and then returns back to 0.8 pu. Write a MATLAB code to plot the power angle (δ) as a function of P_e where $\Delta P_e = 0.01$ pu. Show the plot in radian and degree.

5

Directional Overcurrent Relays and the Importance of Relay Coordination

Aim

The goal of this chapter is to give a brief idea about relay coordination. This subject is explained by covering the general concepts of relay grading using different types of overcurrent relays (OCRs); including definite-time overcurrent relays (DTOCRs), definite-current overcurrent relays (DCOCRs), and inverse-time overcurrent relays (ITOCRs). Starting from current and time gradings, the importance of inverse-time grading is addressed via solving different radial systems. Then, some relatively harder networks are covered to answer the question: *Is the coordination of protective relays an iterative problem?* Through this journey, the technical issue associated with the primitive technique used to optimize relay settings is discussed. Also, the minimum break-point set(MBPS) concept is explained. Thus, this chapter builds the foundation required to design an applicable optimal relay coordinator, which is placed in the next chapter.

From Chapter 2, the time-current characteristic curves (TCCCs) of instantaneous overcurrent relays or definite-current overcurrent relays, definite-time overcurrent relays , and inverse-time overcurrent relays (namely IOCRs/DCOCRs, DTOCRs, and ITOCRs) have been covered. Some relays could have mixed curves where one of these curves is achieved by merging inverse-time and definite-time characteristics to have what are known as inverse definite minimum time (IDMT) characteristics. Also, a comprehensive review has been covered in Chapter 3 for modeling the operating time of ITOCRs under many standards, such as BS142, IEC, IEEE, ANSI, US, CO, IAC, FR, RECT, EDF, and RI as well as RD and other customized/programmable models. This chapter covers some fundamental concepts used in relay coordination. It explains the relay grading strategies used in radial systems, why OCRs should have directional units in non-radial systems, how to find the correct settings of directional ITOCRs, the problem faced in multiple loop networks, and finally why we need to define an upper limit for relay operating times.

5.1 Relay Grading in Radial Systems

In Chapter 2, we have seen different types of OCRs and their TCCCs. For radial systems, the relays can be easily coordinated to ensure that the primary devices operate first. This can be done by selecting proper settings. The three popular relay grading methods are briefly described in the following Sections 5.1.1, 5.1.2, and 5.1.3.

Optimal Coordination of Power Protective Devices with Illustrative Examples, First Edition. Ali R. Al-Roomi.

Companion website: www.wiley.com/go/al-roomi/optimalcoordination

5.1.1 Time Grading

The goal here is to let the circuit breaker (CB) nearest to the fault opens first (El-Hawary, 1983). Thus, the operating time is decreased as the relay is mounted away from the source, and vice versa. To clarify this point, let's take the following example:

Example 5.1 Figure 5.1 represents a very simple radial network. If a fault occurs at the branch between bus 1 and bus 2, calculate the operating time of each relay. Assume that the lowest possible operating time is 0.2 s and the discrimination margin (DM) is 0.4 s.

Solution

Because the first relay (R_1) does not depend on any other protective devices, so we can set it with the lowest possible operating time:

$$\therefore T_1 = 0.2\,\text{s}$$

For the other relays, an increment of 0.4 s is added for each branch until reaching the source:

- For R_2: $T_2 = T_1 + 0.4 = 0.6\,\text{s}$
- For R_3: $T_3 = T_2 + 0.4 = 1.0\,\text{s}$
- For R_4: $T_4 = T_3 + 0.4 = 1.4\,\text{s}$

This can be represented by the bar chart shown in Figure 5.2.

From this example, it is clear that the operating time of R_4 is higher than that of R_1 regardless of the fault location. The problem here is that if a fault occurs near the source, the current will be much higher. Thus, R_4 is expected to take a fast action to isolate this dangerous fault. Instead, R_4 will wait long time before it send the trip signal to its circuit breaker (CB_4). To solve this issue, let's study the next grading method and see its inherent limitation as well.

5.1.2 Current Grading

Because the fault current is proportional to the fault distance, so the idea here is to protect the zones of radial systems based on the value of fault currents. That is, the operating time of protective relays

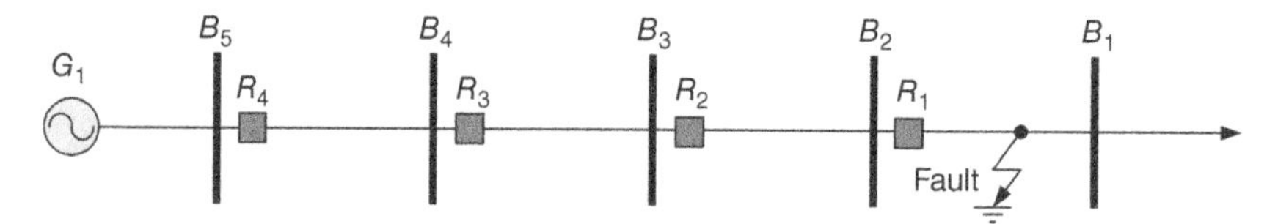

Figure 5.1 Radial system for Example 5.1.

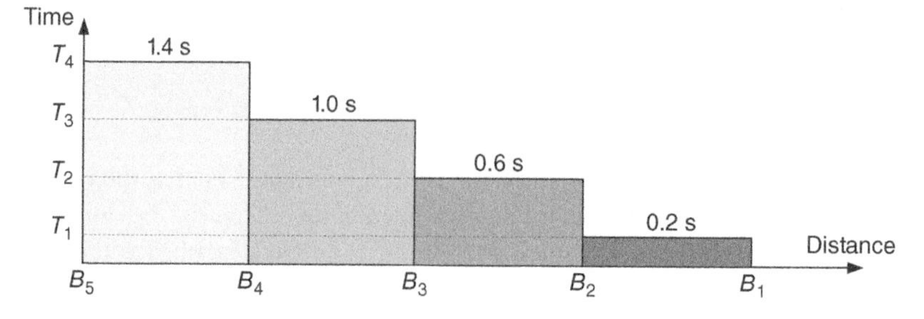

Figure 5.2 Time grading for the radial system of Example 5.1.

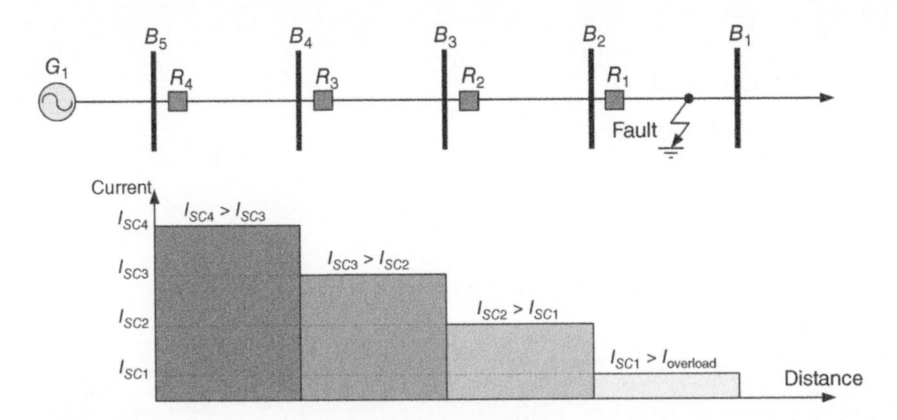

Figure 5.3 Concept of current grading in radial systems.

are set based on their distance from the source. Figure 5.3 clarify the whole idea behind the current grading method. Compared with Figure 5.2, the current magnitude is used here as an indication of the fault severity.

The technical problem associated with this kind of grading is described in Section 2.6.4. However, let's prove it by taking the following example:

Example 5.2 Let's consider the radial network shown in Figure 5.4. Assume lossless lines with reactance of $X_{21} = 0.25\ \Omega$ and $X_{32} = 0.2\ \Omega$. If the source voltage is 11 kV and the second relay has a margin of 130%, analyze the selectivity criterion of this system when:

1. $X_S = 0.4\ \Omega$
2. $X_S = 1.2\ \Omega$

Solution

The procedure is so simple. We need to calculate the 3ϕ short-circuit current based on the fault location. If the total impedance between the source and the fault point is denoted by X_{tot}, then the fault current can be calculated as follows:

$$I_{SC} = \frac{V}{X_{\text{tot}}} = \frac{V_S}{\sqrt{3}\, X_{\text{tot}}} \tag{5.1}$$

- If $X_S = 0.2\ \Omega$:

  For F_A:

$$X_{\text{tot}} = 0.2 + 0.2 + 0.25 = 0.65\ \Omega$$

$$I_{F_A} = \frac{11\ \text{kV}}{\sqrt{3}\ (0.65\ \Omega)} = 9.7705\ \text{kA}$$

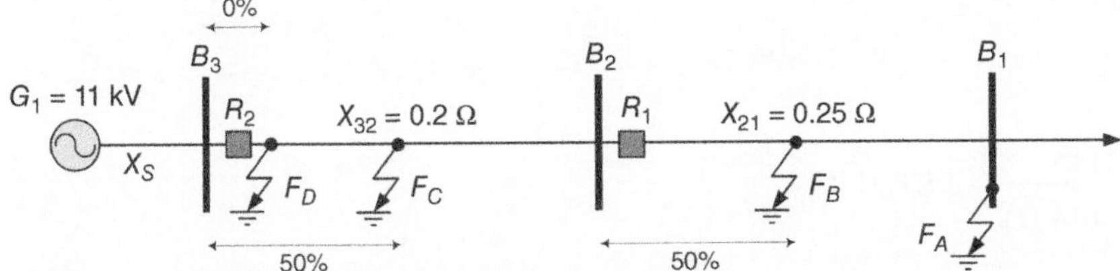

Figure 5.4 Radial system for Example 5.2

For F_B:

$$X_{tot} = 0.2 + 0.2 + 0.125 = 0.525\ \Omega$$
$$I_{F_B} = \frac{11\ \text{kV}}{\sqrt{3}\ (0.525\ \Omega)} = 12.0969\ \text{kA}$$

For F_C:

$$X_{tot} = 0.2 + 0.1 = 0.3\ \Omega$$
$$I_{F_C} = \frac{11\ \text{kV}}{\sqrt{3}\ (0.3\ \Omega)} = 21.1695\ \text{kA}$$

For F_D:

$$X_{tot} = 0.2\ \Omega$$
$$I_{F_D} = \frac{11\ \text{kV}}{\sqrt{3}\ (0.2\ \Omega)} = 31.7543\ \text{kA}$$

The first relay (R_1) should be set[1] with I_{F_A} (i.e. without considering any margin):

$$\therefore I_{R_1} = I_{F_A} = 9.7705\ \text{kA}$$

For R_2, by considering the 30% margin, the definite current can be determined as follows:

$$I_{R_2} = 1.3 I_{F_A} = 12.7017\ \text{kA}$$

By comparing I_{R_1} with I_{F_A} and I_{F_B}, it is clear that R_1 can protect the branch between bus 1 and bus 2 against the faults F_A and F_B. Similarly, by comparing I_{R_2} with I_{F_C} and I_{F_D}, the branch between bus 2 and bus 3 can be protected by R_2 when F_C or F_D occurs.

- If $X_S = 0.4\ \Omega$:

For F_A:

$$X_{tot} = 0.4 + 0.2 + 0.25 = 0.85\ \Omega$$
$$I_{F_A} = \frac{11\ \text{kV}}{\sqrt{3}\ (0.85\ \Omega)} = 7.4716\ \text{kA}$$

For F_B:

$$X_{tot} = 0.4 + 0.2 + 0.125 = 0.725\ \Omega$$
$$I_{F_B} = \frac{11\ \text{kV}}{\sqrt{3}\ (0.725\ \Omega)} = 8.7598\ \text{kA}$$

For F_C:

$$X_{tot} = 0.4 + 0.1 = 0.5\ \Omega$$
$$I_{F_C} = \frac{11\ \text{kV}}{\sqrt{3}\ (0.5\ \Omega)} = 12.7017\ \text{kA}$$

For F_D:

$$X_{tot} = 0.4\ \Omega$$
$$I_{F_D} = \frac{11\ \text{kV}}{\sqrt{3}\ (0.4\ \Omega)} = 15.8771\ \text{kA}$$

---

1 Remember that all the network fault currents must be stepped-down via CTs before sending them to the corresponding protective relays. Here, we are talking about the fault currents measured on the primary side of CTs.

Relay 2 will still detect the faults F_C and F_D when X_S increases from 0.2 to 0.4 Ω because I_{R_2} equals the new value of I_{F_C}. On the opposite side, I_{R_1} is greater than the new values of I_{F_A} and I_{F_B}. Thus, R_1 will not operate when F_A or F_B occurs.

It is clear that the current grading also fails if the source impedance is high or if the impedance between CBs is small.

5.1.3 Inverse-Time Grading

To solve the inherent weakness of the preceding relay grading, the protective relay should have the ability to discriminate between in-zone and out-zone faults. Also, its operating time should be inversely proportional to the fault location. The concept of this grading method can be illustrated in Figure 5.5.

That is, the weaknesses of DTOCRs and DCOCRs can be solved by employing ITOCRs where both the operating time and the current magnitude are considered. In another word, the OCRs used in this type of grading accept both *CTS* and *TDS* for North American relays (or *PS* and *TMS* for European relays). To clarify this point, let's take the following example:

Example 5.3 Figure 5.6 represents a radial system powered by an 11 kV unit. If all the loads have the same power factor, determine the current tap settings and time dial settings of the three OCRs mounted near each busbar. Assume that the relay operating time is governed by the curves shown in Figure 2.14. The discrete current tap settings available for CO8 relay are: 4, 5, 6, 7, 8, 10, and 12 A.

Solution

For 3ϕ systems, the load current can be calculated by the following relation:

$$I = \frac{|S|}{\sqrt{3}V} \tag{5.2}$$

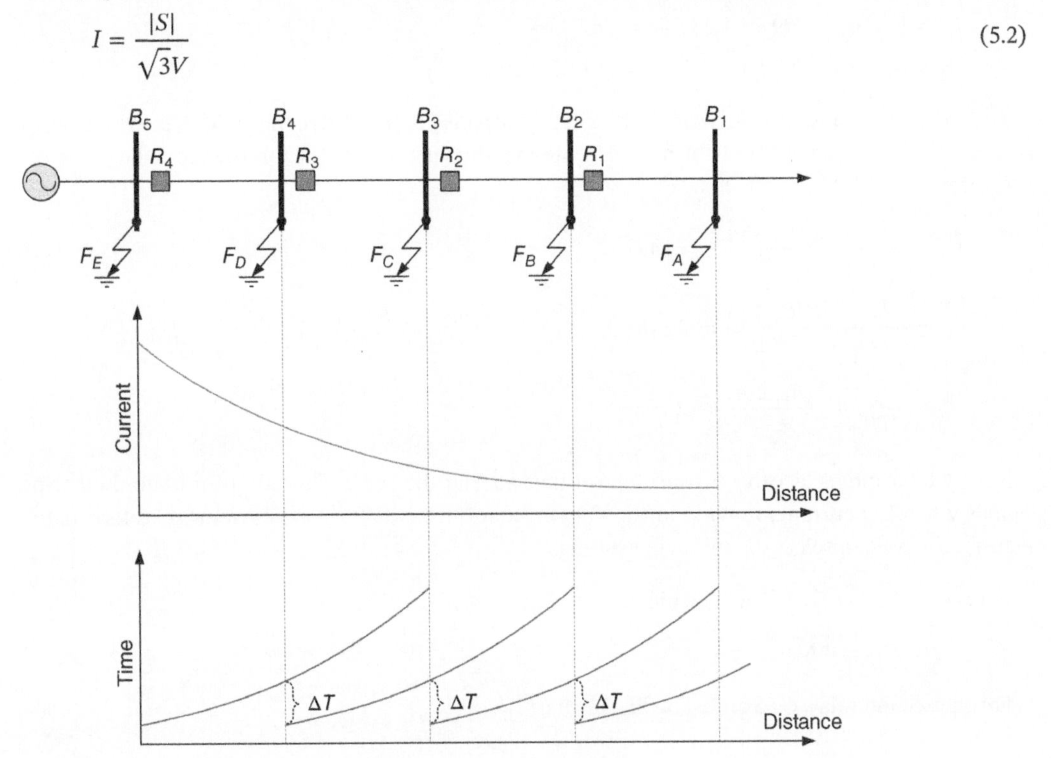

Figure 5.5 Concept of inverse-time grading in radial systems.

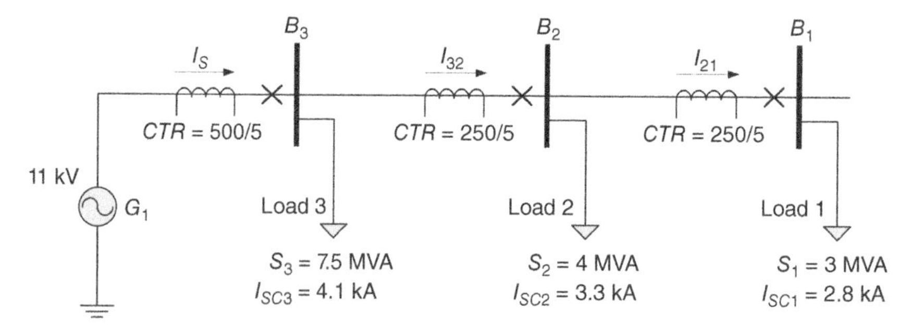

Figure 5.6 A simple radial system used for Example 5.3.

Therefore, the current flowing in each load can be calculated as follows:

$$I_1 = \frac{3 \times 10^6 \text{ VA}}{\sqrt{3}\left(11 \times 10^3 \text{ V}\right)} = 157.46 \text{ A}$$

$$I_2 = \frac{4 \times 10^6 \text{ VA}}{\sqrt{3}\left(11 \times 10^3 \text{ V}\right)} = 209.95 \text{ A}$$

$$I_3 = \frac{7.5 \times 10^6 \text{ VA}}{\sqrt{3}\left(11 \times 10^3 \text{ V}\right)} = 393.65 \text{ A}$$

By applying Kirchhoff's current law (KCL), the pre-fault currents flowing in each branch are:

$$I_{21} = I_1 = 157.46 \text{ A}$$

$$I_{32} = I_2 + I_{21} = 209.95 + 157.46 = 367.40 \text{ A}$$

$$I_S = I_3 + I_{32} = 393.65 + 367.40 = 761.05 \text{ A}$$

As can be seen, these currents cannot be directly supplied to protective relays. This is the purpose of using the three current transformers. The stepped-down currents (i.e. on the secondary side of CTs) are:

$$\bar{I}_{21} = \frac{I_{21}}{CTR_{21}} = \frac{157.46A}{\frac{250}{5}} = 3.1492 \text{ A}$$

$$\bar{I}_{32} = \frac{I_{32}}{CTR_{32}} = \frac{367.40A}{\frac{250}{5}} = 7.3481 \text{ A}$$

$$\bar{I}_S = \frac{I_S}{CTR_S} = \frac{761.05A}{\frac{500}{5}} = 7.6105 \text{ A}$$

Now, we have to select the current tap settings so that the relays should not initiate their trip signals when the currents flowing in the network does not pass the load current. By referring to Figure 2.14, we can select the following settings:

- For the first relay, CTS_1 must be bigger than $\bar{I}_{21}$:

  $$\therefore CTS_1 = 4 \text{ A}$$

- For the second relay, CTS_2 must be bigger than $\bar{I}_{32}$:

  $$\therefore CTS_2 = 8 \text{ A}$$

- For the third relay, CTS_3 must be bigger than $\bar{I}_S$:

$$\therefore CTS_3 = 8\text{ A}$$

As can be seen, the nearest current tap settings are selected.

Since the curves are represented by multiples of pickup current[2] (M), so it is required to determine this setting as well. By referring to (2.1), we first need to calculate the stepped-down short-circuit currents:

$$\bar{I}_{SC1} = \frac{I_{SC1}}{CTR_{21}} = \frac{2.8 \times 10^3 A}{\frac{250}{5}} = 56\text{ A}$$

$$\bar{I}_{SC2} = \frac{I_{SC2}}{CTR_{32}} = \frac{3.3 \times 10^3 A}{\frac{250}{5}} = 66\text{ A}$$

$$\bar{I}_{SC3} = \frac{I_{SC3}}{CTR_S} = \frac{4.1 \times 10^3 A}{\frac{500}{5}} = 41\text{ A}$$

Then, M can be calculated as follows:

$$M_1 = \frac{\bar{I}_{SC1}}{CTS_1} = \frac{56}{4} = 14$$

$$M_2 = \frac{\bar{I}_{SC2}}{CTS_2} = \frac{66}{8} = 8.25$$

$$M_3 = \frac{\bar{I}_{SC3}}{CTS_3} = \frac{41}{8} = 5.125$$

Since the first relay does not depend on other relays, there is no reason to let it wait. Thus, the time dial setting chosen for this relay should be:

$$TDS_1 = \frac{1}{2}$$

Having M_1 and TDS_1, the operating time of the first relay for a fault at bus 1 can be calculated from Figure 2.14 as follows:

$$T_1 \approx 0.1\text{ s}$$

If the first relay fails to operate, the second relay is responsible to clear the preceding fault at bus 1. Thus, the first relay should have enough chance to act before the backup relay operates. Let's assume that the discrimination margin[3] is about 0.4 s (Al-Roomi and El-Hawary, 2019a). Therefore, the operating time of the second relay to isolate the fault at bus 1 is:

$$T_{2_1} = T_1 + 0.4 \approx 0.5\text{ s}$$

We cannot use M_2, because the second relay is working here as a backup relay (i.e. dealing with $\bar{I}_{SC1}$), while M_2 is calculated based on $\bar{I}_{SC2}$. Thus, we need to calculate the following multiples of pickup or tap current:

$$M_{2_1} = \frac{\bar{I}_{SC1}}{CTS_2} = \frac{56}{8} = 7$$

From M_{2_1} and T_{2_1}, we can use Figure 2.14 to determine TDS_2 as follows:

$$TDS_2 \approx 2$$

2 Don't forget, in the IEC standard, this setting is called the plug setting multiplier (*PSM*).
3 It is called coordination time interval (*CTI*), which will be covered in Section 5.3.

The same steps can be carried out to find TDS_3. Because the third relay must be able to operate as a backup relay to clear any fault at bus 2, so we need to calculate the operating time of the second relay when it operates as a primary relay. By referring to Figure 2.14, we can use M_2 and TDS_2 to calculate T_2 as follows:

$$T_2 \approx 0.55 \text{ s}$$

By using the same discrimination margin, the operating time of the third relay, when it operates as a backup relay, is:

$$T_{3_2} = T_2 + 0.4 \approx 0.95 \text{ s}$$

Again, we cannot use M_3 here, because T_{3_2} is calculated when the third relay operates as a backup relay. Thus, we need to find M_{3_2}, which is calculated as follows:

$$M_{3_2} = \frac{\bar{I}_{SC2}}{CTS_3} = \frac{66}{8} = 8.25$$

From M_{3_2} and T_{3_2}, we can use Figure 2.14 to determine TDS_3 as follows:

$$TDS_3 \approx 3.5$$

For our curiosity, the operating time of the third relay to isolate its in-zone fault (i.e. at bus 3) can be determined from the plot by looking for the intersection between M_3 and TDS_3, which is around:

$$T_3 \approx 1.4 \text{ s}$$

We can observe three important points:

1. The calculations are simple and did not consider the load starting currents and overload currents.
2. The values are determined from a log graph, and thus they are not accurate.
3. It is impractical for computer applications, and thus it is crucial to find some mathematical relationships to describe the variability of ITOCRs without referring to any sheet; please, refer to Chapter 3.

Overall, the first and second types, i.e. IOCR/DCOCR and DTOCR, have some major limitations that cause many technical problems if someone tries to use them in mesh and other non-simple radial systems. A good overview is covered in Paithankar and Bhide (2003) and Gers and Holmes (2004). Thus, the last type, which is ITOCR, is the one that is preferred in most ORC problems because its operating time is inversely proportional to the value of the short-circuit current.

5.2 Directional Overcurrent Relays

The non-directional OCRs detect fault currents based on only their magnitudes and then the trip signals are sent to the corresponding **CBs** to clear the faults. To know the major problem of these OCRs, consider the parallel line radial circuit shown in Figure 5.7. In this example, suppose that a fault (F) has occurred on line 2 and near the busbar B. If only non-directional OCRs are used, then the third and fourth relays, i.e. R_3 and R_4, will sense the same fault current magnitude, and thus their trip signals, respectively, sent to CB_3 and CB_4 are initiated at the same time. In addition, CB_2 will be tripped by R_2 after waiting a time delay in order to clear the fault completely from the system.

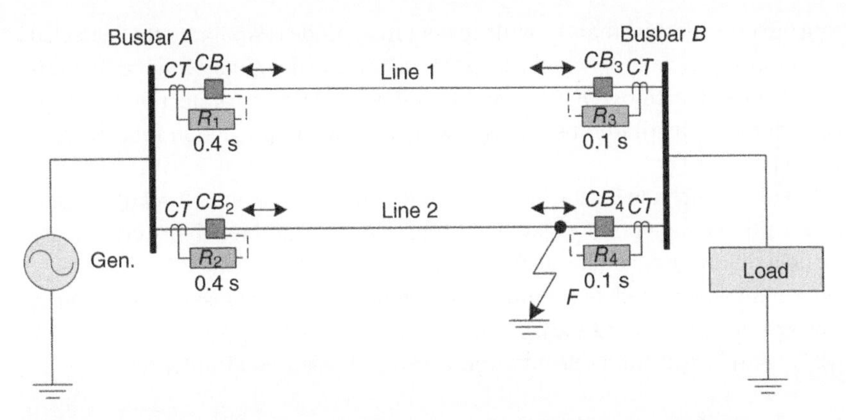

Figure 5.7 Single-end fed power system of parallel feeders with only OCRs.

The problem with this protection scheme is that the load and line 1 will be unnecessarily disconnected and as a result, the protection system is considered unreliable and unselective.

To solve this issue, an additional directional unit is combined with OCRs to identify the fault current directions with respect to a reference signal. Based on that, both the current magnitude and direction are considered to trip the faulty element as fast and selective as possible where the remaining parts of the network can operate normally. This special protective device is called **directional overcurrent relay (DOCR)**, which will be the focus of the next chapters of this book during studying the general ORC problems. The **reference signal**[4] is usually a voltage that can be provided through PT. Current is also used as a polarizing signal in some applications to decrease the total cost. Except for non-inverse-time OCRs, these relays have both *PS* and *TMS* and their TCCCs are similar to those of non-directional OCRs (Paithankar and Bhide, 2003; Gers and Holmes, 2004).

Now, consider that R_3 and R_4 are directional OCRs, i.e. DOCRs, which operate based on both the current magnitude and direction as shown in Figure 5.8. In this case, R_4 will sense both the magnitude and the correct direction of the short-circuit current and will trip CB_4. Similarly, R_2 will trip CB_2 after a time delay, and hence the fault is completely cleared from both ends of line 2. On the other hand, R_3 will restrain (*i.e. not operate*), because the fault current direction is not

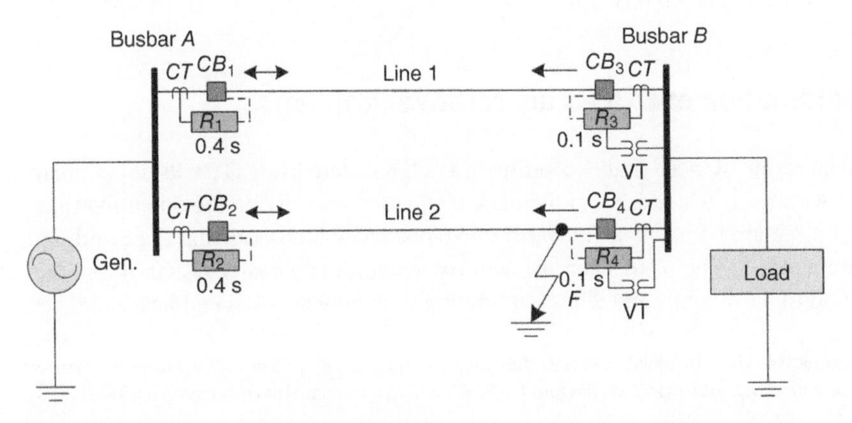

Figure 5.8 Single-end fed power system of parallel feeders with OCRs and DOCRs.

4 It is also known as the **polarizing signal**.

the same as its tripping direction despite that the fault current magnitude is similar to that detected by R_4. Therefore, the continuity of the supply to the load through the healthy line, which is line 1, is preserved. Thus, in practical applications, DOCRs can be used as the primary protection for interconnected sub-transmission and distribution systems, and as the backup protection for transmission systems (Urdaneta et al., 1997).

Now, suppose that R_4 failed to operate for the fault F, then R_1 will act as a remote backup to clear F by tripping CB_1 after waiting enough time delay. Thus, the selectivity and reliability criteria are very important in the field of power system protection.

For other complicated networks such as multi-loop systems, ring feeder systems, or even double-end fed power systems, depending only on non-directional OCRs is a very complicated task and may not satisfy the reliability and selectivity criteria (Paithankar and Bhide, 2003).

5.3 Coordination of DOCRs

The short definition of relay coordination has been given in the introduction of this book. It is a very important stage for any protection design. Correct relay coordination involves selecting the suitable relay setting that assures faults in the protected zone are cleared first by the corresponding primary relays and if they fail, the corresponding backup relays act after a **coordination time interval**[5] (**CTI**), which can be calculated as:

$$CTI = T_{CB} + T_{OS} + T_{SM} \tag{5.3}$$

where T_{CB} is the operating time of the CB after receiving a trip signal from the primary relay, T_{OS} is the over-shoot time,[6] and T_{SM} is a safety margin given to the model to take into account the mismatches due to the relay timing error, CT-ratio error, current magnitude measurement error, etc. (Albasri et al., 2015). The values of CTI range between 0.2 and 0.5 s (Blackburn and Domin, 2006; Anthony, 1995).

Except for some special cases of radial networks, the coordination problem can be considered as a highly constrained MINLP problem, where TMS is continuous and PS is discrete.[7] Such a problem requires an expert protection engineer to solve it analytically where all the fault possibilities, system contingencies, and abnormalities are analyzed and predetermined. Alternatively, it can be easily solved by using optimization methods (Noghabi et al., 2009).

5.4 Is the Coordination of DOCRs an Iterative Problem?

To answer this vital question, we need first to coordinate a radial system using ITOCRs. It is similar to Example 5.3, but with using a standard mathematical equation to estimate the operating time instead of tracing intersection points of semilogarithmic plots. We will observe that the coordination task can be completed in one pass. Then, we will try to coordinate a ring system to see the iterative nature of non-radial systems and why coordinating such networks is a real headache.

---

5 It is also known as a **selective time interval** (**STI**) (Paithankar and Bhide, 2003), as a **coordination delay time** (**CDT**) (Gonen, 2016), as a **grading margin** (**GM**) (So and Li, 2000), and as a **discrimination margin** (**DM**) (Gers and Holmes, 2004).

6 It is limited to the electromechanical relay types, as mentioned before. This time delay is also called the **over-travel delay** and the **coasting time** (Al-Roomi, 2014).

7 In numerical relays they can be provided in almost continuous values (Al-Roomi and El-Hawary, 2019a).

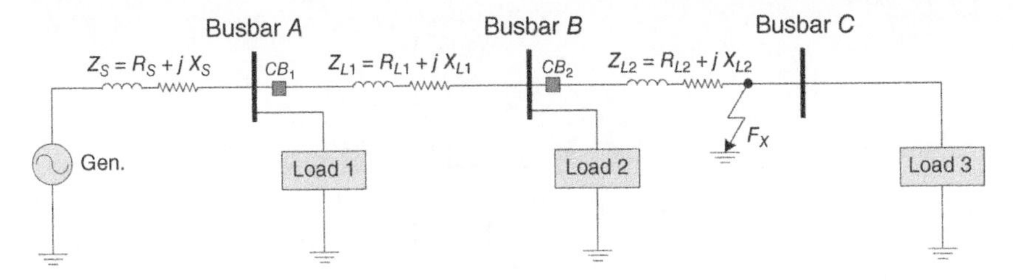

Figure 5.9 Radial network with directional overcurrent relays.

Example 5.4 Figure 5.9 represents a very simple radial system. Suppose a fault (F_x) occurs at busbar C. The two protective devices are non-directional ITOCRs and modeled based on the IEC standard inverse. If the discrimination margin between the primary and backup protection is $CTI = 0.3$ s, coordinate the two relays.

Solution

Because there is no relay after busbar[8] B, so we can take $TMS_2 = 0.1$. For F_x, the second relay (R_2) is the primary protection to clear that fault by tripping its circuit breaker (CB_2) after a delay. The operating time of this relay can be calculated using (3.36) as follows:

$$T_{R_2} = TMS_{R_2} \times \left(\frac{\beta}{PSM_{R_2}^{\alpha} - 1} + \gamma \right) = \left(\frac{0.14 \times 0.1}{PSM_{R_2}^{0.02} - 1} \right) \text{ s}$$

If R_2 fails to operate, R_1 will act as a backup relay to clear F_x. Thus, R_1 should operate after giving enough time to R_2. If R_2 exceeds the maximum allowable chance, then R_1 has to isolate F_x with a minimum operating time of:

$$T_{R_1} = T_{R_2} + CTI = \left(\frac{0.14 \times 0.1}{PSM_{R_2}^{0.02} - 1} + 0.3 \right) \text{ s}$$

That is, we need to adjust the settings of R_1 so that the selectivity criterion ($T_{R_1} \geqslant T_{R_2} + CTI$) is satisfied.

As seen from this radial system, R_2 will never be a backup relay for any fault located after busbar B. Thus, the coordination problem of this simple radial system can be solved easily in one pass.

From this example, we can observe that the coordination of radial systems is not an iterative process. Now, let's try another example. We want to coordinate a very simple non-radial system to see if it can also be solved analytically or iteratively.

Example 5.5 Consider the four-bus ring system shown in Figure 5.10. This test system is taken from (Soman, 2010; Gajbhiye et al., 2005). For simplicity, there is no any operational or topological changes in the network configuration; i.e. a steady-state network is considered. All the predetermined values of PS and the short-circuit current seen by primary and backup relays are tabulated in Table 5.1. Also, assume that these numerical relays have very small lower limits of TMS and there is no limit on their operating times. If the discrimination margin between the primary and backup protection is $CTI = 0.3$ s and the IEC standard inverse model is used, coordinate these eight inverse-time DOCRs.

8 That is, the right side of the radial system.

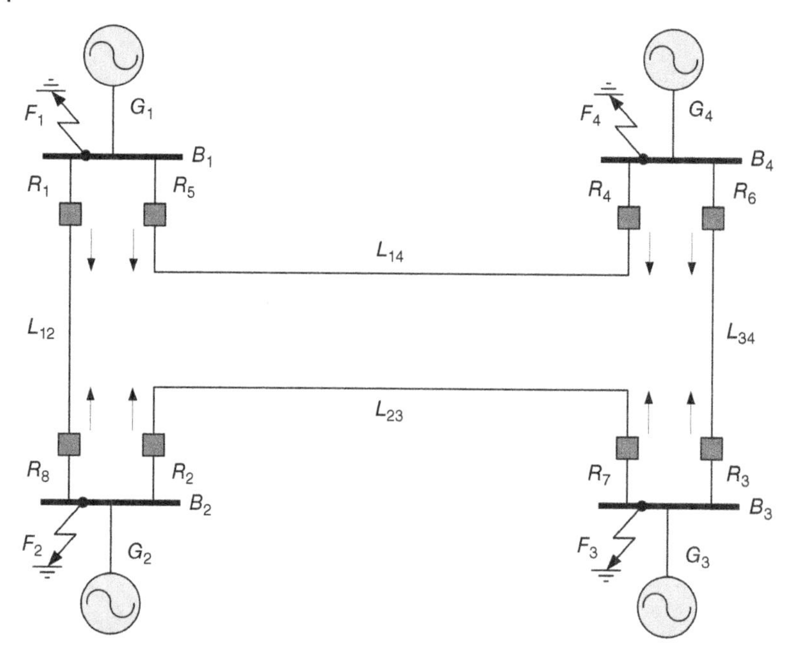

Figure 5.10 Single-line diagram of the four-bus test system used in Example 5.5. Source: Based on Soman (2010); Gajbhiye et al. (2005).

Table 5.1 Data of the four-bus test system used in Example 5.5.

Relay No.	Plug setting (A) Primary side of CT	In-zone Fault current (A)	Out-zone Fault current (A)
R_1	60	1652	152
R_2	80	639	142
R_3	60	937	140
R_4	160	1097	391
R_5	80	553	272
R_6	160	1365	240
R_7	128	868	287
R_8	100	1764	197

Solution

From the figure, DOCRs can detect short-circuit currents only in one direction, so there are two loops:

- **Clockwise Loop:** $R_5 \rightarrow R_6 \rightarrow R_7 \rightarrow R_8 \rightarrow R_5$
- **Anti-Clockwise Loop:** $R_1 \rightarrow R_2 \rightarrow R_3 \rightarrow R_4 \rightarrow R_1$

where the arrows denote the current direction, which can also mean "backup to." The first loop starts with R_5 and ends with R_5 too. Similarly for the second loop, it starts and ends with R_1. This means that the coordination problem is iterative because the devices that are initially considered as primary relays will act as backup relays for some other devices that are initially considered as backup relays. Thus, each DOCR could work as a primary relay as well as a backup relay. Based on this, coordinating relays to operate accurately and selectively is a hard task for protection engineers.

The classical iterative technique is done by randomly selecting one relay from each loop. The second step is to guess an initial operating time by selecting a good starting point for *TMS*. For *PS*, it is predetermined by a good value that lies between the full load and the minimum fault current where some engineering experiences and skills are required. As a result, (3.36) becomes a linear equation with one variable. Because the IEC standard inverse model is used for all the relays, so the operating time of the ith relay can be calculated as follows:

$$T_i = \frac{0.14\ TMS_i}{\left(\frac{I_{f,i}}{PS_i}\right)^{0.02} - 1} = \psi_i\ TMS_i$$

where ψ_i is constant and equals:

$$\psi_i = \frac{0.14}{\left(\frac{I_{f,i}}{PS_i}\right)^{0.02} - 1}$$

Note that, we are not using *CTR* here, because we are referring *PS* to the primary side of CTs directly. Practically, fault currents must be stepped-down before sending them to protective relays.

Now, suppose R_1 and R_5 are selected from each loop and their $TMS^{\text{initial}} = 0.05$. Then, the procedure can be described as follows:

1. **Select the anti-clockwise loop and start from R_1**
   - **Iteration No. 1:** For the fault at bus 2 (i.e. F_2), R_1 should operate as a primary DOCR and its operating time (T_1) can be calculated from (3.36) as:

$$T_1^{\text{primary}} = \frac{\beta \times TMS_1}{\left(\frac{I_{F_1}}{PS_1}\right)^{\alpha} - 1} = \frac{0.14\ (0.05)}{\left(\frac{1652}{60}\right)^{0.02} - 1} = 0.102\,107\ \text{s}$$

If R_1 fails to operate, R_4 will operate after waiting *CTI* as follows:

$$T_4^{\text{backup}} = T_1^{\text{primary}} + CTI = 0.102\,107 + 0.3 = 0.402\,107\ \text{s}$$

So, using this value in (3.36) yields:

$$0.402\,107\ \text{s} = \frac{0.14 \times TMS_4}{\left(\frac{391}{160}\right)^{0.02} - 1} \Rightarrow TMS_5 = 0.051\,789$$

Now, for the fault at bus 1 (i.e. F_1), R_4 should operate as a primary relay with an operating time of:

$$T_4^{\text{primary}} = \frac{0.14\ (0.051\,789)}{\left(\frac{1097}{160}\right)^{0.02} - 1} = 0.184\,707\ \text{s}$$

If R_4 fails to operate, R_3 will operate after waiting *CTI* as follows:

$$T_3^{\text{backup}} = T_4^{\text{primary}} + CTI = 0.184\,707 + 0.3 = 0.484\,707\ \text{s}$$

By using the value of T_3^{backup}, TMS_3 can be calculated from (3.36) as follows:

$$0.484\,707\ \text{s} = \frac{0.14 \times TMS_3}{\left(\frac{140}{60}\right)^{0.02} - 1} \Rightarrow TMS_3 = 0.059\,170$$

Thus, for a fault at bus 4 (i.e. F_4), R_3 should operate as a primary relay with an operating time of:

$$T_3^{\text{primary}} = \frac{0.14\ (0.059\,170)}{\left(\frac{937}{60}\right)^{0.02} - 1} = 0.146\,602\ \text{s}$$

If R_3 fails to operate, R_2 will operate after waiting CTI as follows:

$$T_2^{\text{backup}} = T_3^{\text{primary}} + CTI = 0.146\,602 + 0.3 = 0.446\,602\ \text{s}$$

By using the value of T_2^{backup}, TMS_2 can be calculated from (3.36) as follows:

$$0.446\,602\ \text{s} = \frac{0.14 \times TMS_2}{\left(\frac{142}{80}\right)^{0.02} - 1} \Rightarrow TMS_2 = 0.036\,819$$

Thus, for a fault at bus 3 (i.e. F_3), R_2 should operate as a primary relay with an operating time of:

$$T_2^{\text{primary}} = \frac{0.14\ (0.036\,819)}{\left(\frac{639}{80}\right)^{0.02} - 1} = 0.121\,479\ \text{s}$$

Now, R_1 will act as a backup relay when R_2 fails to operate:

$$T_1^{\text{backup}} = T_2^{\text{primary}} + CTI = 0.121\,479 + 0.3 = 0.421\,479\ \text{s}$$

The new value of TMS_1 is calculated as follows:

$$0.421\,479\ \text{s} = \frac{0.14 \times TMS_1}{\left(\frac{152}{60}\right)^{0.02} - 1} \Rightarrow TMS_1 = 0.056\,492$$

The absolute difference is:

$$\Delta TMS_1 = \left|TMS_1^{\text{new}} - TMS_1^{\text{old}}\right| = |0.056\,492 - 0.05| = 0.006\,492$$

- **Iteration No. 2:** At F_2:

$$T_1^{\text{primary}} = \frac{0.14\ (0.056\,492)}{\left(\frac{1652}{60}\right)^{0.02} - 1} = 0.115\,364\ \text{s}$$

If R_1 fails to operate:

$$T_4^{\text{backup}} = T_1^{\text{primary}} + CTI = 0.115\,364 + 0.3 = 0.415\,364\ \text{s}$$

By using (3.36):

$$0.415\,364\ \text{s} = \frac{0.14 \times TMS_4}{\left(\frac{391}{160}\right)^{0.02} - 1} \Rightarrow TMS_4 = 0.053\,497$$

At F_1:

$$T_4^{\text{primary}} = \frac{0.14\ (0.053\,497)}{\left(\frac{1097}{160}\right)^{0.02} - 1} = 0.190\,797\ \text{s}$$

If R_4 fails to operate:

$$T_3^{\text{backup}} = T_4^{\text{primary}} + CTI = 0.190\,797 + 0.3 = 0.490\,797\ \text{s}$$

By using (3.36):

$$0.490\,797\text{ s} = \frac{0.14 \times TMS_3}{\left(\frac{140}{60}\right)^{0.02} - 1} \Rightarrow TMS_3 = 0.059\,913$$

At F_4:

$$T_3^{\text{primary}} = \frac{0.14\ (0.059\,913)}{\left(\frac{937}{60}\right)^{0.02} - 1} = 0.148\,444\text{ s}$$

If R_3 fails to operate:

$$T_2^{\text{backup}} = T_3^{\text{primary}} + CTI = 0.148\,444 + 0.3 = 0.448\,444\text{ s}$$

By using (3.36):

$$0.448\,444\text{ s} = \frac{0.14 \times TMS_2}{\left(\frac{142}{80}\right)^{0.02} - 1} \Rightarrow TMS_2 = 0.036\,971$$

At F_3:

$$T_2^{\text{primary}} = \frac{0.14\ (0.036\,971)}{\left(\frac{639}{80}\right)^{0.02} - 1} = 0.121\,800\text{ s}$$

If R_2 fails to operate:

$$T_1^{\text{backup}} = T_2^{\text{primary}} + CTI = 0.121\,800 + 0.3 = 0.421\,800\text{ s}$$

By using (3.36):

$$0.421\,800\text{ s} = \frac{0.14 \times TMS_1}{\left(\frac{152}{60}\right)^{0.02} - 1} \Rightarrow TMS_1 = 0.056\,559$$

The absolute difference is:

$$\Delta TMS_1 = \left|TMS_1^{\text{new}} - TMS_1^{\text{old}}\right| = |0.056\,559 - 0.056\,492| = 6.6999 \times 10^{-5}$$

- **Iteration No. 3:** At F_2:

$$T_1^{\text{primary}} = \frac{0.14\ (0.056\,559)}{\left(\frac{1652}{60}\right)^{0.02} - 1} = 0.115\,501\text{ s}$$

If R_1 fails to operate:

$$T_4^{\text{backup}} = T_1^{\text{primary}} + CTI = 0.115\,501 + 0.3 = 0.415\,501\text{ s}$$

By using (3.36):

$$0.415\,501\text{ s} = \frac{0.14 \times TMS_4}{\left(\frac{391}{160}\right)^{0.02} - 1} \Rightarrow TMS_4 = 0.053\,515$$

At F_1:

$$T_4^{\text{primary}} = \frac{0.14\ (0.053\,515)}{\left(\frac{1097}{160}\right)^{0.02} - 1} = 0.190\,860\text{ s}$$

If R_4 fails to operate:

$$T_3^{backup} = T_4^{primary} + CTI = 0.190\,860 + 0.3 = 0.490\,860\text{ s}$$

By using (3.36):

$$0.490\,860\text{ s} = \frac{0.14 \times TMS_3}{\left(\frac{140}{60}\right)^{0.02} - 1} \Rightarrow TMS_3 = 0.059\,921$$

At F_4:

$$T_3^{primary} = \frac{0.14\ (0.059\,921)}{\left(\frac{937}{60}\right)^{0.02} - 1} = 0.148\,463\text{ s}$$

If R_3 fails to operate:

$$T_2^{backup} = T_3^{primary} + CTI = 0.148\,463 + 0.3 = 0.448\,463\text{ s}$$

By using (3.36):

$$0.448\,463\text{ s} = \frac{0.14 \times TMS_2}{\left(\frac{142}{80}\right)^{0.02} - 1} \Rightarrow TMS_2 = 0.036\,973$$

At F_3:

$$T_2^{primary} = \frac{0.14\ (0.036\,973)}{\left(\frac{639}{80}\right)^{0.02} - 1} = 0.121\,985\text{ s}$$

If R_2 fails to operate:

$$T_1^{backup} = T_2^{primary} + CTI = 0.121\,985 + 0.3 = 0.421\,985\text{ s}$$

By using (3.36):

$$0.421\,985\text{ s} = \frac{0.14 \times TMS_1}{\left(\frac{152}{60}\right)^{0.02} - 1} \Rightarrow TMS_1 = 0.056\,560$$

The absolute difference is:

$$\Delta TMS_1 = \left|TMS_1^{new} - TMS_1^{old}\right| = |0.056\,560 - 0.056\,559| = 1 \times 10^{-6}$$

- **Iteration No. 4:** At F_2:

$$T_1^{primary} = \frac{0.14\ (0.056\,560)}{\left(\frac{1652}{60}\right)^{0.02} - 1} = 0.115\,503\text{ s}$$

If R_1 fails to operate:

$$T_4^{backup} = T_1^{primary} + CTI = 0.115\,503 + 0.3 = 0.415\,503\text{ s}$$

By using (3.36):

$$0.415\,503\text{ s} = \frac{0.14 \times TMS_4}{\left(\frac{391}{160}\right)^{0.02} - 1} \Rightarrow TMS_4 = 0.053\,515$$

At F_1:

$$T_4^{\text{primary}} = \frac{0.14\ (0.053\ 515)}{\left(\frac{1097}{160}\right)^{0.02} - 1} = 0.190\ 860\ \text{s}$$

If R_4 fails to operate:

$$T_3^{\text{backup}} = T_4^{\text{primary}} + CTI = 0.190\ 860 + 0.3 = 0.490\ 860\ \text{s}$$

By using (3.36):

$$0.490\ 860\ \text{s} = \frac{0.14 \times TMS_3}{\left(\frac{140}{60}\right)^{0.02} - 1} \Rightarrow TMS_3 = 0.059\ 921$$

At F_4:

$$T_3^{\text{primary}} = \frac{0.14\ (0.059\ 921)}{\left(\frac{937}{60}\right)^{0.02} - 1} = 0.148\ 463\ \text{s}$$

If R_3 fails to operate:

$$T_2^{\text{backup}} = T_3^{\text{primary}} + CTI = 0.148\ 463 + 0.3 = 0.448\ 463\ \text{s}$$

By using (3.36):

$$0.448\ 463\ \text{s} = \frac{0.14 \times TMS_2}{\left(\frac{142}{80}\right)^{0.02} - 1} \Rightarrow TMS_2 = 0.036\ 973$$

At F_3:

$$T_2^{\text{primary}} = \frac{0.14\ (0.036\ 973)}{\left(\frac{639}{80}\right)^{0.02} - 1} = 0.121\ 985\ \text{s}$$

If R_2 fails to operate:

$$T_1^{\text{backup}} = T_2^{\text{primary}} + CTI = 0.121\ 985 + 0.3 = 0.421\ 985\ \text{s}$$

By using (3.36):

$$0.421\ 985\ \text{s} = \frac{0.14 \times TMS_1}{\left(\frac{152}{60}\right)^{0.02} - 1} \Rightarrow TMS_1 = 0.056\ 559$$

The absolute difference is:

$$\Delta TMS_1 = \left|TMS_1^{\text{new}} - TMS_1^{\text{old}}\right| = |0.056\ 559 - 0.056\ 559| = 0$$

It can be clearly seen that the value of TMS_1 did not change from the last iteration, we can stop here.

2. **Select the anti-clockwise loop and start from R_5**
   By repeating the preceding steps, the anti-clockwise loop can be coordinated as well. With the starting point of $TMS_5 = 0.05$, the iterative results are:
   - **Iteration No. 1:**

$$\Rightarrow T_5^{\text{primary}} = 0.177\,557\text{ s} \Rightarrow T_8^{\text{backup}} = 0.477\,557\text{ s} \Rightarrow TMS_8 = 0.046\,572$$

$$\Rightarrow T_8^{\text{primary}} = 0.110\,355\text{ s} \Rightarrow T_7^{\text{backup}} = 0.410\,355\text{ s} \Rightarrow TMS_7 = 0.047\,718$$

$$\Rightarrow T_7^{\text{primary}} = 0.171\,187\text{ s} \Rightarrow T_6^{\text{backup}} = 0.471\,187\text{ s} \Rightarrow TMS_6 = 0.027\,404$$

$$\Rightarrow T_6^{\text{primary}} = 0.087\,578\text{ s} \Rightarrow T_5^{\text{backup}} = 0.387\,578\text{ s} \Rightarrow TMS_5 = 0.068\,594$$

$$\Delta TMS_5 = \left|TMS_5^{\text{new}} - TMS_5^{\text{old}}\right| = |0.068\,594 - 0.05| = 0.018\,594$$

   - **Iteration No. 2:**

$$\Rightarrow T_5^{\text{primary}} = 0.243\,588\text{ s} \Rightarrow T_8^{\text{backup}} = 0.543\,588\text{ s} \Rightarrow TMS_8 = 0.053\,012$$

$$\Rightarrow T_8^{\text{primary}} = 0.125\,614\text{ s} \Rightarrow T_7^{\text{backup}} = 0.425\,614\text{ s} \Rightarrow TMS_7 = 0.049\,493$$

$$\Rightarrow T_7^{\text{primary}} = 0.177\,552\text{ s} \Rightarrow T_6^{\text{backup}} = 0.477\,552\text{ s} \Rightarrow TMS_6 = 0.027\,774$$

$$\Rightarrow T_6^{\text{primary}} = 0.088\,761\text{ s} \Rightarrow T_5^{\text{backup}} = 0.388\,761\text{ s} \Rightarrow TMS_5 = 0.068\,804$$

$$\Delta TMS_5 = \left|TMS_5^{\text{new}} - TMS_5^{\text{old}}\right| = |0.068\,804 - 0.068\,594| = 2.100 \times 10^{-4}$$

   - **Iteration No. 3:**

$$\Rightarrow T_5^{\text{primary}} = 0.244\,332\text{ s} \Rightarrow T_8^{\text{backup}} = 0.544\,332\text{ s} \Rightarrow TMS_8 = 0.053\,084$$

$$\Rightarrow T_8^{\text{primary}} = 0.125\,786\text{ s} \Rightarrow T_7^{\text{backup}} = 0.425\,786\text{ s} \Rightarrow TMS_7 = 0.049\,513$$

$$\Rightarrow T_7^{\text{primary}} = 0.177\,624\text{ s} \Rightarrow T_6^{\text{backup}} = 0.477\,624\text{ s} \Rightarrow TMS_6 = 0.027\,778$$

$$\Rightarrow T_6^{\text{primary}} = 0.088\,774\text{ s} \Rightarrow T_5^{\text{backup}} = 0.388\,774\text{ s} \Rightarrow TMS_5 = 0.068\,806$$

$$\Delta TMS_5 = \left|TMS_5^{\text{new}} - TMS_5^{\text{old}}\right| = |0.068\,806 - 0.068\,804| = 2.000 \times 10^{-6}$$

   - **Iteration No. 4:**

$$\Rightarrow T_5^{\text{primary}} = 0.244\,340\text{ s} \Rightarrow T_8^{\text{backup}} = 0.544\,340\text{ s} \Rightarrow TMS_8 = 0.053\,085$$

$$\Rightarrow T_8^{\text{primary}} = 0.125\,788\text{ s} \Rightarrow T_7^{\text{backup}} = 0.425\,788\text{ s} \Rightarrow TMS_7 = 0.049\,513$$

$$\Rightarrow T_7^{\text{primary}} = 0.177\,625\text{ s} \Rightarrow T_6^{\text{backup}} = 0.477\,625\text{ s} \Rightarrow TMS_6 = 0.027\,778$$

$$\Rightarrow T_6^{\text{primary}} = 0.088\,774\text{ s} \Rightarrow T_5^{\text{backup}} = 0.388\,774\text{ s} \Rightarrow TMS_5 = 0.068\,806$$

$$\Delta TMS_5 = \left|TMS_5^{\text{new}} - TMS_5^{\text{old}}\right| = |0.068\,806 - 0.068\,806| = 0$$

   Because TMS_5 did not change during the fourth iteration, so we can stop here.

Table 5.2 Data of the IEEE three-bus test system used in Example 5.6.

Relay No.	Plug setting (A) "Primary side of CT"	In-zone Fault current (A)	Out-zone Fault current (A)
R_1	180	1978.90	617.22
R_2	100	1525.70	145.34
R_3	100	1683.90	384.00
R_4	150	1815.40	545.00
R_5	120	1499.66	175.00
R_6	120	1766.30	466.17

We can see that the Gauss–Seidel (GS) method can be used to optimally coordinate DOCRs installed in a simple ring network. To fully grasp the idea, let's solve another problem by MATLAB.

Example 5.6 Consider the IEEE three-bus ring system shown in Figure F.1. Again, for simplicity, the steady-state condition is considered. All the predetermined values of *PS* and the short-circuit current seen by primary and backup relays are tabulated in Table 5.2. Also, assume that these numerical relays have very small lower limits of *TMS* and there is no limit on their operating times. The discrimination margin between the primary and backup protection is $CTI = 0.3$ s and the IEC standard inverse model is used. Write a MATLAB code to coordinate these six inverse-time DOCRs by using the Gauss–Seidel method and then plot TCCCs of R_4/R_6 relay pair with respect to I_{relay} and *PSM* on x-axis.

Solution

The code given in `Ring_3BusSystem.m` is:

```
1   clc
2   clear
3   format short
4   IterMax = 100; % maximum iterations
5   Tol = 1e-16; % tolerance (minimum acceptable error)
6   Err1 = 10; % any value bigger than Tol
7   Err2 = 10; % any value bigger than Tol
8   PS = [180, 100, 100, 150, 120, 120]; % plug setting (in A)
9   If_pr = [1978.90, 1525.70, 1683.90, 1815.40, 1499.66, 1766.30]; % ...
        fault current seen by primary relays
10  If_bc = [617.22, 145.34, 384.00, 545.00, 175.00, 466.17]; % fault ...
        current seen by backup relays
11  CTI = 0.3; % coordination time interval
12  % TCCC coefficients
13  alpha = 0.02;
14  beta = 0.14;
15  gamma = 0;
```

```
16      % initial starts for clockwise and anti-clockwise loops
17      TMS = ones(1, 6);
18      TMS(1) = 0.05; % for R1 (ant-clockwise loop)
19      TMS(2) = 0.05; % for R5 (clockwise loop)
20      T = ones(1, 6); % relay operating times
21      %% coordinating clockwise-loop relays
22      for i = 1:IterMax
23          T(1) = TMS(1) * (beta/((If_pr(1)/PS(1))^alpha - 1) + gamma);
24          T(5) = T(1) + CTI;
25          TMS(5) = T(5) / (beta/((If_bc(5)/PS(5))^alpha - 1) + gamma);
26          T(5) = TMS(5) * (beta/((If_pr(5)/PS(5))^alpha - 1) + gamma);
27          T(3) = T(5) + CTI;
28          TMS(3) = T(3) / (beta/((If_bc(3)/PS(3))^alpha - 1) + gamma);
29          T(3) = TMS(3) * (beta/((If_pr(3)/PS(3))^alpha - 1) + gamma);
30          MAT1(i,:) = [i TMS(1) TMS(3) TMS(5) T(1) T(3) T(5)];
31          T(1) = T(3) + CTI;
32          TMS1_new = T(1) / (beta/((If_bc(1)/PS(1))^alpha - 1) + gamma);
33          Err1 = abs(TMS(1) - TMS1_new);
34          TMS(1) = TMS1_new; % update TMS(1) for the next loop
35          if Tol ≥ Err1
36              break
37          end
38      end
39      Tab1 = table(MAT1(:,1), MAT1(:,2), MAT1(:,3), MAT1(:,4), MAT1(:,5), ...
            MAT1(:,6), MAT1(:,7));
40      Tab1.Properties.VariableNames = {'Iteration', 'TMS1', 'TMS3', 'TMS5', ...
            'T1', 'T3', 'T5'};
41      fprintf('For the clockwise loop:\n\n')
42      disp(Tab1)
43      %% coordinating anti-clockwise-loop relays
44      for j = 1:IterMax
45         T(2) = TMS(2) * (beta/((If_pr(2)/PS(2))^alpha - 1) + gamma);
46         T(4) = T(2) + CTI;
47         TMS(4) = T(4) / (beta/((If_bc(4)/PS(4))^alpha - 1) + gamma);
48         T(4) = TMS(4) * (beta/((If_pr(4)/PS(4))^alpha - 1) + gamma);
49         T(6) = T(4) + CTI;
50         TMS(6) = T(6) / (beta/((If_bc(6)/PS(6))^alpha - 1) + gamma);
51         T(6) = TMS(6) * (beta/((If_pr(6)/PS(6))^alpha - 1) + gamma);
52         MAT2(j,:) = [j TMS(2) TMS(4) TMS(6) T(2) T(4) T(6)];
53         T(2) = T(6) + CTI;
54         TMS2_new = T(2) / (beta/((If_bc(2)/PS(2))^alpha - 1) + gamma);
55         Err2 = abs(TMS(2) - TMS2_new);
56         TMS(2) = TMS2_new; % update TMS(2) for the next loop
57         if Tol ≥ Err2
58             break
59         end
60      end
61      Tab2 = table(MAT2(:,1), MAT2(:,2), MAT2(:,3), MAT2(:,4), MAT2(:,5), ...
            MAT2(:,6), MAT2(:,7));
62      Tab2.Properties.VariableNames = {'Iteration', 'TMS2', 'TMS4', 'TMS6', ...
            'T2', 'T4', 'T6'};
63      fprintf('\n\nFor the anti-clockwise loop:\n\n')
64      disp(Tab2)
```

Running this code will generate the following tables:

For the clockwise loop:

Iteration	TMS1	TMS3	TMS5	T1	T3	T5
1	0.05	0.07105	0.023942	0.14252	0.17121	0.064698
2	0.083981	0.073809	0.029182	0.23939	0.17786	0.07886
3	0.085166	0.073906	0.029365	0.24276	0.17809	0.079354
4	0.085208	0.073909	0.029372	0.24288	0.1781	0.079371
5	0.085209	0.073909	0.029372	0.24289	0.1781	0.079372
6	0.085209	0.073909	0.029372	0.24289	0.1781	0.079372
7	0.085209	0.073909	0.029372	0.24289	0.1781	0.079372
8	0.085209	0.073909	0.029372	0.24289	0.1781	0.079372
9	0.085209	0.073909	0.029372	0.24289	0.1781	0.079372
10	0.085209	0.073909	0.029372	0.24289	0.1781	0.079372
11	0.085209	0.073909	0.029372	0.24289	0.1781	0.079372

For the anti-clockwise loop:

Iteration	TMS2	TMS4	TMS6	T2	T4	T6
1	0.05	0.079344	0.10165	0.12497	0.21724	0.25754
2	0.029893	0.069961	0.0966	0.074715	0.19155	0.24475
3	0.029207	0.069641	0.096428	0.073	0.19068	0.24432
4	0.029184	0.06963	0.096422	0.072942	0.19065	0.2443
5	0.029183	0.06963	0.096422	0.07294	0.19064	0.2443
6	0.029183	0.06963	0.096422	0.07294	0.19064	0.2443
7	0.029183	0.06963	0.096422	0.07294	0.19064	0.2443
8	0.029183	0.06963	0.096422	0.07294	0.19064	0.2443
9	0.029183	0.06963	0.096422	0.07294	0.19064	0.2443
10	0.029183	0.06963	0.096422	0.07294	0.19064	0.2443
11	0.029183	0.06963	0.096422	0.07294	0.19064	0.2443

From the last two examples, we can conclude that the coordination of DOCRs in non-radial systems is an iterative process. These two iterative ORC problems can be easily solved, by using the Gauss–Seidel method, because the networks are not complicated and many assumptions have been considered to simplify the problems. The complexity increases when many relays are imposed and both *TMS* and *PS* are taken as variables in (3.36) where *TMS* is available either in continuous or discrete form, *PS* is available within a limited set of discretized values, and the minimum and maximum allowable limits of *TMS*, *PS*, and *T* are restricted. Moreover, this classical iterative method cannot be applied to multi-loop networks in which some of the relays are shared between these loops. As a result, the concept of the **minimum break-point set (MBPS)** has been suggested in the literature to tackle this stiff issue; *please, refer to our literature review.*

Example 5.7 Using the data given in Example 5.6, plot the characteristic curves of R_4/R_6 relay pair with respect to *PSM* on *x*-axis.

Solution

The code given in `DiscriminationMarginP4B6.m` is:

```
1   clc
2   clear
3   I = 100:0.01:2500;
4   for i = 1:length(I)
5       T4(i) = (0.14*0.069647)/(((I(i)/150)^0.02)-1); % Primary Relay
6       M4(i) = I(i)/150;
7       T6(i) = (0.14*0.096431)/(((I(i)/120)^0.02)-1); % Backup Relay
8       M6(i) = I(i)/120;
9   end
10  loglog(M4, T4, 'b', M6, T6, '--r', 'LineWidth', 2)
11  grid on; grid minor; axis([1 max([M4, M6]) 0 100])
12  legend({'Primary Relay (i.e. R_4)', 'Backup Relay (i.e. R_6)'})
13  set(gca, 'FontSize', 13)
14  xlabel('Plug Setting Multiplier, PSM (unitless)')
15  ylabel('Operating Time (s)')
```

The plot shown in Figure 5.11 can be generated when this code is executed, which can also be accessed by opening `DiscriminationMarginP4B6PSM.fig`.

As can be clearly seen, if the fault currents are divided by *PS* of each relay, then the characteristic curves can be represented with respect to *PSM*.

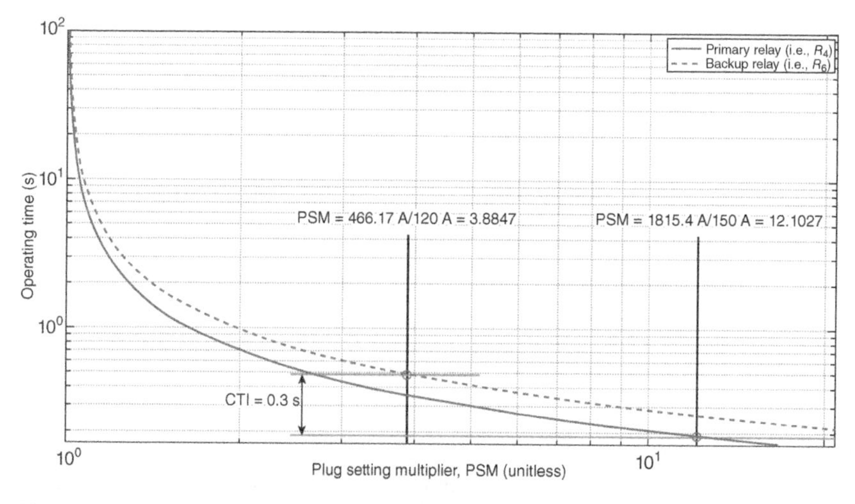

Figure 5.11 TCCCs of R_4/R_6 relay pair with respect to *PSM* on *x*-axis.

5.5 Minimum Break-Point Set

In Section 5.4, we have seen that although ring systems are simple, the coordination problem cannot be solved without using iterative methods. Now, imagine the situation with mesh/interconnected networks where multiple clockwise and anti-clockwise loops exist! For this, we will see that these loops share some relays. This means that the optimal settings achieved for one loop could not be optimal or feasible for other loops. Thus, to eliminate this effect, we have to search for the minimum number of relays that when their CBs are open the clockwise and anti-clockwise loops will vanish. These relays are called minimum break-point relays (MBPR) or minimum break-point set (MBPS). There are multiple choices to find these relays. The goal here is to find one common acceptable setting for these relays so that their coordination in individual loops is achievable (Soman, 2010).

Example 5.8 Consider the network shown in Figure 5.12. Find the following:

1. clockwise loops
2. anti-clockwise loops
3. how many relays in each MBPS
4. identify one possible MBPS

Solution

- **Clockwise Loops:**
  Loop no. 1: $R_1 \rightarrow R_5 \rightarrow R_6 \rightarrow R_1$
  Loop no. 2: $R_2 \rightarrow R_3 \rightarrow R_4 \rightarrow R_2$
  Loop no. 3: $R_1 \rightarrow R_2 \rightarrow R_3 \rightarrow R_4 \rightarrow R_5 \rightarrow R_6 \rightarrow R_1$
- **Anti-Clockwise Loops:**
  Loop no. 1: $R_7 \rightarrow R_{12} \rightarrow R_{11} \rightarrow R_7$
  Loop no. 2: $R_8 \rightarrow R_{10} \rightarrow R_9 \rightarrow R_8$
  Loop no. 3: $R_7 \rightarrow R_{12} \rightarrow R_{11} \rightarrow R_{10} \rightarrow R_9 \rightarrow R_8 \rightarrow R_7$
- **Relays in Each MBPS:** For the clockwise direction, the first and second loops (i.e. the simple loops) do not have any mutual relay. The same thing for the two simple loops of the anti-clockwise direction. Also, there is no any mutual relay between the simple loops of the clockwise direction and the anti-clockwise direction. That is, each simple loop of each direction has its own set. Thus, we are forced to select four relays to break all the clockwise and anti-clockwise loops.

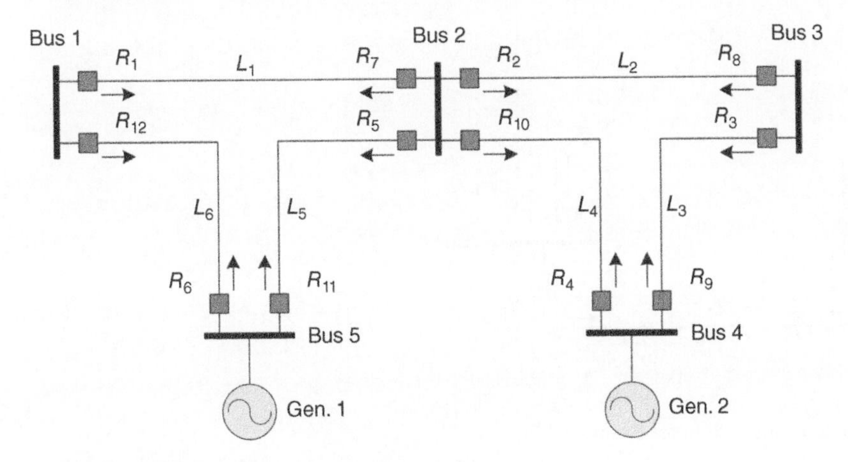

Figure 5.12 A simple five-bus system for Example 5.8.

- **Identify One MBPS:** We can select the following:
  - R_1 to break the first clockwise loop
  - R_2 to break the second clockwise loop
  - R_7 to break the first anti-clockwise loop
  - R_8 to break the second anti-clockwise loop

  The third clockwise loop can be broken by either R_1 or R_2, and the third anti-clockwise loop can be broken by either R_7 or R_8. Thus, to break all the loops of both clockwise and anti-clockwise directions, we can use this set (i.e. $\{R_1, R_2, R_7, R_8\}$), which represents MBPS.

Example 5.9 Consider the network shown in Figure 5.13. Find the following:

1. clockwise loops
2. anti-clockwise loops
3. identify all the possible MBPS

Solution

- **Clockwise Loops:**
  Loop no. 1: $R_1 \rightarrow R_7 \rightarrow R_6 \rightarrow R_1$
  Loop no. 2: $R_4 \rightarrow R_5 \rightarrow R_{10} \rightarrow R_4$
  Loop no. 3: $R_1 \rightarrow R_7 \rightarrow R_{10} \rightarrow R_4 \rightarrow R_1$
- **Anti-Clockwise Loops:**
  Loop no. 1: $R_2 \rightarrow R_5 \rightarrow R_8 \rightarrow R_2$
  Loop no. 2: $R_3 \rightarrow R_9 \rightarrow R_6 \rightarrow R_3$
  Loop no. 3: $R_2 \rightarrow R_3 \rightarrow R_9 \rightarrow R_8 \rightarrow R_2$
- **Identify All the Possible MBPS:**
  Before listing all the possible MBPS, we can observe the following:
  - There are two mutual relays for the clockwise and anti-clockwise simple loops (i.e. the first two loops of each direction), and these two relays are R_5 and R_6.

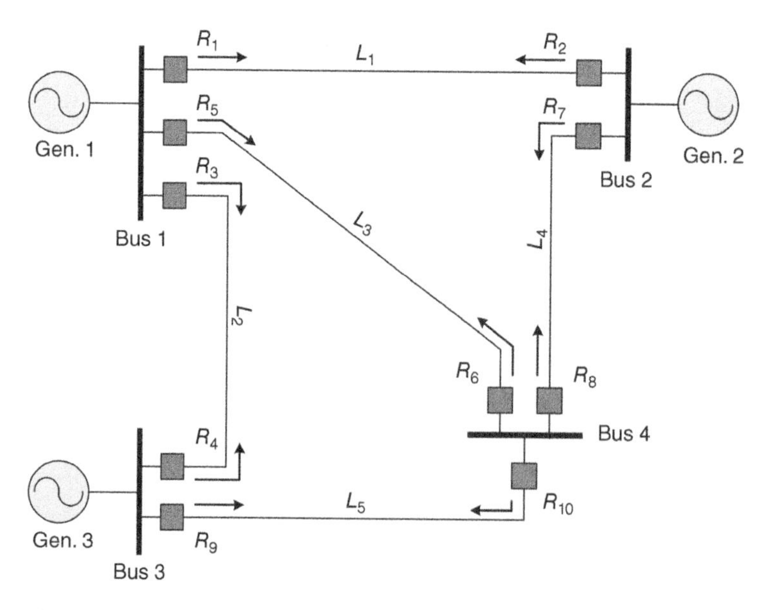

Figure 5.13 A simple four-bus system for Example 5.9.

- Because the third clockwise and anti-clockwise loops do not have R_5 nor R_6, so we have to select only one of them; i.e. either R_5 or R_6. If both relays are selected, then we will be forced to select two more relays to break the third loop of each direction, which ends up with a set bigger than MBPS.
- If R_5 is selected to break the second clockwise loop and the first anti-clockwise loop, then we can select either R_1 or R_7 to break the first and third clockwise loops and either R_3 or R_9 to break the second and third anti-clockwise loops.
- If R_6 is selected to break the first clockwise loop and the second anti-clockwise loop, then we can select either R_4 or R_{10} to break the second and third clockwise loops and either R_2 or R_8 to break the first and third anti-clockwise loops.

Thus, we have eight possible MBPS, which can be collected into two different groups as follows:

$$\left\{R_5, \left\{\begin{matrix} R_1 \\ R_7 \end{matrix}\right\}, \left\{\begin{matrix} R_3 \\ R_9 \end{matrix}\right\}\right\}, \quad \left\{R_6, \left\{\begin{matrix} R_4 \\ R_{10} \end{matrix}\right\}, \left\{\begin{matrix} R_2 \\ R_8 \end{matrix}\right\}\right\}$$

The full list can be extracted from the aforementioned two dense groups as follows:

$$\{R_5, R_1, R_3\},\ \{R_5, R_1, R_9\},\ \{R_5, R_7, R_3\},\ \{R_5, R_7, R_9\},\ \{R_6, R_4, R_2\},$$
$$\{R_6, R_4, R_8\},\ \{R_6, R_{10}, R_2\},\ \text{and}\ \{R_6, R_{10}, R_8\}.$$

From the last two examples, we see that MBPS can be obtained by hand. This is true for small simple networks. For real-world applications, the mesh networks are highly interconnected where some branches could be taken out of service or returned back to service. Thus, identifying MBPS is not an easy task. The literature offers many techniques to do that automatically. For example, we can use graph theory (Dwarakanath and Nowitz, 1980), functional dependency (Jenkins et al., 1992), polynomial time approximation algorithms (Gajbhiye et al., 2005), integer linear programming (ILP) (Gajbhiye et al., 2007), series of matrix computations (Liu and Fu, 2017), and many flavors of combinatorial optimization algorithms (Soman, 2010); especially with meta-heuristic optimization algorithms. Such sophisticated MBPS identification algorithms are used for high dimensional test systems. For example, the study reported in Matthews et al. (2019) show that there are 565 distinct breakers in the MBPS for the Utility 3120-bus test system. How to optimally coordinate these interconnected networks is another story. To do that, different computerized methods are presented in the literature. One of these methods is described in Soman (2010).

5.6 Summary

We have seen that ITOCRs can solve the inherent limitations of DCOCRs and DTOCRs. However, non-directional OCRs are non-selective protective devices. To protect mesh and non-radial networks,[9] DOCRs are preferred (Urdaneta et al., 1997; Paithankar and Bhide, 2003; Gers and Holmes, 2004). They are used as primary relays for interconnected sub-transmission and distribution systems, and as local backup relays for transmission systems. For radial systems, the optimal coordination can be obtained in one pass of calculations. On the opposite side, ring systems require an iterative process to optimize *TMS*. One of the classical ways is to use the Gauss–Seidel method. This optimization technique requires to select one relay from clockwise and anti-clockwise loops and then guess an initial value for its *TMS*. This initial start should

9 For example, multi-loop systems, ring feeder systems, double-end fed power systems, and single-end fed power systems of parallel feeders (Albasri et al., 2015).

not be close to TMS^{min} to avoid violating that side constraint. However, it does not guarantee to reach feasible solutions. Furthermore, PS of all relays are predetermined. This user-defined vector could be improper or could lead to infeasible or/and non-optimal solutions. Moreover, we cannot use this primitive optimization technique to optimally coordinate highly interconnected mesh networks. One way to tackle this issue is to identify the MBPS and then coordinating the relays based on this set.

This book aims to get rid of all these weary computations and highly complicated steps. That is, we want to optimally coordinate all the relays without fixing PS, guessing proper initial values for TMS, identifying clockwise and anti-clockwise loops, or identifying MBPS by complicated algorithms. This is the goal of the next chapters.

Problems

Written Exercises

W5.1 Repeat Example 5.1 when T^{min} and CTI are equal to 0.1 and 0.3 s, respectively.

W5.2 Repeat Example 5.2 with taking $X_{21} = X_{32} = 0.3\ \Omega$ and $X_S = 0.95\ \Omega$. The second relay has a margin of 135%.

W5.3 Repeat Example 5.3 when a numerical CO8 relay is used with $CTI = 0.3$ s. Use the general formula given in (3.36) to compute the time dial setting and trip time of relays. Don't change the values of $\{CTS_1, CTS_2, CTS_3\}$. Take $TDS_1 = 0.5$ and the step-size resolution of $\Delta TDS = 0.001$.

W5.4 Repeat Example 5.3 when a solid-state North American relay is used with $CTI = 0.35$ s. Use the formula given in (3.27) and the modified very inverse time characteristic. Take $TDS_1 = 0.5$ and the step-size resolution of $\Delta TDS = 0.025$.

W5.5 State the other names of the "coordination time interval."

W5.6 State the other names of the "over-shoot time."

W5.7 If the margin between primary and backup electromechanical relays is 0.4 s and the speed of the circuit breaker is 0.15 s from the moment it receives the trip signal from the primary relay. Calculate the over-shoot time of the primary relay if:
- There is no safety margin.
- The safety margin is 0.1 s.

W5.8 Consider Example 5.4. If $PSM_{R_2} = 3$, calculate the trip time of R_1.

W5.9 Repeat Exercise W5.8 with the IEC extremely inverse and short-time inverse models.

W5.10 Consider Exercise W5.8. If the two relays are numerical and manufactured in North America, calculate the trip time of R_1 when the IEC very inverse model is implemented. If $TMS \in [0.025, 1.5]$ and $TDS \in [0.1, 12.5]$, find the operating time of R_1 using:

1. fixed divisor method
2. linear interpolation method

W5.11 Consider the six-bus system shown in Figure 5.14. Find all the clockwise and anti-clockwise loops.

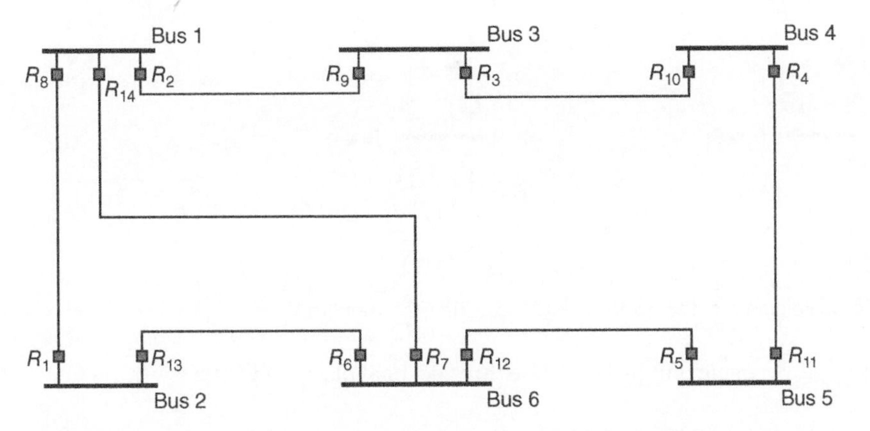

Figure 5.14 Single-line diagram of the six-bus system used in Exercise W5.11.

W5.12 Consider the six-bus system shown in Figure 5.15. Find all the clockwise and anti-clockwise loops.

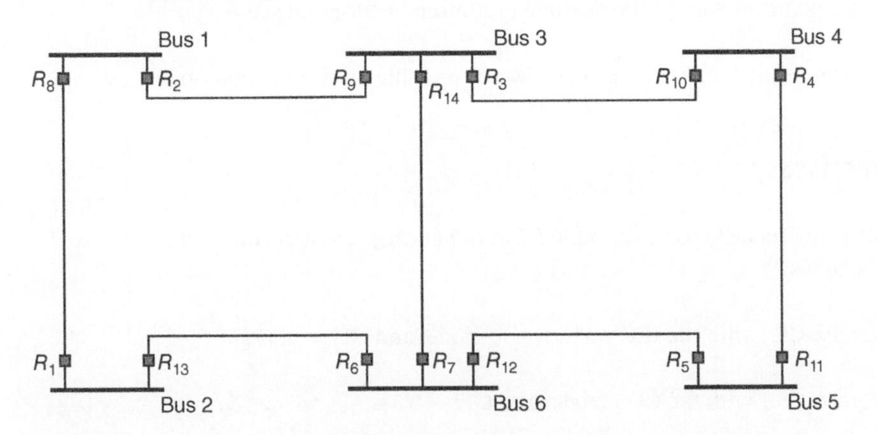

Figure 5.15 Single-line diagram of the six-bus system used in Exercise W5.12.

W5.13 Consider the five-bus system shown in Figure 5.12. Assume that a new branch is established between bus 4 and bus 5 and protected by two DOCRs (R_{13} and R_{14}). The first device is installed close to the end connected to bus 4 and the other is installed close to the other end of the line. The modified five-bus system is shown in Figure 5.16. Find all the clockwise and anti-clockwise loops.

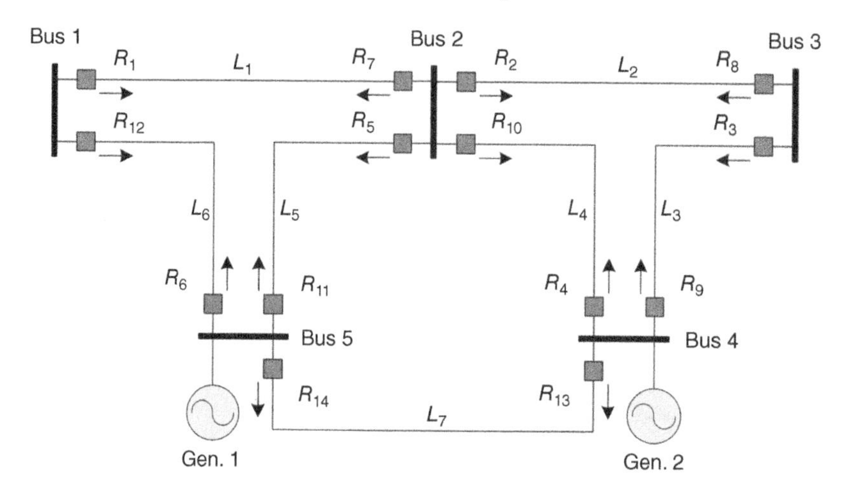

Figure 5.16 Single-line diagram of the five-bus system used in Exercise W5.13.

W5.14 Solve Example 5.6 again, but by hand. Use three decimal places in your answer.

W5.15 For the system shown in Figure 5.14, how many relays are required to achieve the minimum break-point set. List all the possible minimum break-point sets.

W5.16 For the system shown in Figure 5.15, how many relays are required to achieve the minimum break-point set. List all the possible minimum break-point sets.

W5.17 For the system shown in Figure 5.16, show one possible minimum break-point set.

Computer Exercises

C5.1 Design a MATLAB code to solve Example 5.5 automatically. Use a tolerance of $\varepsilon = 1 \times 10^{-16}$ to break the loops.

C5.2 Repeat Exercise C5.1 with the IEC very inverse model and $CTI = 0.35$ s.

C5.3 Repeat Exercise C5.1 with the CO5 model and $CTI = 0.3$ s.

C5.4 Modify the code given in Example 5.6 to coordinate the six relays when the IEEE long time inverse model is used.

C5.5 If the six relays shown in Example 5.6 are recently manufactured in North America and you, as a protection engineer, have been asked to program the IAC 51 model with $CTI = 0.325$ s. Design a MATLAB code to coordinate these relays. Use the formula given in (3.34).

C5.6 If the six relays shown in Example 5.6 are recently manufactured in Europe and you, as a protection engineer, have been asked to program the IAC 51 model with $CTI = 0.325$ s. Design a MATLAB code to coordinate these relays. Use the formula given in (3.34).

6

General Mechanism to Optimally Coordinate Directional Overcurrent Relays

Aim

The goal of this chapter is to give the reader the knowledge required to optimally coordinate DOCRs. It can be considered as the first step to understand the main concept of ORC problems and how to model them mathematically. Thus, the reader must know the content of this chapter before jumping to the next more advanced chapters. Of course, knowing this chapter requires some background in power system protection and classical/modern optimization algorithms, which were the goals of the preceding chapters 1 and 2.

From the literature review, the historical timeline of the ORC problem has been presented. Also, some of the fundamentals have been covered in Chapter 2, which are very essential for those who want to understand the optimal coordination of DOCRs.

This problem can be solved numerically, and the optimal (*or at least near optimal*) solution could be obtained by using many optimization techniques; which was covered in detail in Chapter 1.

The content of this chapter will focus on how to formulate the coordination problem of DOCRs into a mathematical model so that it can be solved later by any available n-dimensional optimization technique.

General Program Requirements: To build a program as a numerical ORC solver to optimally coordinate all the relays between each other, the stages shown in the flowchart[1] of Figure 6.1 must be satisfied. The following Sections 6.1, 6.2, 6.3, 6.4, and 6.5 describe each stage of this flowchart.

6.1 Constructing Power Network

This stage is responsible to provide all the required information about the given network, such as its type (*radial network, double-line network, ring network, etc.*), number of buses, number of lines, line parameters (*resistance, inductance, capacitance, and conductance*), number of generators and loads with their rated values, types of the protective devices and their TCCCs, types of faults and their locations, network's topology (*fixed or dynamic*), the status of circuit breakers (CBs) (*energized or de-energized*), etc.

1 This flowchart shows how to coordinate DOCRs with a steady-state network. The adaptive version of the ORC solver will be discussed later in Chapter 14.

Optimal Coordination of Power Protective Devices with Illustrative Examples, First Edition. Ali R. Al-Roomi.

Companion website: www.wiley.com/go/al-roomi/optimalcoordination

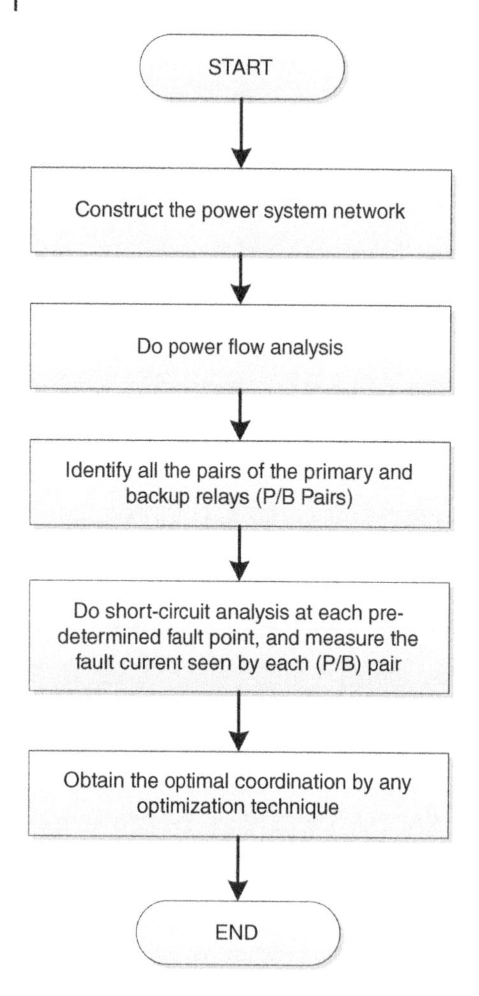

Figure 6.1 General flowchart to optimally coordinate DOCRs.

This can be considered as the first step of any proposed ORC problem because without satisfying this stage there is no other choice to proceed with the next stages. For example, if ETAP[2] is used with some missed information on this stage, then an error message will pop-up when the run button is pressed.

6.2 Power Flow Analysis

The term "**power flow (PF)**" as in some references (Momoh, 2001b; Wang et al., 2008; Zhu, 2009; Wood and Wollenberg, 2013; Grainger and Stevenson, 1994; Murthy, 2014; Saadat, 1999; Glover et al., 2012; Das, 2002; Sivanagaraju and Sreenivasan, 2010; Bergen, 1986; Nasar, 1990), or "**load flow (LF)**" as in others (Stagg and El-Abiad, 1968; Arrillaga and Arnold, 1990; Arrillaga and Watson, 2001; Kirtley, 2010; Das, 2006; El-Hawary, 1983; Kothari and Nagrath, 2003; Wang et al., 2008; Debs, 1988; Das, 2002; Lynn Powell, 2005), is frequently used as a subject of one of the most important tools in electric power systems engineering. Here are some nice sentences taken from textbook

2 It is a commercial electrical software.

authors; Saadat (1999) in page 189: ***"Power flow studies are the backbone of power system analysis and design,"*** Bergen (1986, p. 150): ***"It is an integral part of studies in system planning and operation and is, in fact, the most common of power system computer calculations,"*** and finally by Grainger and Stevenson (1994) in page 329: ***"Power flow studies are of great importance in planning and designing the future expansion of power systems as well as in determining the best operation of existing systems."*** This repeated meaning gives a solid conclusion that all electric power systems need some sorts of PF studies to have the ability to measure, monitor, analyze, estimate, predict, and control many variables and parameters to maintain these systems secure and at their optimal conditions (Debs, 1988). It is clearly highlighted by El-Hawary (2008, p. 319): ***"An ubiquitous EMS[3] application software is the PF program, which solves for network state given specified conditions throughout the system."***

Therefore, to understand the importance of PF studies, the following questions should be raised first: *What does "power flow" mean? What are the techniques used to solve it? What are the pros and cons of each one of these techniques?*

It is well known that modern electric power systems are highly interconnected. These systems are represented by branches and nodes with some **injected sources** and some **consumption points**. The injected sources represent the generating units (such as nuclear, thermal,[4] tidal, hydro, wind farms, solar stations, etc.) connected to the grid. The branches are called powerlines (transmission, sub-transmission, and distribution lines), which are connected between each other through some nodes called busbars. The consumption points are defined as loads; they could be sub-transmission customers (26–69 kV), primary customers (13–4 kV), secondary customers (120 V/60 Hz "**American Standard**" or 240 V/50 Hz "**European Standard**"), or even power station auxiliary plants (air and gas compressors, lube oil cooling systems, lightings, etc.). Batteries, **ultracapacitors**,[5] and **flywheels** are special power components. These bi-directional elements act as loads when there is enough power flowing in the grid, and act as power sources in case there is a shortage in the production of electricity.

After modeling these electric power components from their physical structures to some mathematical expressions, they can then be represented as a power electric circuit. Although network equations can be formulated in different forms, the most common one used for power system analysis is called the **node voltage method** (El-Hawary, 2008; Das, 2006). If the given network is formulated in a **nodal admittance form**, then by the node currents the network can be expressed as **linear algebraic equations**. But practically, electric systems are represented by power values instead of currents, which results in a set of **nonlinear algebraic equations** called "**power flow equations**." These equations can be solved by **iterative techniques** (Saadat, 1999; Murthy, 2014; Husain, 2007). That is, solving PF equations leads to knowing the voltage magnitude $|V|$ and its phase angle δ at each busbar, and the real power P and reactive power Q flowing through each branch. Moreover, from these essential data, much other information can be easily calculated in some sub-algorithms embedded within EMS, which can be used later for many other studies (Grainger and Stevenson, 1994).

Based on this brief introduction, it can be said that PF analysis is carried out to ensure that the following requirements are satisfied (Glover et al., 2012; Murthy, 2014; Gers and Holmes, 2004):

1. Each bus voltage magnitude is close to its rated value:

$$|V_i| \approx V_i^{\text{rated}} \tag{6.1}$$

---

3 EMS: Energy Management System.
4 Some types are gas turbines (GTs), steam turbines (STs), gas generators, diesel generators, etc.
5 Also called **supercapacitors (SCs)**.

2. The total power generation (P_G) should meet the total consumed power, i.e. the summation of power demand (P_D) and power losses (P_L), as follows:

$$P_G = P_D + P_L \tag{6.2}$$

3. All the generators should not exceed the specified lower and upper limits of real and reactive power:

$$P_G^{\min} \leq P_G \leq P_G^{\max} \tag{6.3}$$

$$Q_G^{\min} \leq Q_G \leq Q_G^{\max} \tag{6.4}$$

4. Lines and transformers are not overloaded:

$$I \leq I_L^{\max} \times OLF, \quad \text{where} \quad OLF = 1.25 \text{ to } 1.5 \tag{6.5}$$

Thus, the independent variables here are the voltage magnitude ($|V|$), the angle (δ), the real power (P), and the reactive power (Q) (Glover et al., 2012; Husain, 2007). In PF, each bus has two **known (specified) variables** and two **unknown (unspecified) variables**. The busbar type depends upon the known variables and can be summarized in Table 6.1.

6.2.1 Per-Unit System and Three-to-One-Phase Conversion

Solving PF with real quantities can complicate the solution because real interconnected power systems have different voltage levels and thus require the programmers and operation engineers to transform the entire network to have one voltage level (Saadat, 1999). Instead of this tedious and non-preferable approach, all the actual quantities of voltage (V), current (I), impedance (Z), admittance (Y), and complex power (S) can be normalized with respect to some chosen reference or base values (El-Hawary, 1983). Based on this, all these actual quantities can then be expressed to a one standardized quantity known as a **per-unit quantity** (or just **pu**) (Murthy, 2014). One of the main advantages of using per unit values is that no more complicated circuits are needed to represent power transformers where the values of simplified ideal transformers are same by referring to any side of the transformers (Glover et al., 2012). Inverse transformation can be done at any time to return the network's per-unit values back to their real formats (i.e. V, A, Ω, ℧, VA, W, VAR).

After accomplishing this point, the next important one is about simplifying the entire system from being a 3ϕ network to a simple 1ϕ network. The reason behind this action is because the PF analysis is supposed to be carried out on an interconnected system operated under a normal balanced steady-state condition. Thus, dealing with a per-unit positive-sequence single-line diagram is relatively quite simple compared with a polyphase network containing different levels of voltage.

Table 6.1 Bus types and their known/unknown PF variables.

Busbar type	Known	Unknown
Swing bus[a]	$\|V_i\|$ and δ_i	P_i and Q_i
Generator bus[b]	P_i and $\|V_i\|$	Q_i and δ_i
Load bus[c]	P_i and Q_i	$\|V_i\|$ and δ_i

a) Also called **slack** or **reference bus.**
b) Also called **voltage-controlled** or **PV bus.**
c) Also called **PQ bus.**

For modeling transmission lines, there are two approaches[6]:

1. The **lumped parameter** model.[7]
2. The **distributed parameter** model.

The last approach is also called the *lumped component model* and the *lumped element model.* The distributed parameter approach is more accurate, which is used when the length is longer than 250 km (~155 miles). If the length is less, then the lumped parameter models can be used. These simplified models are categorized as follows (Saadat, 1999):

- The short-length transmission line model, which is used when $\mathfrak{L} \leqslant 80$ km (~50 miles).
- The medium-length transmission line models, which are used when 80 km (~50 miles) $< \mathfrak{L} \leqslant 250$ km (~155 miles).

The most popular medium-length transmission line circuits are called the Γ-circuit, the ⅂-circuit, the T-circuit, and the Π-circuit. The most accurate medium-length transmission line model is derived based on the M-circuit[8] (Al-Roomi and El-Hawary, 2020d).

Compared with the Γ-model and ⅂-model, the T-model and Π-model are more accurate. As a rule of thumb, the T-model is not preferred in most cases because it creates a virtual node between the sending-end and receiving-end buses on each branch[9] (Arrillaga and Watson, 2001; Kirtley, 2010).

6.2.2 Power Flow Solvers

Nowadays, there are many PF techniques proposed in the literature. Some of them are listed in the following text:

1. **Jacobi Method (JM)** (Lynn Powell, 2005)
2. **Gauss Family**[10]: Conventional Gauss–Seidel (GS),[11] Gauss–Seidel with an acceleration factor called α (Stagg and El-Abiad, 1968; Bergen, 1986; Grainger and Stevenson, 1994; Saadat, 1999; Momoh, 2001b; Das, 2002; Kothari and Nagrath, 2003; Lynn Powell, 2005; Das, 2006; Husain, 2007; Murthy, 2014; Zhu, 2009; Kirtley, 2010; Glover et al., 2012)
3. **Z-Matrix Methods** (Brown et al., 1963; Stagg and El-Abiad, 1968; Das, 2002; Lynn Powell, 2005; Murthy, 2014)
4. **Newton–Raphson (NR)**: Rectangular, Polar, and Hybrid Forms (Stagg and El-Abiad, 1968; El-Hawary, 1983; Bergen, 1986; Debs, 1988; Arrillaga and Arnold, 1990; Grainger and Stevenson, 1994; Saadat, 1999; Arrillaga and Watson, 2001; Momoh, 2001b; Das, 2002; Kothari and Nagrath, 2003; Lynn Powell, 2005; Das, 2006; Murthy, 2014; Wang et al., 2008; Zhu, 2009; Glover et al., 2012)

6 Hybrid approaches could also be derived to compromise between simplicity and accuracy (Al-Roomi, 2020).
7 It is also called the **lumped component** model and the **lumped element** model.
8 There are other possible medium-length transmission line models, such as the Equals-Sign-model, the $\subset$-model, the $\supset$-model, the I-model, the O-model, and the ∞-model (Al-Roomi, 2020).
9 However, the T-model is useful in some studies, such as modeling non-ideal transformers.
10 The main objective of using this algorithm is to have indirect solution of inverse matrices per each iteration, especially when dealing with radial networks where many matrices' elements are zero (i.e. sparse matrices). One of the direct solutions is to use Gauss elimination by lower-upper (LU) triangular factorization technique with forward and backward substitution and optimal ordering of the network's nodes (Glover et al., 2012; Debs, 1988), which is a very effective tool used with Newton–Raphson solver.
11 It is similar to JM. The improvement here is that GS utilizes the newly updated independent variables to iterate the remaining variables within the same iteration (Saadat, 1999; Husain, 2007).

5. **Decoupled Family**: Decoupled Newton and fast-decoupled load flow (FDLF) (Bergen, 1986; Debs, 1988; Arrillaga and Arnold, 1990; Grainger and Stevenson, 1994; Saadat, 1999; Arrillaga and Watson, 2001; Momoh, 2001b; Das, 2002; Kothari and Nagrath, 2003; Lynn Powell, 2005; Das, 2006; Murthy, 2014; Wang et al., 2008; Zhu, 2009; Glover et al., 2012)
6. **Jacobian-Free Newton–Krylov (JFNK) Methods** (Knoll and Keyes, 2004; De Rubira, 2012)
7. **Optimal Multiplier Load Flow** (De Rubira, 2012)
8. **Continuous Newton Load Flow** (De Rubira, 2012)
9. **Sequential Conic Programming** (De Rubira, 2012)
10. **Holomorphic Embedding Method (HEM)** (Trias, 2012; De Rubira, 2012)
11. **Quadratized Power Flow** (Lynn Powell, 2005)
12. **DC Load Flow** (Debs, 1988; Momoh, 2001b; Das, 2002; Lynn Powell, 2005; Wang et al., 2008; Zhu, 2009; Glover et al., 2012)
13. **AC-DC Load Flow** (Debs, 1988)
14. **Tellegen's Theorem Based Power Flow** (Debs, 1988; Ferreira, 1990)
15. **Multiport Compensation-Based Power Flow** (Shirmohammadi et al., 1988)
16. **Forward Sweep Algorithm** (Babu et al., 2010)
17. **Three-Order Convergence Based Algorithm** (Sun et al., 2009)
18. **Load Flow Technique Based on Load Transfer and Load Buses Elimination** (Chung et al., 1997)
19. **Branch-to-Node Matrix-Based Power Flow (BNPF)** (Aravindhababu et al., 2001)
20. **Complex Power-Complex Voltage (S-E) Graph Technique** (Ilic-Spong and Zaborszky, 1982)
21. **S-E Oriented Load Flow** (Khodr et al., 2006)
22. **Hybrid Power Flow Methods** (Wang et al., 2005)
23. **Bulirsch–Stoer Method** (Tostado-Véliz et al., 2019)
24. **S-Iteration Process** (Tostado-Véliz et al., 2020a)
25. **Romberg's Integration Scheme** (Tostado-Véliz et al., 2020b)
26. **Meta-Heuristic Based Power Flow Techniques** (Elrayyah et al., 2013; Ou et al., 2013; Wong et al., 2003)
27. **Many Others** (Chang et al., 1994; Mbamalu et al., 1995; Meliopoulos et al., 2003; Kumar and Rao, 2014)

In terms of accuracy, the worst method is the DC load flow, which becomes the best method in terms of processing time. This method is just used in some special applications, like contingency analysis and quick optimal pricing calculations. Also, it is good for getting a general figure or initial point to estimate some online scenarios where the processing time is the most critical factor and at the same time some decimal places of tolerance can be sacrificed. It is important to say that, for accurate and precise calculations, this method is totally discarded (Conejo, 2011). If Tellegen's theorem is applied here instead, as in Ferreira (1990) and Debs (1988), then it might provide good results with very limited usage of memory.[12] The JM and GS methods have simple calculation steps, which make them easy to program. Also, they require less memory and processing time. However, they are sensitive to the slack bus selection. Further, as the network size increases the algorithms utilize more iterations, which is the case faced with real power networks. This phenomenon creates a bold usage limitation (Husain, 2007). The most popular one is the NR method. Some of its main

12 The other choice is to involve AI to achieve highly accurate PF results within a very short time (Al-Roomi and El-Hawary, 2020a).

advantages are its high accuracy and quick convergence rate without depending on the network size or the slack bus selection.

6.2.3 How to Apply the Newton–Raphson Method

Before ending this section, it is interesting to give a brief overview about how to involve the NR method in order to solve the load flow problems. Thus, it can give a nice figure about this essential stage of ORC. Let us consider the typical bus shown in Figure 6.2.

Applying the **Kirchhoff's current law (KCL)** to this bus yields:

$$\begin{aligned} I_i &= y_{i0}V_i + y_{i1}\left(V_i - V_1\right) + y_{i2}\left(V_i - V_2\right) + \cdots + y_{in}\left(V_i - V_n\right) \\ &= \left(y_{i0} + y_{i1} + y_{i2} + \cdots + y_{in}\right)V_i - y_{i1}V_1 - y_{i2}V_2 - \cdots - y_{in}V_n \end{aligned} \tag{6.6}$$

We can re-arrange (6.6) to have:

$$I_i = V_i\sum_{j=0}^{n} y_{ij} - \sum_{j=1}^{n} y_{ij}V_j \quad j \neq i \tag{6.7}$$

If the ith bus is included in the jth vector, then (6.7) can be re-written in terms of the bus admittance matrix as follows:

$$I_i = \sum_{j=1}^{n} Y_{ij}V_j \tag{6.8}$$

The term Y_{ii} will contain all the admittances connected to the ith bus, including the line charging admittance (y_{i0}); *see* (6.6). The polar form of (6.8) is:

$$I_i = \sum_{j=1}^{n} |Y_{ij}||V_j|\angle\theta_{ij} + \delta_j \tag{6.9}$$

Because the complex power at the ith bus is:

$$\begin{aligned} P_i - jQ_i &= V_i^* I_i \\ &= |V_i|\angle - \delta_i\sum_{j=1}^{n} |Y_{ij}||V_j|\angle\theta_{ij} + \delta_j \end{aligned} \tag{6.10}$$

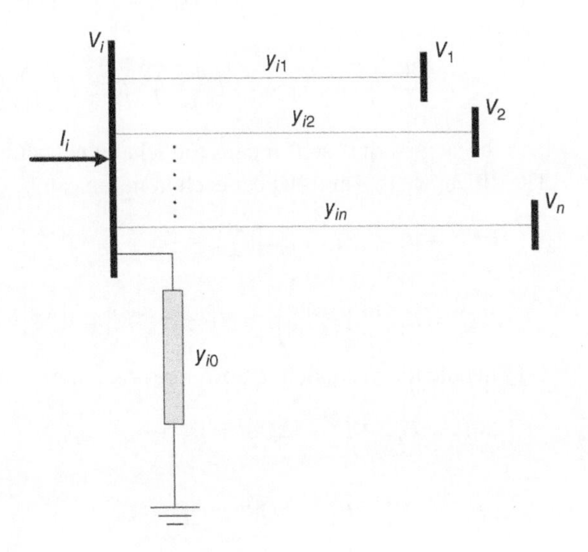

Figure 6.2 Interconnection between the ith busbar and other busbars.

Therefore, by separating the real and imaginary parts, we can get the following two expressions:

$$P_i = \Re\left\{V_i^* I_i\right\} = \sum_{j=1}^{n} |V_i||V_j||Y_{ij}| \cos\left(\theta_{ij} - \delta_i + \delta_j\right) \tag{6.11}$$

$$Q_i = -\Im\left\{V_i^* I_i\right\} = -\sum_{j=1}^{n} |V_i||V_j||Y_{ij}| \sin\left(\theta_{ij} - \delta_i + \delta_j\right) \tag{6.12}$$

With neglecting the first bus, which is considered as a slack bus, the aforementioned derivation can be summarized in the following two points:

- For PQ-buses, we have two equations; (6.11) and (6.12).
- For PV-buses, we have only one equation; (6.11).

Expanding (6.11) and (6.12) by **Taylor's series**, will get something like the following:

$$\left[\begin{array}{c} \Delta P^{(k)} \\ \hline \Delta Q^{(k)} \end{array}\right] = \left[\begin{array}{c|c} J_1^{(k)} & J_2^{(k)} \\ \hline J_3^{(k)} & J_4^{(k)} \end{array}\right] \cdot \left[\begin{array}{c} \Delta\delta^{(k)} \\ \hline \Delta|V^{(k)}| \end{array}\right] \tag{6.13}$$

where k indicates the iteration number, which is stopped once the acceptable tolerance (ε) is reached. The four Js are sub-matrices, which contain partial derivative equations. These four sub-matrices are grouped to construct what is called the **Jacobian matrix (J)**. The unknowns here are the voltage magnitude ($|V|$) and phase angles (δ) for PQ-buses, and just phase angles for PV-buses. All the unknown reactive power Qs can be calculated later by (6.12) once $|V|$ and δ of all PQ- and PV-buses are numerically determined.

To see how (6.13) has been derived, let's expand (6.11) and (6.12) by Taylor's series to have:

$$\begin{array}{l}
P_2^{(k)} + \left(\frac{\partial P_2}{\partial \delta_2}\right)^{(k)} \Delta\delta_2^{(k)} + \cdots + \left(\frac{\partial P_2}{\partial \delta_n}\right)^{(k)} \Delta\delta_n^{(k)} + \left(\frac{\partial P_2}{\partial |V_2|}\right)^{(k)} \Delta|V_2^{(k)}| + \cdots + \left(\frac{\partial P_2}{\partial |V_n|}\right)^{(k)} \Delta|V_2^{(k)}| = P_2^{\mathrm{sch}} \\
\vdots \qquad \vdots \qquad \ddots \qquad \vdots \qquad \vdots \qquad \ddots \qquad \vdots \qquad = \vdots \\
P_n^{(k)} + \left(\frac{\partial P_n}{\partial \delta_2}\right)^{(k)} \Delta\delta_2^{(k)} + \cdots + \left(\frac{\partial P_n}{\partial \delta_n}\right)^{(k)} \Delta\delta_n^{(k)} + \left(\frac{\partial P_n}{\partial |V_2|}\right)^{(k)} \Delta|V_2^{(k)}| + \cdots + \left(\frac{\partial P_n}{\partial |V_n|}\right)^{(k)} \Delta|V_2^{(k)}| = P_n^{\mathrm{sch}} \\
Q_2^{(k)} + \left(\frac{\partial Q_2}{\partial \delta_2}\right)^{(k)} \Delta\delta_2^{(k)} + \cdots + \left(\frac{\partial Q_2}{\partial \delta_n}\right)^{(k)} \Delta\delta_n^{(k)} + \left(\frac{\partial Q_2}{\partial |V_2|}\right)^{(k)} \Delta|V_2^{(k)}| + \cdots + \left(\frac{\partial Q_2}{\partial |V_n|}\right)^{(k)} \Delta|V_2^{(k)}| = Q_2^{\mathrm{sch}} \\
\vdots \qquad \vdots \qquad \ddots \qquad \vdots \qquad \vdots \qquad \ddots \qquad \vdots \qquad = \vdots \\
Q_n^{(k)} + \left(\frac{\partial Q_n}{\partial \delta_2}\right)^{(k)} \Delta\delta_2^{(k)} + \cdots + \left(\frac{\partial Q_n}{\partial \delta_n}\right)^{(k)} \Delta\delta_n^{(k)} + \left(\frac{\partial Q_n}{\partial |V_2|}\right)^{(k)} \Delta|V_2^{(k)}| + \cdots + \left(\frac{\partial Q_n}{\partial |V_n|}\right)^{(k)} \Delta|V_2^{(k)}| = Q_n^{\mathrm{sch}}
\end{array}$$

where the superscript "sch" means the scheduled value.

The difference in δ and $|V|$ per each iteration can be calculated as follows:

$$\begin{array}{l}
\Delta\delta_2^{(k)} = \delta_2^{(k+1)} - \delta_2^{(k)} \;,\; \Delta|V_2^{(k)}| = |V_2^{(k+1)}| - |V_2^{(k)}| \\
\vdots \qquad\qquad \vdots \qquad\qquad \vdots \\
\Delta\delta_n^{(k)} = \delta_n^{(k+1)} - \delta_n^{(k)} \;,\; \Delta|V_n^{(k)}| = |V_n^{(k+1)}| - |V_n^{(k)}|
\end{array}$$

Let's denote the mismatch in power per each iteration by the following expressions:

$$\begin{array}{l}
\Delta P_2^{(k)} = P_2^{\mathrm{sch}} - P_2^{(k)} \;,\; \Delta Q_2^{(k)} = Q_2^{\mathrm{sch}} - Q_2^{(k)} \\
\vdots \qquad\qquad \vdots \qquad\qquad \vdots \\
\Delta P_n^{(k)} = P_n^{\mathrm{sch}} - P_n^{(k)} \;,\; \Delta Q_n^{(k)} = Q_n^{\mathrm{sch}} - Q_n^{(k)}
\end{array}$$

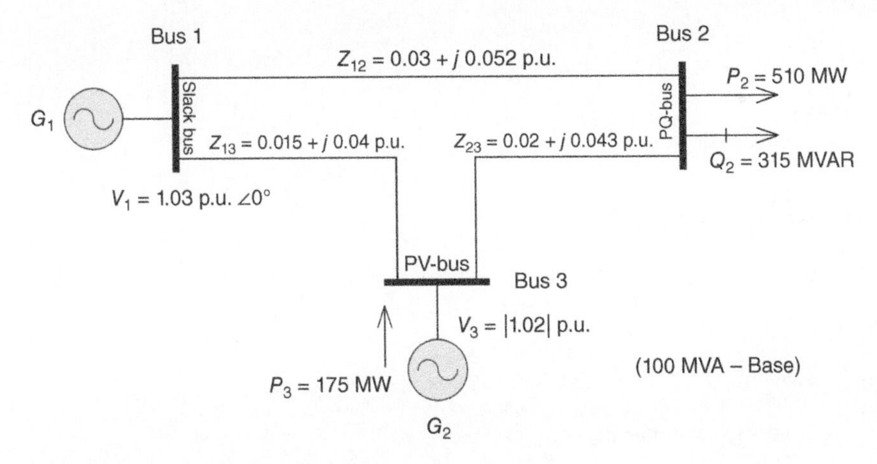

Figure 6.3 A simple three-bus system used in Example 6.1.

Then, the above equations can be represented using the matrix form as follows:

$$\begin{bmatrix} \delta_2^{(k+1)} \\ \vdots \\ \delta_n^{(k+1)} \\ \hline |V_2^{(k+1)}| \\ \vdots \\ |V_n^{(k+1)}| \end{bmatrix} = \begin{bmatrix} \delta_2^{(k)} \\ \vdots \\ \delta_n^{(k)} \\ \hline |V_2^{(k)}| \\ \vdots \\ |V_n^{(k)}| \end{bmatrix} + \left[\begin{array}{ccc|ccc} \left(\frac{\partial P_2}{\partial \delta_2}\right)^{(k)} & \cdots & \left(\frac{\partial P_2}{\partial \delta_n}\right)^{(k)} & \left(\frac{\partial P_2}{\partial |V_2|}\right)^{(k)} & \cdots & \left(\frac{\partial P_2}{\partial |V_n|}\right)^{(k)} \\ \vdots & \ddots & \vdots & \vdots & \ddots & \vdots \\ \left(\frac{\partial P_n}{\partial \delta_2}\right)^{(k)} & \vdots & \left(\frac{\partial P_n}{\partial \delta_n}\right)^{(k)} & \left(\frac{\partial P_n}{\partial |V_2|}\right)^{(k)} & \vdots & \left(\frac{\partial P_n}{\partial |V_n|}\right)^{(k)} \\ \hline \left(\frac{\partial Q_2}{\partial \delta_2}\right)^{(k)} & \cdots & \left(\frac{\partial Q_2}{\partial \delta_n}\right)^{(k)} & \left(\frac{\partial Q_2}{\partial |V_2|}\right)^{(k)} & \cdots & \left(\frac{\partial Q_2}{\partial |V_n|}\right)^{(k)} \\ \vdots & \ddots & \vdots & \vdots & \ddots & \vdots \\ \left(\frac{\partial Q_n}{\partial \delta_2}\right)^{(k)} & \vdots & \left(\frac{\partial Q_n}{\partial \delta_n}\right)^{(k)} & \left(\frac{\partial Q_n}{\partial |V_2|}\right)^{(k)} & \vdots & \left(\frac{\partial Q_n}{\partial |V_n|}\right)^{(k)} \end{array}\right]^{-1} \cdot \begin{bmatrix} P_2^{sch} - P_2^{(k)} \\ \vdots \\ P_n^{sch} - P_n^{(k)} \\ \hline Q_2^{sch} - Q_2^{(k)} \\ \vdots \\ Q_n^{sch} - Q_n^{(k)} \end{bmatrix}$$

For small power networks, applying NR by hand is possible although it is weary. Practically, NR should be coded in numerical programming languages. Example 6.1 shows a MATLAB code used to solve PF problem of a simple three-bus system.

Example 6.1 Consider the network shown in Figure 6.3 where bus 1 is the reference bus. Design a MATLAB code to solve the load flow problem using NR. Set the maximum allowable iterations to 100 and the minimum acceptable error to $\varepsilon = 10^{-5}$. Take the initial start with $|V_2^{(0)}| = 1.0$ pu, $|V_3^{(0)}| = 1.02$ pu, and $\delta_2 = \delta_3 = 0°$. The remaining information is given in the network.

Solution

The code given in `NR3BusSystem.m` is:

```
1   clc
2   clear
3   V=[1.03; 1.0; 1.02]; % Bus voltages
4   d=[0; 0; 0];         % Bus Δ
5   Ps=[-5.1; 1.75];     % Supplied active power
6   Qs=-3.15;            % Supplied reactive power
7   % Lines impedances
8   z12 = 0.03 + 1i*0.052;
9   z13=0.015 + 1i*0.04;
10  z23=0.02 + 1i*0.043;
```

```
11      % Mutual admittances
12      y12 = 1 / z12;
13      y13 = 1 / z13;
14      y23 = 1 / z23;
15      % Self admittances
16      y11 = y12 + y13;
17      y22 = y12 + y23;
18      y33 = y13 + y23;
19      % Bus-Admittance Matrix
20      Ybus=[y11 -y12 -y13; -y12 y22 -y23; -y13 -y23 y33];
21      % Extracting the magnitudes and angles from Ybus
22      Y = abs(Ybus); t = angle(Ybus);
23      iter = 0;
24      PWRaccur = 0.00001; % Power accuracy
25      PMR = 100;          % Set the maximum power residual to a high value
26      while max(abs(PMR)) > PWRaccur
27          iter = iter+1
28          % Calculating real and imaginary parts of power
29          P2 = V(2) * V(1) * Y(2,1) * cos(t(2,1)-d(2)+d(1)) + V(2)^2 * ...
                Y(2,2) * cos(t(2,2)) + V(2) * V(3) * Y(2,3) * ...
                cos(t(2,3)-d(2)+d(3));
30          P3 = V(3) * V(1) * Y(3,1) * cos(t(3,1)-d(3)+d(1)) + V(3)^2 * ...
                Y(3,3) * cos(t(3,3)) + V(3) * V(2) * Y(3,2) * ...
                cos(t(3,2)-d(3)+d(2));
31          Q3 = -V(2) * V(1) * Y(2,1) * sin(t(2,1)-d(2)+d(1)) - V(2)^2 * ...
                Y(2,2) * sin(t(2,2)) - V(2) * V(3) * Y(2,3) * ...
                sin(t(2,3)-d(2)+d(3));
32          % Finding the Jacobian Matrix
33          J(1,1) = V(2) * V(1) * Y(2,1) * sin(t(2,1)-d(2)+d(1)) + V(2) * ...
                V(3) * Y(2,3) * sin(t(2,3)-d(2)+d(3));
34          J(1,2) = -V(2) * V(3) * Y(2,3) * sin(t(2,3)-d(2)+d(3));
35          J(1,3) = V(1) * Y(2,1) * cos(t(2,1)-d(2)+d(1)) + 2 * V(2) * ...
                Y(2,2) * cos(t(2,2)) +  V(3) * Y(2,3) * cos(t(2,3)-d(2)+d(3));
36          J(2,1) = -V(3) * V(2) * Y(3,2) * sin(t(3,2)-d(3)+d(2));
37          J(2,2) = V(3) * V(1) * Y(3,1) * sin(t(3,1)-d(3)+d(1)) + V(3) * ...
                V(2) * Y(3,2) * sin(t(3,2)-d(3)+d(2));
38          J(2,3) = V(3) * Y(2,3) * cos(t(3,2)-d(3)+d(2));
39          J(3,1) = V(2) * V(1) * Y(2,1) * cos(t(2,1)-d(2)+d(1)) + V(2) * ...
                V(3) * Y(2,3) * cos(t(2,3)-d(2)+d(3));
40          J(3,2) = -V(2) * V(3) * Y(2,3) * cos(t(2,3)-d(2)+d(3));
41          J(3,3) = -V(1) * Y(2,1) * sin(t(2,1)-d(2)+d(1)) - 2 * V(2) * ...
                Y(2,2) * sin(t(2,2)) - V(3) * Y(2,3) * sin(t(2,3)-d(2)+d(3));
42          % Calculating power residuals
43          DP = Ps - [P2; P3];
44          DQ = Qs - Q3;
45          PMR = [DP; DQ]; % Power mismatch resolution
46          DX = J \ PMR;
47          d(2) = d(2) + DX(1);
48          d(3) = d(3) + DX(2);
49          V(2) = V(2) + DX(3);
50          PMR, J, V, Δ = 180/pi*d,
51      end
52      % Calculating the remaining unknown variables
53      P1 = V(1)^2 * Y(1,1) * cos(t(1,1)) + V(1) * V(2) * Y(1,2) * ...
            cos(t(1,2)-d(1)+d(2)) + V(1) * V(3) * Y(1,3) * cos(t(1,3)-d(1)+d(3))
54      Q1 = -V(1)^2 * Y(1,1) * sin(t(1,1)) - V(1) * V(2) * Y(1,2) * ...
            sin(t(1,2)-d(1)+d(2)) - V(1) * V(3) * Y(1,3) * sin(t(1,3)-d(1)+d(3))
55      Q3 = -V(3) * V(1) * Y(3,1) * sin(t(3,1)-d(3)+d(1)) - V(3) * V(2) * ...
            Y(3,2) * sin(t(3,2)-d(3)+d(2)) - V(3)^2 * Y(3,3) * sin(t(3,3))
```

If this program is executed in MATLAB or GNU Octave environment, the following results will be obtained just after 5 iterations:

$$V_1 = 1.0300 \text{ pu } \angle 0.0000^\circ$$
$$V_2 = 0.8608 \text{ pu } \angle -6.6004^\circ$$
$$V_3 = 1.0200 \text{ pu } \angle -2.2626^\circ$$
$$S_1 = P_1 + jQ_1 = 3.9706 + j1.6539 \text{ pu} = 397.06 \text{ MW} + j165.39 \text{ MVAR}$$
$$S_2 = -P_2 - jQ_2 = -5.1000 - j3.1500 \text{ pu} = -510.00 \text{ MW} - j315.00 \text{ MVAR}$$
$$S_3 = P_3 + jQ_3 = 1.7500 + j2.6977 \text{ pu} = 175.00 \text{ MW} + j269.77 \text{ MVAR}$$

where $\{|V_1|, \delta_1, |V_3|, P_2, Q_2, P_3\}$ are specified and known from the beginning.

As can be seen, although the system is so simple, this kind of problems should be solved by computers because the branches are lossy. If someone tries to solve it manually by hand, then it will require tens pages. Add to that, many steps could need to be re-calculated again if any value is changed. Instead, the preceding program can find highly accurate solutions for any initial input by just one press.

6.2.4 Sparsity Effect

The Newton–Raphson technique is not preferred in some applications because it consumes a large amount of CPU time and data storage (Husain, 2007); especially with radial systems where most of the Jacobian matrix elements are zero. This happens because each bus is connected to a few number of lines. Thus, the admittance matrix $[Y]$ of such large networks will have few non-zero elements. This phenomenon makes NR-based PF solvers slower and forces them to consume more memory and storage capacity to deal with large $[Y]$ and $[J]$. Modern commercial software take into account the sparsity of electrical networks. These sparsity techniques are employed to have compact storage of system data (Glover et al., 2012).

Example 6.2 Assume that the lower and upper off-diagonal non-zero elements σ are equal or three times that of the diagonal elements (i.e. $\sigma = 1 \to 3$). If the matrix of NR has a size of $[5000 \times 5000]$, determine how many elements are zero is the matrix.

Solution

With a $[5000 \times 5000]$ matrix, the following useless memory can be faced if it is implemented in a practical application:

- The total matrix elements: $m^2 = 5000^2 = 25\ 000\ 000$
- The total non-zero elements: $m + 2\sigma m = (1 + 2\sigma)\, m = 15\ 000$ to $35\ 000$

Imagine! There are 24 965 000 elements, which are saved just as zeros!! This logical astonishment can be clearly seen during coding NR in any specialized numerical programming language. The computer code given in Example 6.3 clarify this by showing the memory usage in MATLAB.

Example 6.3 Assume that the ratio of off-diagonal to diagonal non-zero elements is 300%. Calculate the memory used in MATLAB to deal with $[20 \times 20]$, $[200 \times 200]$, and $[2000 \times 2000]$ matrices.

Solution

The code given in `MatrixSparsity.m` is:

```
1   clc
2   clear
3   n = [20, 200, 2000]; % small, medium, and large dimension matrices
4   a = 3; % ratio of off-diagonal to diagonal non-zero elements
5   NZE = (1+2*a)*n; % non-zero elements
6   display('------------------------------------------------------');
7   for k=1:length(n)
8       A = zeros(n(k), n(k)); % to create [n x n] matrix
9       d = diag(A); % to select diagonal elements of A
10      d = d + rand(length(d), 1); % to fill d with random values
11      A = A + diag(d); % to update diagonal elements of A
12      G = sparse(A);
13      L = length(G(G≠0));
14      while L≠NZE(k)
15          for i=1:n(k)
16              for j=1:n(k)
17                  e = randi(n(k),2,1); % to select random element
18                  if e(1)≠e(2) && A(e(1),e(2))==0 % to select non-zero ...
                        off diagonal element
19                      G = sparse(A);
20                      L = length(G(G≠0));
21                      if L < NZE(k)
22                          A(e(1),e(2)) = rand; % to fill that element
23                      else
24                          break
25                      end
26                  end
27              end
28          end
29          G = sparse(A);
30          L = length(G(G≠0));
31      end
32  disp(['The memory information when the matrix size is: ' ...
        num2str(n(k)) 'x' num2str(n(k))]);
33  display('------------------------------------------------------');
34  disp('The total size of the original matrix is:');
35  whos A
36  B = sparse(A);
37  disp('The total size of the compact matrix (Table of factors) is:');
38  whos B
39  display('------------------------------------------------------');
40  end
```

Running this code will generate the following important information:

```
---------------------------------------------------------------
The memory information when the matrix size is: 20x20
---------------------------------------------------------------
The total size of the original matrix is:
  Name         Size               Bytes  Class     Attributes

  A           20x20               3200  double
```

```
The total size of the compact matrix (Table of factors) is:
  Name          Size                 Bytes  Class     Attributes

  B            20x20                 2408  double    sparse

-------------------------------------------------------------
The memory information when the matrix size is: 200x200
-------------------------------------------------------------
The total size of the original matrix is:
  Name           Size                 Bytes  Class     Attributes

  A            200x200               320000  double

The total size of the compact matrix (Table of factors) is:
  Name           Size                 Bytes  Class     Attributes

  B            200x200               24008  double    sparse

-------------------------------------------------------------
The memory information when the matrix size is: 2000x2000
-------------------------------------------------------------
The total size of the original matrix is:
  Name           Size                    Bytes  Class     Attributes

  A            2000x2000             32000000  double

The total size of the compact matrix (Table of factors) is:
  Name           Size                    Bytes  Class     Attributes

  B            2000x2000               240008  double    sparse

-------------------------------------------------------------
```

As can be clearly seen from the last example, for small applications, taking the inverse matrix has an ignorable effect on the machine memory. But, when large power system networks are simulated, a big portion of the memory will be consumed for nothing! Here, in this example, with $n = 2000 \rightarrow$ the memory reduced from 32 MB down to 240 KB only!! Now, just imagine how the situation would be if the network admittance matrix is a combination of susceptance[13] (B) and conductance (G).

Based on that, other PF techniques have been introduced in order to compromise between accuracy, memory usage, and processing speed. The second most popular PF solvers are the decoupled techniques. The first version was suggested in 1963 by Carpentier (Stagg and El-Abiad, 1968), and it was based on setting the off-diagonal sub-matrices of the Jacobian matrix to zero because the sensitivity of ΔP to $\Delta|V|$ and the sensitivity of ΔQ to $\Delta\delta$ are very small and can be ignored. Then, in 1974, further simplifications to the remaining two sub-matrices of the Jacobian matrix have been suggested by Stott and Alsac to have non-iterative sub-matrices (Saadat, 1999). The last version is

13 The susceptance is also known as permittance, which is coined by Oliver Heaviside—June 1887.

called the fast-decoupled load flow (FDLF) method, which can provide a good solution and convergence rate with a very limited memory usage (Kothari and Dhillon, 2011).

Although some techniques have been recently presented in the literature as competitive PF solvers, still they are not popular as GS, NR, and FDLF methods; especially NR (with activating sparsity technique), which can be seen in most commercial and non-commercial software, such as:

1. ETAP
2. DIgSILENT PowerFactory
3. PowerWorld Simulator
4. EasyPower
5. ASPEN OneLiner
6. CAPE
7. CYME
8. ERACS
9. SKM Systems (DAPPER and CAPTOR)
10. Kalkitech PowerApps
11. PSCAD
12. EuroStag
13. SIEMENS' PSS/E
14. SIMPOW
15. NEPLAN
16. OpenDSS
17. MilSoft Power
18. GE's PSLF
19. RTDS Simulator
20. SPICE
21. ATP-EMTP
22. Nexant
23. Paladin
24. SPARD mp
25. DSATools
26. CAI's Transmission 2000
27. ABB's GridView (power flow)
28. MATLAB/Simulink/Simscape package
29. MATLAB-based power system toolboxes, such as EST, MatEMTP, MatPower, PAT, PSAT, PST, SPS, VST, etc. (Milano, 2005)
30. Python-based power system libraries, such as PandaPower

Example 6.4 Consider the eight-bus test system shown in Figure F.10 of Appendix F. If the source G_3 is disconnected, simulate the test system in PowerWorld Simulator. Find the admittance matrix of the system. Solve the load flow problem using GS.

Solution

The model is given in `8Bus.PWB`. To simulate the network, we have to follow the steps shown in Figure 6.4.

The load flow solution can be found by following the steps shown in Figure 6.5, and the solution is shown in Figure 6.6.

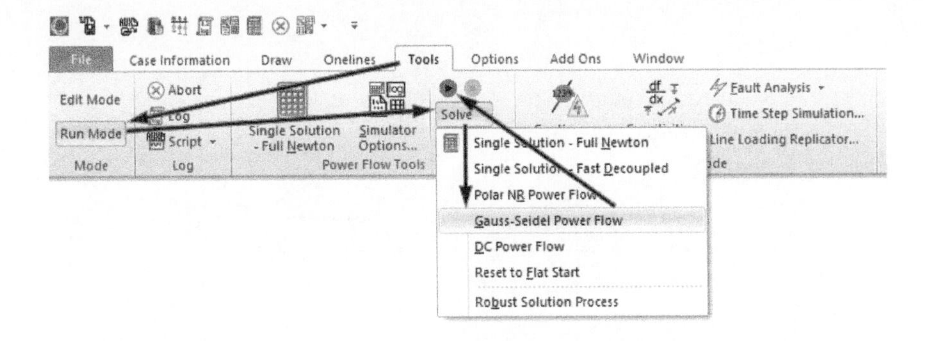

Figure 6.4 Steps required to solve load flow problems in PowerWorld Simulator. Source: PowerWorld Corporation.

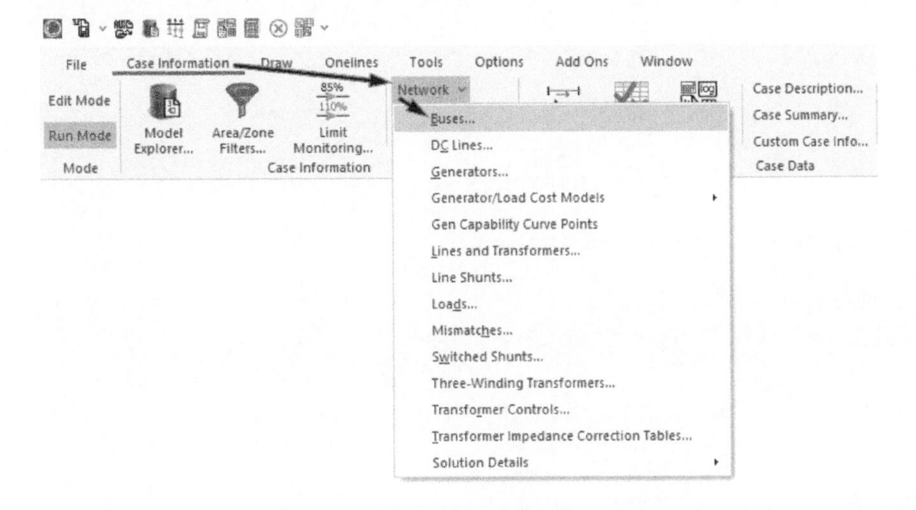

Figure 6.5 Steps required to extract load flow solutions in PowerWorld Simulator. Source: PowerWorld Corporation.

Bus	BusNum	BusName	AreaName	BusNomVolt	BusPUVolt	BusKVVolt	BusAngle	BusLoadMW	BusLoadMVR	BusGenMW	BusGenMVR	BusSS
1	1	1	1	150.00	0.96491	144.736	-2.72					
2	2	2	1	150.00	0.95715	143.573	-3.42	40.00	20.00			
3	3	3	1	150.00	0.91503	137.254	-6.36	60.00	40.00			
4	4	4	1	150.00	0.90476	135.713	-7.19	70.00	40.00			
5	5	5	1	150.00	0.92040	138.060	-5.82	70.00	50.00			
6	6	6	1	150.00	0.96169	144.254	-2.93					
7	7	7	1	10.00	1.00000	10.000	0.00			114.49	90.45	
8	8	8	1	10.00	1.00000	10.000	0.11			127.50	99.15	

Figure 6.6 Load flow solution obtained by PowerWorld Simulator for the eight-bus test system using the GS method. Source: PowerWorld Corporation.

The animation of the power flowing in all the branches of the eight-bus test system is depicted in Figure 6.7.

Then, we have to follow the steps shown in Figure 6.8 so that the admittance matrix can be generated; as shown in Figure 6.9.

For optimal coordination, many research papers have been conducted based on ETAP and DIgSILENT PowerFactory programs where NR, FDLF, and accelerated GS methods are available

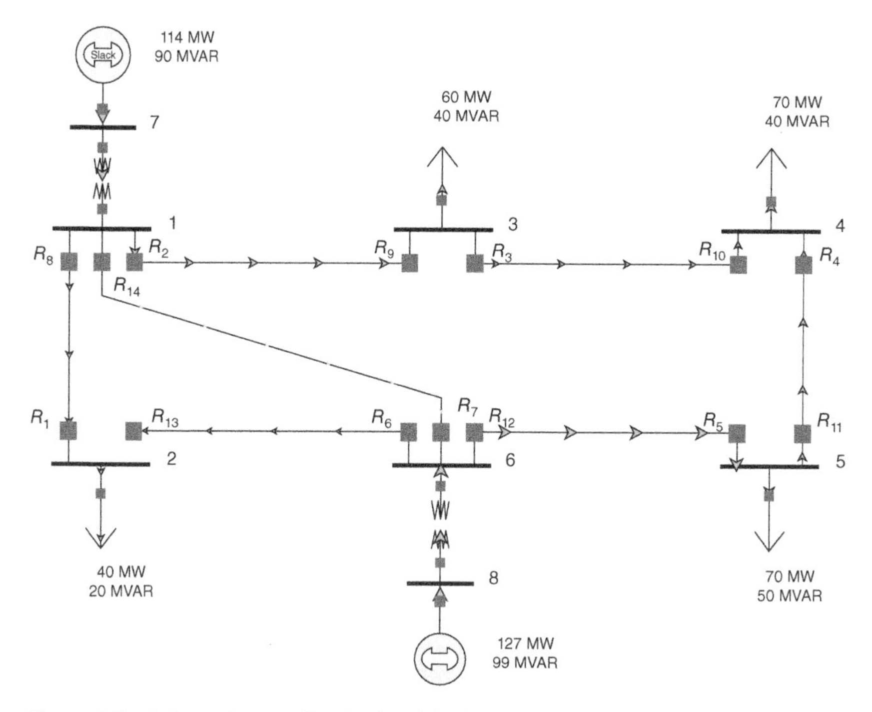

Figure 6.7 Animated power flow in the eight-bus test system.

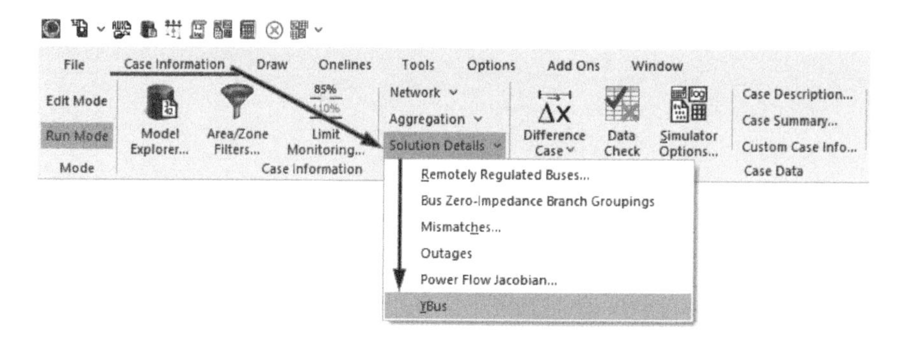

Figure 6.8 Steps required to extract the system admittance matrix in PowerWorld Simulator. Source: PowerWorld Corporation.

	Number	Name	Bus 1	Bus 2	Bus 3	Bus 4	Bus 5	Bus 6	Bus 7	Bus 8
1	1	1	4.6811 - j78.5917	-1.5898 + j19.8728	-1.1110 + j13.9169			-1.9802 + j19.8020	-0.0001 + j25.0000	
2	2	2	-1.5898 + j19.8728	3.3363 - j39.7191				-1.7465 + j19.8463		
3	3	3	-1.1110 + j13.9169		2.6761 - j31.5399	-1.5651 + j17.6230				
4	4	4			-1.5651 + j17.6230	4.0041 - j39.5742	-2.4390 + j21.9512			
5	5	5				-2.4390 + j21.9512	5.0969 - j46.1087	-2.6579 + j24.1574		
6	6	6	-1.9802 + j19.8020	-1.7465 + j19.8463			-2.6579 + j24.1574	6.3846 - j88.8057		-0.0001 + j25.0000
7	7	7	-0.0001 + j25.0000						0.0001 - j25.0000	
8	8	8						-0.0001 + j25.0000		0.0001 - j25.0000

Figure 6.9 Admittance matrix of the eight-bus test system obtained by PowerWorld Simulator. Source: PowerWorld Corporation.

(ETAP Company, 2011; Singh et al., 2013). Therefore, because of the popularity of NR, the researchers still use it as a tool to do accurate/precise calculations for off-line PF analysis. However, if the given network has a dynamic topology, then the ORC problem is considered as an adaptive problem. With this situation, multiple online PF analysis could be carried out within fixed or variable intervals. If NR is used here, then it may cause a technical problem because of its long processing time. Thus, the accuracy and online processing speed are both considered as very important factors in the applications of power system protection.

As a summary, PF analysis can be translated as a "frozen" picture of one moment, condition, or scenario of a dynamic interconnected electric power system (Das, 2002). Now, someone may raise this logical doubt: *If PF analysis is carried out during a steady-state condition, then what is the benefit gained from it if we know that the real interconnected systems are dynamic where power generations and demand are not fixed due to planned outages "maintenance jobs" or forced outages "faults"?!* Actually, this is one of unclear points that could be faced in the literature by many students and researchers, and here we explain it in details. Real power plants are monitored through a real-time automation and control systems.[14] The measurements are taken from very precise and accurate industrial instrument devices mounted on different points of electric power components and send their signals from field to some local control rooms (LCRs) through smart transmitters. Different standard protocols (such as Filedbus, Profibus, HART, DeviceNet, CAN, and ASI) can be used. Then, the data is sent from LCRs to the central or main control room (MCR) through some standard control protocols (such as ControlNet, V-Net, Modbus, and DH+) with the help of fiber-optic and wireless communications for some far RTUs. In MCR, part of the operation crew set in the front of their HMIs where the automation engineers have a separate room with more access level to the entire control system. All these computing and archiving devices are connected to each other through Ethernet protocol. The higher levels of access (or the top layer of the hierarchy diagram) to these DCS and SCADA are given to the plant managers (power production, transmission, distribution) as well as system control department and could be given to the research and development (R&D) department if it is established. These details are important to know the meaning of online operation for both PF and ORC problems.

We should know that the aforementioned real-time measurements are exclusively used for online automation and control. The measurements appear as per existing system and not as per our wish. For example, we can confidently say that nobody can do "cause-and-effect" study by just opening circuit breakers (CBs) of some live busbars! This could jeopardize the entire system and could create some consequent failures to the equipment that might lead to cascading trips till reaching an entire trip "blackout." Therefore, this real-time automation strategy cannot be used for prediction, evaluation, and expansion planning. This is why we need to re-initiate the PF analysis within an adaptive or automatic scheme to re-compute the required information for conducting the necessary analysis. If there are C scenarios that need to be simulated and verified based on the existing situation of the network, then the PF solver should also be executed C times, which in turn a large amount of CPU time is required. This is why there are many techniques available in the literature, which are offered as the best alternatives[15] that can compromise between the solution quality and processing speed.

---

14 Such modern automation systems are distributed control system (DCS), supervisory control and data acquisition (SCADA), programmable logic controllers (PLCs)/remote terminal units (RTUs) with locally mounted human–machine interfaces (HMIs) "i.e. simple version of SCADA," microSCADA for small power plants as offered by ABB company, or even a combination of different systems. For example, using SCADA from one manufacturer as a subsystem in DCS of another manufacturer through a common communication medium called open platform communications (OPC) server, or it can integrate DCS as a subsystem in SCADA.

15 based on the researchers' opinions.

PF analysis as a subject is very big and it is one of the major courses for undergraduate electrical engineers. Because of the space limitation and to avoid any dispersion, the reader can refer to the preceding references (such as (El-Hawary, 2008; Grainger and Stevenson, 1994; Saadat, 1999; Glover et al., 2012; Husain, 2007; Kothari and Dhillon, 2011)) for detailed descriptions with many illustrative and computer examples as well.

6.3 P/B Pairs Identification

As covered before, coordinating relays means that the selectivity among groups of relays is done correctly and sequentially. So, if a fault occurs in any zone, a set of primary relays (R^{pr}) should operate first, and if any relay fails or exceeds the assigned chance (i.e. $T^{pr} + CTI$), one or a set of backup relays (R^{bc}) should operate immediately. Also, each one of these backup relays will work as a primary relay for another group of relays. Thus, finding the P/B relay pairs of all DOCRs is a very important step for solving the ORC problem. There are many ways to determine these P/B relay pairs. The main three methods are listed in the following text:

6.3.1 Inspection Method

It is the most popular method, which is preferable for small and non-complicated networks. It is also recommended if optimal coordination is done based on one fixed topology (i.e. non-adaptive scheme). The basic idea behind this method is to trace the current direction that DOCR can see, and then try to find the other DOCRs in other zones that can see the same current direction.

Example 6.5 Consider the network shown in Figure 5.12. Identify the backup relay(s) of each primary relay by using the inspection method.

Solution
By applying the preceding steps, if R_1 is considered as R^{pr}, then R_6 will work as R^{bc} after waiting CTI. This can be seen by tracing the arrows near the relays shown in Figure 5.12. Also, if R_2 is considered as R^{pr}, then the group $\{R_1, R_4, R_{11}\}$ will work as R^{bc}. If this process is applied to all 12 DOCRs, then Table 6.2 can be constructed.

6.3.2 Graph Theory Methods

There are different ways to identify the P/B relay pairs of electric power networks. These P/B relay pairs identification techniques are based on graph theory where many of them are designed based on the LINKNET structure (Damborg et al., 1984; Ramaswami et al., 1990; Venkata et al., 1991; Birla et al., 2004; Singh et al., 2013; Birla, 2014). One simple method is described in Elrafie and Irving (1993). This approach can be briefly described as follows:

1. **Numbering All the Relays:**
   - Because each line of the network has two DOCRs mounted on both ends, so the total number of DOCRs mounted on the network is 2ß where ß is the total number of lines or branches.

Table 6.2 Primary/backup relay pairs of the five-bus system shown in Figure 5.12.

R^{pr}	1	2	3	4	5	6	7	8	9	10	11	12
R^{bc}	6	1,4,11	2	3	1,4,8	5	4,8,11	9	10	1,8,11	12	7

- Make two groups of DOCRs with a length equals to $ß$. The first group will contain $\{R_1, R_2, \dots, R_j, \dots, R_ß\}$, and the second group will contain $\{R_{ß+1}, R_{ß+2}, \dots, R_{ß+j}, \dots, R_{2ß}\}$; where $j = 1, 2, \dots, ß$.
- From each line, the relay of the first selected end will be numbered by j, and the relay of the other end will be numbered by $(ß + j)$. Continue this process until reaching the last line where R_j becomes $R_ß$ and $R_{ß+j}$ becomes $R_{2ß}$.

2. **Constructing the Nodal and Augmented Incidence Matrices:**
   - Construct the nodal incidence matrix (**A**) where the number of rows is equal to the number of buses ($\aleph$), and the number of columns is equal to the number of lines ($ß$):

$$\mathbf{A} = [\aleph \times ß] = \begin{bmatrix} e_{1,1} & e_{1,2} & \cdots & e_{1,j} & \cdots & e_{1,ß} \\ e_{2,1} & e_{2,2} & \cdots & e_{2,j} & \cdots & e_{2,ß} \\ \vdots & \vdots & \ddots & \vdots & \ddots & \vdots \\ e_{i,1} & e_{i,2} & \cdots & e_{i,j} & \cdots & e_{i,ß} \\ \vdots & \vdots & \ddots & \vdots & \ddots & \vdots \\ e_{\aleph,1} & e_{\aleph,2} & \cdots & e_{\aleph,j} & \cdots & e_{\aleph,ß} \end{bmatrix} \tag{6.14}$$

   - Each ijth element ($e_{i,j}$) can be determined as follows:
     - For each ith bus, look to the end of each line connected to that bus.
     - If that end contains R_j, then give it "+1."
     - Otherwise, if it contains $R_{ß+j}$, then look to the second end and give it "-1" on the same row of the ith bus.
   - Once the construction of **A** is completed, the augmented incidence matrix ($\mathbf{A_d}$) can be constructed by duplicating **A** and multiplying the second one by −1 to have:

$$\mathbf{A_d} = [\mathbf{A}, -\mathbf{A}] \tag{6.15}$$

3. **Identifying P/B Pairs:**
   - From $\mathbf{A_d}$, the positive element of each column will represent $R_{j'}^{bc}$ where $j' = 1, 2, \dots, j-1, j, j+1, \dots, ß-1, ß, ß+1, \dots, ß+j, \dots, 2ß-1, 2ß$.
   - If $R_{j'}^{bc}$ is selected from the j'th column, see the negative element from the same column and save the new ith row.
   - From that ith row, select all the positive elements as primary relays of that backup relay $R_{j'}^{bc}$; except one positive element which needs to be excluded.
   - If $R_{j'}^{bc}$ is located in the first group (i.e. $1 \leqslant j' \leqslant ß$), then exclude the positive element in the location of $(j' + ß)$, and in the location of $(j' - ß)$ if $R_{j'}^{bc}$ is located in the second group (i.e. $ß + 1 \leqslant j' \leqslant 2ß$).

As seen, many steps are required to identify all the P/B relay pairs based on the graph theory method. To clarify this process, let's see Example 6.6.

Example 6.6 Repeat Example 6.5, but by using the graph theory method. (**Hint**: The answer should be similar to the one tabulated in Table 6.2.)

Solution

Note that these DOCRs are numbered in a way to fit with the method of graph theory. There are six branches, which give 12 DOCRs, and thus $j' = 1, 2, \dots, 12$. If the selected line has $R_{j'}$ with $j' = j$ (i.e. is located between 1 and 6), then the other relay of the second end of that line will be numbered by R_{j+6}. From the network, the nodal and augmented incidence matrices can be respectively determined as follows:

$$\mathbf{A} = \begin{bmatrix} e_{1,1} & e_{1,2} & \cdots & e_{1,j} & \cdots & e_{1,ß} \\ e_{2,1} & e_{2,2} & \cdots & e_{2,j} & \cdots & e_{2,ß} \\ \vdots & \vdots & \ddots & \vdots & \ddots & \vdots \\ e_{i,1} & e_{i,2} & \cdots & e_{i,j} & \cdots & e_{i,ß} \\ \vdots & \vdots & \ddots & \vdots & \ddots & \vdots \\ e_{\aleph,1} & e_{\aleph,2} & \cdots & e_{\aleph,j} & \cdots & e_{\aleph,ß} \end{bmatrix} = \begin{bmatrix} 1 & 0 & 0 & 0 & 0 & -1 \\ -1 & 1 & 0 & -1 & 1 & 0 \\ 0 & -1 & 1 & 0 & 0 & 0 \\ 0 & 0 & -1 & 1 & 0 & 0 \\ 0 & 0 & 0 & 0 & -1 & 1 \end{bmatrix}$$

$$\mathbf{A_d} = \left[\begin{array}{cccccc|cccccc} 1 & 0 & 0 & 0 & 0 & -1 & -1 & 0 & 0 & 0 & 0 & 1 \\ -1 & 1 & 0 & -1 & 1 & 0 & 1 & -1 & 0 & 1 & -1 & 0 \\ 0 & -1 & 1 & 0 & 0 & 0 & 0 & 1 & -1 & 0 & 0 & 0 \\ 0 & 0 & -1 & 1 & 0 & 0 & 0 & 0 & 1 & -1 & 0 & 0 \\ 0 & 0 & 0 & 0 & -1 & 1 & 0 & 0 & 0 & 0 & 1 & -1 \end{array}\right]$$

From $\mathbf{A_d}$, if column no. 1 is selected, then $e_{1,1}$ indicates that R_1 works as a backup relay for some primary relays. To find these relays, let's select the negative element of that column which is located in the second row (i.e. $e_{2,1}$). From row no. 2, the positive elements $\{e_{2,2}, e_{2,5}, e_{2,10}\}$ respectively indicate that $\{R_2, R_5, R_{10}\}$ work as primary relays. Because the backup relay R_1 is listed in the first group, so the positive element $e_{2,1+ß}$ (which is $e_{2,7}$) is excluded. Therefore, R_1 has to backup R_2, R_5, and R_{10}.

To see how to find the P/B relay pairs for the backup relays listed in the second group (i.e. R_7^{bc} to R_{12}^{bc}), let's select column no. 8. The positive element $e_{3,8}$ means that R_8 works as a backup relay for some primary relays. $e_{2,8}$ is the negative element of that column, which is located in the second row. Because R_8 is listed in the second group, so the positive element that has to be rejected is ($e_{2,8-ß} = e_{2,2}$). The remaining positive elements $\{e_{2,5}, e_{2,7}, e_{2,10}\}$ respectively represent the primary relays $\{R_5, R_7, R_{10}\}$. Thus, R_8 is considered as R^{bc} for R_5, R_7, and R_{10}.

By repeating these steps for all the columns of $\mathbf{A_d}$, Table 6.2 will be constructed again.

This approach may be relatively more difficult than the previous method, but it gives a systematic way to identify these P/B relay pairs automatically without any human mistakes as if it is solved classically by hand. Based on this, the graph theory method provides some capabilities to the researchers who want to study the ORC of dynamic networks[16] where all the pairs can be automatically re-identified if any element (bus or line) is added to or removed from the network.

6.3.3 Special Software

The methods presented here are similar to those presented in Section 6.3.2. The only difference is that the P/B relay pairs identification process is done through some ready-made tools. Thus, by using some special software, these P/B relay pairs can be determined easily. It is recommended to use this method for very complicated and large systems. However, the main problem associated with this approach is that it is hard to establish an interface between software and numerical programming languages (such as MATLAB, GNU Octave, Python, and Julia). This is important so important for adaptive ORC tools. Do not forget that the main goal here is to obtain optimal relay settings, which can be done by coding optimization algorithms in numerical programming languages. The same technical problem is faced in short-circuit analysis where doing it in external software requires to establish a special interface.

16 That is networks with many configurations.

6.4 Short-Circuit Analysis

Some basic principles have been covered in Chapter 2. It has been shown that the severity of short-circuit current depends on the type of that fault and its location. The good protection design should detect the highest overload current as well as the lowest severe faults, which are reflected on the allowable limits of *PS*. Thus, to carry out the process of optimal relay coordination, it is important to do first the fault analysis on the given network. The near-end 3ϕ short-circuit fault analysis is commonly used because if the coordination is satisfied under this worst case, then it will be definitely satisfied for the other cases.[17] This phenomenon can be graphically explained by Figure 6.10.

Two Questions Can be Raised Here:

- ***How to determine the minimum and maximum PS of DOCR based on the given maximum overload and minimum fault that can be seen by the ith relay?***
- ***How to select the location of faults on each line of the given network?***

It is better to answer these two questions in Chapter 7 when the side constraints of *PS* are covered because many aspects need to be clarified there. However, we need to know how to calculate the minimum and maximum faults seen by relays. Also, we need to know the tools required to do that. Furthermore, we need to know the standards that our calculations should comply with.

6.4.1 Short-Circuit Calculations

The important thing of this section is to say that if anyone wants to solve the ORC problem of DOCRs, then the hand calculation of fault analysis is highly not recommended. This weary calculation can be done quickly and smoothly within the previous commercial software. In this way, a huge amount of effort and time can be saved and can give the ability to easily re-analyze the system under different assumptions and scenarios. On the other hand, the classical calculation by hand may lead to some mistakes or errors and needs to be started again from the stage where the new

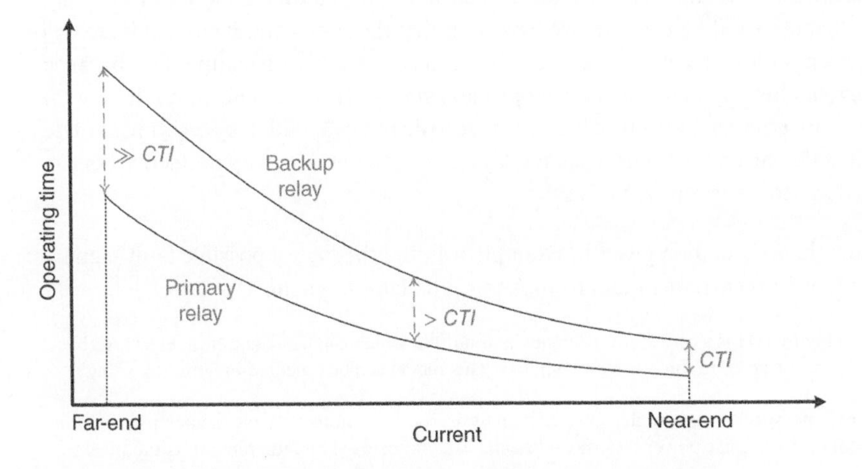

Figure 6.10 Discrimination margin between primary and backup relays.

17 It is valid if only one unified TCCC is used for all DOCRs (Al-Roomi and El-Hawary, 2019b).

assumptions or mistakes have not been considered. For example, the network shown in Figure 6.11 gives a brief idea about how to model 3ϕ electric networks in Simulink.[18]

Although we can model electric power systems in Simulink, it is relatively harder than doing that in other commercial software specialized for electric power systems engineering.

6.4.2 Electric Power Engineering Software Tools

Nowadays, there are many enterprise application software (EAS) that can be used to perform different power system studies including both online and offline. Selecting the proper one depends on many criteria, such as the price, available features and capabilities, speed, user-friendly, animation and interactivity, appearance, documentation and learning resources, popularity, flexibility with other software, customer support, supported formats, available modules,[19] etc.

In Section 6.2, a wide list of software has been shown. Most of them support doing short-circuit fault analysis. For example, ETAP, DIgSILENT PowerFactory, PowerWorld Simulator, EasyPower, CYME, SKM DAPPER, EuroStag, PSS/E, NEPLAN, SIMPOW, MilSoft Power, SPARD mp, and DSATools. For MATLAB, unfortunately, there is no mature third-party toolbox available yet to do short-circuit analysis[20] (Milano, 2005). For , there is one popular library called "PandaPower," which has the ability to do static short-circuit calculation according to the IEC 60909 standard.

Thus, for designing adaptive ORC solvers, it could be viable to rely on rather than MATLAB because the former one is free and much familiar in the industry field. The other option is to establish an interface between the preceding EAS and programming languages (MATLAB, Python, C/C++, etc.) to exchange the data. The third option is to code everything in one program. This means that the programmer should have a very strong experience in programming languages, a deep knowledge in many power system studies, and ensures that the calculations comply with international standards. This is why such big projects are done by establishing specialized research groups.

6.4.2.1 Minimum Short-Circuit Current

As said earlier, to have reliable protection designs, DOCRs should be adjusted in a way that can detect and isolate the smallest fault current. For most cases, single line to ground, which is shown in Figure 2.3a, is the lowest sever short-circuit fault (El-Hawary, 1983; Paithankar and Bhide, 2003). Also, from the last Chapter 5 and Figure 6.10, we have seen that the short-circuit current increases as the distance between fault and source decreases. This is also correct for backup relays because they cannot see currents higher than that seen by primary relays. Thus, to determine the lowest possible short-circuit current seen by each DOCR, we have to do our calculation by creating a single line-to-ground fault at the far-end point from each relay when it acts as a backup protective device. This means that we have to create out-zone faults.

Example 6.7 Using the information given in Example 6.4, find the lowest possible fault seen by each DOCR. The unit of fault current should be in Amps. The fault impedance is zero.

---

18 The system shown in Figure 6.11 is a case study available in Simulink, which can be loaded from MATLAB by just typing `power_3phseriescomp` in the command window. This model can be executed in Simulink using Simscape package.

19 Each module performs one specific task, such as PF, fault analysis, stability, protection, etc. Some programs are offered with default modules to do basic power system studies and ask end-users to buy the remaining modules as per their needs.

20 Even "MatPower," it supports some steady-state studies, such as PF analysis, but not short-circuit analysis. This motivates Cole and Belmans (2011) to create a new toolbox called "MatDyn," based on MatPower, to do some power system dynamic studies. However, still, short-circuit analysis is not supported in MatDyn.

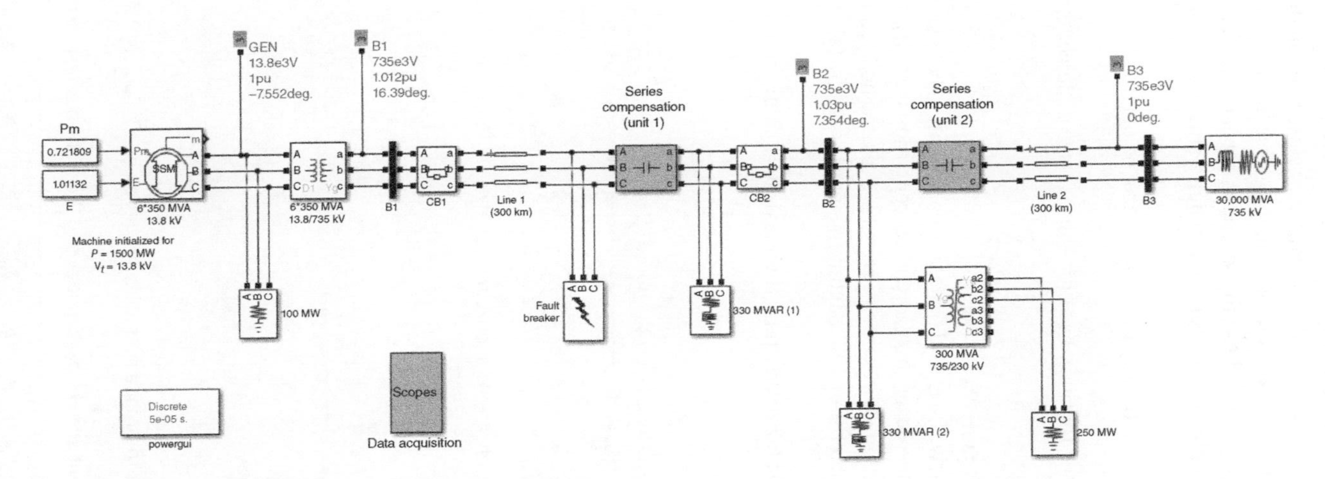

Figure 6.11 Modeling a simple three-bus system in Mathwork's Simulink.

Solution

The model is given in `8Bus.PWB`. To do short-circuit analysis, we have to follow the steps shown in Figure 6.12.

For R_1, it is a backup relay for R_2 and R_{14}. Thus, we have to create out-zone faults near bus 3 and 6 and then see the lowest fault. The word "near" means very close to or on the bus. In PowerWorld Simulator, we can create a fault 99% away from bus 1 and 1% away from bus 3 or bus 6. Also, we can create that fault on bus 3 or bus 6. We will use the second approach; i.e. creating faults on busbars. Figure 6.13 shows the short-circuit current seen by R_1 when a single line-to-ground fault is created on bus 3.

If we continue this process for all the relays, Table 6.3 can be obtained.

From Example 6.7, we can see that it is hard for some relays to detect out-zone far-end single line-to-ground faults.

6.4.2.2 Maximum Short-Circuit Current

At the beginning of this section, we have said that if the coordination between primary and backup DOCRs is satisfied under the most sever short-circuit current, then we will guarantee that the coordination is satisfied under less sever short-circuit currents. Thus, the near-end 3ϕ short-circuit fault analysis is commonly used to determine the maximum current seen by relays.

Example 6.8 Repeat Example 6.7 to compute the maximum short-circuit current seen by each primary relay.

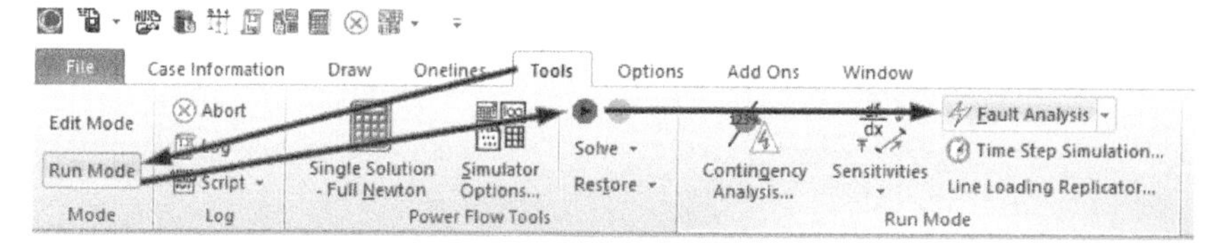

Figure 6.12 Steps required to do 1ϕ far-end short-circuit analysis in PowerWorld Simulator. Source: PowerWorld Corporation.

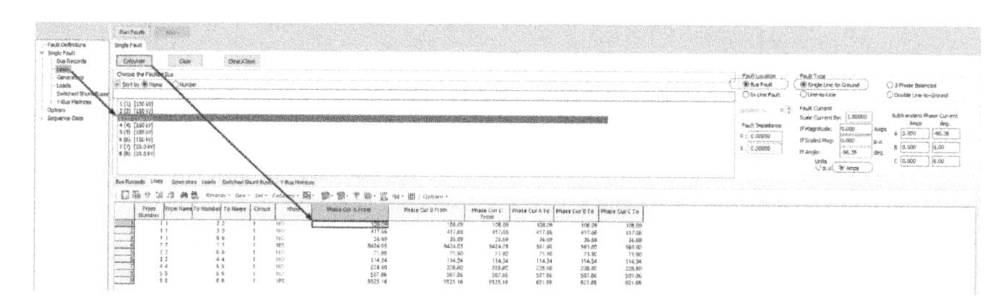

Figure 6.13 Results of single line-to-ground short-circuit fault on bus 3. Source: PowerWorld Corporation.

Table 6.3 Minimum 1ϕ short-circuit current seen by backup relays of the eight-bus test system.

Relay	R_1	R_2	R_3	R_4	R_5	R_6	R_7	R_8	R_9	R_{10}	R_{11}	R_{12}	R_{13}	R_{14}
$I_{f,1\phi}^{\text{bc,min}}$ (A)	108.09	417.66	114.34	228.80	587.86	71.9	36.69	108.09	417.66	114.34	228.80	587.86	71.90	36.69

Solution

Because the question wants to see the short-circuit currents seen by primary relays, so it asks us to calculate the currents for in-zone faults. The maximum short-circuit current occurs at the near-end point; i.e. when the fault location is so close to the ith relay. In this example, we will assume that the near-end or close-in 3ϕ fault is 1% away from each primary relay and occurs in its zone. Thus, for R_1, we have to create a balanced fault at 99% away from bus 1 and 1% away from bus 2. Figure 6.14 explains everything we need to conduct this type of short-circuit analysis in PowerWorld Simulator.

If we continue this process for all the relays, Table 6.4 can be obtained.

6.4.3 Most Popular Standards

For real-word applications, doing short-circuit analysis requires considering many practical points. Thus, the short-circuit calculations should comply with a standard. In the literature, there are more than one standard that can be used in our calculation. The most popular standards are categorized under IEC, IEEE/ANSI, GOST,[21] and VDE.[22] For example, DIgSILENT PowerFactory supports IEC 60909 (including 2016 edition), IEEE 141/ANSI C37, VDE 0102/0103, G74 and IEC 61363 norms and methods. Also, it supports IEC 61660 and ANSI/IEEE 946 for short-circuit currents in DC grids. On the opposite side, ETAP supports ANSI/IEEE C37 & UL 489, IEC 60909, IEC 61363, and GOST R-52735.

The following lines give a brief overview about the most popular standards (ETAP, 2020):

6.4.3.1 ANSI/IEEE Standards C37 & UL 489

In these standards, all machine internal voltage sources and external voltage sources are replaced by an equivalent voltage source at the fault location. This equivalent voltage source equals the pre-fault voltage at that location. Three different impedance networks are formed to calculate momentary,

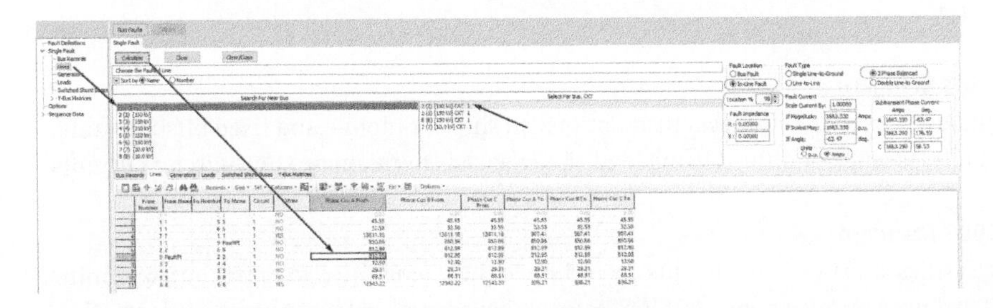

Figure 6.14 Steps required to do 3ϕ near-end short-circuit analysis in PowerWorld Simulator. Source: PowerWorld Corporation.

Table 6.4 Maximum 3ϕ short-circuit current seen by primary relays of the eight-bus test system.

Relay	R_1	R_2	R_3	R_4	R_5	R_6	R_7	R_8	R_9	R_{10}	R_{11}	R_{12}	R_{13}	R_{14}
$I_{f,3\phi}^{pr,max}$ (A)	812.95	1675.55	1033.80	653.10	371.27	1505.63	1239.57	1540.74	3624.32	910.32	1266.99	1658.60	838.73	1300.30

21 GOST standards were originally developed by the government of the Soviet Union, which are now maintained by Russia.

22 Germen standards.

interrupting, and steady-state short-circuit currents, and corresponding fault current duties for various protective devices. These three networks can be categorized as follows:

1. **Sub-transient Network:** 1/2 *cycle network*
2. **Transient Network:** 3/2–4 *cycle network*
3. **Steady-State Network:** *30 cycle network*

ANSI/IEEE standards recommend the use of separate R and X networks to calculate the ratio of X to R. This X/R ratio should be obtained for each individual faulted bus and short-circuit current, and then that ratio is used to determine the multiplying factor to account for the system DC offset.

The following is a list of some useful ANSI/IEEE codes:

- **IEEE C37.04:** *Standard rating structure for AC high-voltage circuit breakers rated on a symmetrical current including supplements: IEEE C37.04f, IEEE C37.04g, IEEE C37.04h, IEEE C37.04i*
- **IEEE C37.010:** *Standard application guide for AC high-voltage circuit breakers rated on a symmetrical current*
- **IEEE C37.010b:** *Standard and emergency load current-carrying capability*
- **IEEE C37.010e:** *Supplement to IEEE C37.010*
- **IEEE C37.13:** *Standard for low-voltage AC power circuit breakers used in enclosures*
- **IEEE C37.013:** *Standard for AC high-voltage generator circuit breakers rated on a symmetrical current basis*
- **IEEE C37.20.1:** *Standard for metal enclosed low-voltage power circuit breaker switchgear*
- **IEEE 399:** *IEEE recommended practice for power system analysis (IEEE Brown Book)*
- **IEEE 141:** *IEEE recommended practice for electric power distribution for industrial plants (IEEE Red Book)*
- **IEEE 242:** *IEEE recommended practice for protection and coordination of industrial and commercial power systems (IEEE Buff Book)*
- **UL 489_9:** *Standard for safety for molded-case circuit breakers, molded-case switches, and circuit breaker enclosure*

6.4.3.2 IEC 61363 Standard

This special standard is used for electrical installations on ships or mobile and fixed offshore units. It includes pre-fault loading conditions using load-flow analysis to calculate short-circuit currents.

6.4.3.3 IEC 60909 Standard

This standard classifies short-circuit currents according to their magnitudes (maximum and minimum) and fault distances from generators. Similar to what we have explained before, this standard uses maximum short-circuit currents to determine equipment ratings, and it uses minimum currents for the settings of protective devices.

The IEC 60909 standard is commonly used in ORC (Amraee, 2012), which is also used in short-circuit analysis module of PowerWorld Simulator.

The following is a list of some useful IEC codes:

- **IEC 62271-100:** *High-Voltage Switchgear and Controlgear, Part 100: High-Voltage Alternating-Current Circuit Breakers*
- **IEC 62271-200:** *High-Voltage Switchgear and Controlgear, Part 200: AC Metal-Enclosed Switchgear and Controlgear for Rated Voltages Above 1 kV and up to and including 52 kV*
- **IEC 62271-203:** *High-Voltage Switchgear and Controlgear, Part 203: Gas-Insulated Metal-Enclosed Switchgear for Rated Voltages Above 52 kV*

- **IEC 60282-2:** *High-Voltage Fuses, Part 2: Expulsion Fuses*
- **IEC 60909-0:** *Short-Circuit Currents in Three-Phase AC Systems, Part 0: Calculation of Currents Edition 2.0 2016-01*
- **IEC 60909-1:** *Short-Circuit Currents in Three-Phase AC Systems, Part 1: Factors for the Calculation of Short-Circuit Currents According to IEC 60909-0*
- **IEC 60909-2:** *Electrical Equipment – Data for Short-Circuit Current Calculations in Accordance with IEC 909*
- **IEC 60909-4:** *Short-Circuit Currents in Three-Phase AC Systems, Part 4: Examples for the Calculation of Short-Circuit Currents*
- **IEC 60947-1:** *Low Voltage Switchgear and Controlgear, Part 1: General Rules*
- **IEC 60947-2:** *Low Voltage Switchgear and Controlgear, Part 2: Circuit Breakers*
- **IEC 61363-1:** *Electrical Installations of Ships and Mobile and Fixed Offshore Units, Part 1: Procedures for Calculating Short-Circuit Currents in Three-Phase AC*
- **IEC 60781:** *Application guide for calculation of short-circuit currents in low-voltage radial systems*

Example 6.9 Consider the Western System Coordinating Council (WSCC) nine-bus test system shown in Figure 6.15. This is one of popular PF test systems. It consists of nine buses, three generators, three power transformers, six lines, and three loads. The base voltage levels are 13.8, 16.5, 18, and 230 kV. The line complex powers are around hundreds of MVA each (Al-Roomi, 2015c). Fortunately, this test system and many others are pre-programmed in DIgSILENT PowerFactory.[23] It can be easily imported by pressing on `File` and then selecting `Examples`. Then, the window shown in Figure 6.16 will be popped-up. Follow the steps shown in that figure to successfully import the WSCC nine-bus test system. After importing it, do short-circuit analysis to find the minimum and maximum fault currents.

Solution

To call short-circuit analysis program, we have to follow the steps shown in Figure 6.17.

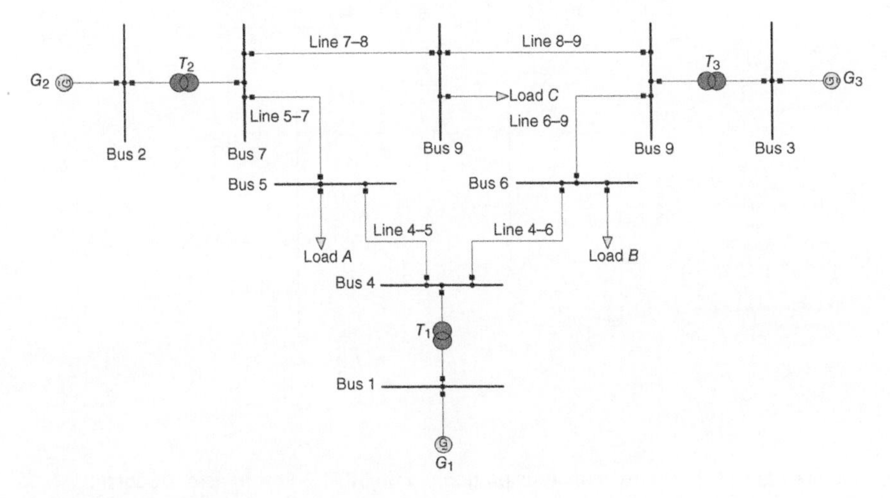

Figure 6.15 The WSCC nine-bus test system as built in DIgSILENT PowerFactory.

23 The demo version can be requested by filling the form given in:

https://www.digsilent.de/en/demo-request.html

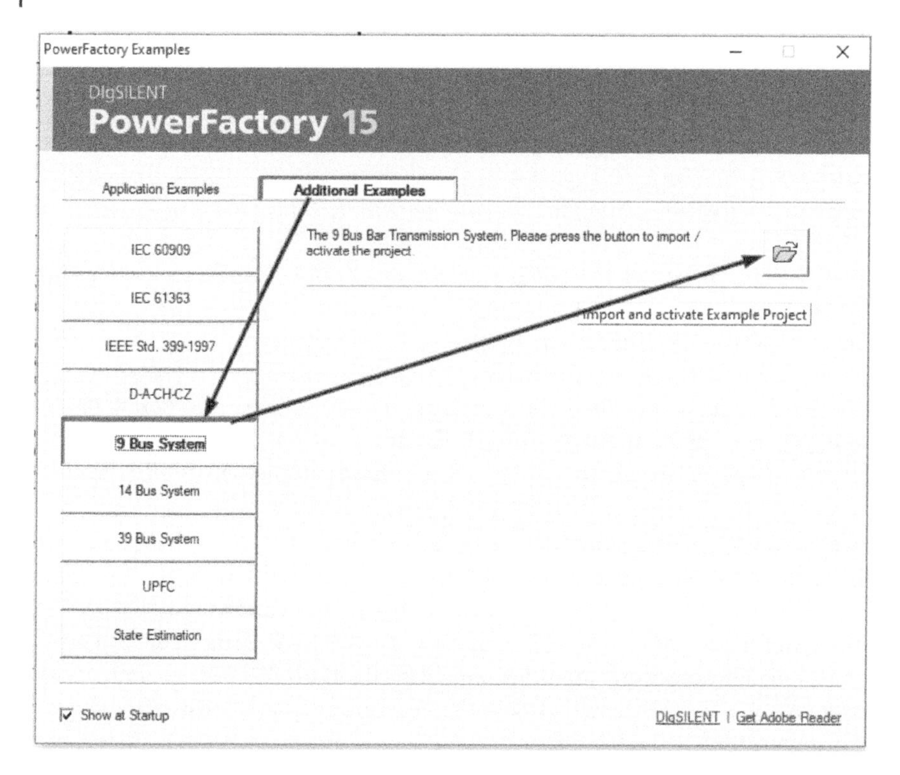

Figure 6.16 Steps required to import the WSCC nine-bus test system in DIgSILENT PowerFactory. Source: DIgSILENT.

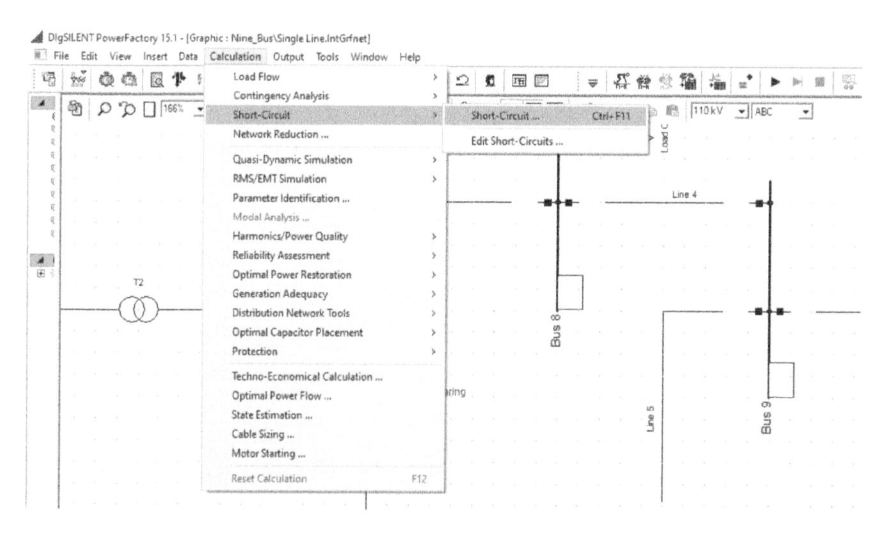

Figure 6.17 Steps required to call short-circuit analysis program in DIgSILENT PowerFactory. Source: DIgSILENT.

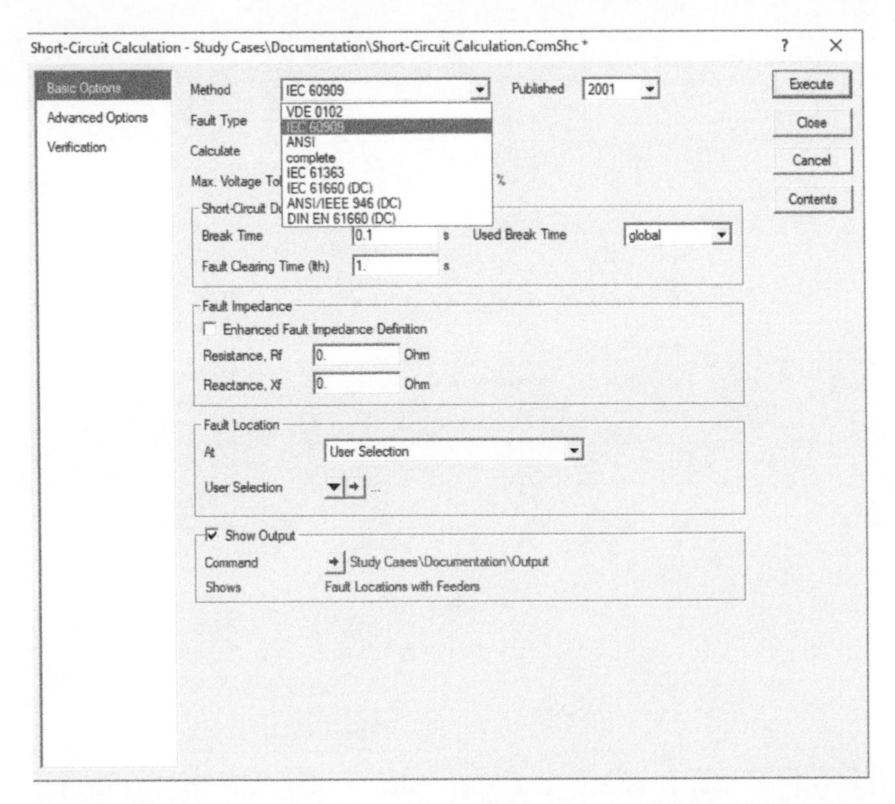

Figure 6.18 Steps required to do short-circuit analysis based on the IEC 60909 standard. Source: DIgSILENT.

Because the question asks about the minimum and maximum short-circuit currents, so we have to use the IEC 60909 standard. Figure 6.18 shows the steps required to do accomplish this analysis.

For calculating minimum short-circuit currents, we will select:

- **Fault Type:** Single Phase to Ground
- **Calculate:** Min. Short-Circuit Currents

If we press on `Execute` button, the following report will be generated:

```
-----------------------------------------------------------------------------------------------------------------
|              |                                                 |  DIgSILENT  | Project:                       |
|              |                                                 | PowerFactory |-------------------------------
|              |                                                 |   15.1.7    | Date:  11/7/2020               |
-----------------------------------------------------------------------------------------------------------------
-----------------------------------------------------------------------------------------------------------------
| Fault Locations with Feeders                                                                                  |
| Short-Circuit Calculation / Method : IEC 60909                      Single Phase to Ground   / Min. Short-Circuit Currents |
-----------------------------------------------------------------------------------------------------------------
| Asynchronous Motors                   | Grid Identification                  | Short-Circuit Duration              |
|   Always Considered                   |   Automatic                          |   Break Time                 0.10 s |
|                                       |                                      |   Fault Clearing Time (Ith)  1.00 s |
|                                       | Conductor Temperature                | c-Voltage Factor                    |
|                                       |   User Defined             No        |   User Defined               No     |
|                                       |                                      |                                     |
-----------------------------------------------------------------------------------------------------------------
```

| Grid: Nine_Bus | System Stage: Nine_Bus | | Annex: / 1 |

		rtd.V. [kV]	Voltage [kV]	Voltage [deg]	c-Factor			Sk" [MVA/MVA]	Ik" [kA/kA]	Ik" [deg]	ip [kA/kA]	Ib [kA]	Sb [MVA]	EFF [-]
Bus 1	A	16.50	0.00	0.00	1.00			557.19 MVA	58.49 kA	-88.02	150.91 kA	58.49	557.19	0.00
	B		9.04	-112.18				0.00 MVA	0.00 kA	0.00	0.00 kA	0.00	0.00	0.96
	C		8.91	114.25				0.00 MVA	0.00 kA	0.00	0.00 kA	0.00	0.00	0.93
T1						Bus 4	A	89.63 MVA	9.41 kA	95.75	24.27 kA			
							B	46.04 MVA	4.83 kA	-84.27	12.47 kA			
							C	43.59 MVA	4.58 kA	-84.23	11.80 kA			
G1							A	467.79 MVA	49.11 kA	-88.74	126.69 kA			
							B	46.04 MVA	4.83 kA	-84.27	12.47 kA			
							C	43.59 MVA	4.58 kA	-84.23	11.80 kA			
Bus 2	A	18.00	0.00	0.00	1.00			500.89 MVA	48.20 kA	-88.66	125.38 kA	48.20	500.89	0.00
	B		9.93	-113.61				0.00 MVA	0.00 kA	0.00	0.00 kA	0.00	0.00	0.96
	C		9.81	114.81				0.00 MVA	0.00 kA	0.00	0.00 kA	0.00	0.00	0.94
T2						Bus 7	A	96.00 MVA	9.24 kA	94.04	24.03 kA			
							B	48.40 MVA	4.66 kA	-85.99	12.11 kA			
							C	47.60 MVA	4.58 kA	-85.94	11.91 kA			
G2							A	405.03 MVA	38.97 kA	-89.30	101.38 kA			
							B	48.40 MVA	4.66 kA	-85.99	12.11 kA			
							C	47.60 MVA	4.58 kA	-85.94	11.91 kA			

| Grid: Nine_Bus | System Stage: Nine_Bus | | Annex: / 2 |

		rtd.V. [kV]	Voltage [kV]	Voltage [deg]	c-Factor			Sk" [MVA/MVA]	Ik" [kA/kA]	Ik" [deg]	ip [kA/kA]	Ib [kA]	Sb [MVA]	EFF [-]
Bus 3	A	13.80	0.00	0.00	1.00			373.01 MVA	46.82 kA	-87.96	120.61 kA	46.82	373.01	0.00
	B		7.80	-115.71				0.00 MVA	0.00 kA	0.00	0.00 kA	0.00	0.00	0.99
	C		7.65	117.62				0.00 MVA	0.00 kA	0.00	0.00 kA	0.00	0.00	0.95
T3						Bus 9	A	96.62 MVA	12.13 kA	94.30	31.24 kA			
							B	49.41 MVA	6.20 kA	-85.71	15.98 kA			
							C	47.21 MVA	5.93 kA	-85.68	15.26 kA			
G3							A	276.50 MVA	34.70 kA	-88.75	89.40 kA			
							B	49.41 MVA	6.20 kA	-85.71	15.98 kA			
							C	47.21 MVA	5.93 kA	-85.68	15.26 kA			
Bus 4	A	230.00	0.00	0.00	1.00			517.33 MVA	3.90 kA	-86.40	9.93 kA	3.90	517.33	0.00
	B		116.08	-95.57				0.00 MVA	0.00 kA	0.00	0.00 kA	0.00	0.00	0.88
	C		115.24	96.77				0.00 MVA	0.00 kA	0.00	0.00 kA	0.00	0.00	0.87
T1						Bus 1	A	362.97 MVA	2.73 kA	91.47	6.97 kA			
							B	55.98 MVA	0.42 kA	95.90	1.07 kA			
							C	54.63 MVA	0.41 kA	96.01	1.05 kA			
Line 1						Bus 5	A	81.70 MVA	0.62 kA	97.48	1.57 kA			
							B	29.63 MVA	0.22 kA	-85.08	0.57 kA			
							C	28.80 MVA	0.22 kA	-85.01	0.55 kA			
Line 6						Bus 6	A	73.53 MVA	0.55 kA	99.79	1.41 kA			
							B	26.36 MVA	0.20 kA	-83.00	0.51 kA			
							C	25.84 MVA	0.19 kA	-82.86	0.50 kA			

```
|                                                                                                                    |
|Bus 5          A  230.00     0.00     0.00  1.00        323.80 MVA    2.44 kA  -83.30      5.93 kA    2.44     323.80  0.00 |
|               B            120.12  -109.67               0.00 MVA    0.00 kA    0.00      0.00 kA    0.00       0.00  0.91 |
|               C            123.96   109.60               0.00 MVA    0.00 kA    0.00      0.00 kA    0.00       0.00  0.93 |
|                                                                                                                    |
| Line 1            Bus 4                            A   196.93 MVA    1.48 kA   95.07      3.60 kA                         |
|                                                    B     6.36 MVA    0.05 kA   89.74      0.12 kA                         |
|                                                    C     6.05 MVA    0.05 kA   89.47      0.11 kA                         |
|                                                                                                                    |
| Line 2            Bus 7                            A   127.08 MVA    0.96 kA   99.22      2.33 kA                         |
|                                                    B     6.36 MVA    0.05 kA  -90.26      0.12 kA                         |
|                                                    C     6.05 MVA    0.05 kA  -90.53      0.11 kA                         |
|                                                                                                                    |
------------------------------------------------------------------------------------------------------------------------
------------------------------------------------------------------------------------------------------------------------
| Grid: Nine_Bus            System Stage: Nine_Bus       |                                   | Annex:          / 3      |
------------------------------------------------------------------------------------------------------------------------
|                 rtd.V.      Voltage      c-              Sk"                Ik"                ip         Ib       Sb    EFF |
|                 [kV]    [kV]     [deg] Factor      [MVA/MVA]      [kA/kA]     [deg]     [kA/kA]      [kA]    [MVA]   [-] |
------------------------------------------------------------------------------------------------------------------------
| Load A                                             A     0.00 MVA    0.00 kA    0.00      0.00 kA                         |
|                                                    B     0.00 MVA    0.00 kA    0.00      0.00 kA                         |
|                                                    C     0.00 MVA    0.00 kA    0.00      0.00 kA                         |
|                                                                                                                    |
|                                                                                                                    |
|Bus 6          A  230.00     0.00     0.00  1.00        306.06 MVA    2.30 kA  -81.26      5.41 kA    2.30     306.06  0.00 |
|               B            119.51  -110.22               0.00 MVA    0.00 kA    0.00      0.00 kA    0.00       0.00  0.90 |
|               C            125.14   109.87               0.00 MVA    0.00 kA    0.00      0.00 kA    0.00       0.00  0.94 |
|                                                                                                                    |
| Line 5            Bus 9                            A   116.82 MVA    0.88 kA  100.44      2.06 kA                         |
|                                                    B     4.16 MVA    0.03 kA  -83.10      0.07 kA                         |
|                                                    C     4.12 MVA    0.03 kA  -82.80      0.07 kA                         |
|                                                                                                                    |
| Line 6            Bus 4                            A   189.33 MVA    1.43 kA   97.70      3.34 kA                         |
|                                                    B     4.16 MVA    0.03 kA   96.90      0.07 kA                         |
|                                                    C     4.12 MVA    0.03 kA   97.20      0.07 kA                         |
|                                                                                                                    |
| Load B                                             A     0.00 MVA    0.00 kA    0.00      0.00 kA                         |
|                                                    B     0.00 MVA    0.00 kA    0.00      0.00 kA                         |
|                                                    C     0.00 MVA    0.00 kA    0.00      0.00 kA                         |
|                                                                                                                    |
|                                                                                                                    |
|Bus 7          A  230.00     0.00     0.00  1.00        519.23 MVA    3.91 kA  -87.15     10.08 kA    3.91     519.23  0.00 |
|               B            116.10   -96.34               0.00 MVA    0.00 kA    0.00      0.00 kA    0.00       0.00  0.88 |
|               C            115.45    97.19               0.00 MVA    0.00 kA    0.00      0.00 kA    0.00       0.00  0.87 |
|                                                                                                                    |
| Line 2            Bus 5                            A    81.38 MVA    0.61 kA   97.92      1.58 kA                         |
|                                                    B    27.92 MVA    0.21 kA  -84.53      0.54 kA                         |
|                                                    C    27.77 MVA    0.21 kA  -84.42      0.54 kA                         |
|                                                                                                                    |
| T2                Bus 2                            A   340.10 MVA    2.56 kA   90.97      6.60 kA                         |
|                                                    B    59.41 MVA    0.45 kA   94.03      1.15 kA                         |
|                                                    C    59.37 MVA    0.45 kA   94.12      1.15 kA                         |
|                                                                                                                    |
| Line 3            Bus 8                            A    98.32 MVA    0.74 kA   95.14      1.91 kA                         |
|                                                    B    31.50 MVA    0.24 kA  -87.25      0.61 kA                         |
|                                                    C    31.61 MVA    0.24 kA  -87.16      0.61 kA                         |
|                                                                                                                    |
|                                                                                                                    |
|Bus 8          A  230.00     0.00     0.00  1.00        362.91 MVA    2.73 kA  -85.04      6.85 kA    2.73     362.91  0.00 |
|               B            119.09  -107.49               0.00 MVA    0.00 kA    0.00      0.00 kA    0.00       0.00  0.90 |
|               C            122.09   107.69               0.00 MVA    0.00 kA    0.00      0.00 kA    0.00       0.00  0.92 |
|                                                                                                                    |
------------------------------------------------------------------------------------------------------------------------
```

```
-----------------------------------------------------------------------------------------------------------------------
| Grid: Nine_Bus              System Stage: Nine_Bus          |                                 | Annex:           / 4    |
-----------------------------------------------------------------------------------------------------------------------
|                   rtd.V.     Voltage      c-          Sk"            Ik"              ip          Ib        Sb    EFF |
|                   [kV]     [kV]   [deg] Factor    [MVA/MVA]    [kA/kA]   [deg]     [kA/kA]      [kA]     [MVA]  [-] |
-----------------------------------------------------------------------------------------------------------------------
|  Line 3          Bus 7                        A     207.25 MVA   1.56 kA   94.70     3.91 kA                         |
|                                               B       1.65 MVA   0.01 kA  108.19     0.03 kA                         |
|                                               C       2.05 MVA   0.02 kA  105.67     0.04 kA                         |
|                                                                                                                     |
|  Line 4          Bus 9                        A     155.67 MVA   1.17 kA   95.30     2.94 kA                         |
|                                               B       1.65 MVA   0.01 kA  -71.81     0.03 kA                         |
|                                               C       2.05 MVA   0.02 kA  -74.33     0.04 kA                         |
|                                                                                                                     |
|  Load C                                       A       0.00 MVA   0.00 kA    0.00     0.00 kA                         |
|                                               B       0.00 MVA   0.00 kA    0.00     0.00 kA                         |
|                                               C       0.00 MVA   0.00 kA    0.00     0.00 kA                         |
|                                                                                                                     |
|                                                                                                                     |
|Bus 9          A  230.00     0.00    0.00  1.00     461.69 MVA   3.48 kA  -86.11     8.79 kA   3.48    461.69  0.00 |
|               B            116.53  -97.12             0.00 MVA   0.00 kA    0.00     0.00 kA   0.00      0.00  0.88 |
|               C            115.57   98.33             0.00 MVA   0.00 kA    0.00     0.00 kA   0.00      0.00  0.87 |
|                                                                                                                     |
|  Line 4          Bus 8                        A     108.73 MVA   0.82 kA   94.98     2.07 kA                         |
|                                               B      34.34 MVA   0.26 kA  -88.14     0.65 kA                         |
|                                               C      33.21 MVA   0.25 kA  -88.13     0.63 kA                         |
|                                                                                                                     |
|  T3              Bus 3                        A     274.87 MVA   2.07 kA   91.57     5.23 kA                         |
|                                               B      60.37 MVA   0.45 kA   94.17     1.15 kA                         |
|                                               C      58.57 MVA   0.44 kA   94.23     1.11 kA                         |
|                                                                                                                     |
|  Line 5          Bus 6                        A      78.85 MVA   0.59 kA  100.48     1.50 kA                         |
|                                               B      26.10 MVA   0.20 kA  -82.79     0.50 kA                         |
|                                               C      25.42 MVA   0.19 kA  -82.70     0.48 kA                         |
|                                                                                                                     |
|Line 2                                                                                                               |
| Fault         A  230.00     0.00    0.00  1.00     446.71 MVA   3.36 kA  -85.28     8.44 kA   3.36    446.71  0.00 |
|               B            115.86 -101.57             0.00 MVA   0.00 kA    0.00     0.00 kA   0.00      0.00  0.88 |
|               C            118.96  101.96             0.00 MVA   0.00 kA    0.00     0.00 kA   0.00      0.00  0.89 |
|                                                                                                                     |
|  Line 2a         Bus 7                        A     356.82 MVA   2.69 kA   93.95     6.74 kA                         |
|                                               B      20.75 MVA   0.16 kA   97.74     0.39 kA                         |
|                                               C      20.68 MVA   0.16 kA   97.85     0.39 kA                         |
|                                                                                                                     |
|  Line 2b         Bus 5                        A      90.05 MVA   0.68 kA   97.77     1.70 kA                         |
|                                               B      20.75 MVA   0.16 kA  -82.26     0.39 kA                         |
|                                               C      20.68 MVA   0.16 kA  -82.15     0.39 kA                         |
|                                                                                                                     |
|  Fault location                               A       0.00 MVA   0.00 kA    0.00     0.00 kA                         |
|                                               B       0.00 MVA   0.00 kA    0.00     0.00 kA                         |
|                                               C       0.00 MVA   0.00 kA    0.00     0.00 kA                         |
|                                                                                                                     |
-----------------------------------------------------------------------------------------------------------------------
```

For calculating maximum short-circuit currents, we will select:

- **Fault Type:** 3-Phase Short-Circuit
- **Calculate:** Max. Short-Circuit Currents

Again, pressing Execute button will generate the following report:

```
-----------------------------------------------------------------------------------------------------------------
|                      |                                                 |  DIgSILENT    | Project:               |
|                      |                                                 |  PowerFactory |------------------------
|                      |                                                 |    15.1.7     | Date:  11/7/2020       |
-----------------------------------------------------------------------------------------------------------------
-----------------------------------------------------------------------------------------------------------------
| Fault Locations with Feeders                                                                                  |
| Short-Circuit Calculation / Method : IEC 60909                  3-Phase Short-Circuit   / Max. Short-Circuit Currents |
-----------------------------------------------------------------------------------------------------------------
| Asynchronous Motors                   | Grid Identification                 | Short-Circuit Duration             |
|   Always Considered                   |   Automatic                         |   Break Time                0.10 s |
|                                       |                                     |   Fault Clearing Time (Ith) 1.00 s |
| Decaying Aperiodic Component (idc)    | Conductor Temperature               | c-Voltage Factor                   |
|   Using Method               B        |   User Defined          No          |   User Defined              No     |
|                                       |                                     |                                    |
-----------------------------------------------------------------------------------------------------------------
```

Grid: Nine_Bus	System Stage: Nine_Bus								Annex:	/ 1		
	rtd.V. [kV]	Voltage [kV]	Voltage [deg]	c-Factor	Sk" [MVA/MVA]	Ik" [kA/kA]	Ik" [deg]	ip [kA/kA]	Ib [kA]	Sb [MVA]	Ik [kA]	Ith [kA]
Bus 1	16.50	0.00	0.00	1.10	1629.33 MVA	57.01 kA	-86.51	147.98 kA	46.90	1340.32	57.01	58.32
T1	Bus 4				393.68 MVA	13.78 kA	95.46	35.76 kA				
G1					1235.96 MVA	43.25 kA	-87.14	112.25 kA				
Bus 2	18.00	0.00	0.00	1.10	1488.65 MVA	47.75 kA	-87.82	124.86 kA	39.00	1216.01	47.75	48.95
T2	Bus 7				428.16 MVA	13.73 kA	94.03	35.91 kA				
G2					1060.81 MVA	34.03 kA	-88.57	88.97 kA				
Bus 3	13.80	0.00	0.00	1.10	1156.02 MVA	48.36 kA	-86.54	125.58 kA	40.79	974.90	48.36	49.48
T3	Bus 9				449.57 MVA	18.81 kA	94.39	48.84 kA				
G3					706.55 MVA	29.56 kA	-87.14	76.75 kA				
Bus 4	230.00	0.00	0.00	1.10	1216.17 MVA	3.05 kA	-86.21	7.88 kA	2.95	1175.29	3.05	3.12
T1	Bus 1				722.19 MVA	1.81 kA	91.67	4.68 kA				
Line 1	Bus 5				260.59 MVA	0.65 kA	95.93	1.69 kA				
Line 6	Bus 6				234.66 MVA	0.59 kA	97.92	1.52 kA				
Bus 5	230.00	0.00	0.00	1.10	898.73 MVA	2.26 kA	-84.69	5.61 kA	2.26	898.73	2.26	2.29
Line 1	Bus 4				528.86 MVA	1.33 kA	94.24	3.30 kA				
Line 2	Bus 7				370.09 MVA	0.93 kA	96.86	2.31 kA				
Bus 6	230.00	0.00	0.00	1.10	854.99 MVA	2.15 kA	-83.28	5.18 kA	2.15	854.99	2.15	2.17
Line 5	Bus 9				338.26 MVA	0.85 kA	97.98	2.05 kA				
Line 6	Bus 4				516.87 MVA	1.30 kA	95.89	3.13 kA				
Bus 7	230.00	0.00	0.00	1.10	1227.43 MVA	3.08 kA	-87.07	8.02 kA	2.93	1167.89	3.08	3.15
Line 2	Bus 5				258.53 MVA	0.65 kA	96.46	1.69 kA				
T2	Bus 2				663.13 MVA	1.66 kA	90.90	4.33 kA				
Line 3	Bus 8				306.77 MVA	0.77 kA	94.37	2.01 kA				

Grid: Nine_Bus	System Stage: Nine_Bus								Annex:	/ 2		
	rtd.V. [kV]	Voltage [kV]	Voltage [deg]	c-Factor	Sk" [MVA/MVA]	Ik" [kA/kA]	Ik" [deg]	ip [kA/kA]	Ib [kA]	Sb [MVA]	Ik [kA]	Ith [kA]
Bus 8	230.00	0.00	0.00	1.10	978.18 MVA	2.46 kA	-85.91	6.26 kA	2.43	967.23	2.46	2.50
Line 3	Bus 7				553.56 MVA	1.39 kA	93.67	3.54 kA				
Line 4	Bus 9				424.65 MVA	1.07 kA	94.63	2.72 kA				
Bus 9	230.00	0.00	0.00	1.10	1104.01 MVA	2.77 kA	-85.95	7.10 kA	2.60	1036.42	2.77	2.83
Line 4	Bus 8				339.97 MVA	0.85 kA	93.85	2.19 kA				
T3	Bus 3				514.14 MVA	1.29 kA	92.08	3.31 kA				
Line 5	Bus 6				250.92 MVA	0.63 kA	98.37	1.61 kA				
Line 2												
Fault	230.00	0.00	0.00	1.10	1118.16 MVA	2.81 kA	-86.10	7.16 kA	2.72	1084.28	2.81	2.86
Line 2a	Bus 7				840.90 MVA	2.11 kA	93.10	5.38 kA				
Line 2b	Bus 5				277.60 MVA	0.70 kA	96.34	1.78 kA				

6.5 Applying Optimization Techniques

From the preceding Sections 6.1, 6.2, 6.3, and 6.4, it has been seen that to do short-circuit analysis the steady-state condition of the network should first be obtained. This can be done by conducting PF analysis, which also depends on the current structure of dynamically changed power network. Knowing the structure is not just important for the PF and short-circuit analysis. It is also used for identifying the correct P/B relay pairs. Once all these stages are completed, the last stage can be initiated to find the best relay settings. Because this stage is vast and considered as the core or heart of this book, Chapter 7 will be dedicated just to explain how to mathematically model the ORC problem of standard inverse overcurrent relays and then solving it by using many approaches. This optimization model will be used as a solid basis to understand the mechanism of other more complicated coordination problems where different protective devices are involved.

Problems

Written Exercises

W6.1 Suppose that all the P/B relay pairs of a dynamic network are given. Do we need to re-identify these pairs if:
1. Power settings of some generators are changed.
2. There are variations in some loads.
3. Some circuit breakers are open.

Why?

W6.2 Is NR preferred for adaptive ORC solvers? Why?

W6.3 Why is the most severe fault commonly used in ORC problems?

W6.4 Consider the IEC 61363 and IEC 60909 standards. Which one should be used to conduct a short-circuit analysis for a large mesh network?

W6.5 Consider the simple three-bus system shown in Figure 6.19. All line admittances are identical and equal to $y_{12} = y_{13} = y_{23} = -j6$ pu. Take bus 1 as the reference bus where $V_1 = 1/0°$ pu. Also, bus 2 is maintained at $|V_2| = 1.05$ pu. Find the following:
1. Bus admittance matrix of the system.
2. Power flow equations.
3. Jacobian matrix.

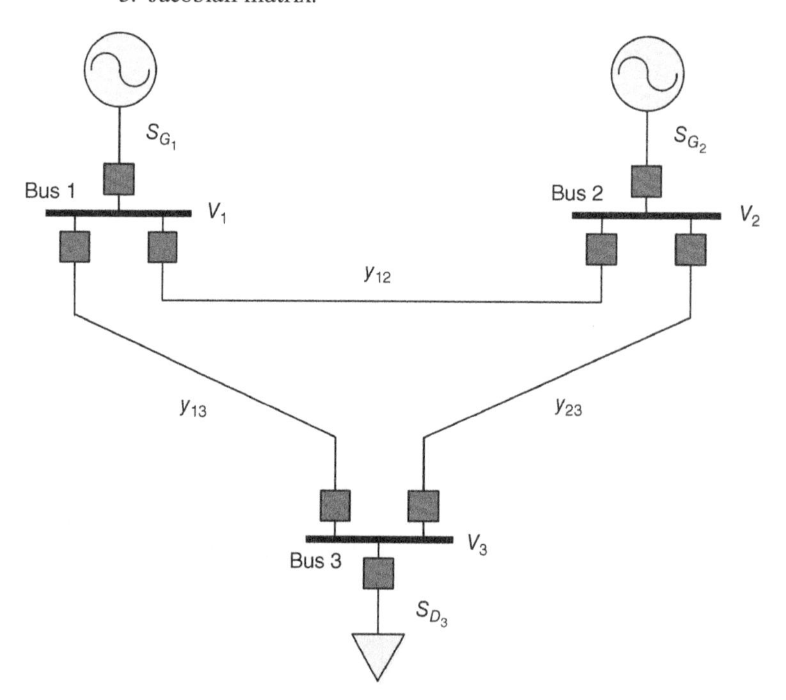

Figure 6.19 Single-line diagram of the three-bus system used in Exercise W6.5.

W6.6 Consider the network given in Figure 6.20. The active power supplied to bus 3 is $P_{G_3} = 300$ MW and the voltage magnitude of that busbar is maintained at $|V_3| = 1.03$ pu. The slack bus is maintained at $V_1 = 1.025/0°$ pu, and the load connected to bus 2 is $S_{D_2} = 400$ MW $+ j200$ MVAR. If the base load is 100 MVA and the converged solution gives $V_2 = 1.000\,57 - j0.036\,69$ pu:

1. Find the bus admittance matrix of the system.
2. Verify S_{D_2}.
3. Find P_{G_1} and Q_{G_1}.

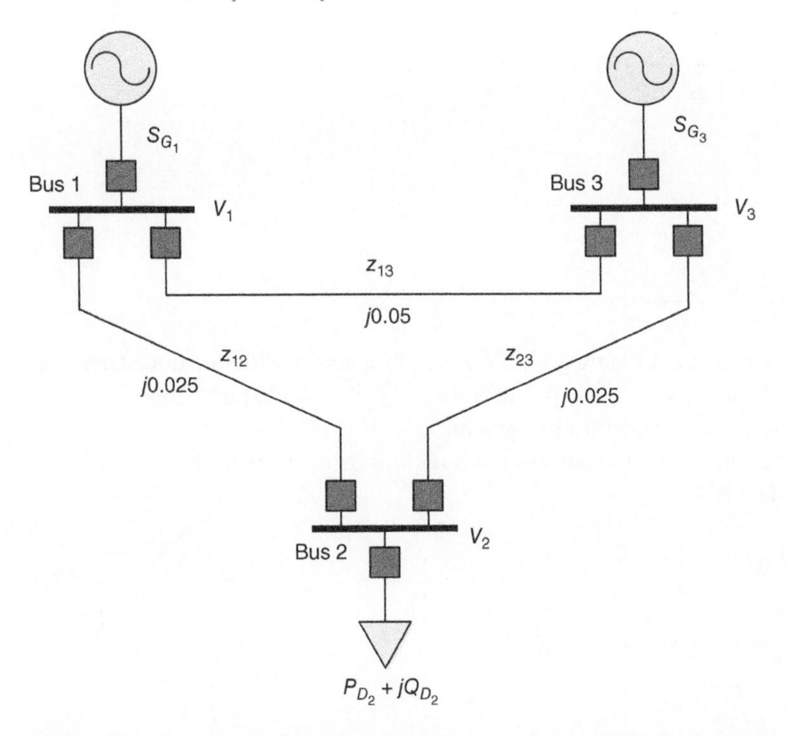

Figure 6.20 Single-line diagram of the three-bus system used in Exercise W6.6.

W6.7 Consider the system given in Figure 6.19. If it is a lossy network with line impedances of $z_{12} = 0.04 + j0.025$ pu, $z_{13} = 0.01 + j0.08$ pu, and $z_{23} = 0.02 + j0.125$ pu:

1. Find the bus admittance matrix of the system.
2. Write down the complete set of power flow equations using the polar form.
3. Complete Table 6.5.

Table 6.5 Table of Exercise W6.7.

	Bus 1	Bus 2	Bus 3
$\lvert V_i \rvert$	1	1.04	0.98
θ_i	0.0°	3.2°	−2.0°
P_{G_i}	?	?	0.0
Q_{G_i}	?	?	0.0
P_{D_i}	0.0	0.0	?
Q_{D_i}	0.0	0.0	?

W6.8 Repeat Exercise W6.7, but now with lossless network. If the new line impedances equal the imaginary part of those given in Exercise W6.7:

1. Find the bus admittance matrix of the system.
2. Write down the complete set of power flow equations using the polar form.
3. Because it is lossless network, verify that $P_{G_1} + P_{G_2} = P_{D_3}$.
4. Complete Table 6.6.

Table 6.6 Table of Exercise W6.8.

	Bus 1	Bus 2	Bus 3
$\lvert V_i \rvert$	1	1.04	0.98
θ_i	0.0°	3.2°	−2.0°
P_{G_i}	?	?	0.0
Q_{G_i}	?	?	0.0
P_{D_i}	0.0	0.0	$P_{G_1} + P_{G_2}$
Q_{D_i}	0.0	0.0	?

W6.9 Consider the system given in Figure 6.19. If it is a lossy network with line impedances of $z_{12} = 0.04 + j0.025$ pu, $z_{13} = 0.01 + j0.08$ pu, and $z_{23} = 0.02 + j0.125$ pu:

1. Find the bus admittance matrix of the system.
2. Write down the complete set of power flow equations using the polar form.
3. Complete Table 6.7.

Table 6.7 Table of Exercise W6.9.

	Bus 1	Bus 2	Bus 3
$\lvert V_i \rvert$	1	1.04	?
θ_i	0.0°	?	?
P_{G_i}	?	1.1	0.0
Q_{G_i}	?	?	0.0
P_{D_i}	0.0	0.0	1.5
Q_{D_i}	0.0	0.0	0.6

W6.10 A small startup power company wants to buy software to do their business in power systems engineering. This includes power flow and short-circuit analysis. After extensive searches, the engineering department minimizes the option to either ETAP or DIgSILENT PowerFactory. If the main customers are German companies where the VDE standard should be followed, which software should be selected? What if the main customers are Russian?

W6.11 What is the built-in command used in MatPower to perform short-circuit analysis?

W6.12 Consider the 6 bus network shown in Figure 6.21. Using the graph theory method explained in Section 6.3, find all the P/B relay pairs of the system.

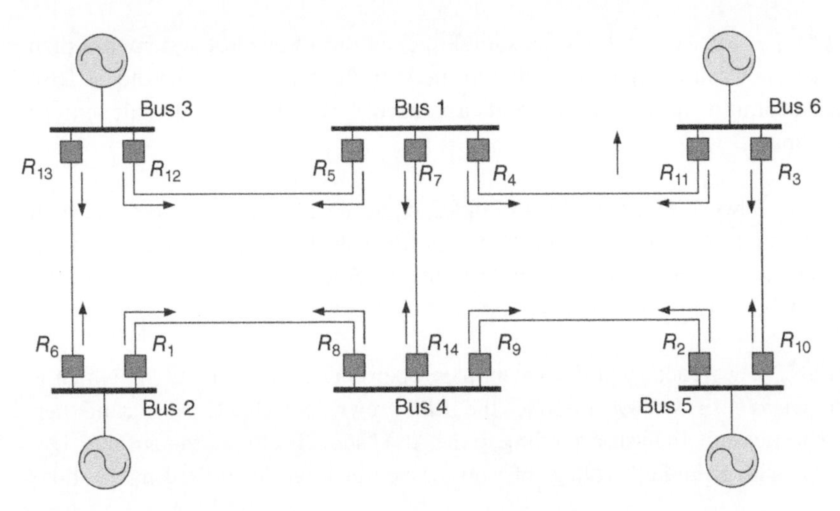

Figure 6.21 Single-line diagram of the six-bus system used in Exercise W6.12.

W6.13 Consider the 8 bus network shown in Figure 6.22. Using the inspection method, find all the P/B relay pairs.

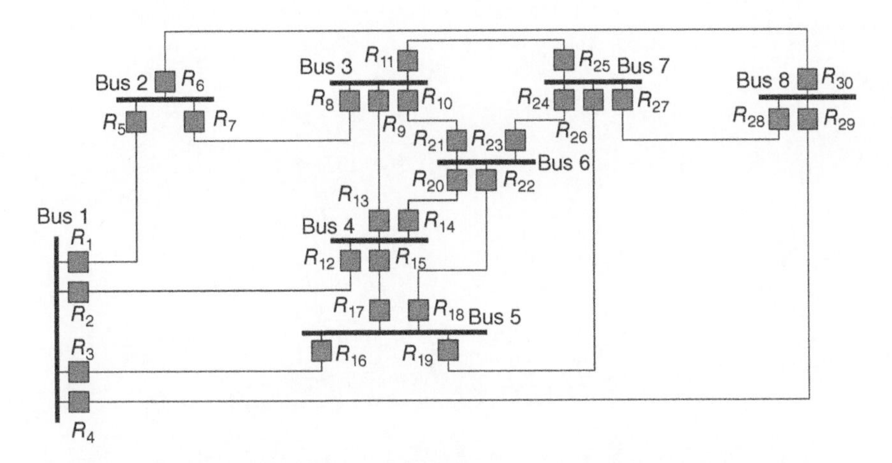

Figure 6.22 Single-line diagram of the eight-bus system used in Exercise W6.13.

Computer Exercises

C6.1 Based on your results obtained in Exercise W6.5, simulate the system in PowerWorld Simulator. Use 138 kV for the nominal voltage and 100 MVA for the base load.

C6.2 Based on your results obtained in Exercise W6.6, simulate the system in PowerWorld Simulator. Use 138 kV for the nominal voltage and 100 MVA for the base load.

C6.3 Based on your results obtained in Exercise W6.9, design a MATLAB code to solve the load flow problem by using NR.

C6.4 Using DIgSILENT PowerFactory built-in examples, open the 14-bus test system and then carry out the short-circuit analysis according to the VDE 0102 standard. Use the 3ϕ fault type with the maximum short-circuit currents approach. Leave the other default settings without any change.

C6.5 Using DIgSILENT PowerFactory built-in examples, open the 39-bus test system and then carry out the short-circuit analysis according to the IEC 60909 standard. Use the 3ϕ fault type with the maximum short-circuit currents approach. Select all the busbars. Leave the other default settings without any change.

C6.6 Using DIgSILENT PowerFactory built-in examples, open the 42-bus test system, which is commonly known as IEEE Std. 399-1997 (see the IEEE brown book (IEEE, 1998)), and then carry out the short-circuit analysis according to the ANSI standard for all the busbars. Use the 3ϕ fault type with a pre-fault voltage of 1 pu. Leave the other default settings without any change.

7

Optimal Coordination of Inverse-Time DOCRs with Unified TCCC

After reading this chapter you should be able to:

- Formulate the mathematical model of the optimal relay coordination (ORC) problem when only directional over-current relays (DOCRs) with one unified time–current characteristic curve (TCCC) are used
- Differentiate between different coordination criteria
- Deal with continuous time multiplier setting (*TMS*) and discrete plug setting (*PS*) vectors
- Know the concept of transient topology
- Express mesh networks in numerical equations
- Solve DOCRs-based ORC problems using meta-heuristic optimization algorithms
- Embed a pre-processing unit as a sub-algorithm to check and correct the selectivity constraints

---

7.1 Mathematical Problem Formulation

For finding the optimal value of a given problem, it needs first to be translated into a mathematical model. This section describes how to formulate the coordination problem so that it can be plugged into any multidimensional optimization algorithm. Also, it has been said that over-current relays (OCRs) can be categorized into three main types: instantaneous over-current relays (IOCRs)/directional over-current relays (DCOCRs), DTOCRs, and inverse time over-current relays (ITOCRs). In this chapter, only the last type will be considered. The other types will be taken later in the Chapters 10 and 17. Also, we will focus here on European relays where time multiplier setting (*TMS*), plug setting (*PS*), and plug setting multiplier (*PSM*) are used. The same mathematical model can be used for North American relays by just replacing *TMS* with time dial setting (*TDS*), *PS* with current tap setting (*CTS*), and *PSM* with *M*. This point has been extensively discussed in Chapter 3.

As per covered in Chapter 1, the design function of any optimization problem should be modeled as an objective function subjected to some design constraints. The following Sections 7.1.1, 7.1.2, 7.1.3, 7.1.4, 7.1.5, and 7.1.6 cover all these parts of classical optimal relay coordination (ORC) problems:

Optimal Coordination of Power Protective Devices with Illustrative Examples, First Edition. Ali R. Al-Roomi.

Companion website: www.wiley.com/go/al-roomi/optimalcoordination

7.1.1 Objective Function

Suppose that an electrical network contains ß branches, and each branch is protected by two CBs mounted on both ends. If each CB is triggered by one directional over-current relay (DOCR), then the preceding network should contain $\varrho = 2ß$ DOCRs. Also, it is known that the operating time of each DOCR depends on *PS* and *TMS*. Thus, the dimension (n) of any DOCR-based ORC problem can be calculated as follows:

$$n = 2\varrho = 4ß \tag{7.1}$$

If an in-zone fault occurs at the location (k), then the operating time of the ith primary relay (R_i) can be mathematically represented as follows:

$$T_{i,k} = f\left(TMS_i, PS_i, I_{i,k}\right), \quad i = 1,2, \ldots, \varrho \tag{7.2}$$

where $I_{i,k}$ is a short-circuit current seen by R_i for a fault occurring at the kth location.

For $T_{i,k}$, as said before, there were many attempts to model ITOCRs mathematically (Pierre and Wolny, 1986; Chan and Maurer, 1992; Almas et al., 2012). The most popular ones are presented in Chapter 3. If the IEC/BS standard model is selected for R_i, then its operating time can be calculated as follows:

$$T_{i,k} = \frac{\beta\ TMS_i}{\left(\frac{I_{i,k}}{PS_i}\right)^{\alpha} - 1} \tag{7.3}$$

The ANSI/IEEE standard model can also be used. It is similar to (7.3), but with considering the γ-coefficient[1,2]:

$$T_{i,k} = TDS_i \times \left[\frac{\beta}{\left(\frac{I_{i,k}}{CTS_i}\right)^{\alpha} - 1} + \gamma\right] \tag{7.4}$$

It has to be said that the lower and upper bounds of TMS_i and TDS_i depend on the relay technology.[3] As said at the beginning of this section, we will focus on (7.3).

The objective of this optimization problem is to find the best values of *PS* and *TMS* of all ϱ DOCRs so that the weighted sum of the operating times, when DOCRs act as primary relays, is minimized for ℓ fault locations as follows:

$$OBJ = \min \sum_{i=1}^{\varrho} \sum_{k=1}^{\ell} w_{i,k} T_{i,k}^{\text{pr}} \tag{7.5}$$

where $w_{i,k}$ represents the fault probability that might happen at the kth location of a branch protected by the ith relay. For the sake of simplicity, all the weights given in (7.5) are considered equal to one (Urdaneta et al., 1988).

To do short-circuit analysis, two options are available. The first option is to do it manually by hand, which is a weary process and highly unrecommended. The other option is to use one of the known commercial and free-distributed software.[4]

1 Please note that there is no subscript i for the coefficients $\{\alpha, \beta, \gamma\}$ because most of the studies consider that all ϱ DOCRs have the same time–current characteristic curve (TCCC). If multiple TCCCs are used, then multi-standard coefficients must be applied, and thus the subscript i must be included to these coefficients. This is also true for multiple user-defined coefficients.

2 The alpha and beta coefficients of (7.4) are different than those of (7.3). Please refer to the general model given in (3.36).

3 This point will be covered when the ORC problem with different relay technologies is considered.

4 A comprehensive list of commercial and free-distributed software is given in Chapter 6.

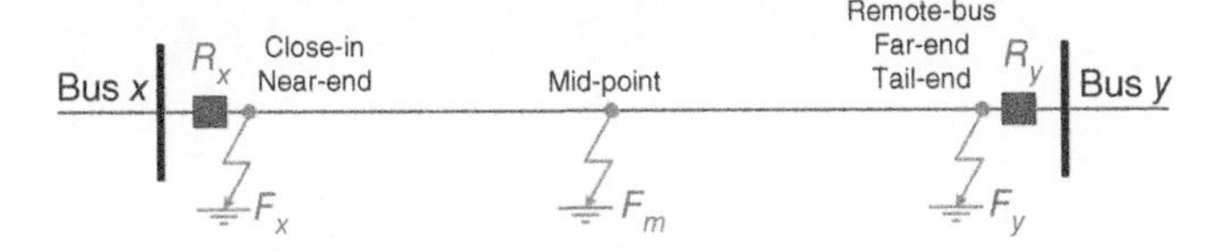

Figure 7.1 Levels of protection design criteria.

For ℓ fault locations on each zone, Damborg et al. (1984) classify three levels of coordination criteria. By referring to Figure 7.1, the first one is called the **desired design criterion**, which considers two fault locations. The first one is called the **near-end** fault[5] and the other one is called the **far-end** fault.[6] For the near-end location, the fault occurs at the nearest possible point of the line where the primary protective relay is installed, and vice versa for the far-end location (Al-Roomi and El-Hawary, 2019b). Please note that the near-end fault for the xth primary DOCR mentioned in Figure 7.1 is considered as the far-end fault for the yth primary DOCR. Because satisfying this criterion for low fault currents is very hard and sometimes impossible, so it could be relaxed to what is called the **minimum criterion** where the relay settings are optimized based on the near-end 3ϕ faults. When larger classes of faults are studied, then one fault at the **mid-point** of each line could be considered. This case is called the **enhanced criterion** (Damborg et al., 1984; Birla et al., 2006b). Some ORC studies considered just the mid-points, such as the nine-bus test system reported in Bedekar and Bhide (2011a) and the three-bus test system reported in Amraee (2012) and Albasri et al. (2015). All these design criteria are covered in Table 7.1 (Al-Roomi and El-Hawary, 2019b).

Based on this detailed classification, (7.5) should be re-expressed as follows:

- **Desired Design Criteria:**

$$OBJ = \begin{cases} \min \left[\sum_{p=1}^{\varrho^{\text{near}}} T_p^{\text{pr,near}} + \sum_{q=1}^{\varrho^{\text{far}}} T_q^{\text{pr,far}} \right], & \text{if } \varrho^{\text{near}} \neq \varrho^{\text{far}} \\ \min \sum_{i=1}^{\varrho} \left(T_i^{\text{pr,near}} + T_i^{\text{pr,far}} \right), & \text{if } \varrho^{\text{near}} = \varrho^{\text{far}} \end{cases} \tag{7.6}$$

- **Minimum Criteria:**

$$OBJ = \min \sum_{i=1}^{\varrho} T_i^{\text{pr,near}} \tag{7.7}$$

- **Enhanced Criteria:**

$$OBJ = \begin{cases} \min \left[\sum_{p=1}^{\varrho^{\text{near}}} T_p^{\text{pr,near}} + \sum_{q=1}^{\varrho^{\text{mid}}} T_q^{\text{pr,mid}} \right], & \text{if } \varrho^{\text{near}} \neq \varrho^{\text{mid}} \\ \min \sum_{i=1}^{\varrho} \left(T_i^{\text{pr,near}} + T_i^{\text{pr,mid}} \right), & \text{if } \varrho^{\text{near}} = \varrho^{\text{mid}} \end{cases} \tag{7.8}$$

Table 7.1 Number of faults considered for coordination criteria.

Coordination criteria	Number and names of fault locations
Desired design Criteria	2-Bolted points: $\{F_x, F_y\}$ for R_x and R_y
Minimum criteria	1-Bolted point: F_x for R_x and F_y for R_y
Enhanced criteria	2-Bolted points: $\{F_x, F_m\}$ for R_x and $\{F_y, F_m\}$ for R_y
Mid-point criteria	1-Bolted points: F_m for both R_x and R_y

Source: Al-Roomi and El-Hawary (2019).

5 It is also known as the **close-in** fault.

6 It is also known as the **remote-bus** fault and the **tail-end** fault.

- **Mid-Point Criteria:**

$$OBJ = \min \sum_{i=1}^{\varrho} T_i^{\text{pr,mid}} \tag{7.9}$$

where the cases ($\varrho^{\text{near}} = \varrho^{\text{far}} = \varrho$) and ($\varrho^{\text{near}} = \varrho^{\text{mid}} = \varrho$) occur when the same number of relays are considered for both fault locations.

Optimizing any one of these objective functions requires to satisfy first the following design constraints:

7.1.1.1 Other Possible Objective Functions

In some references, we could find different objective functions.[7] For example, the operating time of relays when they act as primary and backup protective devices can be summed as follows:

$$OBJ = \min \left[\sum_{k=1}^{\ell} \left(\sum_{i=1}^{\varrho^{\text{pr}}} T_{i,k}^{\text{pr}} + \sum_{j=1}^{\varrho^{\text{bc}}} T_{j,k}^{\text{bc}} \right) \right] \tag{7.10}$$

which can also be classified according to the preceding coordination criteria.

Some references take into account the difference in time between backup and primary relays:

$$OBJ = \min \left[\lambda_1 \cdot \sum_{i=1}^{\varrho} T_i^2 + \lambda_2 \cdot \sum_{j=1}^{\sigma} T_{ji}^2 \right] \tag{7.11}$$

where $\{\lambda_1, \lambda_2\}$ are weights, σ is the number of P/B relay pairs, and T_{ji} is the difference in the operating times of the jth backup and the ith primary relays which can be calculated as follows:

$$T_{ji} = T_j^{\text{bc}} - T_i^{\text{pr}} - CTI \tag{7.12}$$

The same one can be found in Razavi et al. (2008) where CTI is taken into account during minimizing the cost function:

$$OBJ = \min \left\{ \lambda_1 \cdot \sum_{i=1}^{\varrho} (T_i)^2 + \lambda_2 \cdot \sum_{j=1}^{\sigma} \left[T_{ji} - \mu_2 (T_{ji} - |T_{ji}|)^2 \right] \right\} \tag{7.13}$$

where μ_2 is a weighting factor.

An objective function, which is similar to the one given in (7.13), is proposed by Alam et al. (2016):

$$OBJ = \min \left\{ \lambda_1 \cdot \sum_{i=1}^{\varrho} (T_i)^2 + \lambda_2 \cdot \sum_{j=1}^{\sigma} (T_j - CTI)^2 \right\} \tag{7.14}$$

where the sum of λ_1 and λ_2 should be one.[8]

Because fault currents flow in forward and reverse directions, so dual setting DOCRs have been proposed by some manufacturers to act for each direction differently. Zeineldin et al. (2015) present the following objective function:

$$OBJ = \min \left[\sum_{k=1}^{\ell} \left(\sum_{i=1}^{\varrho^{\text{fw}}} T_{i,k}^{\text{fw}} + \sum_{j=1}^{\varrho^{\text{rv}}} T_{j,k}^{\text{rv}} \right) \right] \tag{7.15}$$

where the superscripts fw and rv stand for the forward and reverse directions, respectively.

7 Moreover, some ORC solvers are designed as multi-objective optimizers.
8 Equation (7.14) is implemented in Alam (2019) for adaptive ORC.

7.1.2 Inequality Constraints on Relay Operating Times

To realize the operation of the ith relay, its speed should be bounded between two limits:

$$T_{i,k}^{\min} \leqslant T_{i,k} \leqslant T_{i,k}^{\max} \tag{7.16}$$

where $T_{i,k}^{\min}$ and $T_{i,k}^{\max}$ are, respectively, the minimum and maximum operating times of R_i for a fault occurring at the kth location. $T_{i,k}^{\min}$ depends on the internal components of R_i, whereas $T_{i,k}^{\max}$ depends on the critical clearing time (t_{cr}) required to preserve system stability (Kundur, 1994; Padiyar, 2008; Albasri et al., 2015) and the allowable thermal limit of the protected component[9] (Smith and Jain, 2017).

There is one common mistake that is frequently seen in the literature where this inequality constraint is treated as a side constraint. Since this constraint is applied to a dependent variable, so it is a functional constraint. Also, because it contains "⩽," so it is an inequality functional constraint.[10] Thus, (7.16) can be divided into two inequality constraints as follows:

$$T_{i,k}^{\min} - T_{i,k} \leqslant 0 \tag{7.17}$$

$$T_{i,k} - T_{i,k}^{\max} \leqslant 0 \tag{7.18}$$

7.1.3 Side Constraints on Relay Time Multiplier Settings

Manufacturers of protective relays offer their products with some specifications. One of these specifications is about the lower and upper limits of TMS, which can be expressed as follows[11]:

$$TMS_i^{\min} \leqslant TMS_i \leqslant TMS_i^{\max} \tag{7.19}$$

where $TMS_i^{\min}$ and $TMS_i^{\max}$ are, respectively, the minimum and maximum values of TMS of R_i.

7.1.4 Side Constraints on Relay Plug Settings

The correct range of PS is graphically explained in Figure 7.2. For the ith relay, the lower limit ($PS_i^{\min}$) should be set equal to or greater than the maximum overload current ($I_{OL}^{\max}$), and the upper limit ($PS_i^{\max}$) should be set equal to or less than the minimum fault current.[12] The term $I_{OL}^{\max}$ can be calculated as follows:

$$I_{OL}^{\max} = OLF \times I_L^{\max} \tag{7.20}$$

where $I_L^{\max}$ is the maximum rated current. **OLF** is the **overload factor**, which depends on the element being protected, and it is usually set in the range of 1.25–1.5 (Gers and Holmes, 2004).

Practically speaking, the side constraints on the PSs of all ϱ DOCRs depend on the specification of powerlines[13] and short-circuit analysis. Once these two fundamental steps are successfully done,

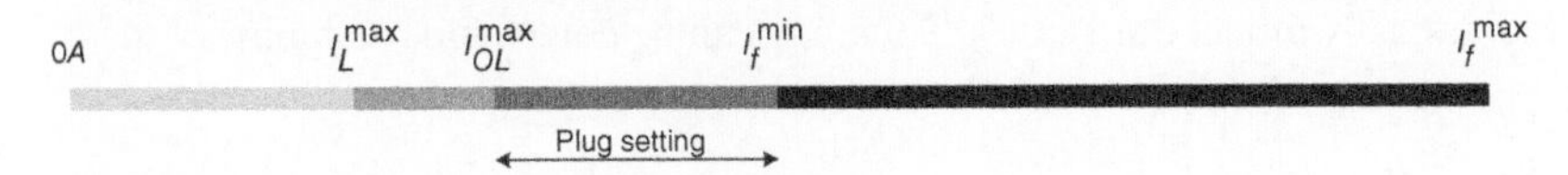

Figure 7.2 The available range of PS.

9 These two factors have been covered in Chapter 4.
10 Please refer to Chapter 1 for more details.
11 It has to be remembered that this variable is unitless (Albasri et al., 2015).
12 In most cases, the minimum fault current is the single-phase (1ϕ) short-circuit current (El-Hawary, 1983).
13 That is transmission, sub-transmission, and distribution lines.

the limits of PS of all ϱ relays can be defined. In general, the side constraint of the ith PS (PS_i) can be mathematically expressed as follows (Al-Roomi, 2014):

$$PS_i^{min} \leqslant PS_i \leqslant PS_i^{max} \tag{7.21}$$

To simplify this constraint, most of the studies presented in the literature consider this side constraint as a fixed vector of discrete values, such as (Noghabi et al., 2009; Amraee, 2012; Albasri et al., 2015). However, some other studies use the practical range associated with each relay, such as (Bedekar and Bhide, 2011a; Albasri et al., 2015). For the sake of clarity, these two bounds can be practically calculated as follows:

$$PS_i^{min} = \frac{OLF \times I_{L,i}^{max}}{CTR_i} \tag{7.22}$$

$$PS_i^{max} = \frac{2}{3CTR_i} I_{f,i}^{min} \tag{7.23}$$

where CTR_i is the **CT-ratio** of the ith relay, and $I_{f,i}^{min}$ is the minimum fault current that must be detected by that relay (Gers and Holmes, 2004).

7.1.5 Selectivity Constraint Among Primary and Backup Relay Pairs

This inequality constraint ensures that the associated backup DOCRs will not operate before their primary DOCRs. This can be accomplished by selecting the proper PS and TMS so that the backup relay(s) can initiate the trip signal to isolate the kth out-zone fault if R_i exceeds the given chance.[14] The mathematical formulation of this constraint can be expressed as follows:

$$T_{j,k} - T_{i,k} \geqslant CTI_i \tag{7.24}$$

where $T_{j,k}$ is the operating time of the jth backup relay for an out-zone fault occurred at the kth location, and CTI_i is the coordination time interval given to the ith primary relay which is also the minimum allowable discrimination margin between R_i and R_j (Gers and Holmes, 2004).

In almost all the ORC studies reported in the literature, the researchers consider only one unified CTI for all ϱ DOCRs. Based on this assumption, (7.24) can be simplified by dropping the subscript i from CTI to have:

$$T_{j,k} - T_{i,k} \geqslant CTI \tag{7.25}$$

Therefore, the operating time of the jth backup relay must be known to check whether the preceding constraint is satisfied or not. This can be easily calculated by using the IEC standard model:

$$T_{j,k} = \frac{\beta \, TMS_j}{\left(\frac{I_{j,k}}{PS_j}\right)^{\alpha} - 1} \tag{7.26}$$

Remember that if North American DOCRs are used, then the ANSI/IEEE standard given in (7.4) must be used to mathematically model the TCCCs. Thus, $T_{j,k}$ can be determined as follows:

$$T_{j,k} = TDS_j \times \left[\frac{\beta}{\left(\frac{I_{j,k}}{CTS_j}\right)^{\alpha} - 1} + \gamma\right] \tag{7.27}$$

14 In other words, it is the sum of the ith relay operating time T_i^{pr} plus its coordination time interval (CTI_i).

where $I_{j,k}$ is a short-circuit current seen by the jth backup relay (R_j) for a fault occurring at the kth location.[15]

Again, remember that if the primary and backup relays are manufactured in different continents (i.e. North American and European DOCRs), then it is recommended to transform between *TDS* and *TMS* in ORC solvers either through the fixed divisor method or the linear interpolation method. This is also very useful when one unified TCCC is used for all North American and European programmable numerical relays. This point has been extensively covered in Chapter 3.

By substituting (7.3) and (7.26) into (7.25), the selectivity constraint can be expressed according to the IEC/BS standard model as follows:

$$\frac{\beta\ TMS_j}{\left(\frac{I_{j,k}}{PS_j}\right)^{\alpha} - 1} - \frac{\beta\ TMS_i}{\left(\frac{I_{i,k}}{PS_i}\right)^{\alpha} - 1} \geqslant CTI \tag{7.28}$$

Similarly, by substituting (7.4) and (7.27) into (7.25), the selectivity constraint can be expressed according to the ANSI/IEEE standard model as follows:

$$TDS_j \times \left[\frac{\beta}{\left(\frac{I_{j,k}}{CTS_j}\right)^{\alpha} - 1} + \gamma\right] - TDS_i \times \left[\frac{\beta}{\left(\frac{I_{i,k}}{CTS_i}\right)^{\alpha} - 1} + \gamma\right] \geqslant CTI \tag{7.29}$$

7.1.5.1 Transient Selectivity Constraint

For having a more realized model, some researchers take into account the case when one of the two-end primary relays operates before the other. At that very short period of time, the network will have a transient topology and thus the fault current will change.

Example 7.1 Consider the simple three-bus system shown in Figure 7.3. The trip speed of R_2 is slower than that of R_1 when a 3ϕ short-circuit fault occurs on point F_x of line 1, and vice versa for point F_y. Draw the transient topology of the network for each fault location.

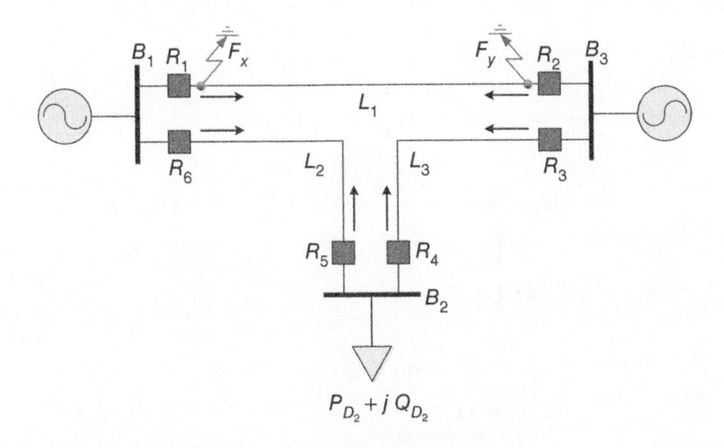

Figure 7.3 A simple three-bus system for Example 7.1.

15 This location belongs to the ith primary relay (R_i). Thus, this fault is considered as an out-zone fault for R_j.

Solution

Here, we have two scenarios:

- **Short-Circuit at F_x:**
  For this fault location, the operating time of the first relay is less than that of the second relay (i.e. $T_1 < T_2$). Thus, the current seen by R_2 will change when the first circuit breaker (CB_1) isolates the end connected to bus 1. This transient topology is shown in Figure 7.4.
- **Short-Circuit at F_y:**
  For this fault location, R_2 will be faster than R_1 (i.e. $T_1 > T_2$). Thus, with assuming identical circuit breakers and similar coordination time intervals, CB_2 will isolate line 1 from being connected to bus 3 before CB_1 isolating the other end. Thus, R_1 will see a different current during that period. This transient topology is shown in Figure 7.5.

Example 7.2 As a numerical example, consider the network given in Figure 6.20. Take the new line impedances as $z_{12} = j0.027$ pu, $z_{13} = j0.035$ pu, and $z_{23} = j0.03$ pu. The active power supplied to bus 3 is $P_{G_3} = 250$ MW and the voltage magnitude of that busbar is maintained at $|V_3| = 1.05$ pu.

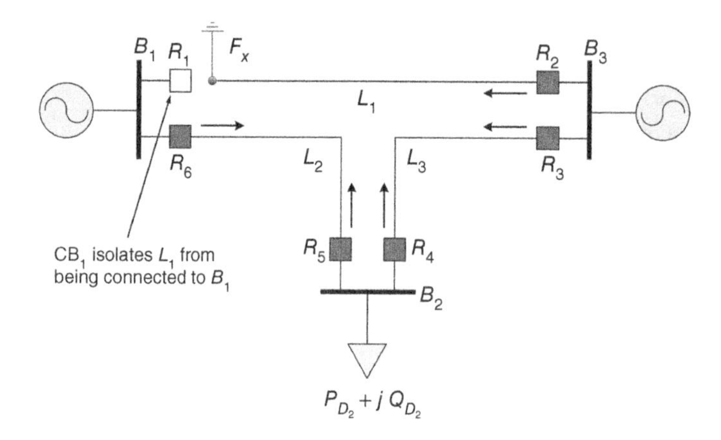

Figure 7.4 Solution of Example 7.1 during the transient topology of F_x.

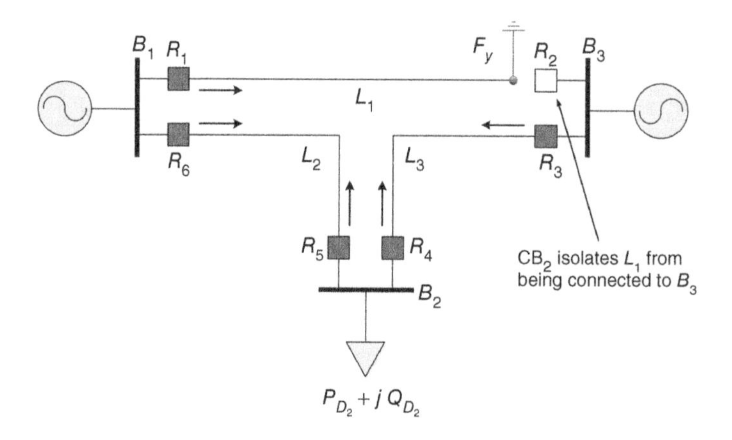

Figure 7.5 Solution of Example 7.1 during the transient topology of F_y.

The slack bus is maintained at $V_1 = 1.02\angle 0^\circ$ pu, and the load connected to bus 2 is $S_{D_2} = 450\text{MW} + j260$ MVAR. The base load is 100 MVA and the nominal voltage is 138 kV. Consider the same two 3ϕ short-circuit faults given in the preceding example, find:

1. The short-circuit current seen by R_2 for F_x before the opening of CB_1.
2. The short-circuit current seen by R_1 for F_y before the opening of CB_2.
3. The transient short-circuit current seen by R_2 for F_x after the opening of CB_1.
4. The transient short-circuit current seen by R_1 for F_y after the opening of CB_2.

Solution

For the first two parts of the question, the network does not have any transient topology (i.e. CB_1 and CB_2 operate at the same moment). Thus, we can use the procedure explained in Chapter 6 for calculating 3ϕ short-circuit currents using PowerWorld Simulator directly without modifying anything. Figure 7.6 shows the system modeled in PowerWorld Simulator. It can be browsed by opening the file `3Bus.PWB`. For the first two parts of the question, the short-circuit currents are:

$$I_{R_2,F_x} = 2.688\,37 \text{ pu} = 1124.73 \text{ A}$$
$$I_{R_1,F_y} = 1.355\,95 \text{ pu} = 567.29 \text{ A}$$

For the transient topology, we can assume that the first isolated end is connected to a virtual busbar where no power source or load is connected and then assuming that the fault occurs on that busbar. Let's call it bus 4. This new configuration can be depicted in Figure 7.7 for F_x. It can be

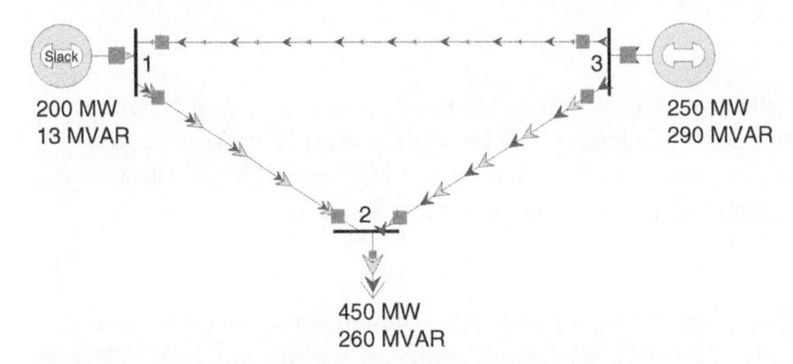

Figure 7.6 Modeling the steady-state topology of the three bus network of Example 7.2 in PowerWorld Simulator.

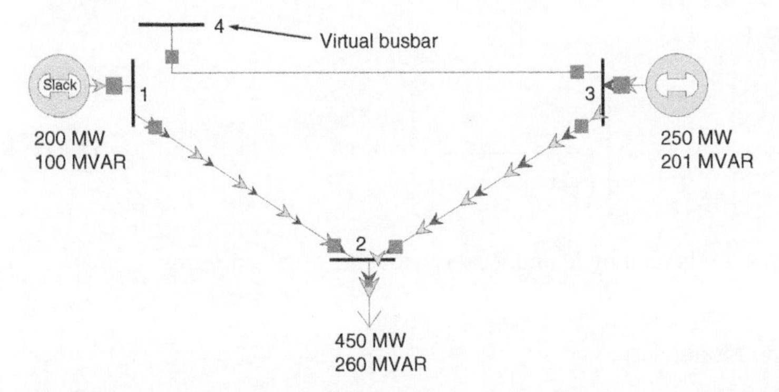

Figure 7.7 Modeling the transient topology of the three bus network of Example 7.2 in PowerWorld Simulator when CB_1 operates before CB_2.

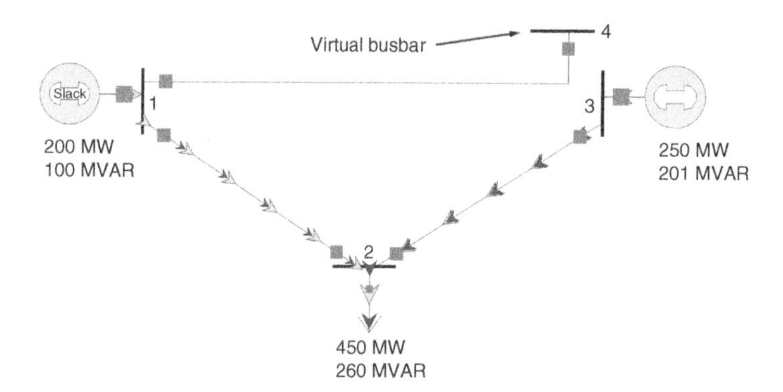

Figure 7.8 Modeling the transient topology of the three bus network of Example 7.2 in PowerWorld Simulator when CB_2 operates before CB_1.

accessed via the file `3Bus_Fx.PWB`. Following the same steps of the first two parts, the transient short-circuit current seen by R_2 during the opening of CB_1 is:

$$I'_{R_2,F_x} = 5.330\ 79 \text{ pu} = 2230.24 \text{ A}$$

The same thing for F_y where Figure 7.8 shows the model saved in the file `3Bus_Fy.PWB`. Thus, the transient short-circuit current seen by R_1 during the opening of CB_2 is:

$$I'_{R_1,F_y} = 5.220\ 31 \text{ pu} = 2184.02 \text{ A}$$

As can be clearly seen from the preceding two examples, the short-circuit currents significantly increase during transient topologies. Therefore, it is important to ensure that the corresponding backup relays will not operate at that moment (Urdaneta et al., 1988; Amraee, 2012; Albasri et al., 2015). This can be achieved by defining the following inequality constraint:

$$T'_{j,k} - T'_{i,k} \geqslant CTI \tag{7.30}$$

where $T'_{i,k}$ and $T'_{j,k}$ are the operating times of the i–jth P/B relay pair under that transient condition.

Again, the constraint given in (7.30) can be respectively expressed according to the IEC/BS and ANSI/IEEE standard models as follows:

$$\frac{\beta\ TMS_j}{\left(\frac{I'_{j,k}}{PS_j}\right)^{\alpha} - 1} - \frac{\beta\ TMS_i}{\left(\frac{I'_{i,k}}{PS_i}\right)^{\alpha} - 1} \geqslant CTI \tag{7.31}$$

$$TDS_j \times \left[\frac{\beta}{\left(\frac{I'_{j,k}}{CTS_j}\right)^{\alpha} - 1} + \gamma\right] - TDS_i \times \left[\frac{\beta}{\left(\frac{I'_{i,k}}{CTS_i}\right)^{\alpha} - 1} + \gamma\right] \geqslant CTI \tag{7.32}$$

where $I'_{i,k}$ and $I'_{j,k}$ are the fault currents seen by R_i and R_j, respectively.

7.1.6 Standard Optimization Model

By collecting all the optimization parts given earlier, the standard optimization model of classical ORC problems can be expressed for European DOCRs as follows:

$$
\begin{aligned}
&\min_{TMS,PS} \; Z(TMS_1, \dots, TMS_\varrho, PS_1, \dots, PS_\varrho) \\
&\text{Subjected to: } T^{\text{pr}}_{i,k} + CTI - T^{\text{bc}}_{j,k} \leqslant 0 \\
&\qquad T^{\text{min}}_{i,k} - T^{\text{pr}}_{i,k} \leqslant 0 \\
&\qquad T^{\text{pr}}_{i,k} - T^{\text{max}}_{i} \leqslant 0 \\
&\qquad TMS^{\text{min}}_i \leqslant TMS_i \leqslant TMS^{\text{max}}_i \\
&\qquad PS^{\text{min}}_i \leqslant PS_i \leqslant PS^{\text{max}}_i
\end{aligned}
\tag{7.33}
$$

where i and j notations are, respectively, used to represent the primary and backup relays, and the notation k represents the fault location.

For North American DOCRs, the following standard optimization model can be used:

$$
\begin{aligned}
&\text{Subjected to: } T^{\text{pr}}_{i,k} + CTI - T^{\text{bc}}_{j,k} \leqslant 0 \\
&\qquad T^{\text{min}}_{i,k} - T^{\text{pr}}_{i,k} \leqslant 0 \\
&\qquad T^{\text{pr}}_{i,k} - T^{\text{max}}_{i} \leqslant 0 \\
&\qquad TDS^{\text{min}}_i \leqslant TDS_i \leqslant TDS^{\text{max}}_i \\
&\qquad CTS^{\text{min}}_i \leqslant CTS_i \leqslant CTS^{\text{max}}_i
\end{aligned}
\tag{7.34}
$$

It has to be remembered that if the transient selectivity constraint expressed in (7.30) is modeled in the ORC problem, then it should also be included in (7.33) and (7.34).

7.2 Optimal Coordination of DOCRs Using Meta-Heuristic Optimization Algorithms

The optimization algorithms presented in Chapter 1 can solve only unconstrained problems, while the simple ORC model shown in Section 7.1 is a highly constrained nonlinear nonconvex optimization problem that needs special care to deal with. In addition, the conventional evolutionary algorithms (EAs) are so slow techniques and thus they are not preferred in some protection designs where the processing speed is a very important factor. Moreover, obtaining a feasible solution is not an easy task, because of the nonlinearity and nonconvexity, having a large number of design constraints, and a limited number of discrete *PS*. Based on these facts, we think that it is essential to reveal some crucial modifications that can be applied to any conventional optimization algorithm so all the preceding issues can be tackled with ensuring having optimal or near-optimal feasible solutions.

7.2.1 Algorithm Implementation

Once the mathematical model given in (7.33) is correctly formulated, any n-dimensional optimization algorithm can be applied to find the optimal values of *TMS* and *PS*. Three possible scenarios could be faced during designing any EA program:

- **TMS and PS Are Discrete → Combinatorial EAs:** This approach could be used to simulate electromechanical DOCRs.
- **TMS Is Continuous and PS Is Discrete → Mixed-Integer[16] EAs:** This approach could be used to simulate solid-state DOCRs.

16 Some references mention that by saying "mixed integer non-linear programming (MINLP) problem."

- **TMS and PS Are Continuous → Real-Coded EAs:** This approach could be used to simulate state-of-the-art numerical DOCRs.

These three categories have a direct implication when the ORC problem is linearized. This part will be covered in Chapter 8 when linear programming (LP) is hybridized with EAs to exploit the optimal solutions.

7.2.2 Constraint-Handling Techniques

The conventional EAs given in Chapter 1 (i.e. biogeography-based optimization [BBO] and differential evolution [DE]) can be applied to any one of the three scenarios listed in Section 7.2.1. For example, the conventional BBO algorithm shown in Figure 1.14 has been successfully used to solve some popular test systems (Al-Roomi, 2014; Albasri et al., 2015). However, the ORC problems are well-known as highly stiff non-convex nonlinear optimization problems where many constraints need to be satisfied.

Various techniques have been proposed to handle and satisfy all the design constraints, such as the **Karush–Kuhn–Tucker** (**KKT**) technique suggested in Urdaneta et al. (1988) and the death penalty function presented in Bedekar and Bhide (2011a). Unfortunately, the literature lacks of enough information about this essential part. It is either not mentioned at all in many papers or insufficient information is given such as the unknown exterior penalty functions reported in Noghabi et al. (2009) and Amraee (2012). Therefore, it is important to carry further investigation on this issue because it directly affects the performance of the optimization algorithm in terms of the overall cost and the processing speed (i.e. CPU time) (Eiben and Smith, 2003; Albasri et al., 2015). The constraint-handling techniques presented in Chapter 1 will be selected here for carrying this investigation.

Example 7.3 The IEEE three-bus test system is shown in Figure F.1. The CT-ratios of R_1 to R_6, the list of P/B relay pairs, the short-circuit currents for 3ϕ fault at the middle of each line, and the rest of information all are given in Appendix F. The *PS* is considered discrete in uniform steps of 0.5 A. Formulate the optimization model given in (7.33) for this test system. Take $T^{\mathrm{pr}} \in [0.1, 0.5]$. Use the IEC standard inverse characteristic curve for all the six relays. Also, neglect the transient topology of the network.

Solution

From Appendix F, we have: $CTI = 0.2$ s, $PS \in [1.5, 5]$ A, and $TMS \in [0.1, 1.1]$. The coordination problem of the three-bus test system with fixed topology and six IEC standard inverse DOCRs will have 12 variables and 30 constraints. Thus, (7.33) can be re-expressed as follows:

$$\min_{TMS,PS} Z(TMS_1, \dots, TMS_6, PS_1, \dots, PS_6)$$

$$\begin{aligned}
\text{Subjected to: } & T^{\mathrm{pr}}_{i,k} + 0.2 - T^{\mathrm{bc}}_{j,k} \leqslant 0 \\
& 0.1 - T^{\mathrm{pr}}_{i,k} \leqslant 0 \\
& T^{\mathrm{pr}}_{i,k} - 0.5 \leqslant 0 \\
& 0.1 \leqslant TMS_i \leqslant 1.1 \\
& 1.5 \leqslant PS_i \leqslant 5
\end{aligned}$$

where the notations i and j respectively represent the primary and backup relays; i.e. P/B pairs.

If the IEC/BS standard model given in (7.3) is selected for all the relays R_i, then (7.33) can be expanded as follows:

$$OBJ = \min \left\{ \frac{0.14\ TMS_1}{\left[\frac{1978.9}{\left(\frac{300}{5}\right)PS_1}\right]^{0.02} - 1} + \frac{0.14\ TMS_2}{\left[\frac{1525.7}{\left(\frac{200}{5}\right)PS_2}\right]^{0.02} - 1} + \frac{0.14\ TMS_3}{\left[\frac{1683.9}{\left(\frac{200}{5}\right)PS_3}\right]^{0.02} - 1} \right.$$

$$\left. + \frac{0.14\ TMS_4}{\left[\frac{1815.4}{\left(\frac{300}{5}\right)PS_4}\right]^{0.02} - 1} + \frac{0.14\ TMS_5}{\left[\frac{1499.66}{\left(\frac{200}{5}\right)PS_5}\right]^{0.02} - 1} + \frac{0.14\ TMS_6}{\left[\frac{1766.3}{\left(\frac{400}{5}\right)PS_6}\right]^{0.02} - 1} \right\}$$

Subjected to:

$$\frac{0.14\ TMS_1}{\left[\frac{1978.9}{\left(\frac{300}{5}\right)PS_1}\right]^{0.02} - 1} + 0.2 - \frac{0.14\ TMS_5}{\left[\frac{175}{\left(\frac{200}{5}\right)PS_5}\right]^{0.02} - 1} \leqslant 0$$

$$\frac{0.14\ TMS_2}{\left[\frac{1525.7}{\left(\frac{200}{5}\right)PS_2}\right]^{0.02} - 1} + 0.2 - \frac{0.14\ TMS_4}{\left[\frac{545}{\left(\frac{300}{5}\right)PS_4}\right]^{0.02} - 1} \leqslant 0$$

$$\frac{0.14\ TMS_3}{\left[\frac{1683.9}{\left(\frac{200}{5}\right)PS_3}\right]^{0.02} - 1} + 0.2 - \frac{0.14\ TMS_1}{\left[\frac{617.22}{\left(\frac{300}{5}\right)PS_1}\right]^{0.02} - 1} \leqslant 0$$

$$\frac{0.14\ TMS_4}{\left[\frac{1815.4}{\left(\frac{300}{5}\right)PS_4}\right]^{0.02} - 1} + 0.2 - \frac{0.14\ TMS_6}{\left[\frac{466.17}{\left(\frac{400}{5}\right)PS_6}\right]^{0.02} - 1} \leqslant 0$$

$$\frac{0.14\ TMS_5}{\left[\frac{1499.66}{\left(\frac{200}{5}\right)PS_5}\right]^{0.02} - 1} + 0.2 - \frac{0.14\ TMS_3}{\left[\frac{384}{\left(\frac{200}{5}\right)PS_3}\right]^{0.02} - 1} \leqslant 0$$

$$\frac{0.14\ TMS_6}{\left[\frac{1766.3}{\left(\frac{400}{5}\right)PS_6}\right]^{0.02} - 1} + 0.2 - \frac{0.14\ TMS_2}{\left[\frac{145.34}{\left(\frac{200}{5}\right)PS_2}\right]^{0.02} - 1} \leqslant 0$$

$$0.1 - \frac{0.14\ TMS_1}{\left[\frac{1978.9}{\left(\frac{300}{5}\right)PS_1}\right]^{0.02} - 1} \leqslant 0$$

$$0.1 - \frac{0.14\ TMS_2}{\left[\frac{1525.7}{\left(\frac{200}{5}\right)PS_2}\right]^{0.02} - 1} \leqslant 0$$

$$0.1 - \frac{0.14\ TMS_3}{\left[\frac{1683.9}{\left(\frac{200}{5}\right)PS_3}\right]^{0.02} - 1} \leqslant 0$$

$$0.1 - \frac{0.14\ TMS_4}{\left[\frac{1815.4}{\left(\frac{300}{5}\right)PS_4}\right]^{0.02} - 1} \leqslant 0$$

$$0.1 - \frac{0.14\ TMS_5}{\left[\frac{1499.66}{\left(\frac{200}{5}\right)PS_5}\right]^{0.02} - 1} \leqslant 0$$

$$0.1 - \frac{0.14\ TMS_6}{\left[\frac{1766.3}{\left(\frac{400}{5}\right)PS_6}\right]^{0.02} - 1} \leqslant 0$$

$$\frac{0.14\ TMS_1}{\left[\frac{1978.9}{\left(\frac{300}{5}\right)PS_1}\right]^{0.02} - 1} - 0.5 \leqslant 0$$

$$\frac{0.14\ TMS_2}{\left[\frac{1525.7}{\left(\frac{200}{5}\right)PS_2}\right]^{0.02} - 1} - 0.5 \leqslant 0$$

$$\frac{0.14\ TMS_3}{\left[\frac{1683.9}{\left(\frac{200}{5}\right)PS_3}\right]^{0.02} - 1} - 0.5 \leqslant 0$$

$$\frac{0.14\ TMS_4}{\left[\frac{1815.4}{\left(\frac{300}{5}\right)PS_4}\right]^{0.02} - 1} - 0.5 \leqslant 0$$

$$\frac{0.14\ TMS_5}{\left[\frac{1499.66}{\left(\frac{200}{5}\right)PS_5}\right]^{0.02} - 1} - 0.5 \leqslant 0$$

$$\frac{0.14\ TMS_6}{\left[\frac{1766.3}{\left(\frac{400}{5}\right)PS_6}\right]^{0.02} - 1} - 0.5 \leqslant 0$$

$$0.1 \leqslant TMS_1 \leqslant 1.1$$
$$0.1 \leqslant TMS_2 \leqslant 1.1$$
$$0.1 \leqslant TMS_3 \leqslant 1.1$$
$$0.1 \leqslant TMS_4 \leqslant 1.1$$
$$0.1 \leqslant TMS_5 \leqslant 1.1$$
$$0.1 \leqslant TMS_6 \leqslant 1.1$$

$$PS_1 \in \{1.5, 2, 2.5, 3, 3.5, 4, 4.5, 5\}$$
$$PS_2 \in \{1.5, 2, 2.5, 3, 3.5, 4, 4.5, 5\}$$
$$PS_3 \in \{1.5, 2, 2.5, 3, 3.5, 4, 4.5, 5\}$$
$$PS_4 \in \{1.5, 2, 2.5, 3, 3.5, 4, 4.5, 5\}$$
$$PS_5 \in \{1.5, 2, 2.5, 3, 3.5, 4, 4.5, 5\}$$
$$PS_6 \in \{1.5, 2, 2.5, 3, 3.5, 4, 4.5, 5\}$$

Example 7.4 Consider the numerical expression given in Example 7.3. If (7.16) is ignored, design a MATLAB code to solve the ORC problem iteratively by using BBO. Use the following initialization parameters: population size of 50, $m_{max} = 0.05$, 4 elite solutions, 1000 generations, and 10 trials. For the constrains, handle them by applying the infinite barrier exterior penalty function. The results and fitness vector of the best run should be exported to MS Excel sheets. Also, plot the best fitness vector (**Hint:** Modify the BBO program code given in Chapter 1.)

Solution
The MATLAB code is given in a folder called `ORC_3Bus_BBO_InfBarrPen`. The program is divided into different files where the main file that can run the whole program is called `Starter.m`. The following optimal relay settings are obtained:

```
The best of Trial # 1 is 1.457
The best of Trial # 2 is 1.4427
The best of Trial # 3 is 1.4779
The best of Trial # 4 is 1.4627
The best of Trial # 5 is 1.5115
The best of Trial # 6 is 1.4874
The best of Trial # 7 is 1.5372
The best of Trial # 8 is 1.4658
The best of Trial # 9 is 1.5451
The best of Trial # 10 is 1.493
The minimum sum of relay operating times = 1.4427
It occurred at run number 2
The optimal setting of TMS1 = 0.10035
The optimal setting of TMS2 = 0.10308
The optimal setting of TMS3 = 0.11773
The optimal setting of TMS4 = 0.10933
The optimal setting of TMS5 = 0.10167
The optimal setting of TMS6 = 0.10311
The optimal setting of PS1 = 2.5
The optimal setting of PS2 = 1.5
The optimal setting of PS3 = 1.5
The optimal setting of PS4 = 1.5
The optimal setting of PS5 = 1.5
The optimal setting of PS6 = 1.5
Mean = 1.488 || Standard Deviation = 0.034178
Elapsed time is 15.351278 seconds.
T1(pr) = 0.26535
```

```
T2(pr) = 0.21585
T3(pr) = 0.23899
T4(pr) = 0.24716
T5(pr) = 0.21406
T6(pr) = 0.26126
T5(bc) - T1(pr) = 0.39238
T4(bc) - T2(pr) = 0.20147
T1(bc) - T3(pr) = 0.25062
T6(bc) - T4(pr) = 0.27754
T3(bc) - T5(pr) = 0.22169
T2(bc) - T6(pr) = 0.54711
```

The plot of the fitness is shown in Figure 7.9, which can also be accessed by opening `BBO_3BusFitnessPlot.fig`.

7.2.3 Solving the Infeasibility Condition

Because of the nature of this highly complex engineering optimization problem, it is very hard to find an optimal and feasible solution without using a large number of iterations[17] and long processing time. Some researchers accept to get good results with violations on some few constraints. This can be seen in Noghabi et al. (2009) and Bedekar and Bhide (2011a).

The causes of this infeasibility condition come from the complexity of the network or/and the lack of diversity of *PS* values. If the network is large and *PS* is provided with a limited number of discrete values, then finding a solution that satisfies all the selectivity constraints of P/B DOCR pairs is a very challenging task where many iterations are required. This motivates many researchers to study the sources and roots of this issue in order to get feasible solutions easily and quickly.

In Karegar et al. (2005), the selectivity constraints are extensively investigated, and the infeasibility range of each P/B relay pair is identified by dividing the overall range into four categories: (i) non-valid, (ii) pre-obtained, (iii) redundant, and (iv) valid. These categories are illustrated in Figure 7.10 where the possible search area (PSA) is highlighted. Based on this, the total number of constraints is reduced, and thus the chance to get feasible solutions increases. Besides, when an

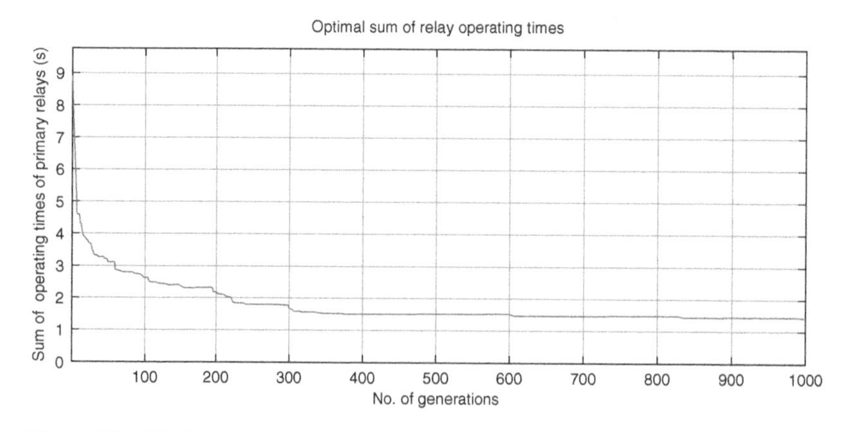

Figure 7.9 The best convergence obtained for Example 7.4. Source: Noghabi et al. (2009).

17 or population size in case of population-based meta-heuristic optimization algorithms.

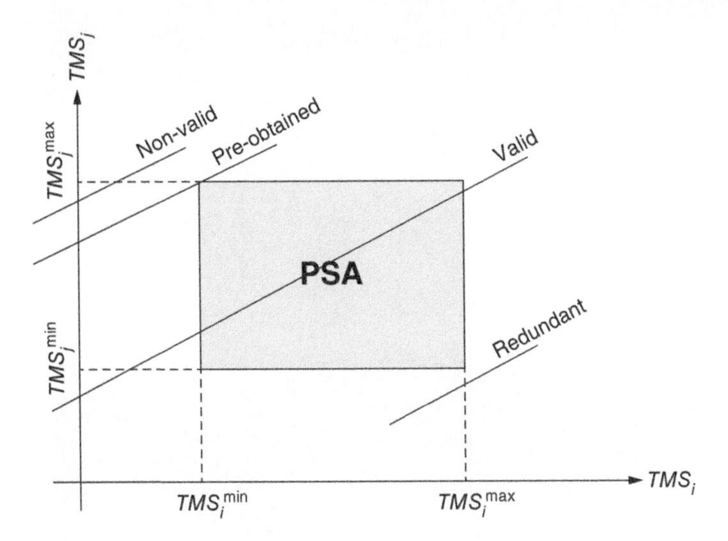

Figure 7.10 Four categories of selectivity constraint lines.

infeasible solution is detected, a search for a feasible solution is carried out through a `while`-loop unit, in which either the relay TCCCs or PSs are changed randomly until a feasible solution is detected. However, this technique cannot be applied if the ORC problem is modeled with one unified TCCC. Although there are other techniques proposed in the literature, such as using complicated equations in Ezzeddine and Kaczmarek (2011a), they are either hard to be implemented or fail to reach feasible search space.

Consequently, the technique proposed in Karegar et al. (2005) is implemented, but with some essential enhancements where both *PS* and *TMS* are randomly generated through the `while`-loop unit shown in Figure 7.10. This pre-processing unit is called the **feasibility checker (FC)**. We will incorporate this FC within BBO[18] to solve highly complex ORC problems. Actually, this is a significant enhancement and a big step in developing superior ORC solvers because this FC has the ability to detect feasible solutions within just a few iterations compared with hundreds and thousands of iterations as with conventional EAs (Noghabi et al., 2009; Bedekar and Bhide, 2011a). To explain it more, suppose that the conventional BBO algorithm shown in Figure 1.14 is used to solve the ORC problem. The mechanism of the FC sub-algorithm can be illustrated in Figure 7.11.

There is a major difference between this FC and random search algorithm (RSA) described in Algorithm 1. In RSA, all the constraints should be satisfied before terminating the `while`-loop. In FC, no need to re-check previously satisfied inequality constraints once FC jumped to the next constraints. In other words, if the selectivity between the i primary relay and the jth backup relay is satisfied, then this inequality constraint is forgotten even if the next P/B relay pairs are satisfied by violating the selectivity of the i–jth P/B relay pair. This strategy can markedly reduce the total CPU time as compared with RSA.

Disabling FC leads to faster algorithms. As discussed earlier, it is a part of a non-zero-sum game. Disabling FC requires to use a huge number of iterations to reach the same solution quality as that obtained by FC, which leads to a longer CPU time. Otherwise, by sticking with the same number of iterations, disabling FC leads to shorter CPU time but with many violations. Further information about enabling and disabling FC is covered in Albasri et al. (2015). To clarify it more, let's see Example 7.5.

18 It can also be incorporated within any conventional or hybrid EA.

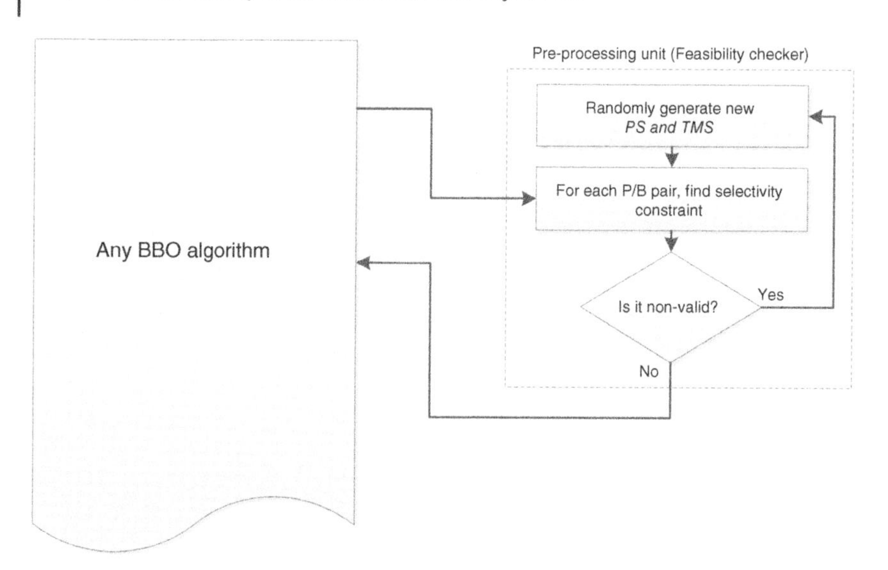

Figure 7.11 Mechanism of the feasibility checker within BBO. This sub-algorithm checks the selectivity constraint of each P/B relay pair and fixes it through a `while`-loop.

Example 7.5 The IEEE eight-bus test system is shown in Figure F.10. The operating times are measured at near-end 3ϕ faults. The relays have limited values of PSs; $PS \in \{0.5, 0.6, 0.8, 1, 1.5, 2, 2.5\}$. The detailed information is given in Appendix F. Although it is a small network, getting feasible solutions for its ORC problem is a very challenging task (Noghabi et al., 2009; Amraee, 2012; Albasri et al., 2015). The goal here is to modify the MATLAB code given in the last example to match with the flowchart shown in Figure 7.11. Show the results with and without the FC. What happen if FC is not activated? Use the binary static exterior penalty function with $r = 10$.

Solution

The MATLAB code is given in a folder called `ORC_8Bus_BBO_BinStaticPen`. Similar to the preceding example, the program is divided into different files where the main file that can run the whole program is called `Starter.m`. The following optimal relay settings are obtained when FC is not activated:

```
The best of Trial # 1 is 31.1038
The best of Trial # 2 is 40.3158
The best of Trial # 3 is 35.6975
The best of Trial # 4 is 30.7207
The best of Trial # 5 is 31.7668
The best of Trial # 6 is 30.1534
The best of Trial # 7 is 31.475
The best of Trial # 8 is 31.4767
The best of Trial # 9 is 29.6065
The best of Trial # 10 is 31.4097
The minimum sum of relay operating times = 29.6065
It occurred at run number 9
The optimal setting of TMS1 = 0.19688
The optimal setting of TMS2 = 0.41575
```

```
The optimal setting of TMS3 = 0.2685
The optimal setting of TMS4 = 0.11038
The optimal setting of TMS5 = 0.43335
The optimal setting of TMS6 = 0.46445
The optimal setting of TMS7 = 0.42663
The optimal setting of TMS8 = 0.39886
The optimal setting of TMS9 = 0.21183
The optimal setting of TMS10 = 0.11581
The optimal setting of TMS11 = 0.30985
The optimal setting of TMS12 = 0.44174
The optimal setting of TMS13 = 0.19864
The optimal setting of TMS14 = 0.56573
The optimal setting of PS1 = 2
The optimal setting of PS2 = 0.5
The optimal setting of PS3 = 0.5
The optimal setting of PS4 = 0.5
The optimal setting of PS5 = 0.5
The optimal setting of PS6 = 0.6
The optimal setting of PS7 = 1.5
The optimal setting of PS8 = 1
The optimal setting of PS9 = 2.5
The optimal setting of PS10 = 0.5
The optimal setting of PS11 = 0.5
The optimal setting of PS12 = 0.6
The optimal setting of PS13 = 1.5
The optimal setting of PS14 = 0.5
Mean = 32.3726 || Standard Deviation = 3.2312
Elapsed time is 31.083081 seconds.
T1(pr) = 0.70895
T2(pr) = 0.71764
T3(pr) = 0.47679
T4(pr) = 0.21627
T5(pr) = 0.98243
T6(pr) = 0.83539
T7(pr) = 0.93999
T8(pr) = 0.83565
T9(pr) = 0.79725
T10(pr) = 0.22515
T11(pr) = 0.61081
T12(pr) = 0.80233
T13(pr) = 0.64292
T14(pr) = 0.90967
T6(bc) - T1(pr) = 0.3039
T1(bc) - T2(pr) = 1.1566
T7(bc) - T2(pr) = 0.69981
T2(bc) - T3(pr) = 0.35321
T3(bc) - T4(pr) = 0.32888
```

```
T4(bc) - T5(pr) = -0.73219
T5(bc) - T6(pr) = 0.45336
T14(bc) - T6(pr) = 0.38109
T5(bc) - T7(pr) = 0.34877
T13(bc) - T7(pr) = 0.42481
T7(bc) - T8(pr) = 0.58179
T9(bc) - T8(pr) = 0.53666
T10(bc) - T9(pr) = -0.53773
T11(bc) - T10(pr) = 0.48314
T12(bc) - T11(pr) = 0.3039
T13(bc) - T12(pr) = 0.3039
T14(bc) - T12(pr) = 0.3039
T8(bc) - T13(pr) = 0.3039
T1(bc) - T14(pr) = 0.3039
T9(bc) - T14(pr) = 0.3039
```

We can see that when FC is not activated we got two violations. Because the binary static exterior penalty function is used with $r = 10$, the actual fitness is 9.6065 and not 29.6065. This can be validated by summing the operating times of primary relays shown in the results. Now, let's activate FC. The following results are obtained:

```
The best of Trial # 1 is 14.9278
The best of Trial # 2 is 17.8551
The best of Trial # 3 is 15.0509
The best of Trial # 4 is 17.8418
The best of Trial # 5 is 15.1494
The best of Trial # 6 is 18.5771
The best of Trial # 7 is 21.2362
The best of Trial # 8 is 17.2401
The best of Trial # 9 is 17.1825
The best of Trial # 10 is 17.3863
The minimum sum of relay operating times = 14.9278
It occurred at run number 1
The optimal setting of TMS1 = 0.14059
The optimal setting of TMS2 = 0.78074
The optimal setting of TMS3 = 0.46958
The optimal setting of TMS4 = 0.67186
The optimal setting of TMS5 = 0.3169
The optimal setting of TMS6 = 0.35175
The optimal setting of TMS7 = 0.629
The optimal setting of TMS8 = 0.49528
The optimal setting of TMS9 = 0.22109
The optimal setting of TMS10 = 0.58774
The optimal setting of TMS11 = 0.31605
The optimal setting of TMS12 = 0.73322
The optimal setting of TMS13 = 0.16582
The optimal setting of TMS14 = 0.51028
The optimal setting of PS1 = 2.5
The optimal setting of PS2 = 0.6
```

```
The optimal setting of PS3 = 2
The optimal setting of PS4 = 0.5
The optimal setting of PS5 = 1.5
The optimal setting of PS6 = 1
The optimal setting of PS7 = 1
The optimal setting of PS8 = 0.5
The optimal setting of PS9 = 2.5
The optimal setting of PS10 = 0.5
The optimal setting of PS11 = 2.5
The optimal setting of PS12 = 0.6
The optimal setting of PS13 = 2.5
The optimal setting of PS14 = 1.5
Mean = 17.2447 || Standard Deviation = 1.9158
Elapsed time is 137.634029 seconds.
T1(pr) = 0.57463
T2(pr) = 1.4164
T3(pr) = 1.3324
T4(pr) = 1.3164
T5(pr) = 1.147
T6(pr) = 0.73633
T7(pr) = 1.2197
T8(pr) = 0.84854
T9(pr) = 0.83211
T10(pr) = 1.1426
T11(pr) = 1.1929
T12(pr) = 1.3317
T13(pr) = 0.71102
T14(pr) = 1.126
T6(bc) - T1(pr) = 0.34789
T1(bc) - T2(pr) = 0.51556
T7(bc) - T2(pr) = 0.32315
T2(bc) - T3(pr) = 0.3179
T3(bc) - T4(pr) = 0.3386
T4(bc) - T5(pr) = 0.37615
T5(bc) - T6(pr) = 1.0879
T14(bc) - T6(pr) = 0.96622
T5(bc) - T7(pr) = 0.60455
T13(bc) - T7(pr) = 1.1008
T7(bc) - T8(pr) = 0.89098
T9(bc) - T8(pr) = 0.58379
T10(bc) - T9(pr) = 0.48491
T11(bc) - T10(pr) = 0.4589
T12(bc) - T11(pr) = 0.34789
T13(bc) - T12(pr) = 0.34789
T14(bc) - T12(pr) = 0.34789
T8(bc) - T13(pr) = 0.34789
T1(bc) - T14(pr) = 0.34789
T9(bc) - T14(pr) = 0.34789
```

We can observe two important things here. First, because FC is built based on a `while`-loop, so the algorithm is slower. In our simulation, it is 442.8% more. Second, activating FC can ensure getting feasible solutions. Thus, even 9.6065 is better than 14.9278, it is infeasible and thus it is worthless.

The example gives us a solid conclusion about the importance of FC in solving ORC problems. However, the additional time consumed by this unit is unacceptable for adaptive ORC solvers. This issue will be solved in Chapter 8.

7.3 Results Tester

The other advantage of FC is to use it as a tester to check the violations in the final results. This effective tool is very useful, especially for the optimal results presented in the literature where no codes are provided. All we need is the information regarding the test system used by the researchers and the final optimal results reported in the literature. The idea is so simple. It is based on (7.25) where the j–ith inequality constraint is violated if the difference between the operation of backup and primary relays is less than CTI:

$$\text{the } j\text{–}i\text{th inequality constraint is} \begin{cases} \text{satisfied}, & \text{if } T_{j,k} - T_{i,k} \geqslant CTI \\ \text{violated}, & \text{otherwise} \end{cases} \tag{7.35}$$

The lower and upper bounds on PS, TMS, and T can be directly checked from the results. Thus, we do not need to embed them in the tester.

Example 7.6 Consider the IEEE eight-bus test system given in the last example. The optimal relay settings reported in Noghabi et al. (2009) for genetic algorithm (GA) and hybrid GA-LP with fixed network topology are tabulated in Table 7.2. Design a MATLAB code to check the violations in the results.

Table 7.2 Optimal relay settings obtained by GA and GA-LP for the IEEE eight-bus test system with fixed network topology.

Primary relay	Conventional GA		Hybrid GA-LP	
	PS	*TMS*	*PS*	*TMS*
R_1	1.0	0.29	1.0	0.3043
R_2	2.5	0.31	2.5	0.2917
R_3	2.5	0.26	2.5	0.2543
R_4	2.5	0.19	2.5	0.1851
R_5	1.5	0.18	1.5	0.1700
R_6	2.5	0.26	2.5	0.2711
R_7	0.5	0.54	0.5	0.5316
R_8	2.5	0.24	2.5	0.2387
R_9	2.0	0.17	2.0	0.1865
R_{10}	2.5	0.19	2.5	0.1895
R_{11}	2.5	0.21	2.5	0.2014
R_{12}	2.5	0.30	2.5	0.2890
R_{13}	1.5	0.23	1.5	0.2297
R_{14}	0.5	0.51	0.5	0.5278

Table 7.3 Violations detected on the optimal relay settings of GA and GA-LP with fixed network topology.

Conventional GA					Hybrid GA-LP				
Backup relay		Primary relay		Violations	Backup relay		Primary relay		Violations
R_j	T_j (s)	R_i	T_i (s)	$T_j - Ti - CTI$ (s)	R_j	T_j (s)	R_i	T_i (s)	$T_j - Ti - CTI$ (s)
6	1.063	1	0.761	0.0021	6	1.108	1	0.798	0.010
1	1.406	2	0.926	0.1801	1	1.476	2	0.871	0.304
7	1.158	2	0.926	**−0.0682**	7	1.140	2	0.871	**−0.032**
2	1.198	3	0.815	0.0830	2	1.127	3	0.797	0.030
3	1.037	4	0.709	0.0282	3	1.015	4	0.691	0.024
4	0.946	5	0.651	**−0.0056**	4	0.921	5	0.615	0.006
5	1.036	6	0.766	**−0.0301**	5	0.979	6	0.799	**−0.120**
14	1.097	6	0.766	0.0304	14	1.135	6	0.799	0.036
5	1.036	7	0.867	**−0.1311**	5	0.979	7	0.854	**−0.175**
13	1.580	7	0.867	0.4130	13	1.578	7	0.854	0.424
7	1.158	8	0.708	0.1498	7	1.140	8	0.704	0.136
9	0.909	8	0.708	**−0.0990**	9	0.997	8	0.704	**−0.007**
10	0.923	9	0.569	0.0541	10	0.920	9	0.624	**−0.004**
11	1.064	10	0.699	0.0651	11	1.021	10	0.697	0.023
12	1.132	11	0.793	0.0397	12	1.091	11	0.760	0.031
13	1.580	12	0.898	0.3823	13	1.578	12	0.865	0.413
14	1.097	12	0.898	**−0.1013**	14	1.135	12	0.865	**−0.030**
8	1.029	13	0.744	**−0.0153**	8	1.024	13	0.743	**−0.020**
1	1.406	14	0.820	0.2862	1	1.476	14	0.849	0.327
9	0.909	14	0.820	**−0.2110**	9	0.997	14	0.849	**−0.151**

Solution
The MATLAB code is given in a folder called `ResultsTester`. There are two files; one for GA and the other for GA-LP. Running these two files will give the results tabulated in Table 7.3 where the bold values mean violations.

Although this tester is useful for researchers, it could fail to detect violations if the final results are rounded with a few decimal places. For example, from *TMS* listed in Table 7.2, GA has fewer decimal places than GA-LP.

Problems

Written Exercises

W7.1 If a mesh network is protected by 40 DOCRs, how many branches it has?

W7.2 Suppose a 120-dimensional optimization algorithm is used to optimally adjust the settings of DOCRs installed in a mesh network. How many relays the network has if all the relays have the same characteristic curve. How many branches in the network?

W7.3 Suppose that a research lab is established in an interdisciplinary department and there are one protection engineer and one data scientist. The lab head assigned these two guys to optimally coordinate the protective devices of a highly interconnected distribution network where only DOCRs with the IEEE extremely inverse characteristic curve are used to minimize the installation and maintenance costs. If the network consists of 200 branches, how many DOCRs the protection engineer found? If the data scientist previously designed a 1600-dimensional optimization algorithm that can solve constrained problems globally, is that algorithm compatible with the ORC problem? If yes, why? and if no, what the data scientist should do to fix that problem?

W7.4 Suppose that a protection engineer wants to design a new complex objective function by combining both the mid-point and desired design criteria. Write down the objective function if you do not know how many fault points are accounted for each DOCR. Repeat it again if all the fault points are accounted for each DOCR.

W7.5 Consider the network given in Figure 5.10. Use (7.33) to formulate the optimization problem of this network. Take $TMS \in [0.025, 1.2]$, $PS \in [1.5, 7.5]$ A, $T \in [0.075, 0.43]$ s, and $CTI = 0.3$ s. Show the concise expression; similar to the first one shown in Example 7.3.

W7.6 Repeat Exercise W7.5 if all DOCRs are manufactured in North America. Take $TDS \in [0.5, 11]$, $CTS \in [1.5, 10]$ A, $T \in [0.1, 0.4]$ s, and $CTI = 0.35$ s.

W7.7 Re-express the long mathematical model derived in Example 7.3 when the IEEE extremely inverse TCCC is used for all the relays. Take $TDS \in [0.25, 12]$, $CTS \in \{1.5, 2, 2.5, \dots, 8\}$ A, $T \in [0.1, 0.4]$ s, and $CTI = 0.3$ s.

W7.8 Apply the steps taken in Example 7.3 to express the ORC problem of the five-bus network shown in Figure 5.12. Use the IEC ultra-inverse model for all the 12 DOCRs.

W7.9 Suppose that you was reading a paper discussing about ORC. From the experimental results, you saw that one P/B relay pair has the following optimal settings: $TMS_i = 0.221$, $TMS_j = 0.283$, $PS_i = 2$ A, $PS_j = 2.5$ A, $CTR_i = 1200/5$, $CTR_j = 800/5$, $I_{i,k} = 3783$ A, $I_{j,k} = 2244$ A, and $CTI = 0.3$ s. All DOCRs are modeled based on the IEC standard inverse. Is the selectivity constraint satisfied?

W7.10 Consider the data given in Exercise W7.9. If the short-circuit currents during the transient topology are $I'_{i,k} = 5117$ A and $I'_{j,k} = 3420$ A, is the selectivity constraint satisfied?

W7.11 If the P/B relay pair has the following settings: $TMS_i = 0.297$, $TMS_j = 0.230$, $PS_i = 2$ A, $PS_j = 2$ A, $CTR_i = 800/5$, $CTR_j = 800/5$, $I_{i,k} = 2150$ A, $I_{j,k} = 1004$ A, and $CTI = 0.35$ s. All DOCRs are modeled based on the IEC standard inverse. The difference between $T_{j,k}$ and $T_{i,k}$ is 0.3213 s, which is less than CTI and thus the selectivity constraint is violated. Can we satisfy this constraint by:

1. Reducing CTI? Explain why.
2. Changing $\{\alpha, \beta, \gamma, \xi\}$ of both TCCCs? Explain why.
3. Adjusting TMS/TDS and PS/CTS of both relays? Explain why.

W7.12 Suppose that a North American backup DOCR operates after awaiting 0.896 s. If $CTS_j =$ 2 A, $TDS_j = 0.5$, and $CTR_j = 80$, calculate the fault current seen by the backup relay ($I_{j,k}$) when the following TCCC model is used:
1. The IEEE long time extremely inverse
2. The IEEE moderately inverse
3. The IEEE short time extremely inverse

W7.13 Based on the MATLAB code provided in Example 7.4 and those developed in Exercise C7.1–C7.10. Use two tables to compare the processing speed of these 10-constraint handling techniques. One table should be devoted to the fixed topology and the other when the transient topology of each DOCR is considered. Set the population size to 50. If the transient topology is not considered, use 85 generations and 10 runs. If it is considered, use 1000 generations and 50 runs. Write down the specification of the computation machine used in this experiment.

W7.14 Consider Exercise W7.13 again. If the mission now is to evaluate the solution quality of each constrain-handling technique. Use the MATLAB codes to export the records. The performance criteria should be the best, worst, mean, standard deviation, and normalized CPU time. Use only the second scenario where both the fixed and transient topologies are considered in the ORC problem.

Computer Exercises

C7.1 Modify the MATLAB code given in Example 7.4 to consider the transient topology of each relay. Refer to Appendix F for the transient short-circuit currents.

C7.2 Modify the MATLAB code given in Example 7.4 and the one developed in Exercise C7.1 to re-solve the IEEE three-bus test system with/without considering the transient topology of each relay. The random search algorithm must be used to handle the design constraints.

C7.3 Repeat Exercise C7.3 with the binary static penalty function as a constraint-handler. The penalty multiplier is $r = 0.15$.

C7.4 Repeat Exercise C7.3 with the superiority of feasible points-type I as a constraint-handler. The penalty settings are $r = 10$ and $\beta = 2$. Please note that these symbols are used in Chapter 1 and not belong to relay equations.

C7.5 Repeat Exercise C7.3 with the superiority of feasible points-type II as a constraint-handler. The penalty settings are $r = 10$ and $\beta = 2$. Please note that these symbols are used in Chapter 1 and not belong to relay equations.

C7.6 Repeat Exercise C7.3 with the eclectic evolutionary algorithm as a constraint-handler. The penalty settings are $K = 10^9$ and $n =$ the number of selectivity constraints. Please note that these symbols are used in Chapter 1 and not belong to relay equations.

C7.7 Repeat Exercise C7.3 with the typical dynamic penalty as a constraint-handler. The penalty settings are $c = 0.5$, $\alpha = 2$, $\beta = 2$, and $r = 1$. Please note that these symbols are used in Chapter 1 and not belong to relay equations.

C7.8 Repeat Exercise C7.3 with the exponential dynamic penalty as a constraint-handler. The penalty settings are $\beta = 1$ and $T = 1/\sqrt{\text{Iter}}$. Please note that these symbols are used in Chapter 1 and not belong to relay equations.

C7.9 Repeat Exercise C7.3 with the adaptive multiplication penalty as a constraint-handler. The penalty settings are $\alpha = 2$, $\beta = 1$, $\varepsilon = 10^{-20}$, $n = 6$ (# of selectivity constraints), $n' = 6$ (# of transient selectivity constraints). Please note that these symbols are used in Chapter 1 and not belong to relay equations. You can modify the penalty function to provide better solutions.

C7.10 Repeat Exercise C7.3 with the self-adaptive penalty as a constraint-handler. You can modify the penalty function to provide better solutions.

C7.11 Suppose that the relays of the IEEE three-bus test system are manufactured in North America. Modify the MATLAB code given in Example 7.4 to optimize *TDS* and *CTS*. Use the IEEE very inverse model for all relay TCCCs. The lower and upper bounds of *TDS* are 0.5 and 11, respectively. Take *CTS* as a discrete variable that moves from 1.5 to 8 in 0.1 steps.

C7.12 Modify the MATLAB code given in Example 7.4 to optimize the relay settings of the IEEE four-bus test system given in Appendix F. Select proper initialization parameters. (**Hint:** Use the special mathematical model given in Appendix F.)

C7.13 Modify the MATLAB code given in Example 7.4 to optimize the relay settings of the IEEE nine-bus test system given in Appendix F. Select proper initialization parameters. (**Hint:** Use the realized equations given in (7.22) and (7.23) for PS^{min} and PS^{max}, respectively.)

C7.14 Modify the MATLAB code given in Example 7.5 to optimize the relay settings of the IEEE 15-bus test system given in Appendix F. Select proper initialization parameters.

C7.15 Modify the MATLAB code given in Example 7.6 to test the feasibility of the optimal relay settings reported in Amraee (2012) for the IEEE eight-bus test system.

C7.16 You are requested to validate the results obtained by DE for the three-bus test system reported in Thangaraj et al. (2010). (**Hint:** You need to design a single file MATLAB code to test the feasibility of the optimal relay settings obtained by the conventional DE algorithm.)

C7.17 You are requested to validate the results obtained by DE for the six-bus test system reported in Thangaraj et al. (2010). (**Hint:** You need to design a single file MATLAB code to test the feasibility of the optimal relay settings obtained by the conventional DE algorithm.)

C7.18 You are requested to validate the results reported in Thangaraj et al. (2010). (**Hint:** Since the question does not mention which result should be checked, so you need to design one or multiple MATLAB codes to test the feasibility of the optimal relay settings obtained by the conventional DE algorithm as well as the five modified versions; MDE1, MDE2, MDE3, MDE4, and MDE5.)

C7.19 You are requested to validate the results obtained by GA and GA-NLP for the nine-bus test system reported in Bedekar and Bhide (2011a). (**Hint:** You need to modify the MATLAB file "`Coordination.m`" given in Exercise C7.13 to show some information behind the final results reported in the literature.)

C7.20 You are requested to validate the results obtained by MINLP and Seeker algorithm for the 15-bus test system reported in Amraee (2012). (**Hint:** You need to modify the MATLAB file "`Coordination.m`" given in Exercise C7.14 to show some information behind the final results reported in the literature.)

8

Incorporating LP and Hybridizing It with Meta-heuristic Algorithms

After reading this chapter you should be able to:

- Linearize the mathematical model of the optimal relay coordination (ORC) problem when only directional over-current relays (DOCRs) with one unified time–current characteristic curve (TCCC) are used
- Deal with both the IEC/BS and the ANSI/IEEE standard models
- Derive the classical linear models by fixing plug setting (*PS*)
- Derive the transformation-based linear models by fixing time multiplier setting (*TMS*)
- Solve these linearized mathematical models by applying linear programming (LP)
- Solve these linearized mathematical models by involving multi-start LP
- Solve these linearized mathematical models by using meta-heuristic optimization algorithms

Based on Chapter 1, modern optimization algorithms are powerful and global solvers for various optimal relay coordination (ORC) problems. However, they are probabilistic- and stochastic-based methods where many iterations are required to reach feasible optimal points. Some researchers suggest to hybridize evolutionary algorithms (EAs) with linear programming (LP) and non-linear programming (NLP) to accelerate the convergence rate and accuracy. Such studies are reported in Noghabi et al. (2009), Bedekar and Bhide (2011a), Papaspiliotopoulos et al. (2014), Albasri et al. (2015), and Al-Roomi and El-Hawary (2017a). However, in Chapter 7, we have seen that the mathematical formulation of ORC is nonlinear. Thus, there is a technical problem when LP is selected as a sub-algorithm in EAs. Detailed information about this problem and how to practically solve it in different ways is given in the following text.

8.1 Model Linearization

In Chapter 1, it has been said that to be able to apply LP as a sub-algorithm,[1] the optimization model should be expressed in a linear form. Referring to the preceding ORC models presented in Chapter 7, it is obvious that the objective function and the functional constraints are nonlinear. Thus, the first essential step that has to be applied, before initiating LP, is to linearize the optimization model. The literature shows one possible way to do that, which is about taking plug setting (*PS*)

1 To fine-tune the individuals supplied from EAs.

Optimal Coordination of Power Protective Devices with Illustrative Examples, First Edition. Ali R. Al-Roomi.

Companion website: www.wiley.com/go/al-roomi/optimalcoordination

supplied from EAs for all ϱ relays as a constant vector, and the time multiplier setting (TMS) vector as an initial point to LP. Actually, there is also another approach that can be applied as well. This novel approach can be achieved by simply linearizing the model as a function of PS instead of TMS. However, a precaution should be given to the PS vector since its ϱ elements are discrete variables (Al-Roomi and El-Hawary, 2020e).

8.1.1 Classical Linearization Approach

This approach can be applied to both the IEC/BS and ANSI/IEEE standards as follows:

8.1.1.1 IEC Curves: Fixing Plug Settings and Varying Time Multiplier Settings

To linearize the IEC/BS standard given in (7.3) for the ith relay, the following expression is used:

$$T_{i,k} = TMS_i \times \frac{\beta}{\left(\frac{I_{i,k}}{PS_i}\right)^{\alpha} - 1} = \psi_{i,k} TMS_i;\ \psi_{i,k} = \frac{\beta}{\left(\frac{I_{i,k}}{PS_i}\right)^{\alpha} - 1} \tag{8.1}$$

where PS_i is constant, which is either initially pre-defined in classical LP algorithms or automatically determined by EAs in hybrid algorithms.

Substituting (8.1) in (7.33) for $T_{i,k}^{\text{pr}}$ and $T_{j,k}^{\text{bc}}$ yields:

$$Z = \min \sum_{i=1}^{\varrho} \psi_{i,k} TMS_i$$

$$\begin{aligned}
\text{Subjected to: } \psi_{i,k} TMS_i + CTI - \psi_{j,k} TMS_j &\leqslant 0\\
T_{i,k}^{\min} - \psi_{i,k} TMS_i &\leqslant 0\\
\psi_{i,k} TMS_i - T_{i,k}^{\max} &\leqslant 0\\
TMS_i^{\min} \leqslant TMS_i &\leqslant TMS_i^{\max}
\end{aligned} \tag{8.2}$$

Example 8.1 Consider the nonlinear mathematical formulation given in Example 7.3. Linearize the expression if the optimal PSs obtained in Example 7.4 are used for the six directional over-current relays (DOCRs).

Solution

The optimal vector of PS obtained in Example 7.4 is:

$$PS = [2.5, 1.5, 1.5, 1.5, 1.5, 1.5]^T$$

Thus, the expression given in Example 7.3 can be linearized as follows:

$$\begin{aligned}
OBJ = \min \{&2.6441TMS_1 + 2.0940TMS_2 + 2.0300TMS_3\\
&+2.2607TMS_4 + 2.1056TMS_5 + 2.5337TMS_6\}
\end{aligned}$$

Subjected to:

$$\begin{aligned}
2.6441TMS_1 - 6.4696TMS_5 &\leqslant -0.2\\
2.0940TMS_2 - 3.8172TMS_4 &\leqslant -0.2\\
2.0300TMS_3 - 4.8788TMS_1 &\leqslant -0.2\\
2.2607TMS_4 - 5.0885TMS_6 &\leqslant -0.2\\
2.1056TMS_5 - 3.7014TMS_3 &\leqslant -0.2\\
2.5337TMS_6 - 7.8422TMS_2 &\leqslant -0.2
\end{aligned}$$

$$-2.6441TMS_1 \leqslant -0.1$$
$$-2.0940TMS_2 \leqslant -0.1$$
$$-2.0300TMS_3 \leqslant -0.1$$
$$-2.2607TMS_4 \leqslant -0.1$$
$$-2.1056TMS_5 \leqslant -0.1$$
$$-2.5337TMS_6 \leqslant -0.1$$

$$2.6441TMS_1 \leqslant 0.5$$
$$2.0940TMS_2 \leqslant 0.5$$
$$2.0300TMS_3 \leqslant 0.5$$
$$2.2607TMS_4 \leqslant 0.5$$
$$2.1056TMS_5 \leqslant 0.5$$
$$2.5337TMS_6 \leqslant 0.5$$

$$0.1 \leqslant TMS_1 \leqslant 1.1$$
$$0.1 \leqslant TMS_2 \leqslant 1.1$$
$$0.1 \leqslant TMS_3 \leqslant 1.1$$
$$0.1 \leqslant TMS_4 \leqslant 1.1$$
$$0.1 \leqslant TMS_5 \leqslant 1.1$$
$$0.1 \leqslant TMS_6 \leqslant 1.1$$

8.1.1.2 IEEE Curves: Fixing Current Tap Settings and Varying Time Dial Settings

Similarly, the ANSI/IEEE standard can be linearized as follows:

$$T_{i,k} = TDS_i \times \left[\frac{\beta}{\left(\frac{I_{i,k}}{CTS_i}\right)^{\alpha} - 1} + \gamma \right] = \varphi_{i,k} TDS_i; \ \varphi_{i,k} = \frac{\beta}{\left(\frac{I_{i,k}}{CTS_i}\right)^{\alpha} - 1} + \gamma \tag{8.3}$$

and, again, CTS_i is held constant for the LP stage.

Thus, by substituting (8.3) in (7.34) for $T^{\text{pr}}_{i,k}$ and $T^{\text{bc}}_{j,k}$, the following standard linear model can be obtained:

$$Z = \min \sum_{i=1}^{\varrho} \varphi_{i,k} TDS_i$$

$$\begin{aligned} \text{Subjected to: } & \varphi_{i,k} TDS_i + CTI - \varphi_{j,k} TDS_j \leqslant 0 \\ & T^{\min}_{i,k} - \varphi_{i,k} TDS_i \leqslant 0 \\ & \varphi_{i,k} TDS_i - T^{\max}_{i,k} \leqslant 0 \\ & TDS^{\min}_i \leqslant TDS_i \leqslant TDS^{\max}_i \end{aligned} \tag{8.4}$$

8.1.2 Transformation-Based Linearization Approach

This approach is tricky and requires to involve the **transformation** technique used in linear regression analysis (Kutner et al., 2004). For the IEC/BS and ANSI/IEEE standards, this tricky approach can be described as follows:

8.1.2.1 IEC Curves: Fixing Time Multiplier Settings and Varying Plug Settings

For the ith relay, the IEC/BS standard given in (7.3) can be linearized as follows:

$$T_{i,k} = \vartheta_i \widetilde{PS}_{i,k}; \quad \vartheta_i = \beta TMS_i \tag{8.5}$$

where TMS_i is constant, which is either initially pre-defined in LP or automatically determined by EAs in hybrid optimizers. $\widetilde{PS}_{i,k}$ is a transformed PS, which is built based on:

$$\widetilde{PS}_{i,k} = \frac{PS_i^{\alpha}}{I_{i,k}^{\alpha} - PS_i^{\alpha}} \tag{8.6}$$

Substituting (8.5) in (7.33) for $T_{i,k}^{\mathrm{pr}}$ and $T_{j,k}^{\mathrm{bc}}$ yields:

$$Z = \min \sum_{i=1}^{o} \vartheta_i \widetilde{PS}_{i,k}$$

$$\begin{aligned} \text{Subjected to: } & \vartheta_i \widetilde{PS}_{i,k} + CTI - \vartheta_j \widetilde{PS}_{j,k} \leqslant 0 \\ & T_{i,k}^{\min} - \vartheta_i \widetilde{PS}_{i,k} \leqslant 0 \\ & \vartheta_i \widetilde{PS}_{i,k} - T_{i,k}^{\max} \leqslant 0 \\ & \widetilde{PS}_{i,k}^{\min} \leqslant \widetilde{PS}_{i,k} \leqslant \widetilde{PS}_{i,k}^{\max} \end{aligned} \tag{8.7}$$

where the lower and upper bounds of $\widetilde{PS}_{i,k}$ can be, respectively, calculated as follows:

$$\widetilde{PS}_{i,k}^{\min} = \frac{\left(PS_i^{\min}\right)^{\alpha}}{I_{i,k}^{\alpha} - \left(PS_i^{\min}\right)^{\alpha}} \tag{8.8}$$

$$\widetilde{PS}_{i,k}^{\max} = \frac{\left(PS_i^{\max}\right)^{\alpha}}{I_{i,k}^{\alpha} - \left(PS_i^{\max}\right)^{\alpha}} \tag{8.9}$$

After fine-tuning the transformed PS ($\widetilde{PS}_{i,k}$) by LP for the ith relay and at the kth fault location, the actual quantity PS_i can be retrieved by applying the following de-transformation formula:

$$PS_i = \left[\frac{\widetilde{PS}_{i,k} I_{i,k}^{\alpha}}{1 + \widetilde{PS}_{i,k}}\right]^{1/\alpha} = I_{i,k} \times \sqrt[\alpha]{\frac{\widetilde{PS}_{i,k}}{1 + \widetilde{PS}_{i,k}}} \tag{8.10}$$

For more information about this transformed optimization model, the full derivation is given in Appendix D.

8.1.2.2 IEEE Curves: Fixing Time Dial Settings and Varying Current Tap Settings

Following the same transformation steps, the ANSI/IEEE standard given in (7.4) can be linearized as follows:

$$T_{i,k} = \vartheta_i \widetilde{CTS}_{i,k} + \xi_i; \quad \vartheta_i = \beta TDS_i, \quad \xi_i = \gamma TDS_i \tag{8.11}$$

where TDS_i is constant, which is either initially pre-defined in LP or automatically determined by EAs in hybrid optimizers. $\widetilde{CTS}_{i,k}$ is a transformed variable, which is expressed as follows:

$$\widetilde{CTS}_{i,k} = \frac{CTS_i^{\alpha}}{I_{i,k}^{\alpha} - CTS_i^{\alpha}} \tag{8.12}$$

Substituting (8.11) in (7.34) for $T_{i,k}^{\text{pr}}$ and $T_{j,k}^{\text{bc}}$ yields:

$$Z = \min \sum_{i=1}^{\varrho} \vartheta_i \widetilde{CTS}_{i,k} + \xi_i$$

$$\begin{aligned}
\text{Subjected to: } \left(\vartheta_i \widetilde{CTS}_{i,k} + \xi_i\right) + CTI - \left(\vartheta_j \widetilde{CTS}_{j,k} + \xi_j\right) &\leqslant 0 \\
T_{i,k}^{\min} - \left(\vartheta_i \widetilde{CTS}_{i,k} + \xi_i\right) &\leqslant 0 \\
\vartheta_i \widetilde{CTS}_{i,k} + \xi_i - T_{i,k}^{\max} &\leqslant 0
\end{aligned}$$

$$\widetilde{CTS}_{i,k}^{\min} \leqslant \widetilde{CTS}_{i,k} \leqslant \widetilde{CTS}_{i,k}^{\max} \tag{8.13}$$

where the lower and upper bounds of $\widetilde{CTS}_{i,k}$ can be, respectively, calculated as follows:

$$\widetilde{CTS}_{i,k}^{\min} = \frac{\left(CTS_i^{\min}\right)^{\alpha}}{I_{i,k}^{\alpha} - \left(CTS_i^{\min}\right)^{\alpha}} \tag{8.14}$$

$$\widetilde{CTS}_{i,k}^{\max} = \frac{\left(CTS_i^{\max}\right)^{\alpha}}{I_{i,k}^{\alpha} - \left(CTS_i^{\max}\right)^{\alpha}} \tag{8.15}$$

Again, once the transformed current tap settings ($\widetilde{CTS}$) of all ϱ DOCRs are fine-tuned by the LP sub-algorithm, the corresponding actual current tap settings (CTS) can be retrieved by de-transforming the former settings via the following formula:

$$CTS_i = \left[\frac{\widetilde{CTS}_{i,k} I_{i,k}^{\alpha}}{1 + \widetilde{CTS}_{i,k}}\right]^{1/\alpha} = I_{i,k} \times \sqrt[\alpha]{\frac{\widetilde{CTS}_{i,k}}{1 + \widetilde{CTS}_{i,k}}} \tag{8.16}$$

The full derivation about this transformed optimization model is given in Appendix D.

Example 8.2 Modify the MATLAB program given in Example 1.6 to optimize the linearized ORC problem given in Example 8.1 numerically via LP. For the internal solver of `linprog`, use `interior-point-legacy`.

Solution

Because of the sparsity, only one or two variables are not zero in the inequality matrix (A). The problem can be mathematically expressed as follows:

$$\min_{TMS} Z\left(TMS_1, TMS_2, \dots, TMS_6\right) \text{ such that } \begin{cases} A \cdot TMS \leqslant b \\ TMS^{\min} \leqslant TMS \leqslant TMS^{\max} \end{cases}$$

where A, b, TMS^{min}, and TMS^{max} are expressed in vector notation as follows:

$$A = \begin{bmatrix} 2.6441 & 0 & 0 & 0 & -6.4696 & 0 \\ 0 & 2.0940 & 0 & -3.8172 & 0 & 0 \\ -4.8788 & 0 & 2.0300 & 0 & 0 & 0 \\ 0 & 0 & 0 & 2.2607 & 0 & -5.0885 \\ 0 & 0 & -3.7014 & 0 & 2.1056 & 0 \\ 0 & -7.8422 & 0 & 0 & 0 & 2.5337 \\ -2.6441 & 0 & 0 & 0 & 0 & 0 \\ 0 & -2.0940 & 0 & 0 & 0 & 0 \\ 0 & 0 & -2.0300 & 0 & 0 & 0 \\ 0 & 0 & 0 & -2.2607 & 0 & 0 \\ 0 & 0 & 0 & 0 & -2.1056 & 0 \\ 0 & 0 & 0 & 0 & 0 & -2.5337 \\ 2.6441 & 0 & 0 & 0 & 0 & 0 \\ 0 & 2.0940 & 0 & 0 & 0 & 0 \\ 0 & 0 & 2.0300 & 0 & 0 & 0 \\ 0 & 0 & 0 & 2.2607 & 0 & 0 \\ 0 & 0 & 0 & 0 & 2.1056 & 0 \\ 0 & 0 & 0 & 0 & 0 & 2.5337 \end{bmatrix}$$

$$b = \begin{bmatrix} -0.2 \\ -0.2 \\ -0.2 \\ -0.2 \\ -0.2 \\ -0.2 \\ -0.1 \\ -0.1 \\ -0.1 \\ -0.1 \\ -0.1 \\ -0.1 \\ 0.5 \\ 0.5 \\ 0.5 \\ 0.5 \\ 0.5 \\ 0.5 \end{bmatrix}, \quad TMS^{min} = \begin{bmatrix} 0.1 \\ 0.1 \\ 0.1 \\ 0.1 \\ 0.1 \\ 0.1 \end{bmatrix}, \quad TMS^{max} = \begin{bmatrix} 1.1 \\ 1.1 \\ 1.1 \\ 1.1 \\ 1.1 \\ 1.1 \end{bmatrix}$$

The following MATLAB code is given in `Linearized_ORC_3Bus_LP.m`:

```
1	clc
2	clear
3	tic;
4	% Linear inequality constraints:
5	A = [2.6441 0 0 0 -6.4696 0;
6	    0 2.0940 0 -3.8172 0 0;
7	    -4.8788 0 2.0300 0 0 0;
8	    0 0 0 2.2607 0 -5.0885;
9	    0 0 -3.7014 0 2.1056 0;
10	    0 -7.8422 0 0 0 2.5337;
11	    -2.6441 0 0 0 0 0;
12	    0 -2.0940 0 0 0 0;
13	    0 0 -2.0300 0 0 0;
14	    0 0 0 -2.2607 0 0;
15	    0 0 0 0 -2.1056 0;
16	    0 0 0 0 0 -2.5337;
17	    2.6441 0 0 0 0 0;
18	    0 2.0940 0 0 0 0;
19	    0 0 2.0300 0 0 0;
20	    0 0 0 2.2607 0 0;
21	    0 0 0 0 2.1056 0;
22	    0 0 0 0 0 2.5337];
23	b = [-0.2 -0.2 -0.2 -0.2 -0.2 -0.2 -0.1 -0.1 -0.1 -0.1 -0.1 -0.1 0.5 ...
	    0.5 0.5 0.5 0.5 0.5]';
24	% Linear Equality constraints:
25	Aeq = [];
26	beq = [];
27	% Lower and Upper Bounds:
28	lb = [0.1 0.1 0.1 0.1 0.1 0.1];
29	ub = [1.1 1.1 1.1 1.1 1.1 1.1];
30	% Objective function "min[Z(TMS1,TMS2,...,TMS6)] = min[sum(T1 + T2 + ...
	    T3 + T4 + T5 + T6)]":
31	Z = [2.6441 2.0940 2.0300 2.2607 2.1056 2.5337];
32	% Solve the linear program
33	options = optimoptions('linprog', 'Algorithm', ...
	    'interior-point-legacy', 'display', 'iter');
34	TMS = linprog(Z, A, b, Aeq, beq, lb, ub, [], options)
35	% Minimal point
36	Obj = 2.6441*TMS(1) + 2.0940*TMS(2) + 2.0300*TMS(3) + 2.2607*TMS(4) + ...
	    2.1056*TMS(5) + 2.5337*TMS(6)
37	toc;
```

If we run this simple program, the following result will be generated in the command window:

```
Residuals:   Primal      Dual        Upper      Duality     Total
             Infeas      Infeas      Bounds     Gap         Rel
             A*x-b       A'*y+z-w-f  {x}+s-ub   x'*z+s'*w   Error
------------------------------------------------------------------
Iter    0:   1.04e+03 1.06e+01 4.87e+02 8.87e+03 1.04e+03
Iter    1:   9.95e+01 4.63e-15 4.65e+01 9.17e+02 9.95e+01
Iter    2:   3.40e-01 4.99e-15 1.59e-01 2.74e+01 1.10e+00
Iter    3:   1.43e-15 1.57e-14 0.00e+00 4.71e+00 1.48e+00
Iter    4:   3.21e-14 2.39e-15 1.57e-16 2.38e+00 1.28e+00
```

```
     Iter    5:   2.39e-15 4.16e-13 0.00e+00 7.42e-01 7.42e-01
     Iter    6:   2.91e-15 2.96e-14 0.00e+00 7.82e-02 7.82e-02
     Iter    7:   2.68e-16 2.22e-15 2.48e-16 4.31e-04 4.31e-04
     Iter    8:   2.45e-16 1.72e-15 1.11e-16 2.18e-08 2.18e-08
     Iter    9:   1.97e-16 8.17e-16 0.00e+00 2.18e-15 2.19e-15
Optimization terminated.

TMS =
    0.1000
    0.1000
    0.1109
    0.1073
    0.1000
    0.1000

Obj =
    1.4054

Elapsed time is 0.028735 seconds.
```

To see the importance of LP in ORC problems, let's compare the final result obtained in Example 8.2 with that of Example 7.4; in terms of solution quality and processing speed. From Example 7.4, the fitness was 1.4427 s and the processing time of the algorithm was 15.351 278 s. After using LP as a fine-tuner, the fitness is reduced to 1.4054 s with a negligible processing time. We saved 99.8128% of the total CPU time consumed by the meta-heuristic optimization algorithm. This is the main reason why LP is so important for adaptive ORC problems where both the solution quality and processing speed are vital factors that should not be tolerated. But, the question that should be raised here is: *How to find the optimal set of plug settings?* Here, we are offering two options. The first one is by involving LP in a multi-start strategy to act as a local optimizer, while the other one is by hybridizing LP with meta-heuristic optimization algorithms to act as a fine-tuner.

8.2 Multi-start Linear Programming

The core of the multi-start strategy is described in Chapter 1. In Example 1.7, the multi-start LP was initiated to explore all the possible values of δ. This method is known as **exhaustive search.**[2] In ORC problems, the situation is different. Instead of optimizing one parameter in the outer-loop, there are many parameters that need to be explored. To be more specific, these parameters are the pre-defined PSs[3] of all DOCRs. Thus, the typical linearized ORC problem will have ϱ parameters. Based on (7.1), the number of parameters equals the number of interconnected branches being protected by DOCRs. The number of parameters could increase in some cases; as will be discussed later in Chapters 10, 11, 12, and 13.

2 It is also known as **brute-force search** and **generate and test.**
3 Or TMSs as with the transformation-based linearization approach.

Thus, it is impractical to generate and test all the possible sets of *PS*. Instead, we can deal with this situation by involving combinatorial optimization where the optimal set of *PS* can be attained using inexpensive computation. There are many classical and meta-heuristic techniques to do that efficiently, which will be covered in Section 8.3. For the multi-start strategy, we can randomly generate discrete values for a specific number of loops using either `rand` with `round` or `randi`; refer to Example 1.4. This basic mechanism is the most primitive way to have a combinatorial optimizer. Understanding the concept behind it will help developing more advanced multi-start optimizers, such as shrinking the original search space to have a very reach spot of optimal or near optimal solutions (Al-Roomi and El-Hawary, 2020g). In Example 8.3, we will use the primitive approach. The procedure used to generate random discrete values will be adopted later for biogeography-based optimization (BBO)-LP.

Example 8.3 Modify the MATLAB program given in Example 8.2 to make LP working in a multi-start mode. For PS^{max}, give more weight to select lower values; let's say 50% for $PS^{max} = 2$ and 50% for $PS^{max} = 5$. Show the final result and plot the fitness vs. all the trials. (**Hint:** Refer to the MATLAB code given in Example 1.7 for the multi-start strategy.)

Solution

The MATLAB code is given in `MultiStart_Linearized_ORC_3Bus_LP.m`. If we run it, the following result will be generated in the command window:

```
Elapsed time is 0.716222 seconds.
Tab =
```

Run	PS1	PS2	PS3	PS4	PS5	PS6	TMS1	TMS2	TMS3	TMS4	TMS5	TMS6	Obj
1	3.5	2.5	1.5	4.5	1.5	3.5	0.1	0.1	0.11092	0.1	0.1	0.1	1.7242
2	2	2	1.5	2	1.5	2	0.10112	0.1	0.11092	0.1	0.1	0.1	1.4471
3	4.5	3	3.5	1.5	1.5	2.5	0.1	0.1	0.1	0.1227	0.1	0.1	1.6897
4	1.5	2	1.5	1.5	2	2	0.12251	0.1	0.11669	0.11278	0.1	0.1	1.5078
5	5	2.5	5	2	2	1.5	0.1	0.1	0.1	0.1	0.1	0.1	1.6716
6	3.5	5	1.5	4	2	4	0.26629	0.11159	0.33537	0.12439	0.25472	0.12242	3.3755
7	2.5	4	4	2.5	1.5	3	0.24239	0.104	0.18633	0.12254	0.17629	0.11924	2.6143
8	2	2	1.5	2	1.5	1.5	0.10112	0.1	0.11092	0.1	0.1	0.1	1.4159
9	1.5	4	2.5	2	3	5	0.34856	0.10341	0.32267	0.12593	0.32208	0.11095	3.558
10	4	1.5	3	1.5	2.5	3.5	0.1	0.1	0.1	0.10725	0.1	0.1	1.6595
11	3.5	2	2	3.5	4	5	0.1	0.1	0.11515	0.1	0.1	0.1	1.88
12	1.5	2.5	3.5	4.5	5	4	0.14453	0.30306	0.12216	0.2622	0.1075	0.24661	3.7144
13	1.5	2	1.5	1.5	2	1.5	0.12251	0.1	0.11669	0.11278	0.1	0.1	1.4766
14	5	2.5	3	1.5	2.5	3.5	0.1	0.1	0.1	0.11787	0.1	0.1	1.7633
15	4	1.5	5	1.5	2	3	0.1	0.1	0.1	0.10725	0.1	0.1	1.674
16	3	5	1.5	4.5	1.5	1.5	0.29794	0.11409	0.35997	0.12607	0.35246	0.14	3.5163
17	2	2	1.5	1.5	1.5	1.5	0.10112	0.1	0.11092	0.11278	0.1	0.1	1.4201
18	1.5	1.5	2	2	1.5	2	0.11857	0.1	0.1	0.1	0.1	0.1	1.4384
19	4	4	4	2	3.5	3	0.2464	0.10394	0.25818	0.12547	0.25576	0.11958	3.3287
20	3	2.5	2	4.5	3.5	4	0.1	0.1	0.11114	0.1	0.1	0.1	1.8341
21	3	1.5	3	1.5	3	2.5	0.1	0.1	0.1	0.10725	0.1	0.1	1.5796
22	2	2	2	1.5	1.5	2	0.10056	0.1	0.1	0.11278	0.1	0.1	1.4476
23	2.5	2.5	4	3.5	4	5	0.10053	0.1	0.1	0.1	0.1	0.1	1.894
24	2	2	2	2	1.5	1.5	0.10056	0.1	0.1	0.1	0.1	0.1	1.4122
25	1.5	1.5	1.5	1.5	2	2	0.12251	0.1	0.11669	0.10725	0.1	0.1	1.4742
26	4	1.5	3.5	4.5	5	2.5	0.12615	0.34132	0.12124	0.24681	0.10664	0.26878	3.5549
27	4.5	4	3	2.5	3.5	2	0.24851	0.10436	0.28845	0.12407	0.25468	0.12858	3.3569
28	2	1.5	2	2	1.5	1.5	0.10056	0.1	0.1	0.1	0.1	0.1	1.3911
29	1.5	1.5	1.5	2	1.5	2	0.11923	0.1	0.11092	0.1	0.1	0.1	1.4422
30	3	4	4.5	2	2.5	5	0.25571	0.1034	0.21651	0.12569	0.23071	0.11085	3.1157
31	3.5	2	4	3	4.5	2	0.12464	0.31028	0.11724	0.2662	0.10154	0.28198	3.3544
32	5	5	4.5	2	4	2	0.27562	0.11408	0.29038	0.13331	0.24356	0.13753	3.7481
33	2	1.5	2	1.5	1.5	1.5	0.10056	0.1	0.1	0.10725	0.1	0.1	1.3828
34	4.5	1.5	3.5	3.5	1.5	2.5	0.1	0.1	0.1	0.1	0.1	0.1	1.6709

```
35     2     2   1.5     2   1.5     2   0.10112       0.1   0.11092       0.1       0.1       0.1   1.4471
36   1.5     2   1.5     2   1.5   1.5   0.11923       0.1   0.11092       0.1       0.1       0.1   1.4321
37     3     4   3.5   3.5     3     4   0.20657   0.10346   0.17262   0.11748   0.16179   0.11238   2.6396
38     2   1.5     2   1.5   1.5   1.5   0.10056       0.1       0.1   0.10725       0.1       0.1   1.3828
39     4     2   1.5     5     5   1.5   0.12801   0.22036   0.13146   0.19593   0.10686   0.26101    2.964
40     2   1.5     2   1.5     2     2   0.10056       0.1       0.1   0.10725       0.1       0.1   1.4353
41   1.5     2   1.5   1.5   1.5     2   0.11923       0.1   0.11092   0.11278       0.1       0.1   1.4675
42     3     3   2.5     2   2.5     5       0.1       0.1       0.1    0.1028       0.1       0.1   1.7681
43     4     2   1.5     2     3     3       0.1       0.1   0.12704       0.1       0.1       0.1    1.678
44     2     2     2   1.5     2   1.5   0.10056       0.1       0.1   0.11278       0.1       0.1   1.4377
45     3   1.5   4.5     4     4   2.5       0.1       0.1       0.1       0.1       0.1       0.1   1.7598
46   2.5   3.5   3.5     4   1.5     4       0.1       0.1       0.1       0.1       0.1       0.1   1.7774
47   1.5   1.5     2     2   1.5   1.5   0.11857       0.1       0.1       0.1       0.1       0.1   1.4072
48   1.5     2   1.5     2     2     2   0.12251       0.1   0.11669       0.1       0.1       0.1   1.5036
49   4.5   3.5   1.5   3.5   4.5   1.5   0.12496   0.23618   0.14698   0.25787    0.1016   0.33975   3.4127
50   1.5   1.5   2.5     2     2     3   0.12366       0.1       0.1       0.1       0.1       0.1   1.5483
Best_Start =
    33
Global_TMSopt =
    0.1006    0.1000    0.1000    0.1073    0.1000    0.1000
Best_Fitness =
    1.3828
Best_PS =
    2.0000    1.5000    2.0000    1.5000    1.5000    1.5000
```

The fitness is shown in Figure 8.1, which can also be opened in MATLAB via `MultiStartLP_3Bus.fig`.

As can be seen, the multi-start strategy can provide good results within a very short time. The values of *TMS* are fine-tuned for the best set of *PS* found during the 50 trials. Despite the solution quality and processing time, we were forced to give more weight to lower values of *PS* because it is hard for a stochastic process to settle on the optimal set of *PS*. This will be a real problem for larger and more complex mesh networks. We could enhance the performance of this primitive multi-start program by borrowing the shrinking concept presented in Al-Roomi and El-Hawary (2020g). Instead, we will hybridize LP with one meta-heuristic optimization algorithm to smartly select the best set of *PS*, which is given in Section 8.3.

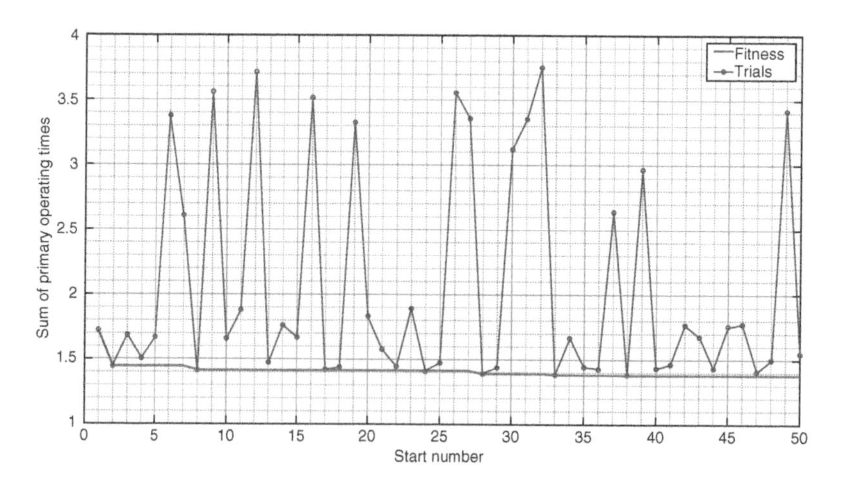

Figure 8.1 The trials and best convergence obtained for Example 8.3.

8.3 Hybridizing Linear Programming with Population-Based Meta-heuristic Optimization Algorithms

To solve the inherent limitation of multi-start LP schemes, the outer loop with random generated initial points can be replaced with more intelligent tools, such as meta-heuristic optimization algorithms.

8.3.1 Classical Linearization Approach: Fixing PS/CTS and Varying TMS/TDS

Let's assume that the linearized ORC model given in (8.2) is solved by LP and this classical optimizer is hybridized with BBO. Then, by taking into account the feasibility checker shown in Figure 7.11, this hybrid approach can be depicted by the flowchart shown in Figure 8.2 where BBO is used as the main algorithm.

It can be seen that the elite solution of BBO, which contains the best *TMS* and *PS* of all ϱ DOCRs, is fine-tuned by LP and then the improved solution is injected into the place of the worst solution available in the population of BBO. If there is no improvement, then the new solution can be rejected from occupying any existing solution in the population. If the elitism stage is defined with more than one elite solution, then we can fine-tune all these solutions. Also, we can fine-tune all the individuals (i.e. including elite and bad solutions) or only some selected individuals of the population. However, this will significantly increase the processing time, and thus making the algorithm unsuitable for adaptive ORC schemes. On the opposite side, we can try implementing the **jumping rate** (**Jr**) technique, used in the **opposition-based learning** (**OBL**) algorithm (Tizhoosh, 2005; Simon, 2013), to activate LP only for some specific generations of EAs. This can effectively improve the processing speed of the proposed hybrid algorithm. Because LP acts as a local optimizer, so it might be difficult to reach the global optimal point. One approach is to let EA works normally without LP for the first quarter, half, or any user-defined range of the total generations, and then activating LP for the rest of the generations. Based on this, both the processing speed and the exploration level can be improved. Also, do not forget the drawback of the death penalty function where the useful information in some infeasible solutions is erased. Thus, we can also try recycling some random solutions, including bad ones, in LP. Some of them could provide very good results.

8.3.2 Transformation-Based Linearization Approach: Fixing TMS/TDS and Varying PS/CTS

Now, assume that the vector of *PS* is the one that needs to be optimized and the vector of *TMS* is considered as an initial guess to LP.[4] That is, the linearized model given in (8.7) is used. Figure 8.3 graphically explains the major difference with the last BBO-LP flowchart.

Example 8.4 Consider the numerical expression given in Example 8.1. Modify the BBO program given in Example 7.4 by embedding LP as a sub-algorithm just after the migration stage. This sub-algorithm should fine-tune the elite solutions supplied from BBO. Based on your MATLAB

---

4 In new versions of MATLAB, the `linprog` package does not require initial points anymore. This is very important news because we can reduce the dimension of EAs by half and letting them working as combinatorial optimization algorithms where the objective is to find the best set of *PS*. The optimal values of *TMS* can then be obtained through LP. This arrangement will lead to a significant reduction in CPU time.

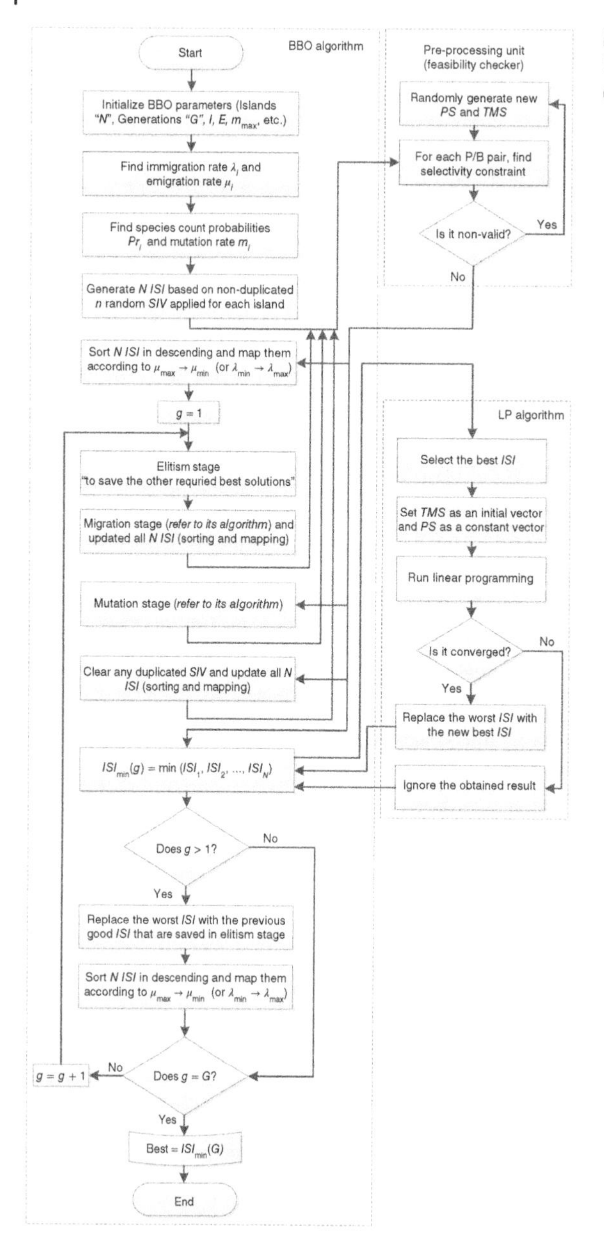

Figure 8.2 Flowchart of the *TMS*-based BBO-LP algorithm reinforced by the feasibility checker sub-algorithm.

version, `linprog` could ask you to remove the initial point (i.e. the vector of *TMS*). Because LP can accelerate converging to the optimal solution, it does not require to use many generations and large population size. For this, use the following initialization parameters: population size of 8, $m_{\max} = 0.05$, 4 elite solutions,[5] 100 generations, and 1 trial. Use the binary static penalty function as a constraint-handler with a penalty multiplier of $r = 0.15$. Show the final result.

5 Using many elite solutions is not recommended for classical meta-heuristic optimization algorithms because this practice will force the algorithm to settle on local optima rather than the global one. In general, using a few elites (like 0 or 1) will lose useful information on the other good individuals that may guide the algorithm to settle on better solutions. Similarly, using many elites will reduce the exploration level by premature convergence, and thus

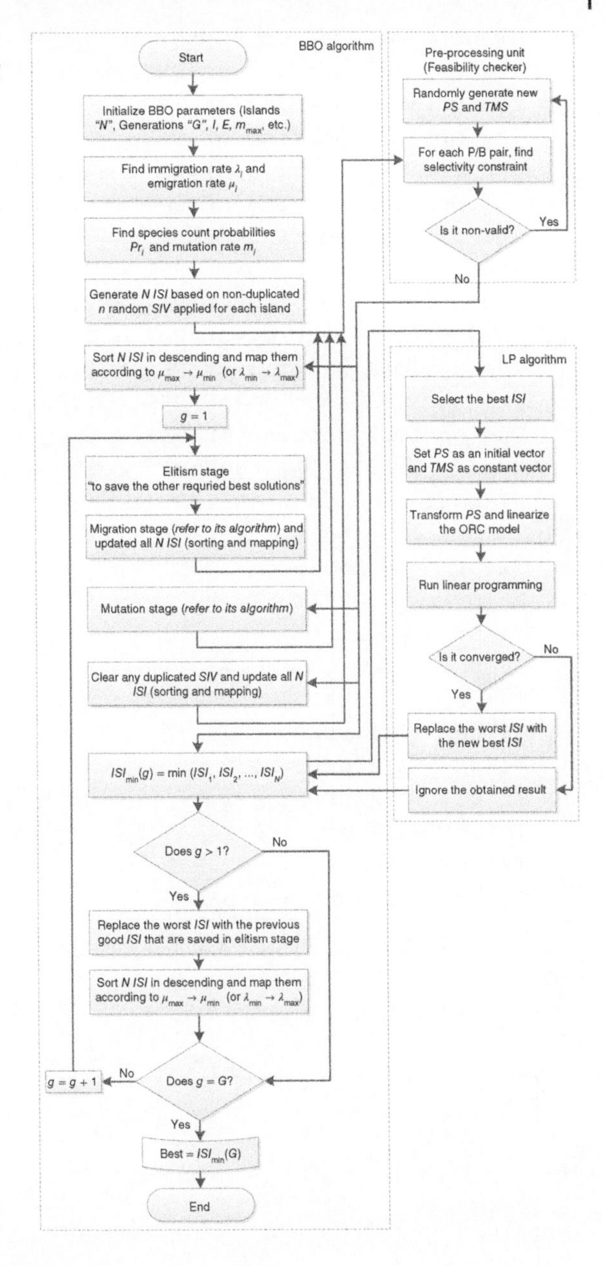

Figure 8.3 Flowchart of the *PS*-based BBO-LP algorithm reinforced by the feasibility checker sub-algorithm.

Solution

The MATLAB files are given in a folder called `LinORC_3Bus_BBOLP_BinStat`. If we run it, the following result will be generated in the command window:

```
The best of Trial # 1 is 1.3828
The minimum sum of relay operating times = 1.3828
```

---

the performance worsens (Can et al., 2008; Simon et al., 2009; Al-Roomi and El-Hawary, 2016b). However, here the story is different! Although a small population size is used, half of these individuals are set as elite solutions. The reason is that we want to recycle these solutions into LP for fine-tuning. This can increase the chance to get better solutions by excessive exploitation. The remaining half of the population is dedicated to exploration. This ratio is not constant as the network dimension, complexity, and algorithm initialization parameters are changed.

```
It occurred at run number 1
The optimal setting of TMS1 = 0.10056
The optimal setting of TMS2 = 0.1
The optimal setting of TMS3 = 0.1
The optimal setting of TMS4 = 0.10725
The optimal setting of TMS5 = 0.1
The optimal setting of TMS6 = 0.1
The optimal setting of PS1 = 2
The optimal setting of PS2 = 1.5
The optimal setting of PS3 = 2
The optimal setting of PS4 = 1.5
The optimal setting of PS5 = 1.5
The optimal setting of PS6 = 1.5
Mean = 1.3828 || Standard Deviation = 0
Elapsed time is 2.081415 seconds.
T1(pr) = 0.24418
T2(pr) = 0.2094
T3(pr) = 0.22282
T4(pr) = 0.24247
T5(pr) = 0.21056
T6(pr) = 0.25337
T5(bc) - T1(pr) = 0.40278
T4(bc) - T2(pr) = 0.2
T1(bc) - T3(pr) = 0.2
T6(bc) - T4(pr) = 0.26639
T3(bc) - T5(pr) = 0.22873
T2(bc) - T6(pr) = 0.53085
```

The fitness is shown in Figure 8.4, which can also be opened in MATLAB via `LinORC_3Bus_BBOLP.fig`.

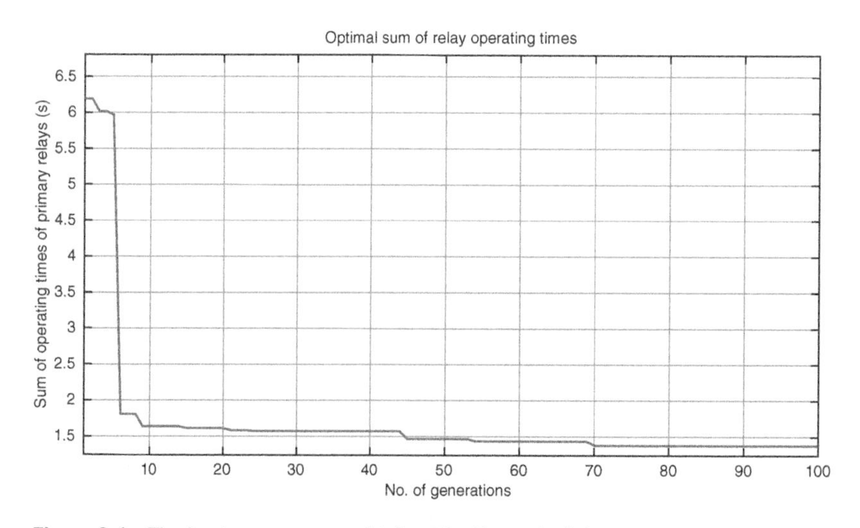

Figure 8.4 The best convergence obtained for Example 8.4.

As can be clearly seen, the hybrid BBO-LP can provide the same global optima obtained by the multi-start-based LP. It does not need to guess the location of the best set of *PS* to give more weight on it. This is very useful, especially for larger and more complex networks. For the processing speed, BBO-LP consumed 2.081 415 s compared with just 0.028 735 s in Example 8.3. However, this is nothing if we compare it with the conventional BBO given in Example 7.4 where the fitness and processing time were 1.4427 s and 15.351 278 s, respectively. This means that we can have better solution(s) with saving around 86.44% of the total CPU time consumed by the conventional BBO algorithm. This fantastic feature presents the hybrid schemes between LP and EAs as superior optimization algorithms that could be used for solving ORC problems with dynamic networks; i.e. adaptive ORC solvers.

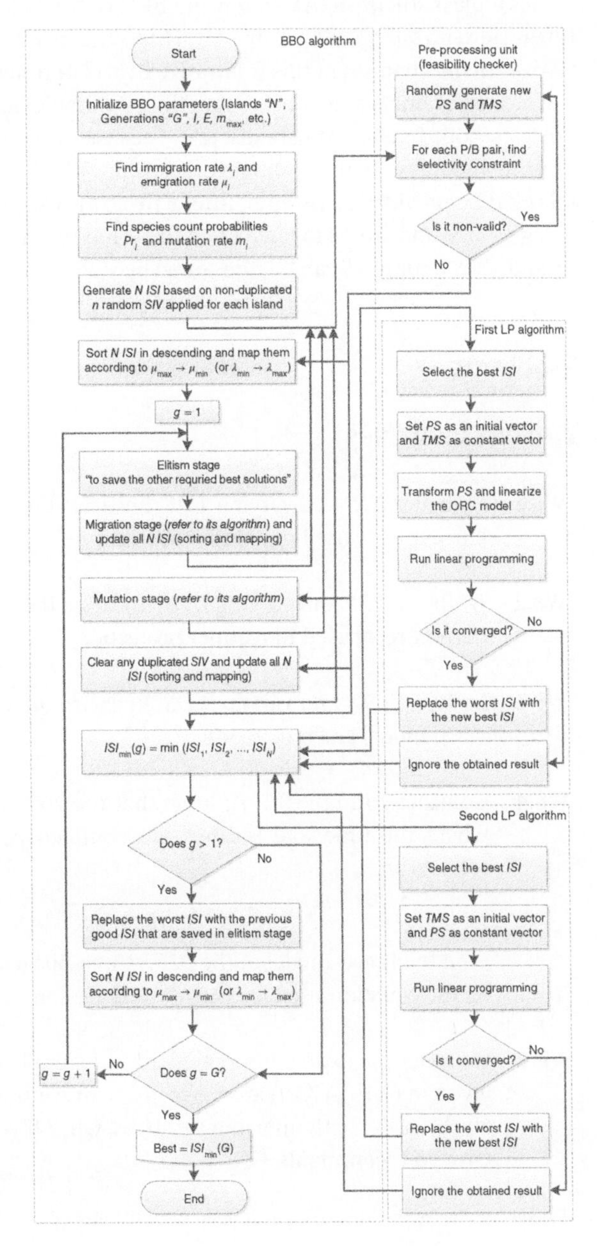

Figure 8.5 Flowchart of the *TMS/PS*-based BBO-2LP algorithm reinforced by the feasibility checker sub-algorithm.

8.3.3 Innovative Linearization Approach: Fixing/Varying TMS/TDS and PS/CTS

This is the most advanced hybrid approach between EAs and LP. It incorporates the advantages of both algorithms shown in Figures 8.2 and 8.3. That is, two separated LP sub-algorithms are coded in two different places of BBO for both *TMS* and *PS*. Thus, by this hybridization strategy, the user can ensure that both settings are fine-tuned. However, there are some couple questions that are still open and require detailed answers! For example: *Which relay setting should be fine-tuned first, TMS or PS? Is it required to fine-tune both settings in every iteration? What is about the exploration level of the overall algorithm? Is it useful to embed the Jr technique used? What is the suitable value of Jr? How fast is it with/without Jr?*

This topic is still fresh (Al-Roomi and El-Hawary, 2020e). Actually, based on our extensive literature review, no one yet talked about it. This is the first time ever. The flowchart shown in Figure 8.5 reveals the mechanism of this innovative hybrid optimization algorithm.

It has to be always remembered that the *PS* variables are mostly discrete, and for some relay technologies, both *PS* and *TMS* are discrete. Because the strategy here is to fine-tune these settings in a separate mode,[6] so both LP sub-algorithms or only the LP sub-algorithm responsible to fine-tune *PS* could be provided as **integer linear programming (ILP)**. Based on the nature of relay technologies, it is impossible to have LP to fine-tune *PS* and ILP to fine-tune *TMS*. That is, either both *PS* and *TMS* or only *PS* can be fine-tuned by ILP.

Problems

Written Exercises

W8.1 By fixing *CTS* and varying *TDS*, linearize the characteristic curve model of the UK overcurrent relay used in rectifier protection.

W8.2 By fixing *TDS* and varying *CTS*, linearize the characteristic curve model of the UK overcurrent relay used in rectifier protection.

W8.3 By fixing *CTS* and varying *TDS*, linearize the characteristic curve model of the IAC and ANSI model given in (3.34).

W8.4 Consider Example 8.1. Assume that the six relays are manufactured in North America where the time dial settings are bounded between 0.5 and 11, and the current tap settings are:

$$CTS = [2.5, 1.5, 1.5, 1.5, 1.5, 1.5]$$

Find the linear model of the IEEE three-bus test system if all DOCRs are modeled based on the equation used for IAC-66. (**Hint:** Refer to (3.34) and Table 3.9.)

W8.5 Consider the expression presented in Example 8.1. If the inequality and side constraints given in (7.16)–(7.19) are neglected, can we solve the linearized optimization algorithm algebraically without using LP? If no, why? If yes, how and what is the optimal solution? Give your comments.

6 This is because of the linearization done in the ORC model.

W8.6 Why are the transformed plug and current tap settings (i.e. $\widetilde{PS}_{i,k}$ and $\widetilde{CTS}_{i,k}$) functions of fault location (k) while the actual ones (i.e. PS_i and CTS_i) are not?

W8.7 Can we use the transformation-based linearized model given in (8.7) with LP if OCRs are numerical? Explain.

W8.8 Can we use the transformation-based linearized model given in (8.7) to optimize electromechanical DOCRs? Explain.

W8.9 What is the main difference between the classical multi-start strategy and the hybrid differential evolution (DE)/LP algorithms? Which one is preferred for small networks and which one is preferred for large networks?

W8.10 State one possible method to improve the performance of the classical multi-start strategy.

W8.11 If we want to hybrid LP with EAs to fine-tune *TMS* of numerical DOCRs, when can we use mixed-integer EAs and when can we use combinatorial EAs?

Computer Exercises

C8.1 Modify the MATLAB code given in Example 8.2 to coordinate the relays when the IEC very inverse type is used.

C8.2 Modify the MATLAB code given in Example 8.2 to coordinate the relays when the IEEE very inverse type is used. the lower and upper bounds of the time dial setting should be 0.5 and 11, respectively.

C8.3 Modify the MATLAB code given in Example 8.2 to consider the transient topology of each relay. Refer to Appendix F for the transient short-circuit currents. Use the following pre-defined plug settings:

$$PS = [2.5, 2,3, 2.5, 2.5, 1.5]$$

Do not consider $T^{\min}$ and $T^{\max}$. Compare your result with that obtained by BBO-LP in Albasri et al. (2015).

C8.4 Repeat Exercise C8.4, but now with considering $T^{\min} = 0.1$ and $T^{\max} = 0.5$. What is the new objective function?

C8.5 Based on the model developed in Exercise W8.4, find the optimal current tap settings by using LP with `dual-simplex` algorithm. (**Hint:** Modify the MATLAB program given in Example 8.1.)

C8.6 Modify the MATLAB program developed in Exercise C8.5 to run in multi-start modes.

C8.7 Modify the MATLAB program developed in Example 8.4 by considering the transient selectivity constraints. Refer to Appendix F for the transient short-circuit currents.

C8.8 Repeat Exercise C8.7, but with fine-tuning just the first best solution. (**Hint:** Do not fine-tune the remaining elite solutions in LP.)

C8.9 Modify the MATLAB code given in Example 8.4 to optimize the relay settings of the IEEE four-bus test system. Select proper initialization parameters. (**Hint:** Use the special mathematical model given in Appendix F.)

C8.10 Modify the MATLAB code given in Example 8.4 to optimize the relay settings of the IEEE eight-bus test system given in Appendix F. Select proper initialization parameters. (**Hint:** Useful information is given in Example 7.5.)

C8.11 Modify the MATLAB code given in Example 8.4 to optimize the relay settings of the IEEE nine-bus test system given in Appendix F. Select proper initialization parameters. (**Hint:** Use the realized equations given in (7.22) and(7.23) for $PS^{\min}$ and $PS^{\max}$, respectively.)

C8.12 Modify the MATLAB code given in Example 8.4 to optimize the relay settings of the IEEE 15-bus test system given in Appendix F. Select proper initialization parameters.

9

Optimal Coordination of DOCRs With OCRs and Fuses

After reading this chapter you should be able to:

- Express the mathematical model of ORC problems when both directional and non-directional overcurrent relays (DOCRs and OCRs) are imposed
- Express the mathematical model of ORC problems when overcurrent relays and fuses are imposed
- Differentiate the selectivity constraints when DOCRs are integrated with other protective devices
- Code and solve different linearized ORC problems using linear programming
- Code and solve different ORC problems using meta-heuristic optimization algorithms

In Chapters 7 and 8, we have learned lots of things regarding optimal relay coordination from DOCRs perspective. This fundamental step lets us grasp the tools and knowledge required to quickly understand more complex ORC problems. Thus, the content of this chapter will be explained mainly through illustrative and computer examples.

9.1 Simple Networks

Coordinating DOCRs need to trace the current direction on each branch. On the opposite side, OCRs can see the current from both directions, which means that there is a challenge to achieve the selectivity criteria between primary and backup OCRs. In Chapters 2 and 5, we have seen that OCRs are not preferred in mesh and interconnected networks. However, we could see them in some locations, like protecting generators by installing OCRs on the receiving-end of PV busbars. For instance, we could see OCRs installed between the step-up transformers and the PV busbars of the 8-bus system shown in Figure F.10, which by the way are not considered during solving the ORC problem. In some simple applications, OCRs could be the main protective devices that could be used to achieve different performance criteria.[1]

9.1.1 Protecting Radial Networks by Just OCRs

For radial networks, OCRs can be used alone because the first primary relay does not act as a backup relay for other OCRs. This is true for the rest of OCRs until reaching the last OCR that is installed

1 Such as the simplicity, cost, reliability, etc. Please refer to Chapter 2 for more details.

Optimal Coordination of Power Protective Devices with Illustrative Examples, First Edition. Ali R. Al-Roomi.

Companion website: www.wiley.com/go/al-roomi/optimalcoordination

near the source. Thus, the coordination is easy and the optimal coordination can be attained analytically.

Example 9.1 Consider the system shown in Figure 9.1. This very basic radial system is taken from (Ralhan and Ray, 2013; Gokhale and Kale, 2014). Both OCRs are non-directional and modeled based on the IEC standard inverse. The near-end short-circuit currents for R_1 and R_2 are 4000 A and 3000 A, respectively. These fault currents are stepped-down through two CTs before being sent to the corresponding protective relays. The CT-ratios used in this problem are 300:1 and 100:1 for R_1 and R_2, respectively. The relay technology supports *TMS* with a lower bound of 0.1 and a minimum allowable operating time of 0.2 second. If $PS_1 = PS_2 = 1$ A and $CTI = 0.57$ s, derive the mathematical expression of this ORC problem.

Solution

Table 9.1 shows the data required to find the optimal relay settings. The objective function of this simple model is:

$$Z = \min \sum_{i=1}^{2} T_i^{\text{pr}} = \min \left(T_1^{\text{pr}} + T_2^{\text{pr}}\right)$$

Because the plug settings of the two relays are given, we can linearize the objective function to be as follows:

$$Z = \min \left(\psi_1 \, TMS_1 + \psi_2 \, TMS_2\right)$$

This objective function is subjected to the selectivity constraint defined in (7.25) as well as the minimum operating time inequality constraint given in (7.17) and the lower bound portion of the time multiplier setting side constraint given in (7.19). Therefore, we have the following new expressions:

- To satisfy (7.25), the operating time of the jth backup relay should be known. By taking ψ_j as the coefficient when the jth relay acts as a backup OCR,[2] then the preceding selectivity constraint can be expressed as follows:

$$\psi_j \, TMS_j - \psi_i \, TMS_i \geqslant CTI \tag{9.1}$$

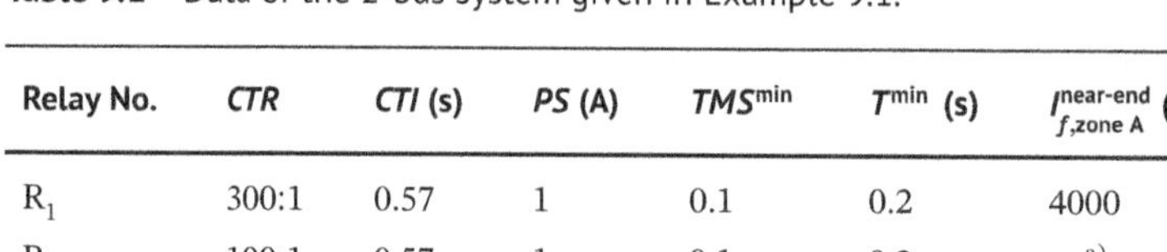

Figure 9.1 Single-line diagram of the 2-bus system given in Example 9.1.

Table 9.1 Data of the 2-bus system given in Example 9.1.

Relay No.	*CTR*	*CTI* (s)	*PS* (A)	$TMS^{\min}$	$T^{\min}$ (s)	$I_{f,\text{zone A}}^{\text{near-end}}$ (A)	$I_{f,\text{zone B}}^{\text{near-end}}$ (A)
R_1	300:1	0.57	1	0.1	0.2	4000	3000
R_2	100:1	0.57	1	0.1	0.2	—[a]	3000

a) "—" means R_2 cannot see the fault happens in the zone of R_1

2 Please refer to (7.26).

- To satisfy (7.17), the following expression is used:

$$\psi_i \; TMS_i \geqslant T_i^{\min} \tag{9.2}$$

- To satisfy the lower bound portion of (7.19), the following expression is used:

$$TMS_i \geqslant TMS_i^{\min} \tag{9.3}$$

By using the data tabulated in Table 9.1, the P/B relay pairs' coefficients a_i and a_j can be determined. Table 9.2 shows the results.

Thus, the problem can be solved using the classical LP as follows:

Minimize $2.63\,303\ TMS_1 + 1.98\,889\ TMS_2$

subjected to:

$2.97\,060\ TMS_1 - 1.98\,889\ TMS_2 \geqslant 0.57$

$2.63\,303\ TMS_1 \geqslant 0.2$

$1.98\,889\ TMS_2 \geqslant 0.2$

$TMS_1 \geqslant 0.1$

$TMS_2 \geqslant 0.1$

9.1.2 Protecting Double-Line Networks by OCRs and DOCRs

Now, let's formulate the ORC problem of a simple system when both OCRs and DOCRs are involved. Example 9.1 shows a coordination problem similar to that taken in Figure 5.8. In this example, we will see how to linearize this ORC problem.

Example 9.2 Consider the system shown in Figure 9.2, which is taken from (Gokhale and Kale, 2014). This system is protected by both OCRs and DOCRs. As can be seen from the arrows, R_1 and R_4 are non-directional OCRs, while R_2 and R_3 are DOCRs. This is the second very basic linear ORC problem, which is an example of parallel feeders with a single-end fed systems[3]. The pre-defined

Table 9.2 The coefficients of the linearized relay operating times of the 2-bus system given in Example 9.1.

Relay constants	Fault location	
	Just beyond bus 1	Just beyond bus 2
ψ_1	$\dfrac{0.14}{\left(\frac{4000}{300}\right)^{0.02} - 1} = 2.63\,303$	$\dfrac{0.14}{\left(\frac{3000}{300}\right)^{0.02} - 1} = 2.97\,060$
ψ_2	—[a)]	$\dfrac{0.14}{\left(\frac{3000}{100}\right)^{0.02} - 1} = 1.98\,889$

a) "—" means R_2 cannot see the fault occurs in the zone of R_1

3 It has to be known that the non-directional overcurrent relays R_1 and R_4 do not have backup relays. If anyone of these two primary relays fail to isolate the corresponding in-zone fault then the network is supposed to be protected by a relay installed in the left side of the generator busbar; which is not shown here. Thus, there are only two selectivity constraints that should be satisfied for R_2 and R_3.

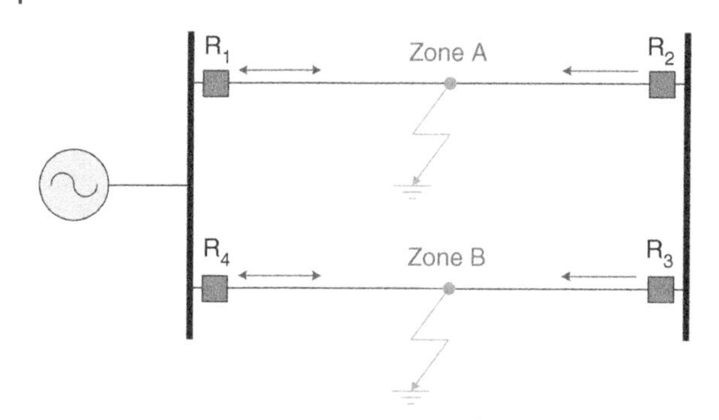

Figure 9.2 Single-line diagram of the 2-bus system given in Example 9.2.

plug settings of all the relays are set equal to 1 A. These four relays are modeled based on the IEC standard inverse characteristic curve. A mid-point short-circuit fault is considered for each zone. These fault currents are stepped-down through four CTs, with a CT-ratio of 300:1, before being sent to the corresponding protective relays. The total fault current for each point is assumed to be 4000 A, and it is divided as follows:

- For the mid-point of zone A:
  R_1 will see 10 A, while R_2 and R_4 will see 10/3 A. Therefore, $(10 + 10/3) \times 300 = 4000$ A. Because R_3 has a directional unit, it will not see that fault.
- For the mid-point of Zone B:
  R_4 will see 10 A, while R_1 and R_3 will see 10/3 A. Therefore, $(10 + 10/3) \times 300 = 4000$ A. Because R_2 has a directional unit, so it will not see that fault.

The minimum allowable operating time of these relays is 0.2 s and the lower bound of time multiplier setting is 0.1. Table 9.3 summarizes all the required parameters and coefficients. Express this ORC problem mathematically as a linear programming problem.

Solution

Similar to the previous test system, the linearized objective function that needs to be minimized is:

$$Z = \min\left(\psi_1\ TMS_1 + \psi_2\ TMS_2 + \psi_3\ TMS_3 + \psi_4\ TMS_4\right)$$

The four constants $\{\psi_1, \psi_2, \psi_3, \psi_4\}$ can be calculated using (8.1). The same thing can be done for ψ_j when the jth relay acts as a backup. These backup relay coefficients are required to express the selectivity constraints of the ORC problem. By using the data tabulated in Table 9.3, the coefficients of each P/B relay pair (i.e. ψ_i and ψ_j) can be determined. These values are tabulated in Table 9.4.

Table 9.3 Data of the 2-bus system given in Example 9.2.

Relay No.	Type	*CTR*	*CTI* (s)	*PS* (A)	TMS^{min}	T^{min} (s)	$I^{near\text{-}end}_{f,zone\ A}$ (A)	$I^{near\text{-}end}_{f,zone\ B}$ (B)
R_1	OCR	300:1	0.57	1	0.1	0.2	3000	1000
R_2	DOCR	300:1	0.57	1	0.1	0.2	1000	—
R_3	DOCR	300:1	0.57	1	0.1	0.2	—[a]	1000
R_4	OCR	300:1	0.57	1	0.1	0.2	1000	3000

a) "—" means R_2 cannot see the fault happens in the zone of R_1

Table 9.4 The coefficients of the linearized relay operating times of the 2-bus system given in Example 9.2.

Relay constants	Fault location	
	Mid-point of zone A	Mid-point of zone B
ψ_1	$\dfrac{0.14}{\left(\frac{3000}{300}\right)^{0.02} - 1} = 2.9706$	$\dfrac{0.14}{\left(\frac{1000}{300}\right)^{0.02} - 1} = 5.7444$
ψ_2	$\dfrac{0.14}{\left(\frac{1000}{300}\right)^{0.02} - 1} = 5.7444$[a]	—
ψ_3	—[b]	$\dfrac{0.14}{\left(\frac{1000}{300}\right)^{0.02} - 1} = 5.7444$
ψ_4	$\dfrac{0.14}{\left(\frac{1000}{300}\right)^{0.02} - 1} = 5.7444$	$\dfrac{0.14}{\left(\frac{3000}{300}\right)^{0.02} - 1} = 2.9706$

a) In Gokhale and Kale (2014), the value is rounded to 5.749 because it is calculated after being stepped down through *CTR*, which is set to 3.33 A instead of dealing with all the fractions of 3.3 333 333· · ·.
b) "—" means R_2 cannot see the fault happens in the zone of R_1

Now, by employing all the information shown in Tables 9.3 and 9.4, the classical LP model can be formulated as follows:

Minimize $2.9706\ TMS_1 + 5.7444\ TMS_2 + 5.7444\ TMS_3 + 2.9706\ TMS_4$

subjected to:

$5.7444\ TMS_4 - 5.7444\ TMS_2 \geqslant 0.57$

$5.7444\ TMS_3 - 5.7444\ TMS_1 \geqslant 0.57$

$2.9706\ TMS_1 \geqslant 0.2$

$5.7444\ TMS_2 \geqslant 0.2$

$5.7444\ TMS_1 \geqslant 0.2$

$2.9706\ TMS_2 \geqslant 0.2$

$TMS_1 \geqslant 0.1$

$TMS_2 \geqslant 0.1$

$TMS_3 \geqslant 0.1$

$TMS_4 \geqslant 0.1$

9.2 Little Harder Networks

In Section 9.1, we have seen how to coordinate OCRs of a simple radial network and OCRs/DOCRs of a simple double-line system. The mathematical expressions were so easy, which can be solved numerically as well as analytically. This section deals with some little harder ORC problems.

9.2.1 Combination of OCRs and DOCRs

In Example 9.2, we can see that R_4 will work as a backup relay for R_2. Thus, for the mid-point fault of zone A, the power will still be delivered to the other bus because R_1 is faster than R_4, and vice versa for the mid-point of zone B. In the next example, we will see that the ORC problem becomes a little harder when another bus is inserted between the two buses.

Example 9.3 Consider the system shown in Figure 9.3, which is taken from (Bedekar and Bhide, 2011b; Singh et al., 2011). A bolted 3ϕ short-circuit fault is considered at the mid-point of all the lines, except the point D where the fault is located between bus 3 and load 2. Please, note that there is no parallel lines between bus 1 and bus 2. These six relays are based on the IEC standard inverse TCCC. The ORC problem is linearized by taking *PS* equal to 1. The base quantities are 25 MVA and 11 kV with line impedances of $Z_{L_1} = Z_{L_2} = 0.08 + j1$ pu and $Z_{L_3} = 0.16 + j2$ pu. The P/B relay pairs, of all the relays shown in Figure 9.3, are listed in Table 9.5. The relay type, technology, TCCC, and *CTR* are shown in Table 9.6. The minimum allowable operating time of all the relays is $T^{\min} = 0.1$ s. The time multiplier settings of all the relays are bounded between 0.025 and 1.2. The time delay between P/B relay pairs is $CTI = 0.3$ s. Express the ORC problem of this electric network mathematically as a linear programming problem. Use the objective function given in (7.10).

Solution

As said before, this system is linearized, because *PS* is not considered as a design variable anymore. Thus, the operating time of each ith relay can be calculated as follows:

$$T_i = \psi_i \times TMS_i$$

where the coefficient ψ_i of the ith primary relay can be calculated as follows:

$$\psi_i = \frac{0.14}{I_{R_i}^{0.02} - 1}$$

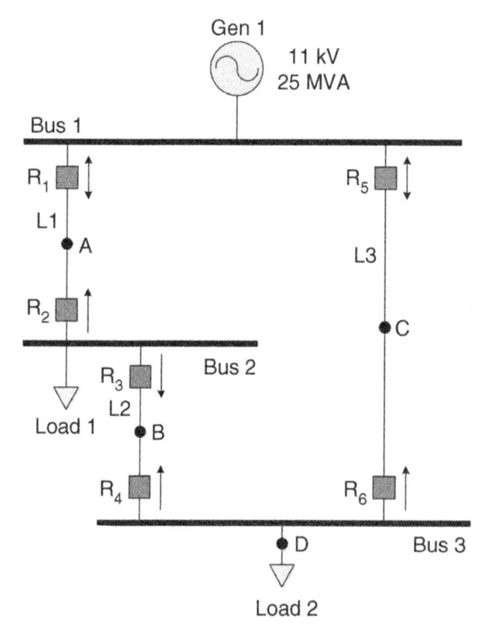

Figure 9.3 Single-line diagram of the 3-bus system given in Example 9.3.

Table 9.5 Primary/backup relay pairs of the 3-bus system given in Example 9.3.

Fault point	Primary relay R_i	Backup relay R_j
A	R_1	—
	R_2	R_4
B	R_3	R_1
	R_4	R_5
C	R_5	—
	R_6	R_3
D	R_3	R_1
	R_5	—

Table 9.6 Relay type, technology, IEC model, and *CTR* of the 3-bus system given in Example 9.3.

Relay No.	Type	Technology	Operating time (s)	CT-ratio
R_1	OCR	Numerical	$\frac{0.14TMS}{PSM^{0.02}-1}$	1000:1
R_2	DOCR	Numerical	$\frac{0.14TMS}{PSM^{0.02}-1}$	300:1
R_3	DOCR	Numerical	$\frac{0.14TMS}{PSM^{0.02}-1}$	1000:1
R_4	DOCR	Numerical	$\frac{0.14TMS}{PSM^{0.02}-1}$	600:1
R_5	OCR	Numerical	$\frac{0.14TMS}{PSM^{0.02}-1}$	600:1
R_6	DOCR	Numerical	$\frac{0.14TMS}{PSM^{0.02}-1}$	600:1

Similarly, the previous equation can also be used to calculate the coefficient ψ_j of the *j*th backup relay when the out-zone fault current is considered instead. The short-circuit current seen by each relay and all the model coefficients are tabulated in Table 9.7 where only the notation i is used for both P/B relay pair coefficients[4].

In this system, we use a different objective function. We want to minimize the sum of operating times of all the relays when they act as primary and backup protective devices; as stated in (7.10). Thus, by using the pre-defined plug settings, the linearized objective function is:

$$Z = \min \sum_{i=1}^{6} \sum_{k=A}^{D} \psi_i^k \ TMS_i, \quad \text{where:} \quad k = \{A, B, C, D\}$$

4 i.e. based on the P/B relay pairs shown in Table 9.5, the primary or backup relay coefficient can be easily determined. These coefficients will be further clarified when the final mathematical expression is shown at the end of this example.

Table 9.7 Fault currents seen by the six IEC standard inverse OCRs/DOCRs relays and their ψ coefficients considered in the 3-bus system given in Example 9.3.

Fault point		Relay					
		R_1	R_2	R_3	R_4	R_5	R_6
A	I_{R_i}	6.579	3.13	—	1.565	1.565	—
	ψ_i	3.6462	6.0651	—	15.5591	15.5591	—
B	I_{R_i}	2.193	—[a]	2.193	2.193	2.193	—
	ψ_i	8.8443	—	8.8443	8.8443	8.8443	—
C	I_{R_i}	1.0965	—	1.0965	—	5.4825	1.8275
	ψ_i	75.9152	—	75.9152	—	4.0443	11.5397
D	I_{R_i}	1.6447	—	1.6447	—	2.7412	—
	ψ_i	13.9988	—	13.9988	—	6.8720	—
$\sum_{k=A}^{D} \psi_i^k$		102.4045	6.0651	98.7583	24.4034	35.3197	11.5397

a) "—" means that the corresponding relay cannot see that fault.

Because the dimension is only 6, so it can be further simplified as follows:

$$Z = \min \sum_{i=1}^{6} \left[\left(\psi_i^A + \psi_i^B + \psi_i^C + \psi_i^D \right) \times TMS_i \right]$$

Thus, the classical LP model can be formulated as follows:[5]

$$\textbf{Minimize} \ 102.4045\ TMS_1 + 6.0651\ TMS_2 + 98.7583\ TMS_3 + 24.4034\ TMS_4 + 35.3197\ TMS_5 + 11.5397\ TMS_6$$

subjected to:

$$15.5591\ TMS_4 - 6.0651\ TMS_2 \geqslant 0.3$$

$$8.8443\ TMS_1 - 8.8443\ TMS_3 \geqslant 0.3$$

$$8.8443\ TMS_5 - 8.8443\ TMS_4 \geqslant 0.3$$

$$75.9152\ TMS_3 - 11.5397\ TMS_6 \geqslant 0.3$$

$$13.9988\ TMS_1 - 13.9988\ TMS_3 \geqslant 0.3$$

$$3.6462\ TMS_1 \geqslant 0.1$$

$$6.0651\ TMS_2 \geqslant 0.1$$

$$8.8443\ TMS_3 \geqslant 0.1$$

$$8.8443\ TMS_4 \geqslant 0.1$$

$$4.0443\ TMS_5 \geqslant 0.1$$

$$11.5397\ TMS_6 \geqslant 0.1$$

5 Please, see the last row of Table 9.7.

$$0.025 \leqslant TMS_1 \leqslant 1.2$$

$$0.025 \leqslant TMS_2 \leqslant 1.2$$

$$0.025 \leqslant TMS_3 \leqslant 1.2$$

$$0.025 \leqslant TMS_4 \leqslant 1.2$$

$$0.025 \leqslant TMS_5 \leqslant 1.2$$

$$0.025 \leqslant TMS_6 \leqslant 1.2$$

As can be clearly seen, the problem complexity significantly increases with just inserting one additional busbar. In Chapters 10, 12, 13, 16, and 17, we will see that the complexity could be further increased when different relay technologies and characteristic curves are considered.

9.2.2 Combination of Fuses, OCRs, and DOCRs

A similar thing can be applied if a combination of fuses, OCRs, and DOCRs is faced in the same network. We can think about fuses as OCRs, but with different rating, operating speed, and characteristic curves. This can be graphically described by looking at the coordination of the simple radial system shown in Figure 9.4 (Madureira and Vieira, 2018). We can see that both types of protection devices have inverse-time characteristic curves, but they are not the same and thus we cannot mimic the operating time of fuses by the equations used for overcurrent relays. The main problem faced in this coordination scheme is that there is no general characteristic curve equation that can be used to calculate the operating time of fuses at different fault currents. Each manufacturer has its own characteristic curves and curve fitted equations. Also, there is inherent uncertainty in

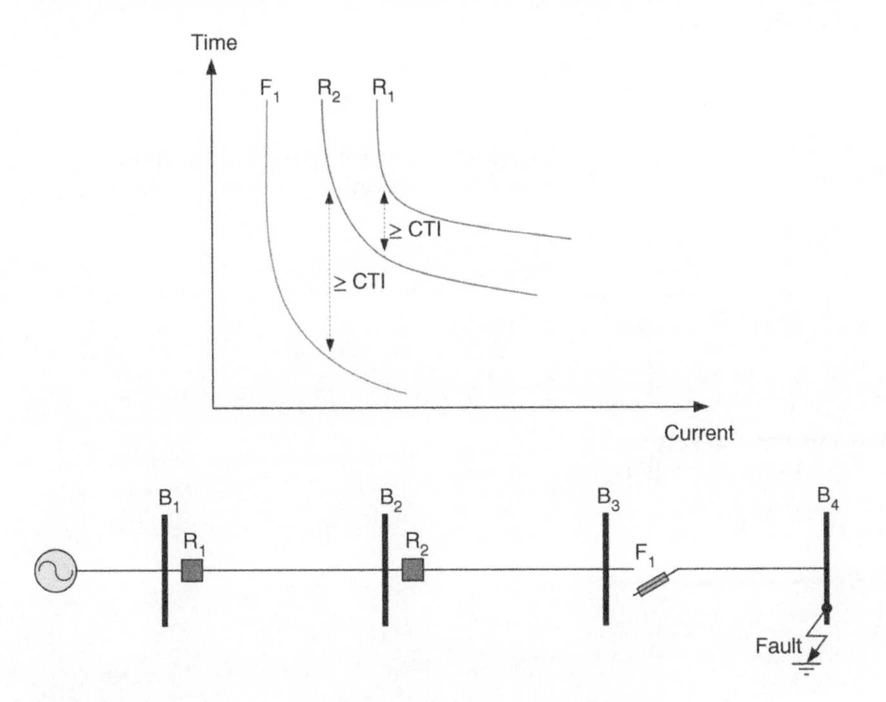

Figure 9.4 Coordination between OCRs and fuses in a radial system.

the fusible links characterized by the existence of maximum and minimum operating times. This uncertainty was represented in Soares (2009) with a range of ±15% around the average melting current of the inverse time curve provided by the manufacturer.

It has to be said that coordinating fuses with other protective devices is not an easy task. Some technical problems are reported in Dudor and Padden (1994) and Gómez et al. (2017). There are some practical guidelines and recommendations that need to be followed during installing such devices (Gers and Holmes, 2004; Das, 2012). Chapter 2 gives a quick introduction. This section covers some ideas used to optimally coordinate fuses with overcurrent relays in non-radial systems.

There are a few attempts to solve optimal coordination problems of networks having both fuses and overcurrent relays so that the best relay settings and fuse types can be obtained. Askarian et al. (2007) used the following objective function to coordinate the protective devices shown in Figure 9.5:

$$\textbf{Minimize } \lambda_1\left(\sum T_r + \sum T_f\right)^2 + \lambda_2\left\{\sum\left[\Delta T_{rr} + \mu_2\left(\Delta T_{rr} - \left|\Delta T_{rr}\right|\right)\right]^2 + \sum\left[\Delta T_{rf} + \mu_2\left(\Delta T_{rf} - \left|\Delta T_{rf}\right|\right)\right]^2 + \sum\left[\Delta T_{ff} + \mu_2\left(\Delta T_{ff} - \left|\Delta T_{ff}\right|\right)\right]^2\right\} \tag{9.4}$$

and the variables $\{\Delta T_{rf}, \Delta T_{ff}, \Delta T_{rr}\}$ can be calculated as follows:

$$\Delta T_{rf} = T_{Br} - 0.75T_{Pf} \tag{9.5}$$

$$\Delta T_{ff} = T_{Bf} - T_{Pf} - 0.35 \tag{9.6}$$

$$\Delta T_{rr} = T_{Br} - T_{Pr} - 0.3 \tag{9.7}$$

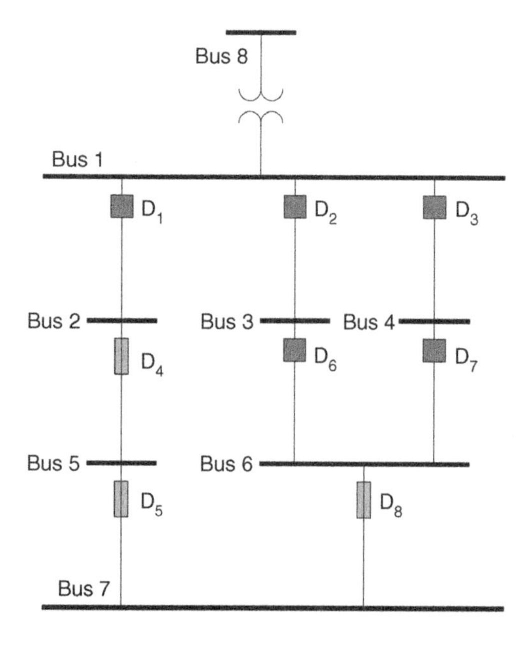

Figure 9.5 Coordinating protective devices in a simple 8-bus system.

where

- T_r and T_f are backup relay and main fuse operating time for fault close to the main fuse;
- T_{Bf} and T_{Pf} are backup and main fuse operating time for fault close to the main fuse; and
- T_{Br} and T_{Pr} are backup relay and main relay operating time for fault close to the main relay.

The other attempt is reported in Madureira and Vieira (2018). The network under study is shown in Figure 9.6. The authors used the classical objective function given in () with an addition penalty function to handle the selectivity constraint. The modified objective function used in their ORC solver is:

$$\min Z = \sum_{i=1}^{\varrho} T_{i,k} + \lambda \sum_{j=1}^{\sigma} \left[\max\left(0, CTI - \Delta T_{ji}\right)\right] \tag{9.8}$$

where ΔT_{ji} is the time interval between the operation of the primary and backup protective devices.

To consider the uncertainty around the average melting current of fuses, a tolerance of ±15% for the operating time of the fuse links is considered. Thus, ΔT_{ji} is calculated as follows:

- Relay-Relay:

$$\Delta T_{ji} = T_j - T_i \tag{9.9}$$

- Relay-Fuse:

$$\Delta T_{ji} = 0.85T_j - Ti \tag{9.10}$$

- Fuse-Relay:

$$\Delta T_{ji} = T_j - 1.15T_i \tag{9.11}$$

- Fuse-Fuse:

$$\Delta T_{ji} = 0.85T_j - 1.15T_i \tag{9.12}$$

Now, just imagine the complexity when the ORC problem contains fuses, OCRs, and DOCRs where these relays could be electromechanical, solid-state, digital, or numerical, and their characteristic curves could be definite-time, definite-current, inverse-time, or a combination of them. This is one reason why ORC problems are so hard compared with many other optimization-based power system studies. Jalilzadeh Hamidi et al. (2019) suggested to use ITOCRs with multistage TCCCs where the fuse operation is emulated based on a model fitted by a logarithmic-based nonlinear regression.

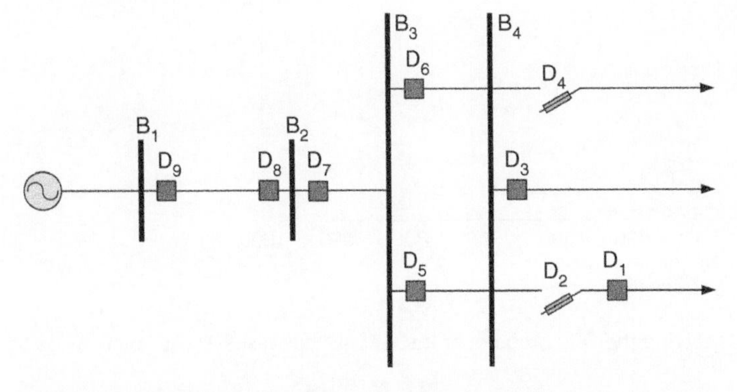

Figure 9.6 Coordinating protective devices in a simple 4-bus system.

9.3 Complex Networks

In Section 9.2, we have seen that the problem complexity exponentially increases as the network size increases regardless of whether the network contains fuses or not.

Example 9.4 Consider the IEEE 42-bus test system given in Appendix F. It is considered one of the largest ORC test systems available in the literature. This test system is protected by a mixture of 97 directional and non-directional OCRs. Thus, the dimension is 194 (i.e. 97 variables of type *TMS* and 97 variables of type *PS*) when only one unified TCCC is used for all the relays. Assume that the IEC standard inverse characteristic curve is used. Use the BBO algorithm programmed earlier to optimally coordinate these OCRs and DOCRs. The program should be initiated using the following

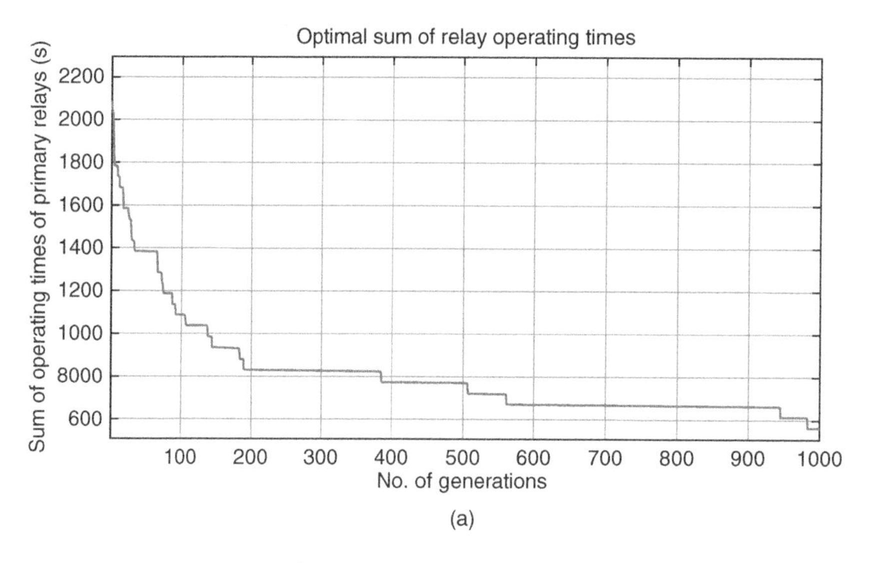

(a)

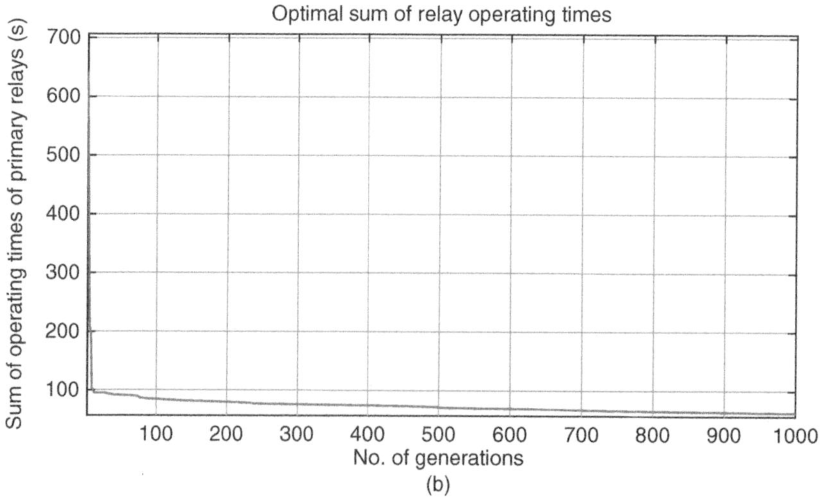

(b)

Figure 9.7 Fitness curves of BBO for solving the ORC problem of the IEEE 42-bus test system. (a) FC is deactivated. (b) FC is activated.

settings: $m_{max} = 0.1$, 1000 generations, 1 trial, population size of 60, and 4 elite solutions. Check the CPU time and the selectivity constraints when the feasibility checker (FC) is activated/deactivated.

Solution

The program files are placed in a folder called `BBO_42BusSystem_IECStadInv`. We can activate FC by setting the switch `ValidSW` to 1. The following results are obtained[6] by running `Starter.m` file:

- FC is not activated:

```
Fitness = 63.47901639 s
Elapsed Time = 28.060432 s
Number of Violations = 10 unselective P/B pairs
```

- FC is activated:

```
Fitness = 62.6059 s
Elapsed Time = 85.511550 s
Number of Violations = 0 unselective P/B pairs
```

Figure 9.7a shows the fitness curve of BBO when FC is deactivated, and vice versa for Figure 9.7b.

The last example shows the importance of the FC sub-algorithm in finding feasible solutions. However, it also reveals its main drawback, which is concentrated in consuming a significant amount of CPU time. Also, the MATLAB program shows the complexity of this ORC problem.

Problems

Written Exercises

W9.1 Solve the optimization model of Example 9.1 analytically.

W9.2 Solve the optimization model of Example 9.2 analytically.

W9.3 Consider the 6-bus system shown in Figure 6.21. Does it contain OCRs? If no why? If yes where?

W9.4 Consider the 3-bus system given in Example 9.3. Solving its ORC problem by LP gives the following optimal time multiplier settings:

$$TMS^* = [0.0589, 0.0250, 0.0250, 0.0290, 0.0629, 0.0250]$$

while the optimal settings obtained by GA in Bedekar and Bhide (2011b) is:

$$TMS^* = [0.0765, 0.0340, 0.0339, 0.0360, 0.0711, 0.0294]$$

Calculate the fitness of each optimal setting. Which is better?

6 Please, note that the processing speed depends on the computing machine used in the simulation. Also, because the program is based on a meta-heuristic optimization algorithm, so the results will change in every run.

W9.5 Again, consider the 3-bus system given in Example 9.3. Assume that all the European relays are numerical and can be programmed by any IEC or IEEE TCCC. If the IEEE moderately inverse characteristic curve is used for all the relays, express the ORC model mathematically.

W9.6 Consider the 3-bus system shown in Figure 9.8. There are two parallel lines between bus 1 and bus 2, and two additional faults at load 1 and load 2 with different numbering. The P/B relay pairs and the total fault currents,[7] of all the eight relays, are tabulated in Table 9.8. The other important thing about this test system is that it is a linearized optimization problem with $PS = 1$ for all the eight relays (i.e. $PS_i = 1\ \forall\ i \in [1,8]$). Add to that, all these eight relays have the same CR-ratio of 100:1 (i.e. $CTR_i = 100{:}1\ \forall\ i \in [1,8]$). The relay types, technology, IEEE model, and CT-ratio are listed in Table 9.9. All the relays are manufactured in Europe. They are numerical and can be programmed by any IEC or IEEE TCCC. The minimum allowable operating time of all the relays is $T^{\min} = 0.1$ s. The time multiplier setting is bounded between 0.025 and 1.2. The coordination time interval (or time delay) between each P/B relay pair is $CTI = 0.3$ s. The coefficients of all the P/B relay pairs are tabulated in Table 9.10. Express the linearized ORC problem using the same objective adopted in Example 9.3.

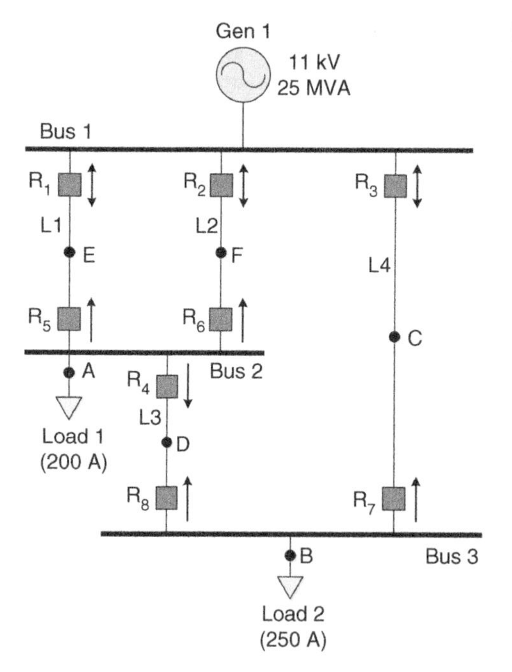

Figure 9.8 Single-line diagram of the 3-bus system used in Exercise W9.6.

Computer Exercises

C9.1 Consider the linearized mathematical expression given in Example 9.1. Design a MATLAB program to solve this problem numerically via LP. Use the `dual-simplex` algorithm.

7 The term "the total fault currents" means the summation of short-circuit currents seen by corresponding primary relays of each fault location before being stepped down through their CTs. For example, Table 9.10 shows the total current seen by the primary relays for the fault location A (i.e. R_1, R_2, and R_8) is: $I_{R_1} + I_{R_2} + I_{R_8} = 10 + 10 + 3.3 = 23.3$ A (after being stepped down) or $23.3 \times CTR(= 100) = 2330$ A (before being stepped down).

Table 9.8 The total fault currents and the P/B relay pairs of the 3-bus system given in Exercise W9.6.

Fault Point	Total Fault current (A)	Primary Relay R_i	Backup Relay R_j
A	2330	R_1	—
		R_2	—
		R_8	R_3
B	1200	R_3	—
		R_4	R_1, R_2
C	1400	R_3	—
		R_7	R_4
D	1400	R_4	R_1, R_2
		R_8	R_3
E	2800	R_1	—
		R_5	R_2, R_8
F	2800	R_2	—
		R_6	R_1, R_8

Table 9.9 Relay types, technology, IEEE model, and CT-ratio of the 3-bus system given in Exercise W9.6.

Relay No.	Type	Technology	Operating Time (s)	CT-Ratio
R_1	MI OCR	Numerical	$TMS \times \left[\frac{0.0515}{PSM^{0.02}-1}+0.114\right]$	100:1
R_2	MI OCR	Numerical	$TMS \times \left[\frac{0.0515}{PSM^{0.02}-1}+0.114\right]$	100:1
R_3	MI OCR	Numerical	$TMS \times \left[\frac{0.0515}{PSM^{0.02}-1}+0.114\right]$	100:1
R_4	MI DOCR	Numerical	$TMS \times \left[\frac{0.0515}{PSM^{0.02}-1}+0.114\right]$	100:1
R_5	MI DOCR	Numerical	$TMS \times \left[\frac{0.0515}{PSM^{0.02}-1}+0.114\right]$	100:1
R_6	MI DOCR	Numerical	$TMS \times \left[\frac{0.0515}{PSM^{0.02}-1}+0.114\right]$	100:1
R_7	MI DOCR	Numerical	$TMS \times \left[\frac{0.0515}{PSM^{0.02}-1}+0.114\right]$	100:1
R_8	MI DOCR	Numerical	$TMS \times \left[\frac{0.0515}{PSM^{0.02}-1}+0.114\right]$	100:1

C9.2 Consider the linearized mathematical expression given in Example 9.2. Design a MATLAB program to solve this problem numerically via LP. Use the `interior-point` algorithm.

C9.3 Consider the linearized mathematical expression given in Example 9.3. Design a MATLAB program to solve this problem numerically via LP. Use the `interior-point-legacy` algorithm.

Table 9.10 The ψ coefficients considered for the eight OCRs/DOCRs of the 3-bus system given in Exercise W9.6.

Fault Point		Relay							
		R$_1$	R$_2$	R$_3$	R$_4$	R$_5$	R$_6$	R$_7$	R$_8$
A	I_{R_i}	10.0	10.0	3.30	—	—	—	—	3.30
	ψ_i	1.2068	1.2068	2.2451	—	—	—	—	2.2451
B	I_{R_i}	3.45	3.45	5.10	6.90	—	—	—	—
	ψ_i	2.1677	2.1677	1.6689	1.4216	—	—	—	—
C	I_{R_i}	2.00	2.00	10.0	4.00	—	—	4.00	—
	ψ_i	3.8032	3.8032	1.2068	1.9458	—	—	1.9458	—
D	I_{R_i}	5.00	5.00	4.00	10.0	—	—	—	4.00
	ψ_i	1.6883	1.6883	1.9458	1.2068	—	—	—	1.9458
E	I_{R_i}	20.0	6.00	2.00	—	8.00	—	—	2.00
	ψ_i	0.9481	1.5255	3.8032	—	1.3267	—	—	3.8032
F	I_{R_i}	6.00	20.0	2.00	—	—	8.00	—	2.00
	ψ_i	1.5255	0.9481	3.8032	—	—	1.3267	—	3.8032
$\sum_{k=A}^{F} \psi_i^k$		11.3396	11.3396	14.6730	4.5742	1.3267	1.3267	1.9458	11.7973

C9.4 Design a MATLAB program to optimize the mathematical model developed in Exercise W9.5.

C9.5 Design a MATLAB program to optimize the mathematical model developed in Exercise W9.6.

C9.6 Modify the MATLAB program given in Example 9.4 if all OCRs and DOCRs are European and have the IEC long-time inverse characteristic curve.

C9.7 If all the relays are manufactured in North America, repeat Exercise C9.6 with using the IEEE moderately inverse characteristic curve. Use $TDS \in [0.5, 12]$, $CTS \in [1.5, 5]$, and $\Delta CTS = 0.25$.

C9.8 Repeat Exercise C9.6, but with using the BNP-EDF characteristic curve.

C9.9 In Example 9.4, we have seen that the main disadvantage of using FC is the high CPU time. The goal here is to solve this time-consuming optimization algorithm within just one second or less! Modify the MATLAB programs given in Example 8.4 and Example 9.4 by embedding LP to act as the one described in Figure 8.2. The LP sub-algorithm should be reconfigurable to the following:

1. Accepts the vector of *TMS* supplied by BBO as an initial point. (**Hint:** Use `optimset` with `LargeScale` option.)

```
1    options=optimset('LargeScale', 'on','Display','off');
2    [Ti,fi]=linprog(f,MAT,b,[],[],lb,ub,x0,options);
```

2. Fine-tunes the vector of *PS* without an initial point. (**Hint:** Use `optimoptions` with any LP algorithm; such as `dual-simplex`, `interior-point`, or `interior-point-legacy`.)

```
1    options=optimoptions('linprog', 'Algorithm', ...
         'interior-point-legacy', 'display', 'off');
2    [Ti,fi]=linprog(f,MAT,b,[],[],lb,ub,[],options);
```

10

Optimal Coordination with Considering Multiple Characteristic Curves

After reading this chapter you should be able to:

- Solve ORC problems when ITOCRs have different TCCCs
- Apply this approach when the network contains only DOCRs or both OCRs/DOCRs
- Apply this approach when the ORC problem is linearized
- Know the inherent weaknesses and technical issues of this approach

In Chapters 7 and 8, it has been assumed that all the relays have one unified TCCC. The questions that might be raised here are: *What if each relay has a characteristic curve that not necessarily be similar to others? Can we enhance the solution quality by using multiple TCCCs? What is about the problem complexity? Is this approach appropriate for adaptive relay coordination?* This chapter tries to answer these essential questions.

10.1 Introduction

In Chapter 9, we have seen how to coordinate DOCRs and OCRs/DOCRs with/without fuses. By comparing Figure 9.3 with Figure 9.8, the complexity significantly increases by adding just one line in parallel to that connected between bus 1 and bus 2. In some applications, we could see different relay technologies with different TCCCs. Taking into account these considerations will definitely increase the complexity of the ORC problem.

Example 10.1 Consider the system shown in Figure 9.8, which represents a little hard network. Do a literature survey to show that the complexity of this ORC problem can be more if its relays are modeled using different characteristic curves and technologies.

Solution
Bedekar and Bhide (2011a) successfully solved the ORC problem of the network shown in Figure 10.1, which has the same topology of the one shown in Figure 9.8 with some little differences. Although the system contains only 3 busbars, the power is distributed from bus 1 to bus 2 through double lines. Thus, 8 relays are used instead of 6; as faced in Example 9.3. The P/B relay pairs are tabulated in Table 10.1. Also, these overcurrent relays are a combination

Optimal Coordination of Power Protective Devices with Illustrative Examples, First Edition. Ali R. Al-Roomi.

Companion website: www.wiley.com/go/al-roomi/optimalcoordination

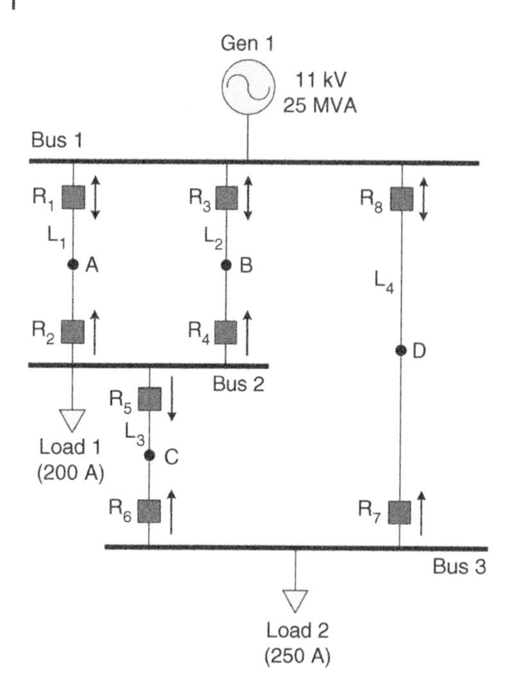

Figure 10.1 Single-line diagram of a 3-bus system containing different types and technologies of OCRs and DOCRs.

Table 10.1 Primary/backup relay pairs of the 3-bus system shown in Figure 10.1.

Fault point	Primary relay R_i	Backup relay R_i
A	R_1	—[a)]
	R_2	R_3, R_6
B	R_3	—
	R_4	R_1, R_6
C	R_5	R_1, R_3
	R_6	R_8
D	R_7	R_5
	R_8	—

a) "—" means there is no backup relay.

of (OCRs and DOCRS), (numerical and electromechanical), and (instantaneous "or definite-current," definite-time, and inverse-time TCCCs), which in turn increases the overall complexity. This special ORC problem requires a special programming code to solve it numerically. Table 10.2 lists the type, technology, IEC model, and CT-ratio of each relay.

The following sections show how to optimally coordinate DOCRs and OCRs/DOCRs with multiple inverse-time TCCCs. Knowing this concept will help the reader to understand how to embed

Table 10.2 Relay type, technology, IEC model, and CT-ratio of each relay shown in Figure 10.1.

Relay No.	Type	Technology	Operating time (s)	CT-Ratio
R_1	Standard inverse OCR	Numerical	$\frac{0.14\ TMS}{PSM^{0.02}-1}$	200:1
R_2	Instantaneous DOCR	Numerical	0.08	100:1
R_3	Standard inverse OCR	Numerical	$\frac{0.14\ TMS}{PSM^{0.02}-1}$	200:1
R_4	Instantaneous DOCR	Numerical	0.08	100:1
R_5	Very inverse time DOCR	Numerical	$\frac{13.5\ TMS}{PSM-1}$	300:1
R_6	Definite time DOCR	Numerical	0.28	100:1
R_7	Extremely inverse time DOCR	Numerical	$\frac{80\ TMS}{PSM^2-1}$	100:1
R_8	Standard inverse OCR	Electromechanical	$\frac{0.14\ TMS}{PSM^{0.02}-1}$	300:1

any other TCCCs. This includes all the standard TCCCs covered in Chapter 3, IOCRs/DCOCRs, DTOCRs, mixed TCCCs,[1] and user-defined TCCCs.

10.2 Optimal Coordination of DOCRs with Multiple TCCCs

If each DOCR has its own TCCC, then different curves have to be used. This means that the relay parameters $\{\alpha, \beta, \gamma\}$ are not constant anymore. Thus, instead of dealing with just two decision variables in each DOCR (i.e. *PS* and *TMS*), they will become five, which in turn increases the ORC problem dimension by 250%.

Based on the relay technology,[2] TCCC could be restricted to some European and North American standards, or even user-defined curves by optimizing the parameters $\{\alpha, \beta, \gamma\}$ or the model itself.[3] To ensure having feasible solutions, five main corrections have to be applied to optimization algorithms, which are:

- If someone wants to apply the preceding hybrid optimization algorithms shown in Figures 8.2, 8.3, and 8.5, then these three parameters have to be set as constants in the LP sub-algorithm(s). This can be easily done by fixing the latest optimal values of $\{\alpha, \beta, \gamma\}$ that were detected by BBO or any other EA.
- In the EA sub-algorithm, the problem dimension (n) can be automatically calculated based on the number of branches being protected by DOCRs. This can be done by modifying (7.1) as follows:

$$n = 5\varrho = 10ß \tag{10.1}$$

However, if standard TCCCs are used, then their parameters $\{\alpha, \beta, \gamma\}$ are linked together. Thus, instead of using three variables of scalar type, it is possible to use only one variable of vector type. For this case, n can be effectively reduced down to:

$$n = 3\varrho = 6ß \tag{10.2}$$

1 Please refer to Chapter 2 for more details.
2 As said before, it can be electromechanical, solid-state, digital "hardware-based," or numerical.
3 Please refer to Chapter 3 for more details.

The other option is to use one scalar discrete variable. This variable should act as an index of a pre-defined matrix with a dimension of $[ind \times 3]$ where ind is the number of standard TCCCs defined in the program and the three columns represent the parameters $\{\alpha, \beta, \gamma\}$. To clarify it more, let's consider all TCCCs tabulated in Table 3.10 except the last one.[4] If $ind = 10$, then the IEEE moderately inverse TCCC should be used; and so on. To summarize these three options, we call them: (i) *three-scalar-variables approach*, (ii) *one-vector-variable approach*, and (iii) *one-scalar-variable approach*.

- Because we have programmable and non-programmable relays, so $\{\alpha, \beta, \gamma\}$ could be fixed vectors for some standard TCCCs or could be defined by users. Thus, there are two possibilities during solving ORC problems using multiple TCCCs:
  1. **All the Relays Are Non-programmable:**
     The one-scalar-variable approach is the best choice. In MATLAB, the lower bound of ind must equal to 1, and the upper bound must equal the number of standard TCCCs given in the ORC problem. For example, if RI and U2 are excepted from Table 3.10, then the upper bound of ind will equal 27. Thus, $ind \in [1, 27]$, which represents the index of that reduced table. In some numerical programming languages, such as Python, the indexing start from 0 instead of 1, and thus $ind \in [0, 26]$.
  2. **All or Some Relays Are Programmable:**
     The three-scalar-variables approach is the best choice for user-defined $\{\alpha, \beta, \gamma\}$. The lower and upper bounds can be set by user. As a quick adjustment, we can set these side constraints based on the corresponding minimum and maximum values shown in Table 3.10. Thus, the default side constraints can be set as follows:
     - If RI is not considered (i.e. $\xi = 1$): $\alpha \in [0.02, 5.6]$, $\beta \in [0.00\ 342, 45\ 900]$, and $\gamma \in [0, 1.967]$.
     - If RI is considered (i.e. $\xi \neq 1$): $\alpha \in [-1, 5.6]$, $\beta \in [-4.2373, 45\ 900]$, $\gamma \in [0, 1.967]$, and $\xi \in [1, 1.43\ 644]$.
     - If RECT, EDF, and RI are not considered: $\alpha \in [0.02, 2.5]$, $\beta \in [0.00\ 342, 315.2]$, and $\gamma \in [0, 1.967]$.

     We can also set the default side constraints based on only European TCCCs, North American TCCCs, IEC TCCCs, IEEE TCCCs, IEC/IEEE TCCCs, etc.
- The lower and upper bounds of each TMS_i or TDS_i depend on the standard TCCC adopted for the ith relay. For the European standard TCCCs, TMS_i should be used with a typical side constraint of $TMS_i \in [0.1, 1.1]$ unless a specific relay model is given. For the North American standard TCCCs, TDS_i should be used with a typical side constraint of $TDS_i \in [0.5, 11]$ unless a specific relay model is given. Thus, to match the IEC/BS and the ANSI/IEEE standard inverse-time TCCC models in the same optimizer, we have two possible options. The first one is to use the fixed divisor approach given in (3.40)–(3.42) and the other one is to use the linear interpolation approach given in (3.43)–(3.50). Here, we have four possible scenarios for non-programmable numerical relays:
  1. If all ϱ DOCRs are modeled based on only TMS, then we have to use (3.36) directly with $TDM = TMS$.
  2. If all ϱ DOCRs are modeled based on only TDS, then we have to use (3.36) directly with $TDM = TDS$.
  3. If DOCRs are modeled based on both TMS and TDS and the random process of the optimization algorithm is based on the lower and upper bounds of TMS, then we need to transform TMS to TDS for North American TCCC models. Thus, we have to use (3.42) if the fixed divisor

4 i.e. the RI characteristic curve is not considered. Thus, no need to use ξ since all TCCCs have $\xi = 1$.

approach is selected, or we have to substitute (3.50) into (3.36) for *TDM* if the linear interpolation approach is selected.

4. If DOCRs are modeled based on both *TMS* and *TDS* and the random process of the optimization algorithm is based on the lower and upper bounds of *TDS*, then we need to transform *TDS* to *TMS* for European TCCC models. Thus, we have to use (3.41) if the fixed divisor approach is selected, or we have to substitute (3.49) into (3.36) for *TDM* if the linear interpolation approach is selected.

For programmable numerical relays, the operation is so smooth. We can use *TDS* with European standard TCCC models and *TMS* with North American Models. The flowchart shown in Figure 10.2 describes the steps required to model TCCCs in ORC solvers. Moreover, with

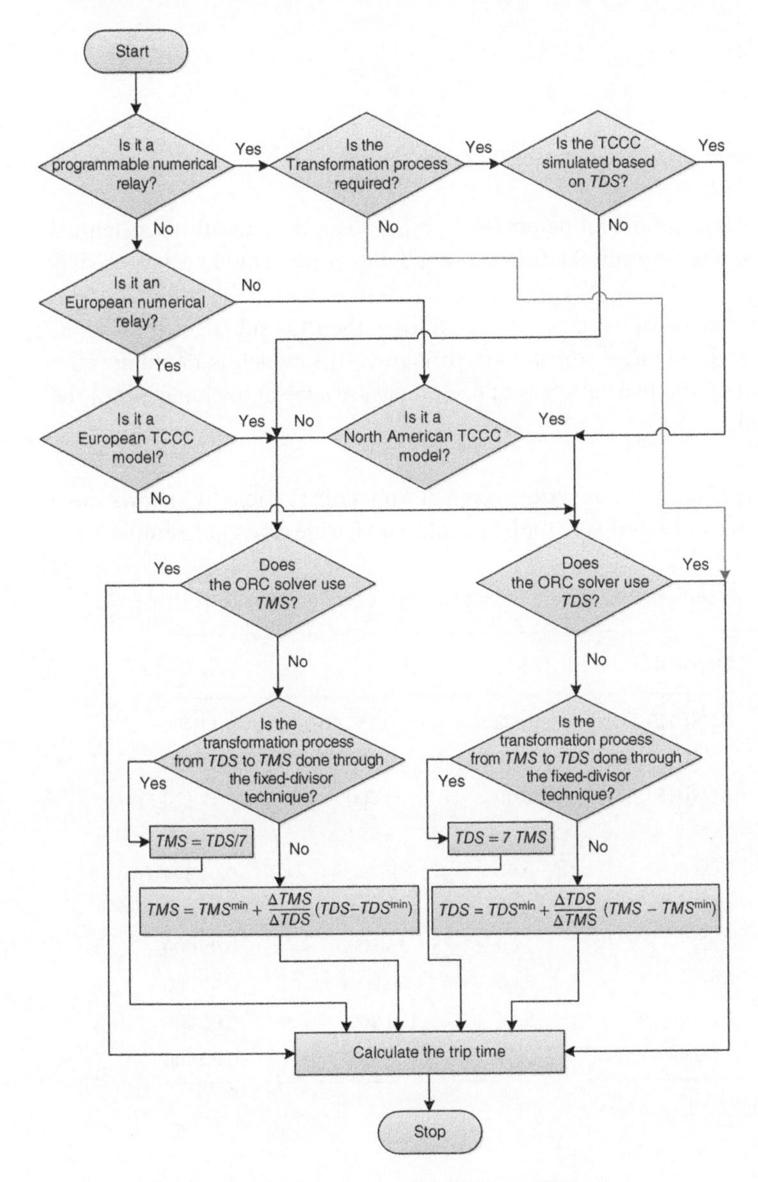

Figure 10.2 The steps required to model TCCCs in ORC solvers.

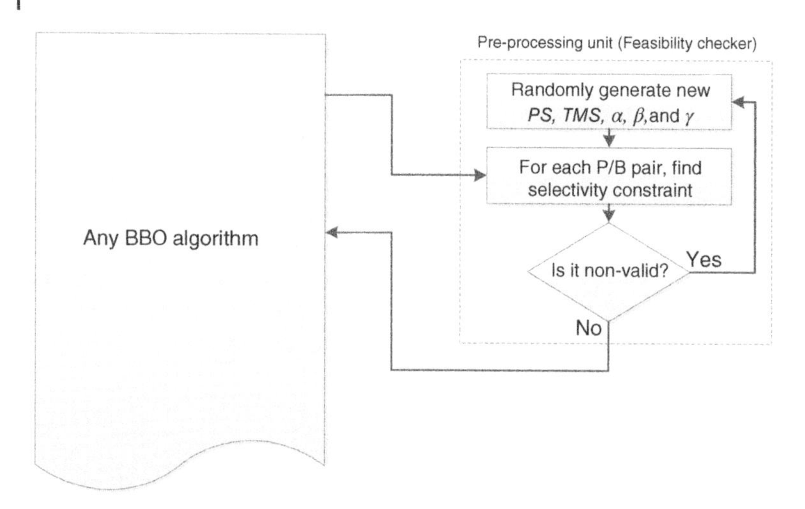

Figure 10.3 Mechanism of FC within BBO when different TCCCs are considered.

programmable numerical relays, the model parameters $\{\alpha, \beta, \gamma\}$ can be set with user-defined values.[5] Furthermore, the relation between the trip time and fault current could be governed by user-defined models.

- Since the parameters $\{\alpha, \beta, \gamma\}$ are considered as new variables in the BBO sub-algorithm, so FC should be modified to deal with this new adjustment. In Figure 10.3, which is modified from Figure 7.11, it is obvious that all the five variables of each relay are subject to change until the i-jth P/B relay pair is satisfied.

Example 10.2 Consider the IEEE 15-bus test system given in Appendix F. Table 10.3 shows some standard TCCCs of those available in Table 3.10. Modify the MATLAB code given in Example 7.5 to

Table 10.3 Standard values of inverse-time overcurrent relay constants used in Example 10.2.

Type of curve	Standard	α	β	γ
Moderately inverse	ANSI/IEEE	0.02	0.05 150	0.11 400
Very inverse	ANSI/IEEE	2.00	19.6100	0.49 100
Extremely inverse	ANSI/IEEE	2.00	28.2000	0.12 170
Inverse	CO8	2.00	5.95 000	0.18 000
Short-time inverse	CO2	0.02	0.02 394	0.01 694
Standard inverse (IDMT[a])	IEC	0.02	0.14 000	0.00 000
Very inverse	IEC	1.00	13.5000	0.00 000
Extremely inverse	IEC	2.00	80.0000	0.00 000
Long-time inverse	AREVA / UK	1.00	120.000	0.00 000
Short-time inverse	AREVA	0.04	0.05 000	0.00 000

a) IDMT: Inverse Definite Minimum Time

5 We can even set ξ as a user-defined value; see the parameters of the RI characteristic curve given in Table 3.10.

solve the ORC problem of this network by considering all the TCCCs listed in Table 10.3. Assume that all the relays are programmable numerical where *TDS* can be used with European TCCCs and *TMS* can be used with North American TCCCs.

Solution

The MATLAB files of this modified BBO program are available in a folder called `BBO_15Bus_MultipleTCCCs`. After running the file `Starter.m`, Table 10.4 can be attained.[6]

Table 10.4 Optimal coordination of the 15-bus system with multiple TCCCs.

Primary R_i	DOCRs settings		TCCCs values			TCCCs standard names
	TMS	*PS*	α	β	γ	
R_1	0.182 885	1.5	2	80	0	IEC Extremely Inverse
R_2	0.232 583	2.5	2	5.95	0.18	CO8 Inverse
R_3	0.623 956	2.5	2	28.2	0.1217	IEEE Extremely Inverse
R_4	0.147 104	2.5	2	28.2	0.1217	IEEE Extremely Inverse
R_5	0.540 057	2.5	0.04	0.05	0	AREVA Short-Time Inverse
R_6	0.553 597	1.5	2	28.2	0.1217	IEEE Extremely Inverse
R_7	0.339 854	1.5	2	80	0	IEC Extremely Inverse
R_8	0.748 261	1.5	2	5.95	0.18	CO8 Inverse
R_9	0.503 833	2.5	2	5.95	0.18	CO8 Inverse
R_{10}	0.319 966	1.5	2	28.2	0.1217	IEEE Extremely Inverse
R_{11}	0.490 257	2.0	2	5.95	0.18	CO8 Inverse
R_{12}	0.518 064	2.5	2	5.95	0.18	CO8 Inverse
R_{13}	0.243 115	1.5	2	80	0	IEC Extremely Inverse
R_{14}	0.845 047	1.0	2	28.2	0.1217	IEEE Extremely Inverse
R_{15}	0.797 605	2.5	2	28.2	0.1217	IEEE Extremely Inverse
R_{16}	0.231 673	2.5	2	28.2	0.1217	IEEE Extremely Inverse
R_{17}	0.331 655	2.5	2	28.2	0.1217	IEEE Extremely Inverse
R_{18}	0.488 519	1.5	2	28.2	0.1217	IEEE Extremely Inverse
R_{19}	0.693 861	2.0	0.04	0.05	0	AREVA Short-Time Inverse
R_{20}	0.219 173	1.5	2	80	0	IEC Extremely Inverse
R_{21}	0.160 622	2.0	2	28.2	0.1217	IEEE Extremely Inverse
R_{22}	0.906 792	1.0	2	28.2	0.1217	IEEE Extremely Inverse
R_{23}	0.182 915	1.5	1	13.5	0	IEC Very Inverse
R_{24}	0.601 498	1.0	2	28.2	0.1217	IEEE Extremely Inverse
R_{25}	1.047 118	0.5	0.02	0.02 394	0.01 694	CO2 Short-Time Inverse
R_{26}	0.803 415	1.0	2	80	0	IEC Extremely Inverse
R_{27}	0.481 713	2.0	2	28.2	0.1217	IEEE Extremely Inverse

(Continued)

6 Please note that the results are not necessarily identical in each run because the algorithm has a probabilistic process, and thus there is no unique answer.

Table 10.4 (Continued)

Primary R_i	DOCRs settings		TCCCs values			TCCCs standard names
	TMS	*PS*	α	β	γ	
R_{28}	1.008 869	2.5	2	5.95	0.18	CO8 Inverse
R_{29}	0.426 727	2.5	2	5.95	0.18	CO8 Inverse
R_{30}	0.479 480	1.0	2	80	0	IEC Extremely Inverse
R_{31}	0.786 416	1.5	2	80	0	IEC Extremely Inverse
R_{32}	0.806 506	2.0	2	5.95	0.18	CO8 Inverse
R_{33}	0.421 003	2.0	2	28.2	0.1217	IEEE Extremely Inverse
R_{34}	0.861 915	1.5	2	80	0	IEC Extremely Inverse
R_{35}	0.465 239	1.0	2	19.61	0.491	IEEE Very Inverse
R_{36}	0.254 528	2.0	2	19.61	0.491	IEEE Very Inverse
R_{37}	0.265 497	1.5	2	80	0	IEC Extremely Inverse
R_{38}	0.931 085	2.5	2	5.95	0.18	CO8 Inverse
R_{39}	0.773 385	0.5	1	13.5	0	IEC Very Inverse
R_{40}	0.550 922	1.5	2	28.2	0.1217	IEEE Extremely Inverse
R_{41}	0.384 480	1.5	2	80	0	IEC Extremely Inverse
R_{42}	0.445 843	2.5	0.04	0.05	0	AREVA Short-Time Inverse
Best (s)	7.621 750					

If we use the same population size and the number of generations reported in (Amraee, 2012), the program will take around 3.8 minutes compared with around 1 minute if only one unified TCCC is used for all the relays. However, the fitness can be significantly enhanced when multiple TCCCs are employed.

10.3 Optimal Coordination of OCRs/DOCRs with Multiple TCCCs

The same concept can be applied when a combination of directional and non-directional OCRs are involved. The only difference is in the way to define the selectivity constraints among primary and backup relay pairs. This has been discussed in Chapter 9 by covering different examples with a gradient of difficulties.

Example 10.3 Consider the IEEE 42-bus test system given in Appendix F, which represents a large network containing both OCRs and DOCRs. In Example 9.4, a BBO program was provided to solve this network via MATLAB. The mission now is to modify that program to accept all the characteristic curves listed in Table 10.3. Assume programmable numerical relays where *TDS* can be used with European TCCCs and *TMS* can be used with North American TCCCs.

Solution

The MATLAB files of this modified BBO program are available in a folder called `BBO_42Bus_MultipleTCCCs`. The program files are placed in a folder called `BBO_42BusSystem_IEC`

`StadInv`. We can activate FC by setting the switch `ValidSW` to 1. The following results are obtained by running `Starter.m` file:

- FC is not activated:

```
Fitness = 26.5773855 s
Elapsed Time = 1191.951870 s
Number of Violations = 23 unselective P/B pairs
```

- FC is activated:

```
Fitness = 38.9723 s
Elapsed Time = 1355.657432 s
Number of Violations = 0 unselective P/B pairs
```

We can see that the feasible fitness can be reduced from 62.61 to 38.97 s if multiple TCCCs are used. However, the processing time increased from 85.51 s (which is unacceptable) to 1355.66 s (which is highly unacceptable). Also, if FC is not activated, then 13 additional violations will be faced, which is an indication that the ORC problem becomes much harder if multiple TCCCs are considered.

10.4 Inherent Weaknesses of the Multi-TCCCs Approach

Based on some recent investigations about the performance of this approach, it has been found that there are many unforgiven weaknesses and technical issues that make this approach impractical (Al-Roomi and El-Hawary, 2017a). We have discussed some of them in Examples 10.2 and 10.3. These complications can be summarized as follows:

- The problem dimension is higher than the normal one by 150% to 250%[7].
- As a consequence of that increase, the algorithm becomes slower.
- Based on the last point, this ORC solver is inappropriate for adaptive coordination schemes.
- Because of that increase in the dimension, the algorithm requires more generations and/or population size to explore the search space for the optimal or near optimal solutions.
- To deal with these additional variables, the algorithm requires many modifications in its main structure.
- Moreover, using many TCCCs could satisfy all the design constraints, but they might also force the optimizer to settle on unselective settings. This phenomenon could happen based on the fact that satisfying optimal coordination at the near-end 3ϕ fault could not be correct for other fault locations. That is, because different TCCCs are imposed in the model, so the smallest *CTI* does not always equal the value measured at the highest fault current. This can be illustratively described in Figure 10.4. Thus, dealing with many TCCCs means that the selectivity constraint between each P/B relay pair must be satisfied by considering a gradient of fault locations to cover all the length of each branch from the near-end to the far-end bus. Therefore, continuing to use the multiple TCCCs approach means that the most exhausting objective function given in (7.5) must be used again to prevent having unselective operation at some spots.

7 It is 250% if each one of the parameters $\{\alpha, \beta, \gamma\}$ occupies one decision element in the optimization algorithm (i.e. when the three-scalar-variables approach is used), and that dimension can be reduced down to just 150% if the preceding three parameters are optimized together as a vector of length 3 (i.e. using the one-vector-variable approach) or as a scalar variable by liking them via their table index (i.e. using the one-scalar-variable approach).

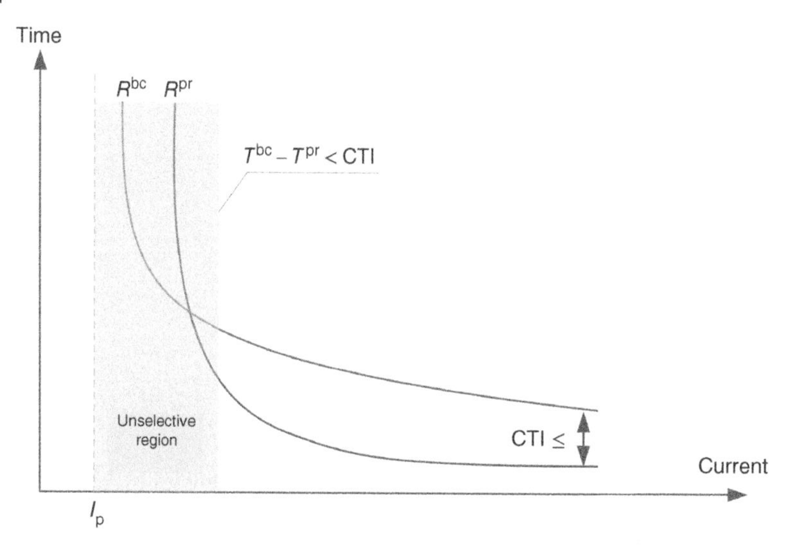

Figure 10.4 Unselective region that could be faced with multiple TCCCs approach.

Problems

Written Exercises

W10.1 Suppose that a network is protected by DOCRs. How many DOCRs does it have if there are 35 branches? What is the dimension of the ORC solver if the one-scalar-variable approach is implemented to optimally coordinate these DOCRs by using multiple TCCCs?

W10.2 A protection engineer wants to protect a large distribution network by a population-based evolutionary algorithm. After conducting a comprehensive study, it has been found that the cheapest way is to use a combination of inverse-time OCRs and DOCRs. What is the dimension of the optimizer if the engineer is thinking to employ more than one TCCC to optimally coordinate 350 relays using the three-scalar-variables approach? Without AI, is this optimizer appropriate to act as an adaptive ORC solver?

W10.3 Comparing the three-scalar-variables approach and the one-scalar-variable approach, which one should be used for user-defined $\{\alpha, \beta, \gamma\}$ and which one should be used for standard $\{\alpha, \beta, \gamma\}$?

W10.4 Referring to Table 3.10, find the side constraints of $\{\alpha, \beta, \gamma\}$ if only US and CO standards are used in an ORC problem modeled based on multiple TCCCs.

W10.5 Consider the flowchart shown in Figure 10.2. A meta-heuristic optimizer is designed based on the typical side constraints of *TMS*. If possible, you want to use *TMS* with one of the IEEE TCCC models directly without transforming it to *TDS*. What should you do if you have:

1. Programmable numerical relays manufactured in Europe?
2. Non-programmable numerical relays manufactured in North America?

If the transformation is required, use the linear-interpolation technique.

W10.6 Consider the feasibility checker shown in Figure 10.3. Its mechanism seems like that of RSA explained in Algorithm 1. Is it slow like RSA? Explain.

W10.7 In Examples 10.2 and 10.3, we have seen that when the ORC solver is designed based on multiple TCCCs it takes a longer time to reach the optimal solution. By referring to Chapters 7, 8, and 9, suggest one way to accelerate its processing speed. Does it solve all the inherent weaknesses and technical issues associated with multiple TCCCs?

Computer Exercises

C10.1 Modify the MATLAB program given in Example 10.2 to optimally coordinate the 42 DOCRs based on the following groups of Table 10.3:

- Only the North American standard TCCCs
- Only the European standard TCCCs
- Only the IEEE and IEC standard TCCCs
- Only the IEEE standard TCCCs
- Only the IEC standard TCCCs
- Only manufacturers' standard TCCCs

C10.2 Suppose that the model parameters $\{\alpha, \beta, \gamma\}$ should be optimally defined by user. Modify the preceding MATLAB program to deal with these new continuous variables. Take the following side constraints: $\alpha \in [0.02, 2]$, $\beta \in [0.05, 120]$, and $\gamma \in [0, 0.491]$.

C10.3 Repeat Exercise C10.2, but considering continuous *PS*.

C10.4 In Example 10.3, it has been seen that the processing speed is totally unacceptable. The mission now is to embed LP as a fine-tuning stage; similar to the one described in Figure 8.2 but with deactivating the initial guess of *TMS*. For the BBO sub-algorithm, use a population size of 20, 4 elites, 30 iterations, and $m_{\max} = 0.1$. For the LP sub-algorithm, use `interior-point-legacy`. Compare the result of this new optimizer with that obtained in Example 10.3.

C10.5 Modify the MATLAB program given in Example 10.3 by taking $\{\alpha, \beta, \gamma\}$ as user-defined variables where their lower and upper bounds are: $\alpha \in [0.02, 2]$, $\beta \in [0.05, 120]$, and $\gamma \in [0, 0.491]$.

C10.6 Repeat Exercise C10.5, but with considering BBO-LP developed in Exercise C10.4. For BBO, use a population size of 20, 2 elites, 200 iterations, and $m_{\max} = 0.1$. For LP, use `interior-point-legacy`. Compare the result of this new optimizer with that obtained in Exercise C10.5.

11

Optimal Coordination with Considering the Best TCCC

After reading this chapter you should be able to:

- Solve the technical issues associated with the multi-TCCCs approach
- Optimally coordinate ITOCRs by searching for the best standard or user-defined TCCC
- Express the ORC optimizer using three different mechanisms
- Know one of the main technical issues faced in distribution systems where the short-circuit currents of P/B relay pairs are not monotonically change

It has been seen in Chapter 10 how to implement multiple TCCCs in solving ORC problems. Although that approach can effectively minimize the objective function, it has many unforgiven weaknesses that make it worthless; unless many weary adjustments are applied to the main structure of the optimizer. Instead, dealing with only one TCCC unified to all ϱ ITOCRs makes the mission much simpler than before. For one TCCC approach, most of the papers presented in the literature use the International Electrotechnical Commission (IEC) standard inverse TCCC. Unfortunately, the other European and North American standards get less attention, although they might provide better solutions. Of course, there are some recent trials, such as the studies reported in Albasri et al. (2015), Jalilzadeh Hamidi et al. (2019), Korashy et al. (2020), and Alam et al. (2020). However, recent numerical relays have many undiscovered built-in time-current characteristic curves, such as those listed in Table 3.10. Also, for modern programmable numerical relays, the end-users can adjust their own TCCCs. This feature can increase the chance to get better solutions by finding just one TCCC unified to all the relays with satisfying both the selectivity and optimality criteria (Al-Roomi and El-Hawary, 2017a).

11.1 Introduction

The problem occurs when the margin between the operating times of the ith primary and jth backup relays is less than Coordination Time Interval CTI, which means that the selectivity constraint given in (7.25) is violated. Consider Figure 10.4. It shows that selectivity among the i-jth P/B relay pair is satisfied at the higher short-circuit currents. For low fault currents, the time interval becomes less than the minimum acceptable value (i.e. CTI). It might end up with a negative value, which means that the backup relay is faster than the primary relay.

Optimal Coordination of Power Protective Devices with Illustrative Examples, First Edition. Ali R. Al-Roomi.

Companion website: www.wiley.com/go/al-roomi/optimalcoordination

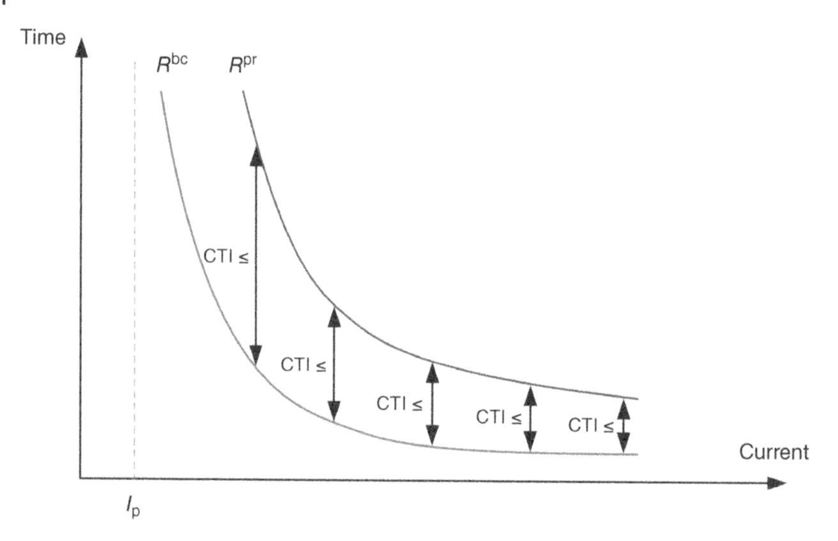

Figure 11.1 Solving selectivity issue if only one TCCC is unified to all ITOCRs.

The root problem comes from using incompatible TCCCs. One of the possible ways to solve this issue is by smartly using correct TCCCs where the selectivity constraint is always satisfied. This could be achieved by employing a combination of inverse-time, definite-time, and definite-current characteristic curves; similar to those covered in Chapter 2. The other one is by employing a multistage inverse-time TCCC, which is suggested in Jalilzadeh Hamidi et al. (2019).

However, the preceding approaches are not easy and create another wall of challenges, especially if we use meta-heuristic optimization algorithms where the goal is to get rid of all weary steps and complicated mathematical equations. Thus, the simplicity criterion is not satisfied. The other concern is about the processing speed of these optimization schemes. Dealing with many variables and lots of if-statements will definitely have a bad impact on the overall CPU time, which turns them into inapplicable/impractical ORC solvers for real-world problems where dynamic changes always exist.

The simpler way is to stick with the conventional approach where only one unified TCCC is used for all relays. This is valid for both standard and user-defined inverse-time TCCCs (Al-Roomi, 2020). If we apply this concept to the ith primary and jth backup relays, then the unselective region of Figure 10.4 will be solved as depicted in Figure 11.1.

11.2 Possible Structures of the Optimizer

In this book, two scenarios are considered. The first one focuses on searching for one unified standard TCCC, and the other scenario focuses on searching for one user-defined TCCC. The first scenario can be easily satisfied by selecting one standard TCCC at each simulation run and then displaying the best type. This technique is shown in Figure 11.2a. The second scenario can be satisfied by two different techniques:

1. Executing meta-heuristic optimization algorithms[1] many times with randomly generated $\{\alpha, \beta, \gamma\}$. Thus, the problem dimension given in (7.1) is kept without any change. But, this

1 In this chapter, we are using Biogeography-Based Optimization BBO/BBO-LP.

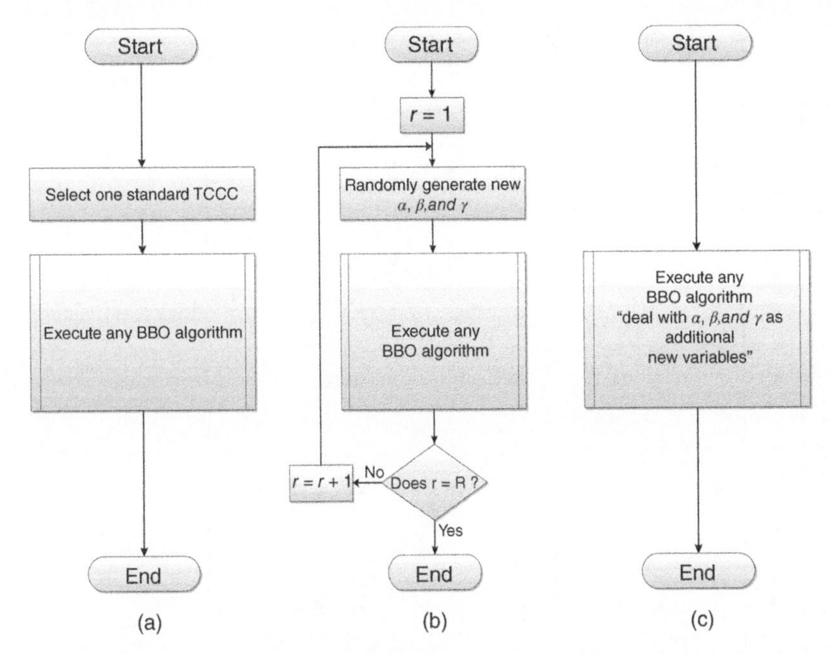

Figure 11.2 Using different techniques to change TCCC within meta-heuristic optimization algorithms. (a) The first technique, (b) the second technique, and (c) the third technique.

technique needs a specific number of simulation runs or trials to optimize the preceding parameters as shown in Figure 11.2b.

2. The other technique treats these parameters as three additional variables, so the problem dimension becomes:

$$n = 3 + 2\varrho = 3 + 4ß \tag{11.1}$$

Thus, the mission is given to the optimizer to find the best user-defined values of $\{\alpha, \beta, \gamma\}$. This technique is shown in Figure 11.2c. Although the problem dimension is slightly increased, the third technique is very smart and much faster than the second technique shown in Figure 11.2b.

Example 11.1 Consider the IEEE 15-bus test system given in Appendix F and the standard TCCCs given in Table 10.3 of Example 10.2. Modify the MATLAB code given in Example 7.5 to solve the ORC problem of this network by searching for the best standard TCCC. Use the simple structure given in Figure 11.2a. Two different scenarios should be taken in this experiment. The first scenario uses discrete *PS* to simulate the working principle of electromechanical relays, while the second scenario uses continuous *PS* as a one of many advantages of modern numerical relays. Both BBO and BBO-LP should be used to optimize $\sum T_i$ of all ϱ DOCRs for each scenario.

Solution

The MATLAB files of the modified BBO program are available in a folder called `BBO_15Bus_OptimalTCCC` and those of the modified BBO-LP program are available in a folder called `BBOLP_15Bus_OptimalTCCC`. After running the file `Starter.m` of each program in a computing machine similar to the one reported in Al-Roomi and El-Hawary (2017a), Table 11.1 is attained.

Table 11.1 Performance comparison between BBO and BBO-LP. population size and number of generations for BBO and BBO-LP are respectively ($N = 120$ and $G = 385$) and ($N = 20$ and $G = 30$).

Type of TCCC	Discrete plug setting (i.e. electromechanical DOCRs)						Continuous plug setting (i.e. numerical DOCRs)					
	BBO + FC			BBO-LP + FC			BBO + FC			BBO-LP + FC		
	Time	Violations	CPU	Time	Violations	CPU	Time	Violations	CPU	Time	Violations	CPU
IEEE Moderately Inverse	21.40	**0**	649.97	12.60	**0**	15.10	21.86	**0**	321.67	12.82	**0**	10.93
IEEE Very Inverse	7.86	**0**	435.64	5.02	**0**	12.86	8.32	**0**	201.20	5.17	**0**	9.72
IEEE Extremely Inverse	3.33	**0**	333.92	**2.28**	**0**	12.09	3.27	**0**	147.98	2.31	**0**	9.27
Inverse "CO8"	5.39	**0**	1031.71	4.47	**0**	23.60	5.68	**0**	623.02	4.95	**0**	18.80
Short-Time Inverse "CO2"	18.20	8	1060.25	10.95	**0**	20.71	18.82	11	559.25	12.01	**0**	14.55
IEC Standard Inverse (IDMT)	33.84	**0**	488.83	13.76	**0**	13.58	32.81	**0**	231.65	13.75	**0**	10.33
IEC Very Inverse	7.54	**0**	320.26	5.24	**0**	11.04	7.00	**0**	134.64	5.00	**0**	8.64
IEC Extremely Inverse	**2.09**	**0**	**273.89**	2.45	**0**	**10.56**	**2.01**	**0**	119.35	**2.26**	**0**	8.66
AREVA/UK Long-Time Inverse	32.11	**0**	284.83	32.22	**0**	10.75	39.15	**0**	**114.92**	33.32	**0**	**8.16**
AREVA Short-Time Inverse	15.60	9	1026.76	10.17	**0**	20.47	16.49	6	538.95	11.55	**0**	14.36

These results are extracted after 10 independent simulation runs.

Table 11.1 shows the results of BBO and BBO-LP algorithms when *PS* is treated as a discrete/continuous variable. From the first look, it can be clearly observed that the violations happen if short-time inverse TCCCs are used; for both European and North American standards. This infeasibility problem can be avoided if the hybrid BBO-LP is used. As the inversion of TCCC becomes slow, the algorithm requires more CPU time to satisfy the selectivity constraints between P/B relay pairs, and the obtained solutions become worse. In addition, it is obvious that when discrete *PS* is used, the feasibility checker (FC) requires more CPU time to find a feasible solution within the limited elements of the vector of *PS*. This phenomenon happens with both BBO and BBO-LP versions. However, most of the results obtained by the discrete mode of *PS* show better solution quality than that obtained by the continuous mode of *PS*, for both BBO and BBO-LP. The reason is that the continuous mode needs a large number of generations to tune the settings of DOCRs. In this contest, the IEC extremely inverse TCCC always scores the best position as compared with the others, except the first scenario with the LP stage where the IEEE extremely inverse TCCC is the winner.

In order to see the benefits of using the technique proposed in this study (i.e. using only one optimal TCCC), Table 11.2 shows the results obtained by a combination of different North American and European TCCCs[2] together in one solution. In this situation, $\{\alpha, \beta, \gamma\}$ are not constants anymore, and thus the original dimension increases by 250% if the three-scalar-variables approach is used. This means that the algorithm deals with 210 variables instead of 84. From Table 11.2, there are three problems; two of them can be clearly addressed while the third one is almost hidden. The first problem is about the complexity of the algorithms if they are built based on 210 variables instead of 84. The second problem is about the processing speed issue because with extra 126 variables the optimizer requires more CPU time and hence becomes very slow compared with itself when it is initiated by only 84 variables. The third problem is about the feasibility. Apparently, the solutions obtained by different TCCCs are feasible, but they could violate some selectivity constraints for some fault locations. That is, if multiple TCCCs are used to optimize the coordination problem of DOCRs, then the obtained solution is feasible under the given short-circuit currents calculated for the identified P/B relay pairs. As per the data given in Amraee (2012), the short-circuit analysis of this system was carried out under 3ϕ near-end faults on all the lines. Thus, the feasibility of the ORC problem, when it is solved using multiple TCCCs, may not hold for other fault locations, such as at the middle and far-end points as shown in Figure 10.4. Based on that, using only one optimal TCCC for all ϱ ITOCRs can satisfy the feasibility, optimality, simplicity, and processing speed performance criteria.

11.3 Technical Issue

The monotonic curve of the *j*th backup relay shown in Figure 11.1 is guaranteed in radial systems. The story is not always the same in mesh systems. There are multiple sources that feed multiple loads through interconnected branches. Thus, the short-circuit current seen by the *j*th backup relay is not always proportional to the location of the out-zone fault (Al-Roomi and El-Hawary, 2019b).

In Chapter 7, it has been seen that the selectivity constraint can be satisfied by giving enough time to each primary relay before initiating its backup relay(s). This checking process is done at some user-defined 3ϕ faults. The standard fault locations are shown in Figure 7.1 and listed in Table 7.1. It seems that everything works smoothly without any practical problem. That is, if the discrimination

2 The standard TCCCs listed in Table 10.3.

Table 11.2 Performance comparison between "DOCRs with multiple TCCCs" and "DOCRs with only the IEC standard inverse TCCC."

Comparison Criteria	IEC Standard Inverse			Standard TCCCs					
	Seeker (Amraee, 2012)	BBO	BBO-LP	BBO[1]	BBO[2]	BBO[3]	BBO[4]	BBO[5]	BBO[6]
Population Size	120								
TCCC Standards	Only IEC IDMT			[1]All	[2]N. American	[3]European	[4]IEEE/IEC	[5]IEEE	[6]IEC
PS Mode	Discrete								
Generation No.	385								
Variables No.	84			210					
Min. OBJ (s)	12.227	33.842	12.609	7.622	5.941	7.311	5.683	5.833	**5.260**

margin between each P/B relay pair is satisfied at the highest short-circuit current (i.e. 3ϕ fault in most cases (El-Hawary, 1983)), then that margin will definitely be larger for lesser currents as depicted in Figure 6.10. However, it has been found that some backup relays see non-monotonic changes in short-circuit currents when the bolted point is gradually shifted from the near-end point to the far-end point. This phenomenon reveals a very important fact that the existing techniques used in solving ORC problems do not assure the feasibility of their solutions. The following example covers this phenomenon through a numerical experiment.

Example 11.2 Consider the IEEE 8-bus test system given in Appendix F. Assume that different 3ϕ short-circuit faults are created between bus 1 and bus 6. For this faulty branch, the primary DOCR mounted on bus 1 is R_{14} and its backup DOCRs are R_1 and R_9. Similarly, R_7 is the primary DOCR of bus 6 and its backup DOCRs are R_5 and R_{13}. Table 11.3 shows the 3ϕ short-circuit currents fed to these six DOCRs and the operating times measured from them when the IEC standard inverse TCCC is used. The optimal settings of these two P/B relay pairs obtained by BBO-LP are tabulated in Table 11.4. Plot the operating times of these two P/B relay pairs.

Solution

The operating times of both P/B relay pairs, at different fault locations, are depicted in Figure 11.3. From both plots, it is very clear that the selectivity constraint is satisfied at the near-end fault of each primary relay (i.e. 1% and 99% away from bus 1 for R_{14} and R_7, respectively). It happens because this test system is built based on the minimum criterion. If some faults are created on different locations of that line, then there is no guarantee that the selectivity constraint will be satisfied. This can be observed in the preceding plots. For instance, the backup relay R_9 operates before the primary relay R_{14} at the far-end point where the fault is 99% away from bus 1 and 1% away from bus 6. Similarly, the backup relay R_{13} operates before R_7 at the far-end point where the fault is 1% away from bus 1 and 99% away from bus 6.

This numerical example can be generalized for other coordination criteria listed in Table 7.1. Even if all the three standard points (i.e. near-end, middle, and far-end points) are considered, the suspicion still exists. This can be easily proved by looking at the crossover between each P/B relay pair shown in Figure 11.3. For R_{14}, the crossover happens with R_9 at about 10% away from bus 6. For R_7, the crossover happens with R_{13} at about 20% away from bus 1. It is obvious that there are other

Table 11.3 The near-end 3ϕ fault currents and operating times of the P/B relay pairs of branch 1-6 of the IEEE 8-bus test system.

Fault Location k (%)	Primary Relays				Backup Relays							
	R_7		R_{14}		R_1		R_5		R_9		R_{13}	
	$I_{7,k}$ (A)	$T_{7,k}$ (s)	$I_{14,k}$ (A)	$T_{14,k}$ (s)	$I_{1,k}$ (A)	$T_{1,k}$ (s)	$I_{5,k}$ (A)	$T_{5,k}$ (s)	$I_{9,k}$ (A)	$T_{9,k}$ (s)	$I_{13,k}$ (A)	$T_{13,k}$ (s)
1	907.51	0.8475	1291.33	0.4948	443.60	0.8621	703.08	1.1136	824.74	0.7954	449.43	0.6099
5	933.78	0.8187	1255.12	0.5052	424.89	0.9092	709.50	1.0986	815.64	0.8049	431.67	0.6325
10	967.10	0.7859	1210.68	0.5191	402.41	0.9766	717.80	1.0799	804.68	0.8168	410.28	0.6636
15	1001.01	0.7561	1167.45	0.5339	381.00	1.0550	726.42	1.0614	794.19	0.8287	389.85	0.6980
20	1035.53	0.7289	1125.37	0.5497	360.72	1.1472	735.39	1.0431	784.14	0.8406	370.41	0.7361
25	1070.66	0.7040	1084.38	0.5666	341.65	1.2560	744.70	1.0249	774.51	0.8525	352.04	0.7784
30	1106.43	0.6810	1044.45	0.5848	323.89	1.3849	754.37	1.0069	765.26	0.8643	334.81	0.8251
35	1142.86	0.6598	1005.51	0.6044	307.57	1.5377	764.41	0.9891	756.37	0.8762	318.84	0.8763
40	1179.99	0.6401	967.56	0.6257	292.85	1.7172	774.84	0.9714	747.84	0.8880	304.26	0.9315
45	1217.84	0.6217	930.55	0.6488	279.91	1.9239	785.68	0.9540	739.62	0.8998	291.21	0.9900
50	1256.47	0.6046	894.47	0.6740	268.94	2.1530	796.93	0.9368	731.71	0.9116	279.88	1.0495
55	1295.90	0.5885	859.33	0.7016	260.16	2.3890	808.62	0.9198	724.09	0.9234	270.45	1.1070
60	1336.20	0.5734	825.11	0.7320	253.76	2.6030	820.77	0.9030	716.74	0.9351	263.13	1.1577
65	1377.41	0.5591	791.84	0.7656	249.94	2.7532	833.39	0.8865	709.65	0.9469	258.09	1.1963
70	1419.59	0.5457	759.56	0.8028	248.80	2.8019	846.51	0.8702	702.79	0.9587	255.50	1.2174
75	1462.80	0.5329	728.30	0.8442	250.43	2.7328	860.15	0.8541	696.17	0.9705	255.47	1.2177
80	1507.12	0.5207	698.14	0.8903	254.80	2.5653	874.35	0.8382	689.75	0.9823	258.07	1.1964
85	1552.64	0.5092	669.15	0.9419	261.85	2.3390	889.13	0.8225	683.54	0.9940	263.28	1.1566
90	1599.28	0.4981	641.52	0.9995	271.48	2.0944	904.45	0.8071	677.49	1.0059	271.10	1.1027
95	1647.32	0.4876	615.30	1.0637	283.47	1.8613	920.44	0.7919	671.64	1.0178	281.36	1.0412
99	1686.75	0.4795	595.48	1.1201	294.66	1.6924	933.72	0.7799	667.07	1.0273	291.25	0.9898

Table 11.4 Optimal settings of some relays of the IEEE 8-bus test system—near-end 3ϕ faults with the IEC standard inverse TCCC.

Setting	R_1	R_5	R_7	R_9	R_{13}	R_{14}
TMS	0.1040	0.1072	0.1	0.1086	0.1003	0.1
PS	0.8	1.5	2.5	2.0	0.6	2.0

doubtful points that need to be checked in addition to the preceding three standard points. Thus, satisfying any one, or even multiple, of coordination criteria listed in Table 7.1 could be insufficient to ensure the feasibility of the optimal solution.

The more complex, but safer, option is to use a gradient of fault points along each line to satisfy the selectivity constraint. That is, ($k = 1, 2, \cdots, \ell$), where ($\ell >> 2$) and covers all the spots of

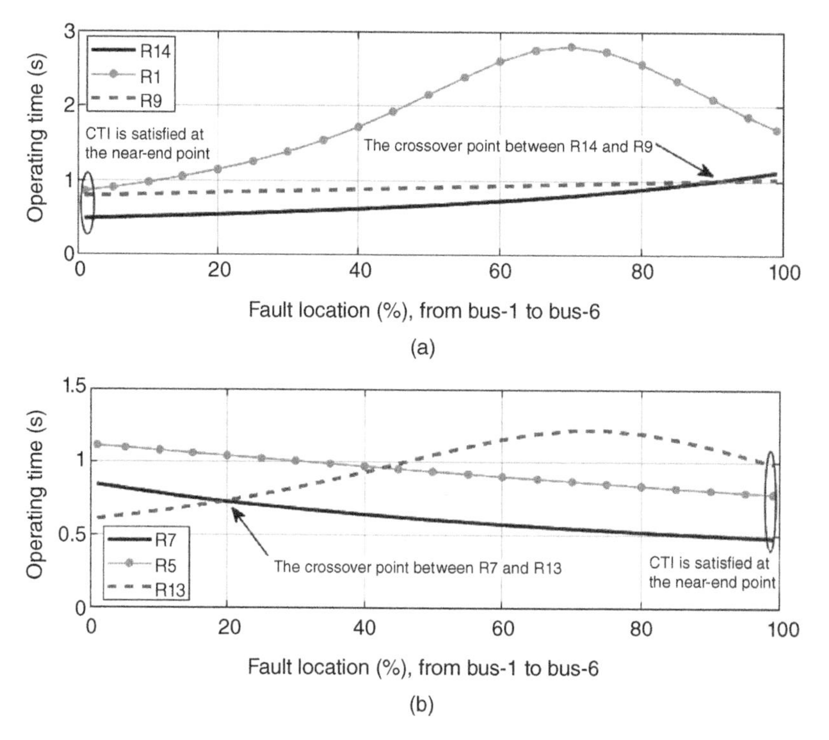

Figure 11.3 Infeasibility of discrimination margins of the P/B relay pairs at some fault locations between bus 1 and bus 6.

each faulty line. This means that the simplification applied to (7.5), regarding ℓ, is not valid anymore. Although this correction will make ORC problems harder than ever, the solutions obtained by this approach are more practical and thus the system reliability can be enhanced. The problem dimension and complexity will dramatically increase if multiple TCCCs are involved (Albasri et al., 2015). Also, if other protective relays and fuses are involved, then the new optimization model of ORC problems will need a very powerful algorithm to deal with this dilemma. Thus, artificial intelligence (AI) could be a must.

Problems

Written Exercises

W11.1 A mesh network consisted of 105 branches is protected by DOCRs. If all the relays have the same TCCC, find the following:

1. The number of relays mounted on the network
2. The dimension of the ORC problem when the 1st technique shown in Figure 11.2a is used
3. The dimension of the ORC problem when the 2nd technique shown in Figure 11.2b is used
4. The dimension of the ORC problem when the 3rd technique shown in Figure 11.2c is used

W11.2 Consider the three techniques shown in Figure 11.2. For a user-defined TCCC, which are the valid techniques? Why?

W11.3 If the preceding user-defined TCCC should be obtained for a dynamic network, then which is the best technique that should be used? Why?

W11.4 Consider the IEEE 42-bus test system where 97 directional and non-directional ITOCRs are used to protect the network. Suppose that the 2nd technique shown in Figure 11.2b is used. What is the dimension of the ORC problem if:

1. The multi-TCCCs approach is used
2. The optimal TCCC approach is used

W11.5 Repeat Exercise W11.4, but with considering 3rd technique shown in Figure 11.2c.

Computer Exercises

Example 11.1 contains four optimization algorithms. Two of them are based on BBO to optimally coordinate the relays using continuous and discrete *PS*. The other two programs do the same task, but they are reinforced by LP to resolve the processing time issue. All these four programs are built based on the 1st technique shown in Figure 11.2a. The following computer exercises rely on the correct program that should be picked up for the modification:

C11.1 Using the 2nd technique shown in Figure 11.2b, design a BBO program to solve the ORC problem of the IEEE 15-bus test system based on user-defined TCCC and continuous *PS*.

C11.2 Repeat Exercise C11.1, but with considering the 3rd technique shown in Figure 11.2c.

C11.3 Using the 2nd technique shown in Figure 11.2b, design a BBO program to solve the ORC problem of the IEEE 15-bus test system based on user-defined TCCC and discrete *PS*.

C11.4 Repeat Exercise C11.3, but with considering the 3rd technique shown in Figure 11.2c.

C11.5 Repeat Exercise C11.1 for the hybrid BBO-LP optimization algorithm.

C11.6 Repeat Exercise C11.2 for the hybrid BBO-LP optimization algorithm.

C11.7 Repeat Exercise C11.3 for the hybrid BBO-LP optimization algorithm.

C11.8 Repeat Exercise C11.4 for the hybrid BBO-LP optimization algorithm.

12

Considering the Actual Settings of Different Relay Technologies in the Same Network

After reading this chapter you should be able to:

- Consider the effect of relay technology in solving ORC problems
- Combine electromechanical, static, digital, and numerical relays in one network
- Incorporate both directional and non-directional ITOCRs

12.1 Introduction

The literature contains many amazing optimization algorithms that have been proposed to solve ORC problems effectively and quickly. Almost all the existing studies are based on a general unrealistic assumption that all ITOCRs have the same model.[1] If someone refers to modern electric power networks, he/she will find that different relay technologies are used in the same network. That is, electromechanical, static, digital "hardware-based" and numerical "software-based" OCRs all could be installed in the same network.[2] Thus, the existing ORC optimizers will not work unless adjusting their classical models to accept different settings of *TMS*, *PS*, *T*, and *CTI* for the same TCCC.[3] Taking this consideration will increase the problem dimension by 1.5× if only one unified TCCC is used for all relays, and by 1.2× if multiple TCCCs are used:[4]

- For one unified TCCC:

$$n = 3\varrho = 6ß \tag{12.1}$$

- For multiple TCCCs:
  - For one-vector-variable and one-scalar-variable approaches:

$$n = 4\varrho = 8ß \tag{12.2}$$

  - For three-scalar-variables approach:

$$n = 6\varrho = 12ß \tag{12.3}$$

1 Be careful, we are talking here about the relay model and not TCCC.
2 In some networks, both European and North American relays could also be faced.
3 The same thing can be applied to *TDS* and *CTS* for North American relays.
4 Regardless of the relay technology used, remember that the ORC problem dimensions using multiple TCCCs are bigger than those when only one unified TCCC is used for all the relays. Thus, 1.2× is the gain applied to the dimensions of ORC problems having multiple TCCCs. This point is clarified in Table 12.1.

Optimal Coordination of Power Protective Devices with Illustrative Examples, First Edition. Ali R. Al-Roomi.

Companion website: www.wiley.com/go/al-roomi/optimalcoordination

Table 12.1 Dimensions of ELD, classical ORC, and realistic ORC problems.

Test system	ELD problem	Classical DOCRs coordination		Mixture of different technologies[a] ✓	
		One TCCC[b]	multiple TCCCs	One TCCC ✓	Multiple TCCCs
3-Bus (Urdaneta et al., 1988)	1-3	12	30	18	36
6-Bus (Noghabi et al., 2009)	2-4	28	70	42 ✓	84
15-Bus (Amraee, 2012)	6-7	84	210	126 ✓	252
30-Bus (Papaspiliotopoulos et al., 2017a)	2-5	60	150	90	180
42-Bus (Albasri et al., 2015)	3	194	485	291 ✓	582

a) ✓: means this case is simulated in this book.
b) Any type of time-current characteristic curves (TCCCs); please refer to Table 3.10.

Table 12.1 gives some examples of five popular test systems where ELD stands for economic load dispatch. This chapter tries to realize the existing ORC model to deal with different relay technologies. To validate its correctness, some test systems are solved by using BBO. We will cover the case when only inverse-time directional and non-directional OCRs are used. Also, we will assume that the correct relay technology is optimized rather than pre-defined by users. The pre-defined relay type approach is more realistic because it is a network-dependent ORC solver. However, the optimized relay type approach is more difficult in programming, and thus knowing it will let the programmers design the other approach smoothly.

12.2 Mathematical Formulation

The new mathematical formulation of this semi-realistic ORC problem can be modeled in any n-dimensional optimization algorithm as follows:

12.2.1 Objective Function

Suppose that a mesh network has ß branches. If each branch is equipped with two DOCRs mounted on both ends, then there are $\varrho = 2ß$ relays in that network. The operating time of the ith relay for a short-circuit fault occurred at the kth location can be mathematically expressed as follows (Albasri et al., 2015):

$$T_{i,k}^{y_i} = f(TMS_i^{y_i}, PS_i^{y_i}, I_{R_i^{y_i},k}), \qquad i = 1, 2, \ldots, \varrho \tag{12.4}$$

where the superscript y_i denotes the technology of the ith relay installed on one end or terminal of a faulty line, and $I_{R_i^{y_i},k}$ is a short-circuit current seen by the ith relay for a fault occurring at the kth location.

Because the current is independent of relay technology, so the superscript y_i can be dropped from the current term to have just $I_{R_i,k}$. Based on (12.4), the operating time of a relay installed on the ith terminal depends on the technology of that relay itself. If that relay is replaced with a relay having different technology, then $T_{i,k}^{y_i}$ could have a different curve. This creates many problems during minimizing the sum of operating times of these ϱ relays by existing mathematical models. The operating time ($T_{i,k}^{y_i}$) can be numerically determined by using any model. For the generalized model of the IEC/BS and ANSI/IEEE characteristic curves:

$$T_{i,k}^{y_i} = TMS_i^{y_i} \left[\frac{\beta}{\left(\frac{I_{R_i,k}}{PS_i^{y_i}} \right)^{\alpha} - 1} + \gamma \right], \quad \begin{array}{l} i = 1, \dots, \varrho \\ k = 1, \dots, \ell \end{array} \tag{12.5}$$

If 3ϕ short-circuit analysis is carried out for ℓ fault locations on each line, then the objective function can be derived by minimizing the sum of operating times of all ϱ DOCRs when they act as primary relays:

$$\text{OBJ} = \min \sum_{i=1}^{\varrho} \sum_{k=1}^{\ell} w_{i,k} T_{i,k}^{y_i,pr} \tag{12.6}$$

where $w_{i,k}$ is the weight given to the fault occurred at the kth location. For the sake of simplicity, a weight of 1 is used for all the fault locations (Urdaneta et al., 1988). Also, most of the studies presented in the literature consider just the fault located at the near-end point of each relay. Thus, (12.6) can be simplified to:

$$\text{OBJ} = \min \sum_{i=1}^{\varrho} T_i^{y_i,pr} \tag{12.7}$$

To ensure that the solutions obtained from minimizing (12.7) are feasible and useful, the following design constraints should be satisfied:

12.2.2 Selectivity Constraint Among Primary and Backup Relay Pairs

The goal of satisfying this constraint is to ensure that each primary protective relay has enough chance to isolate the fault occurred in its zone. If it fails to operate within that period, then one or more backup relays located in the upstream should act. This inequality constraint can be mathematically expressed as follows:

$$T_{j,k}^{u_j} \geqslant T_{i,k}^{y_i} + CTI_i^{y_i} \tag{12.8}$$

where $T_{j,k}^{u_j}$ is the operating time of the jth backup relay (R_j) for a fault occurred at the kth out-zone location of the ith primary relay, and the superscript u_j denotes the technology[5] of R_j. $CTI_i^{y_i}$ is the coordination time interval or the allowable chance given to the ith primary relay (R_i) to clear that fault (Blackburn and Domin, 2006; Anthony, 1995).

5 To check whether the relay is electromechanical, static, digital, or numerical.

From (12.8), it can be clearly seen that *CTI* is not constant anymore. Similar to (12.5), $T_{j,k}^{u_j}$ can be calculated for the fault located at the kth point as follows:

$$T_{j,k}^{u_j} = TMS_j^{u_j}\left[\frac{\beta}{\left(\frac{I_{R_j,k}}{PS_j^{u_j}}\right)^{\alpha} - 1} + \gamma\right], \quad \begin{matrix} j = 1, \ldots, \varrho \\ k = 1, \ldots, \ell \end{matrix} \tag{12.9}$$

where $I_{R_j,k}$ is a short-circuit current seen by R_j for a fault occurring at the kth location.[6]

12.2.3 Inequality Constraints on Relay Operating Times

From (12.5), the minimum and maximum values of $T_{i,k}^{y_i}$ depend on the technology of R_i. Thus, by mentioning the technology type, the operating time can be bounded between two limits as follows:

$$T_{i,k}^{y_i,\min} \leqslant T_{i,k}^{y_i} \leqslant T_{i,k}^{y_i,\max} \tag{12.10}$$

where $T_{i,k}^{y_i,\min}$ and $T_{i,k}^{y_i,\max}$ are, respectively, the minimum and maximum operating times of R_i for a fault occurring at the kth location. The lower bound ($T_{i,k}^{y_i,\min}$) depends on the technology of the ith relay, and the applicable upper bound ($T_{i,k}^{y_i,\max}$) is determined based on the thermal limit of the protected element or the critical clearing time ($T_{i,k}^{y_i,\mathrm{cr.}}$) required to preserve system stability[7] (Kundur, 1994; Padiyar, 2008; Albasri et al., 2015).

Because $T_{i,k}^{y_i}$ is a dependent variable, as can be obviously observed in (12.4)–(12.5), so (12.10) is a functional constraint. Based on this, the constraint must be re-expressed to have the following two inequality constraints (Albasri et al., 2015):

$$T_{ik}^{y_i,\min} - T_{i,k}^{y_i} \leqslant 0 \tag{12.11}$$

$$T_{i,k}^{y_i} - T_{i,k}^{y_i,\max} \leqslant 0 \tag{12.12}$$

12.2.4 Side Constraints on Relay Time Multiplier Settings

For *TMS*, each relay is manufactured with a specific domain and step-size resolution predefined based on y_i. This side constraint is formulated for R_i as follows:

$$TMS_i^{y_i,\min} \leqslant TMS_i^{y_i} \leqslant TMS_i^{y_i,\max} \tag{12.13}$$

where $TMS_i^{y_i,\min}$ and $TMS_i^{y_i,\max}$ are, respectively, the minimum and maximum allowable limits of *TMS* of the ith relay manufactured based on the technology[8] (y_i).

12.2.5 Side Constraints on Relay Plug Settings

Similar to $TMS_i^{y_i}$, the domain and step-size resolution of $PS_i^{y_i}$ depend on the relay technology. However, the same guidance used in Chapter 7 is applied here as well. Thus, this side constraint

6 i.e. belonging to the ith primary relay.
7 These two criteria have been covered in Chapter 4.
8 This is the difference here compared with other similar studies. In this realization, even with the same TCCC, the variable bounds and step-size resolution are not constant anymore.

can be expressed for R_i as follows (Noghabi et al., 2009; Amraee, 2012; Albasri et al., 2015):

$$PS_i^{y_i,\min} \leqslant PS_i^{y_i} \leqslant PS_i^{y_i,\max} \tag{12.14}$$

where $PS_i^{y_i,\min}$ and $PS_i^{y_i,\max}$ are, respectively, the minimum and maximum allowable limits of PS of the ith relay manufactured based on the technology.

By combining the preceding objective function and design constraints, the final mathematical expression of this realized optimization model can be formulated as follows:

$$\textbf{Minimize} \sum_{i=1}^{\varrho} T_{i,k}^{y_i,\text{pr}}$$

$$\begin{aligned} \text{Subjected to: } & T_{i,k}^{y_i} + CTI_i^{y_i} - T_{j,k}^{u_j} \leqslant 0 \\ & T_{i,k}^{y_i,\min} - T_{i,k}^{y_i} \leqslant 0 \\ & T_{i,k}^{y_i} - T_{i,k}^{y_i,\max} \leqslant 0 \\ & TMS_i^{y_i,\min} \leqslant TMS_i^{y_i} \leqslant TMS_i^{y_i,\max} \\ & PS_i^{y_i,\min} \leqslant PS_i^{y_i} \leqslant PS_i^{y_i,\max} \end{aligned} \tag{12.15}$$

12.3 Biogeography-Based Optimization Algorithm

The detailed information about BBO is given in Chapter 1, while Chapters 7–11 show how to employ this meta-heuristic optimization algorithm to solve different ORC problems. This section covers some essential modifications to make BBO applicable to this realized ORC model. The main guidelines can also be applied to any other optimization algorithm.

12.3.1 Clear Duplication Stage

This optional stage can be activated to increase the diversity of islands by avoiding the features of one island being duplicated on other islands. However, the clear duplication stage is completely disabled because the realistic settings of TMS and PS of all the relay technologies are discrete (Al-Roomi and El-Hawary, 2019a). Thus, the collateral damage is more than the expected benefit.

12.3.2 Avoiding Facing Infeasible Selectivity Constraints

Based on the work done in Chapters 7, 8, 11, and 10, the selectivity constraint given in (12.8) for each P/B relay pair is satisfied by initiating a *while*-loop to generate random y_i, $PS_i^{y_i}$, and $TMS_i^{y_i}$ until getting feasible values. This pre-processing unit is known as the feasibility checker (FC), which can accelerate finding feasible solutions to this highly constrained nonlinear non-convex mixed-integer ORC problem.

12.3.2.1 Linear Programming Stage

Because both TMS and PS are discrete variables and each relay has its own realized settings, the linear programming (LP) sub-algorithm proposed in Chapter 8 is deactivated.[9]

9 If this fine-tuning stage must be activated, then LP should be replaced by ILP. Please, refer to Chapter 8.

12.3.3 Linking $PS_i^{y_i}$ and $TMS_i^{y_i}$ with y_i

It is important to note that the settings $PS_i^{y_i}$ and $TMS_i^{y_i}$ are randomly generated based on their technology y_i. This means that the first variable that needs to be randomly generated is y_i, which is a discrete value that lies between 1 and 3 (1 for electromechanical or electromagnetic, 2 for static

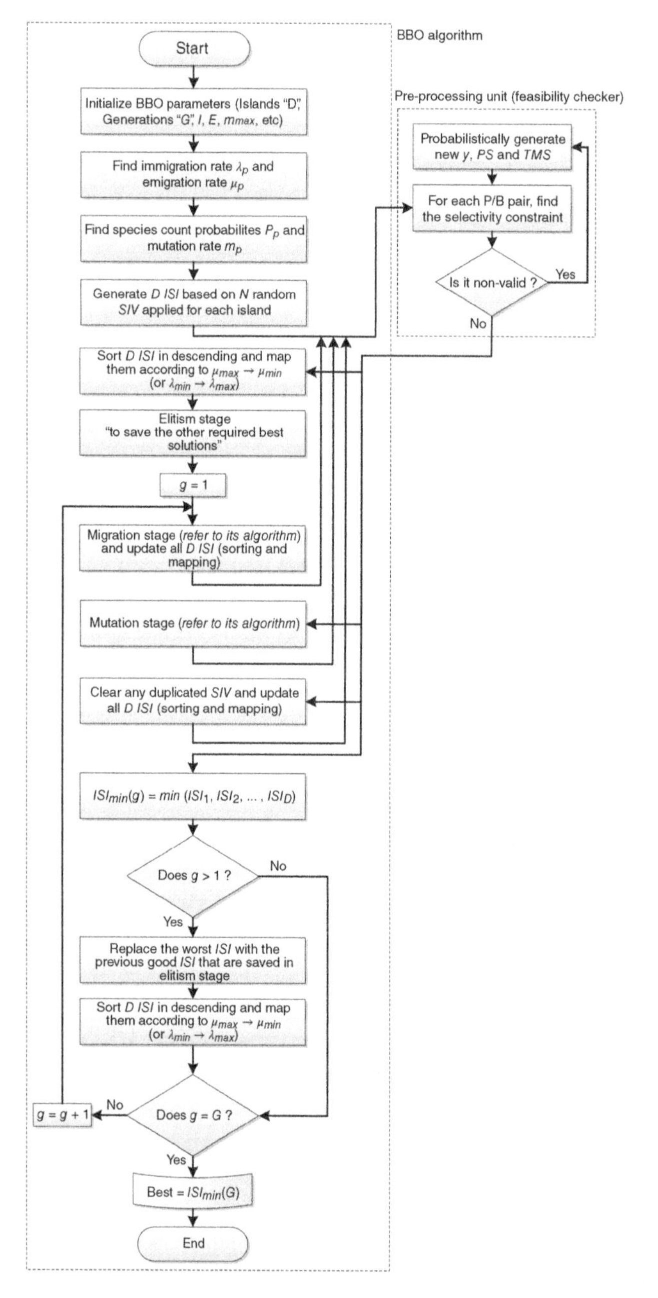

Figure 12.1 Flowchart of the modified BBO algorithm used for optimizing settings of relays with different technologies.

or solid-state, and 3 for digital or numerical). Based on this value, the step-size resolutions and the lower and upper bounds of $PS_i^{y_i}$ and $TMS_i^{y_i}$ are updated.

By integrating all the aforementioned modifications and taking into account the considerations recommended in Alroomi et al. (2013a,b), the final design of the BBO algorithm used to solve this ORC problem is shown in Figure 12.1.

Example 12.1 Consider the IEEE 15-bus test system given in Appendix F. The simulation parameters used in each relay technology are tabulated in Table 12.2. The BBO parameters are: population size of 50, 4 elite solutions, non-death penalty function with a penalty factor of $r = 30$, $m_{max} = 0.1$, 30 trials, and each trial uses 1000 generations. The dimension of this test system is given in Table 12.1. Finally, the IEC standard inverse TCCC is used to model all ϱ DOCRs, which means that the coefficients $\{\alpha, \beta, \gamma\}$ are, respectively, set to $\{0.02, 0.14, 0\}$. By modifying one of the MATLAB programs used in Chapters 7, 8, 11, and 10, solve the ORC problem of this test system by taking into account the differences in relay technologies.

Table 12.2 Simulation parameters used in each relay technology (Ungrad, 1995; ALSTOM, 2002; AREVA, 2011; Amraee, 2012; Zellagui and Abdelaziz, 2015).

	TMS			*PS* (A)			*T* (s)		*CTI*
Technology used	TMS^{min}	TMS^{max}	**Step-size**	PS^{min}	PS^{max}	**Step-size**	T^{min}	T^{max}	**(s)**
Numerical[a)]	0.025	1.5	0.001	0.5	2.5	0.01	0.01	4.0[b)]	0.3
Static	0.05	1.3	0.025	0.5	2.5	0.1	0.03	4.0	0.35
Electromechanical	0.1	1.1	0.05	{0.5, 0.6, 0.8, 1.0, 1.5, 2.0, 2.5}[c)]			0.05	4.0	0.4[d)]

a) The settings of directional and non-directional OCRs of AREVA Micom/P12xy series are used.
b) Although it can go beyond this limit, it is restricted to this value to avoid exceeding the stability critical clearing time.
c) The original settings of the 15-bus test system are given as a vector of steps of 0.5 A between PS^{min} and PS^{max}.
d) The original *CTI* of the 15-bus test system is 0.2 s, which is impractical.

Solution

The MATLAB files of the modified BBO program are available in a folder called `BBO_15Bus_DifferentRelayTechnologies`. By running `Starter.m`, the results tabulated in Table 12.3 are attained.

The optimal solution shows that the relay technology can decide whether the settings and decimal places are feasible or not; even if only the IEC standard inverse model is used.

12.4 Further Discussion

The ORC solver developed in this chapter assumes that the technology of each relay is not predetermined. Thus, it optimizes both the settings and technologies of protective relays. If all the relay technologies are known and fixed before initiating the optimizer, then the process will be easier with less problem dimension.

It is known that there are many manufacturers available in the market for each relay technology. These relays come in different models and versions. Thus, to make the results more realistic, each

Table 12.3 The full results obtained for the IEEE 15-bus test system incorporated with relays having different technologies.

Primary relay	Optimal settings				Primary relay	Optimal settings			
	PS (A)	*TMS*	*T* (s)	Type		*PS* (A)	*TMS*	*T* (s)	Type
R_1	1.1	0.85	1.9087	Static	R_{22}	0.5	0.9	1.5587	Electromechanical
R_2	2.4	0.225	0.7426	Static	R_{23}	1.5	0.75	1.9572	Static
R_3	1.83	0.974	2.5441	Numerical	R_{24}	0.5	0.95	1.7589	Electromechanical
R_4	1.5	0.825	2.2535	Static	R_{25}	1.3	0.85	2.1562	Static
R_5	2	0.8	2.3385	Electromechanical	R_{26}	1.9	0.525	1.5535	Static
R_6	2.3	0.6	1.8161	Static	R_{27}	2.13	0.561	1.8647	Numerical
R_7	1.23	1.144	2.7520	Numerical	R_{28}	2.23	0.607	1.8509	Numerical
R_8	1.8	0.575	1.6471	Static	R_{29}	2.5	0.375	1.0934	Static
R_9	1.25	0.794	1.8122	Numerical	R_{30}	2.1	0.475	1.3908	Static
R_{10}	1.6	0.7	1.8113	Static	R_{31}	0.78	1.192	2.3559	Numerical
R_{11}	2.31	0.368	1.2260	Numerical	R_{32}	0.64	1.264	2.5989	Numerical
R_{12}	1.48	0.539	1.4907	Numerical	R_{33}	1.54	0.981	2.6531	Numerical
R_{13}	1.83	0.808	2.2500	Numerical	R_{34}	1	0.8	1.7716	Electromechanical
R_{14}	1.2	0.7	1.7190	Static	R_{35}	1.3	0.9	2.3630	Static
R_{15}	1.55	0.979	2.6312	Numerical	R_{36}	0.7	1.125	2.2534	Static
R_{16}	0.6	1.05	2.0697	Electromechanical	R_{37}	1.62	1.04	2.7890	Numerical
R_{17}	1.2	0.875	2.0002	Static	R_{38}	2	0.925	2.9179	Static
R_{18}	2.13	0.364	0.9880	Numerical	R_{39}	0.7	1	2.0893	Static
R_{19}	1.14	1.045	2.2970	Numerical	R_{40}	1.2	1.125	2.7400	Static
R_{20}	1.6	0.675	1.6995	Static	R_{41}	2.16	0.849	2.3826	Numerical
R_{21}	0.5	0.95	1.6141	Electromechanical	R_{42}	1.94	0.496	1.4353	Numerical
Best (s)					83.1461				
Worst (s)					94.2674				
Mean (s)					87.8263				
StDev (s)					3.0954				

relay setting could have a slight difference based on its manufacturer, model, and version. The model developed in this section can deal with this extended realistic situation.

Problems

Written Exercises

W12.1 Re-solve Exercise W10.1 again when the protective relays are manufactured based on different technologies.

W12.2 Re-solve Exercise W11.1 again when the protective relays are manufactured based on different technologies.

W12.3 Re-solve Exercise W11.4 again when the protective relays are manufactured based on different technologies.

W12.4 Re-solve Exercise W11.5 again when the protective relays are manufactured based on different technologies.

Computer Exercises

C12.1 Modify the MATLAB program given in Example 12.1 to solve the ORC problem of the IEEE 6-bus test system shown in Figure F.7. All the necessary data and information about the system can be found in Appendix F. Use 200 iterations instead of 1000.

C12.2 Consider the IEEE 42-bus test system given in Appendix F, which contains both directional and non-directional ITOCRs. Modify the MATLAB program given in Example 12.1 for this large test system.

13

Considering Double Primary Relay Strategy

After reading this chapter you should be able to:

- Model main-1 and main-2 relays in ORC problems
- Satisfy the selectivity criteria when both local and remote backup relays are involved
- Solve the coordination problem for the extreme case where the number of main-2 relays (or local backup relays) equals the number of main-1 relays
- Consider only electromechanical, only static, or a mixture of them for main-2 relays
- Incorporate both directional and non-directionals ITOCRs

13.1 Introduction

In Chapter 2, it has been said that the protective relays could be electromechanical (or electromagnetic), solid-state (or static), digital, or numerical. Also, a brief time-line been introduced. This historical account is essential since it emphasizes that real electric power systems may contain many types and models of protective relays at the same time. During **upgrading/retrofit/rehabilitation** phases, there are two possible alternatives that can be selected by protection engineers, either:

- Completely replacing all electromechanical, static, and digital "hardware-based" relays with numerical relays, or;
- Integrating these old relays with numerical relays as a second line of defense.

If the first approach is followed, then the upgrading phase is straightforward without dealing with "**out of stock**" or "**obsolete**" spare parts. Besides, simplicity is preserved, and the classical ORC model can be used. However, this protection design requires removing all outdated infrastructures and replacing gained skill-sets. In contrast, while the second approach requires more maintenance, it allows upgrading existing protection systems and providing additional backup protection. Both approaches can be seen in many realistic electric power networks where electromechanical, static, and/or digital "hardware-based" relays are either completely replaced or partially retained to work in parallel with numerical relays.

This chapter focuses on the second approach where both old and new ITOCRs are assumed to protect electric networks from the same fault. These electromechanical, static, and/or digital "hardware-based" relays (i.e. old relays) mounted at the same points (i.e. the left and right ends of

Optimal Coordination of Power Protective Devices with Illustrative Examples, First Edition. Ali R. Al-Roomi.

Companion website: www.wiley.com/go/al-roomi/optimalcoordination

powerlines) of numerical relays are called "**local backup**" or "**secondary**" relays. Some protection engineers called these new and old relays as "**main-1**" and "**main-2**" relays, respectively, instead of primary and local backup/secondary relays. The logical reason behind that comes from the fact that when there is a fault the corresponding local backup/secondary relays can act without waiting the chance given to numerical relays to operate plus their time delays explained by (5.3) (Phadke et al., 1989; Paithankar and Bhide, 2003; Mysore, 2010). Therefore, these old ITOCRs work independently and they could act before new ITOCRs. Thus, the terms (main-1 and main-2 relays) are more suitable than the terms (primary and local backup/secondary relays). If the protection design based on main-1 and main-2 relays is adopted, then it is called a "**double primary relay strategy (DPRS)**"; and this special case of ORC is the core of this chapter.

In real-world applications, this DPRS is mostly applied to just a few, or even many (but not all), busbars. However, we try to solve the **extreme case** of DPRS. That is, each terminal of powerlines has both main-1 and main-2 ITOCRs where main-1 ITOCRs are always taken as numerical relays and main-2 ITOCRs could be electromechanical or static.[1] This extreme case has two possible solutions:

- If there is no violation, then DPRS can be applied at any P/B relay pair. This phenomenon can be seen in small test systems.
- Because the test systems used with DPRS are feasibly solved in the literature, the violations of this optimization problem have a different meaning. They give important information about the spots where DPRS is not applicable.[2] This phenomenon can be seen in large test systems.

Solving this extreme case means that the problem dimension is increased by 100% (i.e. the problem is duplicated). If all the relays are manufactured based on the same technology, then the dimension of the ORC problem can be calculated as follows:

- For one unified TCCC:

$$n = 4\varrho = 8\text{ß} \tag{13.1}$$

- For multiple TCCCs:
  - For one-vector-variable and one-scalar-variable approaches:

$$n = 6\varrho = 12\text{ß} \tag{13.2}$$

  - For three-scalar-variables approach:

$$n = 10\varrho = 20\text{ß} \tag{13.3}$$

If different relay technologies are used, then (12.1)–(12.3) should be modified to match with DPRS as follows:

- For one unified TCCC:

$$n = 6\varrho = 12\text{ß} \tag{13.4}$$

1 The digital relays can also be considered as main-2 relays, but they are not included because their settings are almost identical to those of numerical relays. The objective here is to analyze the feasibility when different relay settings and step-size resolutions are involved. Please, refer to Table 12.2.

2 In other words, these violations can tell which busbars are incompatible with DPRS.

- For multiple TCCCs:
  - For one-vector-variable and one-scalar-variable approaches:

$$n = 8\varrho = 16ß \tag{13.5}$$

  - For three-scalar-variables approach:

$$n = 12\varrho = 24ß \tag{13.6}$$

Therefore, if the IEEE 3-bus test system shown in Figure 13.1 is used, then the number of variables, i.e. *PS* and *TMS*, increases from being 12 variables (for 6 DOCRs) to 24 variables (for 12 DOCRs). The complexity of this problem increases considerably if the TCCCs of these DOCRs are not identical (i.e. using multiple North American and/or European standards, such as inverse, very inverse, extremely inverse, and even - in some cases - user-defined TCCCs). Thus, the coefficients $\{\alpha, \beta, \gamma\}$ are not constants anymore, which means that each DOCR has 5 variables instead of 2. Based on this, if these DOCRs (i.e. the relays installed in the IEEE 3-bus test system) are equipped with asymmetrical TCCCs, then the dimension of that ORC problem further increases from being 24 variables (for one unified TCCC) to 60 variables (for multiple TCCCs). Table 13.1 compares this special ORC problem with the classical one. If the IEEE 42-bus test system (IEEE Std. 399-1997) is selected, then the dimension of its ELD problem is just 3 (i.e. three units), while it has 194 *TMS* variables and 194 *PS* variables (total is 388) for this special ORC problem. If multiple TCCCs are used to coordinate these DOCRs, then the dimension increases to 970 variables. This chapter uses one unified TCCC for all ITOCRs. The other highly complicated scenarios, which are based on multiple TCCCs and relays with different technologies, can be solved by following the steps described in Chapters 10 and 12.

To solve this special ORC problem with one unified TCCC for all ITOCRs, the hybrid BBO/DE optimization algorithm presented in Chapter 1 is applied here, but after modifying it to act as a combinatorial optimization algorithm. Again, this algorithm is reinforced by a modified FC to avoid any infeasible settings of P/B relay pairs. Solving the extreme case of this ORC problem ensures detecting which P/B relay pairs accept DPRS when this strategy is not adopted for all circuit breakers. The proposed combinatorial BBO/DE algorithm is evaluated using numerical/electromechanical,

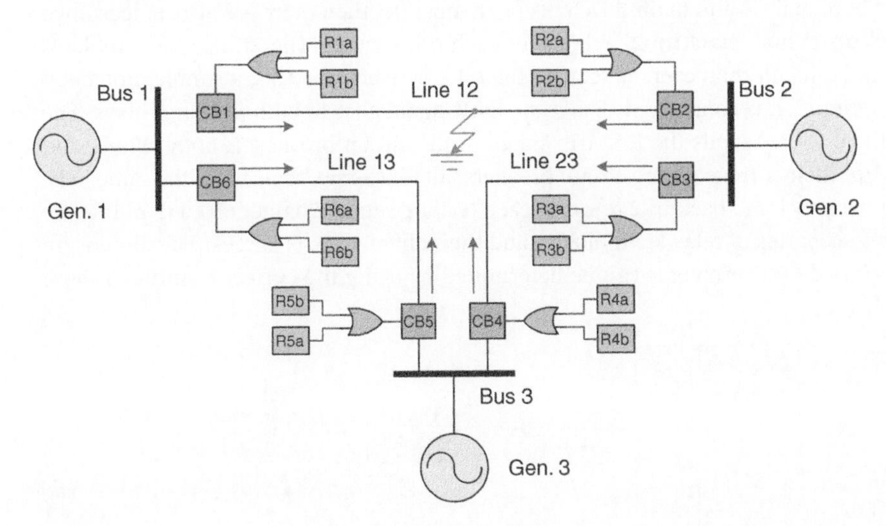

Figure 13.1 Single-line diagram of the IEEE 3-bus test system equipped with main-1 "a" and main-2 "b" DOCRs.

Table 13.1 Dimensions of ELD, classical ORC, and DPRS-based ORC problems.

Test system	ELD problem	Classical DOCRs coordination		Double relays strategy[a] ✓	
		One TCCC[b]	Multiple TCCCs	One TCCC ✓	Multiple TCCCs
3-Bus (Urdaneta et al., 1988)	1–3	12	30	24	60
6-Bus (Noghabi et al., 2009)	2–4	28	70	56 ✓	140
15-Bus (Amraee, 2012)	6–7	84	210	168 ✓	420
30-Bus (Papaspiliotopoulos et al., 2017a)	2–5	60	150	120	300
42-Bus (Albasri et al., 2015)	3	194	485	388 ✓	970

a) ✓: means this case is simulated in this book.
b) Any type of time-current characteristic curves (TCCCs); for example, IDMT standard.

numerical/static, and numerical/electromechanical-static sets as main-1/main-2 ITOCRs. Digital ITOCRs are not considered because their hardware features are almost similar to numerical ITOCRs.

13.2 Mathematical Formulation

Figure 13.1 shows how the extreme case of DPRS can be illustrated on the IEEE 3-bus test system, given in Urdaneta et al. (1988), while assuming that a fault takes place on branch 1–2. In the classical ORC problem, R_1 and R_2 are allowed to act as primary relays to isolate that fault. If R_1 fails to operate, then R_5 operates as a backup relay. Similarly, R_4 acts as a backup relay for R_2. With DPRS, the operational philosophy is completely different because each circuit breaker is initiated by a trip signal that might come from its main-1 (a numerical) or main-2 (an electromechanical or a static) DOCR. If both main-1 and main-2 DOCRs fail to operate, then there are also, at least, two DOCRs (i.e. "**backup-1**" and "**backup-2**" relays) for each primary set. Therefore, ORC problems with DPRS are more difficult than ever; as seen in Table 13.1. To achieve this, a simple protection logic, shown in Figure 13.2, is considered. The same fault signal goes to both primary relays. The fastest device is the one that sends the first trip signal to the circuit breaker. If both main-1 and main-2 relays and/or their corresponding circuit breakers fail to operate, then again the same logic is applied to their backup-1 and backup-2 relays of each backup set. In Chapters 10, 11, and 12, we have covered many scenarios of relay technologies and their characteristic curves. The dimension (n) of such DPRS-based ORC problems can be determined by using the correct formula of those

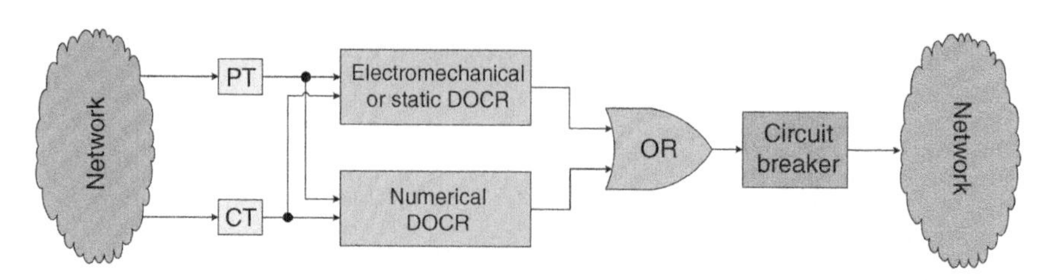

Figure 13.2 A simple protection logic of the double primary relay strategy.

listed in (13.1)–(13.6). These equations should be used for the extreme case where each circuit breaker is connected to two protective relays. If all ITOCRs are directional, then this special ORC problem can be mathematically modeled in any n-dimensional optimization algorithm as follows:

13.2.1 Objective Function

Assume that a network contains ϱ main-1 and ς main-2 DOCRs. The general objective function can be defined as follows:

$$\text{OBJ} = \min \sum_{k=1}^{l} \left[\sum_{i=1}^{\varrho} \left(w_{ia,k} T_{ia,k}^{\text{pr}} \right) + \sum_{j=1}^{\varsigma} \left(w_{jb,k} T_{jb,k}^{\text{pr}} \right) \right] \tag{13.7}$$

where $T_{ia,k}^{\text{pr}}$ and $T_{jb,k}^{\text{pr}}$ are, respectively, the operating times of the ith main-1 "a" and the jth main-2 "b" DOCRs when they act as primary protective relays for the kth fault location. The coefficients $w_{ia,k}$ and $w_{jb,k}$ are the kth fault location probability weights assigned to each relay, respectively.

If $\varsigma \neq \varrho$, then it means that some terminals of branches are not equipped with DPRS.[3] Based on this, the problem dimension is reduced, and hence becomes easier to solve. For the extreme case, both ϱ and ς are equal, so (13.7) becomes:

$$\text{OBJ} = \min \sum_{k=1}^{l} \left[\sum_{i=1}^{\varrho} \left(w_{ia,k} T_{ia,k}^{\text{pr}} + w_{ib,k} T_{ib,k}^{\text{pr}} \right) \right] \tag{13.8}$$

For simplicity, all the weights given in (13.8) are set equal to one (Urdaneta et al., 1988). Also, if only one common TCCC is used for all 2ϱ DOCRs, then the ORC problem can be solved using only 3ϕ near-end faults (i.e. $\ell = 1$) where the other less severe fault types and locations can also be achieved within that solution[4] (Birla et al., 2006b; Albasri et al., 2015). Therefore, (13.8) is further simplified to:

$$\text{OBJ} = \min \sum_{i=1}^{\varrho} \left(T_{ia}^{\text{pr}} + T_{ib}^{\text{pr}} \right) \tag{13.9}$$

The operating time of the xth DOCR of the ith primary protective set can be computed using the generalized model as follows:

$$T_{ix}^{\text{pr}} = TMS_{ix} \left[\frac{\beta}{\left(\dfrac{I_{R_{ix}}}{PS_{ix}} \right)^{\alpha} - 1} + \gamma \right], \quad \begin{array}{l} i = 1, \ldots, \varrho \\ x = a \text{ or } b \end{array} \tag{13.10}$$

where TMS_{ix} and PS_{ix} are the two independent variables of the xth DOCR of the ith primary protective set (x is either a or b), and $I_{R_{ix}}$ is the fault current seen by the xth relay. Based on the most studies conducted in the literature, the IEC/BS standard inverse TCCC is used, so the coefficients $\{\alpha, \beta, \gamma\}$ are, respectively, set equal to $\{0.02, 0.14, 0\}$ for all 2ϱ DOCRs. By referring to Figure 13.2, the same short-circuit current is seen by both main-1 and main-2 DOCRs through one common CT, so $I_{R_{ix}}$ is replaced by I_{R_i}. Therefore, (13.10) is modified for the IEC standard inverse model as follows:

$$T_{ix}^{\text{pr}} = \frac{0.14\, TMS_{ix}}{\left(\dfrac{I_{R_i}}{PS_{ix}} \right)^{0.02} - 1}, \quad \begin{array}{l} i = 1, \ldots, \varrho \\ x = a \text{ or } b \end{array} \tag{13.11}$$

3 i.e. the extreme case is not applicable here.

4 Please, do not forget the note presented in Section 11.3 of Chapter 11.

Now, after defining the terms T_{ia}^{pr} and T_{ib}^{pr} of (13.9) by (13.11), this objective function should satisfy the following design constraints:

13.2.2 Selectivity Constraint

If T_{ja}^{bc} and T_{jb}^{bc} are the jth new (i.e. numerical) and old (i.e. electromechanical or static) backup relays assigned to T_{ia}^{pr} and T_{ib}^{pr} of the ith primary protective set, then the operating time of the fastest backup relay should be equal or longer than the slowest sum of the operating time and the coordination time interval of the xth primary relay:

$$\min\left[T_{ja}^{\text{bc}}, T_{jb}^{\text{bc}}\right] \geq \max\left[T_{ia}^{\text{pr}} + CTI_{ia}, T_{ib}^{\text{pr}} + CTI_{ib}\right] \tag{13.12}$$

where CTI_{ia} and CTI_{ib} are the coordination time intervals defined for the ath and bth DOCRs of the ith primary protective set to isolate their in-zone faults. Because different technologies are involved, so the typical value of CTI is between 0.2 and 0.5 s (Blackburn and Domin, 2006; Anthony, 1995). Table 12.2 shows the practical range of different relay technologies.

The operating time of the xth backup relay of the jth protective set[5] can be calculated for an out-zone fault as follows:

$$T_{jx}^{\text{bc}} = \frac{0.14\, TMS_{jx}}{\left(\dfrac{I_{R_j}}{PS_{jx}}\right)^{0.02} - 1}, \quad \begin{array}{l} j = 1, \ldots, \varrho \\ x = a \text{ or } b \end{array} \tag{13.13}$$

where I_{R_j} is the fault current that is supposed to be cleared by either the ith main-1 or main-2 relay and seen by both the jth backup-1 and backup-2 relays.

13.2.3 Inequality Constraints on Relay Operating Times

The operating time of each ixth relay is practically bounded between two limits:

$$T_{ix}^{\min} \leqslant T_{ix} \leqslant T_{ix}^{\max} \tag{13.14}$$

where $T_{ix}^{\min}$ is the minimum operating time that the ixth DOCR can achieve, and $T_{ix}^{\max}$ is the maximum operating time allowed to reach by the ixth DOCR without losing the system stability or exceeding the thermal limit of the protected element. From (13.10) and (13.11), T_{ix} is a dependent variable, so (13.14) must be divided into two functional inequality constraints as follows:

$$T_{ix}^{\min} - T_{ix} \leqslant 0 \tag{13.15}$$

$$T_{ix} - T_{ix}^{\max} \leqslant 0 \tag{13.16}$$

13.2.4 Side Constraints on Relay Time Multiplier Settings

Based on the technology of each relay, the side constraints of TMS can be mathematically expressed as follows:

$$TMS_{ix}^{\min} \leqslant TMS_{ix} \leqslant TMS_{ix}^{\max} \tag{13.17}$$

where $TMS_{ix}^{\min}$ and $TMS_{ix}^{\max}$ are, respectively, the minimum and maximum allowable limits of TMS of the ixth DOCR.

---

5 Each ith protective set contains main-1 and main-2 relays, and each jth protective set contains backup-1 and backup-2 relays.

13.2.5 Side Constraints on Relay Plug Settings

The trip signal of the ixth DOCR is initiated once the current angle is correct and the current magnitude exceeds the given predetermined value. This value is defined as the severity level, which can be measured as follows:

$$PSM_{ix} = \frac{I_{R_i}}{PS_{ix}} \tag{13.18}$$

where PSM_{ix} is the plug setting multiplier of the ixth DOCR. If $PSM_{ix} < 1$, it means that no fault exists in the system.

This equation requires that the value of PS should be carefully selected to sense the lowest possible short-circuit current and, at the same time, it allows the overload current to flow without initiating the trip signal.[6] Therefore, the proper PS of the ixth DOCR should be equal to or less than two-thirds of the minimum fault current and equal to or higher than the overload current. This can be expressed as:

$$PS_{ix}^{\min} \leqslant PS_{ix} \leqslant PS_{ix}^{\max} \tag{13.19}$$

where $PS_{ix}^{\min}$ and $PS_{ix}^{\max}$ are, respectively, the minimum and maximum allowable limits of PS of the ixth DOCR.

To have a more realistic representation, these bounds are determined through some specific equations for each ixth DOCR; as described before in (7.20)–(7.23). However, most of the studies presented in the literature take PS as a predefined vector of discrete settings for all DOCRs.

13.3 Possible Configurations of Double Primary ORC Problems

In Section 13.2, a full mathematical optimization model has been formulated. However, this special ORC problem could also be solved analytically if all main-2 relays are considered as local backup relays with dropping remote backup relays. This protection scheme is not practical and highly risky, but it is one of the following three possible protection schemes:

1. **Only Remote Backup Relays Are Considered:**
   This is the classical ORC problem, which is also the one presented in the literature (Urdaneta et al., 1988, 1997; Birla et al., 2006b; Zeineldin et al., 2006; Mansour et al., 2007; Noghabi et al., 2009; Thangaraj et al., 2010; Bedekar and Bhide, 2011a; Amraee, 2012; Singh et al., 2013; Albasri et al., 2015; Papaspiliotopoulos et al., 2017a; Zellagui and Abdelaziz, 2015).
2. **Both Local/Remote Backup Relays Are Considered:**
   It is the core of this chapter. The protection design based on DPRS could be seen in some practical applications. It is introduced here with the extreme case where all ϱ circuit breakers are initiated either by main-1 numerical relays or main-2 old relays. But, most of (or even all) the time this strategy is partially applied to some selected circuit breakers.
3. **Only Local Backup Relays Are Considered:**
   If all remote backup relays are dropped from the protection design, which is highly not recommended, then all main-2 (electromechanical, static, and digital "hardware-based") relays will

6 Separate overload protection should handle above nominal non-fault currents if the maximum allowable period is exceeded. Thermal overload relays and "slow blow" fuses are some common overload protective devices. Refer to Chapter 4 for more details.

act as local backup relays because the corresponding state-of-the-art numerical relays are always faster than the preceding relays.

If someone selects this protection design, then the exact optimal solution can be analytically obtained by setting both relays to their highest possible speed as follows:

$$T_i^{\text{primary}} = T_{ia}^{\min} \tag{13.20}$$

$$T_i^{\text{local backup}} = T_{ib}^{\min} \tag{13.21}$$

That is, there is no need to add *CTI* to discriminate between the primary and local backup relays. The reason behind this is that each one can operate without waiting for the other. This means that the selectivity constraint described in (13.12) is also dropped.

However, there are many practical problems associated with this odd protection design, such as:

- It is not practical to leave all the relays installed on the other branches without effectively utilizing them as a second line of defense. Even if they are all completely unutilized to clear out-zone faults, there is also some doubt about whether the selectivity criterion is satisfied or not. This phenomenon could be seen in distribution systems having short length branches because short distribution lines have low impedance, and thus there is a chance that the other out-zone numerical relays might act before the in-zone old relays. This could also happen if the in-zone local backup relays are electromechanical and the other out-zone remote backup relays are static or digital because the last two old relay types are faster and thus they could isolate the faulty branch before the respective in-zone local backup relays; especially if $I_{R_j} \approx I_{R_i}$. This means that the protection design could be unselective, which is one of the major weaknesses of this protection scheme.
- For isolating faults, it is risky to depend on the same CTs and PTs for both protective relays.
- Having double relays on each circuit breaker is a very expensive design and requires lots of maintenance. Practically, this strategy is partially applied to some selected circuit breakers. It has been said before that the extreme case is investigated here because when it is solved the other less complicated problems, where DPRS is partially applied, can be easily solved as well.

Therefore, going with the third protection scheme (i.e. only local backup relays) is not feasible.

Example 13.1 Consider the IEEE 6-bus test system shown in Figure F.7 where all the necessary data and information can be found in Appendix F. The settings of *TMS*, *PS*, and *CTI* that should be used in this example are tabulated in Table 12.2. The maximum operating time ($T^{\max}$) is set equal to 1 instead of 4. For this test system, use both the conventional BBO and hybrid BBO/DE algorithms to optimally coordinate the relays when the extreme case of DPRS is applied. The initialization parameters are: 200 generations, population size of 50, and 30 independent runs. The clear duplication stage (of both BBO and BBO/DE) is completely disabled to avoid having infinite loops due to searching within discrete variables of *PS* and *TMS*. The elitism stage is activated by recycling the best four individuals of one iteration into the other. The step-size (F) and crossover rate (C_r) of DE are both equal to 0.5. Also, the binary static-exterior penalty function (BS-EPF) should be used to handle the design constraints. During modeling main-2 DOCRs, first simulate them as electromechanical relays and then simulate them as static relays. Thus, you need to design two conventional BBO and two hybrid BBO/DE optimization algorithms. Use the IEC standard inverse TCCC for all DOCRs. Also, it is assumed that all numerical relays are manufactured based on the same technology. The same assumption is applied to the old relays. Thus, the dimension should be calculated based on (13.1).

Solution

Referring to the network shown in Figure F.7, each circuit breaker can be initiated by a trip signal that might come from either main-1 or main-2 DOCR. If both relays fail to operate or if their circuit breaker does not open, then there is at least one set of backup-1 and backup-2 DOCRs located in the upstream ready to act after awaiting the assigned time delay. From the last row of Table 12.2, both *TMS* and *PS* are discrete where electromechanical DOCRs have very limited settings of *PS*. As stated in Table 13.1, the ELD problem has only four variables for this test system. The classical ORC problem has 28 variables with many constraints that need to be satisfied. Using the extreme case of DPRS means dealing with 56 variables[7] and the search space has different layers assigned to each type of DOCRs. This is the reason why the ORC problems are considered highly constrained nonlinear non-convex mixed-integer optimization problems; especially if the extreme case of DPRS is applied. Besides, this problem has the following constraints:

- 20 inequality constraints for (13.12)
- 28 inequality constraints for (13.15)
- 28 inequality constraints for (13.16)
- 28 side constraints for (13.17)
- 28 side constraints for (13.19)

The MATLAB files of the BBO and BBO/DE programs are available in a folder called `BBO_BBODE_6Bus_UnifiedMain2`. By running `Starter.m` of each program, the final results tabulated in Table 13.2 are attained. As can be seen from the table, two simulations are covered where the ϱ main-1 relays are always selected as numerical DOCRs. Each one of these two simulations is carried out by two optimization algorithms: the conventional BBO and the hybrid BBO/DE; both with FC. The constraints are handled by BS-EPF with a penalty multiplier of $r = 30$. The proposed hybrid BBO/DE with FC gives better results in both simulations. Disabling FC can save a remarkable CPU time, but the probability to get feasible solutions decreases steeply. For example, based on the computing machine used in this example,[8] if electromechanical relays are used as main-2 DOCRs, then BBO without FC can save around 75% of the total CPU time spent when FC is enabled (BBO consumes 12.82 seconds and BBO+FC consumes 52.93 seconds). However, disabling FC generates two violations. Thus, to get feasible solutions without FC, it is required to increase the number of generations and/or population size plus modifying the penalty function and other settings, which leads to higher CPU time; especially with larger test systems. Returning back to Table 13.2, it is obvious that when static DOCRs are selected as main-2 protective devices the optimizer can find better solutions, while electromechanical DOCRs can provide better average and standard deviations. The reason is that the static DOCRs have a wider search space than that of the electromechanical DOCRs; as can be seen in the step-size resolutions of *PS* and *TMS* listed in Table 12.2. Therefore, when the static DOCRs are used the optimizer needs more generations to explore the search space. The fitness curves of the BBO and BBO/DE algorithms are shown in Figure 13.3 for both simulations (i.e. all main-2 relays are either static or electromechanical DOCRs).

---

7 There are 28 DOCRs in the network as an extreme case of DPRS. If each relay has two variables (i.e. *PS* and *TMS* if only the IEC standard inverse TCCC is used for all 2ϱ DOCRs), then the dimension of this ORC problem is 56.
8 The specifications are: ALIENWARE M14x Laptop, 64-bit Windows 10 OS, Intel Core i7-4700MQ CPU @ 2.4 GHz, and 16 GB RAM.

Table 13.2 Simulation results of the IEEE 6-bus test system when the extreme case of DPRS is applied – main-2 relays are either electromechanical or static.

DOCR No.	Main-1: Numerical & Main-2: Electromechanical								Main-1: Numerical & Main-2: Static							
	BBO + FC				Hybrid BBO/DE + FC				BBO + FC				Hybrid BBO/DE + FC			
	Numerical		Electromechanical		Numerical		Electromechanical		Numerical		Static		Numerical		Static	
	PS	*TMS*	*PS*	*TMS*	*PS*	*TMS*	*PS*	*TMS*	*PS*	*TMS*	*PS*	*TMS*	*PS*	*TMS*	*PS*	*TMS*
R_1	1.47	0.522	2.0	0.50	0.63	0.605	0.5	0.70	1.34	0.520	2.3	0.425	1.54	0.540	0.9	0.550
R_2	2.49	0.190	2.5	0.15	1.03	0.383	0.6	0.40	0.80	0.518	1.9	0.325	0.97	0.431	1.0	0.375
R_3	1.43	0.277	1.0	0.40	1.71	0.399	0.8	0.55	1.24	0.447	2.2	0.675	0.73	0.589	1.0	0.475
R_4	1.00	0.357	0.8	0.30	2.27	0.167	0.6	0.35	1.96	0.128	1.0	0.250	2.21	0.091	1.2	0.225
R_5	2.18	0.216	1.0	0.25	0.72	0.317	1.0	0.25	1.12	0.283	1.1	0.350	0.63	0.361	1.7	0.275
R_6	1.91	0.486	2.0	0.25	0.96	0.299	2.0	0.25	1.29	0.239	2.5	0.125	2.25	0.148	2.1	0.150
R_7	1.48	0.570	2.5	0.35	2.17	0.272	1.5	0.40	2.41	0.396	1.7	0.375	1.81	0.292	1.4	0.450
R_8	1.38	0.261	0.8	0.40	0.69	0.251	2.0	0.15	2.00	0.233	1.1	0.200	0.81	0.299	0.6	0.225
R_9	1.22	0.346	0.8	0.40	0.70	0.527	1.5	0.25	1.41	0.377	0.7	0.525	2.31	0.231	1.1	0.375
R_{10}	0.89	0.440	1.0	0.45	0.83	0.542	0.5	0.55	1.58	0.370	2.2	0.250	1.02	0.644	1.8	0.325
R_{11}	2.10	0.385	1.5	0.55	1.02	0.498	2.0	0.30	0.76	0.668	1.8	0.300	2.06	0.290	1.4	0.325
R_{12}	0.87	0.783	2.5	0.60	1.10	0.693	0.6	0.75	0.52	0.985	1.6	0.550	1.78	0.601	2.5	0.500
R_{13}	1.09	0.186	2.5	0.10	2.32	0.187	1.0	0.25	1.16	0.248	1.3	0.150	1.04	0.469	2.1	0.050
R_{14}	2.14	0.343	2.5	0.25	1.40	0.316	0.6	0.45	2.40	0.252	1.0	0.425	0.96	0.423	1.1	0.375
OBJ (s)	20.1840						**19.0082**		20.0758						**18.9457**	
Mean (s)	22.4254						**21.6983**		22.6473						**22.4227**	
StDev (s)	1.4291						**1.3229**		1.4808						**1.4679**	

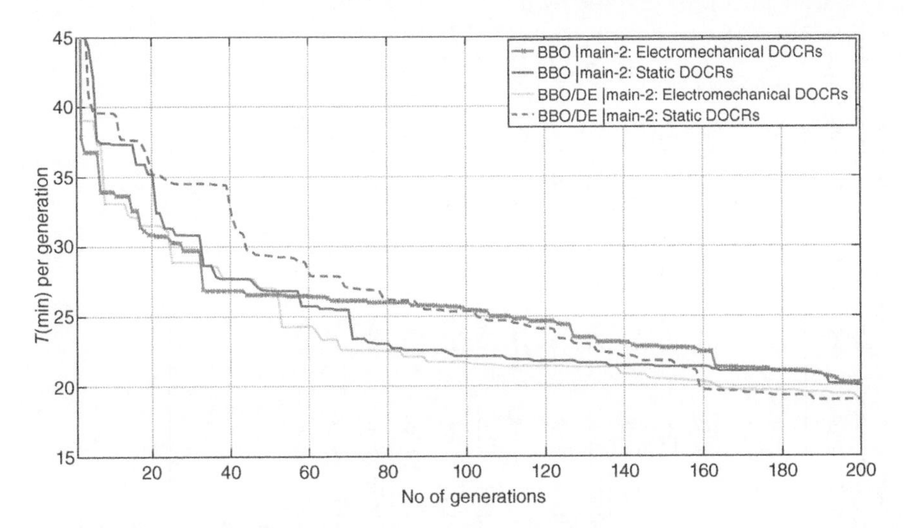

Figure 13.3 Curves of fitness functions of BBO and BBO/DE when the extreme case of DPRS is applied to the IEEE 6-bus test system – main-2 relays are either electromechanical or static.

From Table 12.2, it is clear that the technology determines the lower and upper limits of relay variables as well as their step-size resolutions. Changing these resolutions can affect the feasible search space and making it more difficult to find, but the mechanism of the technique remains without any change. This means that the realistic model presented in this study can be applied to any DOCR-based ORC problem by simply updating the preceding settings. For example, the numerical relay of this study is AREVA Micom/P12xy. However, the SIEMENS relay model 7SJ80 suggested in Costa et al. (2017) can also be used by just updating the corresponding step-size resolutions.

Example 13.2 Repeat Example 13.1 again, but now assume that a mixture of electromechanical and static relays are used for main-2 relays. Use only the BBO/DE program to find the optimal solution for four different random groups of main-2 relay types.

Solution

In this example, a mixture of static and electromechanical DOCRs are used as main-2 relays. They can be selected based on a vector of uniformly distributed pseudorandom integers between 1s and 2s, which are generated by the MATLAB command `randi(`u, υ, ϱ`)`; where u means the maximum discrete value that the vector elements can reach starting from 1, υ means the required number of vectors to be generated, and ϱ is the total number of DOCRs assigned as main-2. These three parameters are set as: $u = 2$, $\upsilon = 4$, and $\varrho = 14$. Also, the preceding penalty factor is updated to $r = 35$ for this example.

The MATLAB files of the BBO/DE program are available in a folder called `BBODE_6Bus_NumStaticElectmech`. By running `Starter.m`, the final results tabulated in Table 13.3 are attained for four random vectors of main-2 relay types. All the solutions obtained for these 4 groups are feasible. The total number of installed relays, their types, models, TCCCs, and locations as well as the initialization parameters of the optimization algorithm all play an important role in converging to better solutions. From Table 13.3, it can be clearly seen that the first group has the lowest standard deviation, while the second group has the lowest fitness and mean. Because the ORC problem is solved with the extreme case of DPRS where the total number of static relays are almost equal to that of electromechanical relays, so all the means are almost equal. The fitness curves obtained by the BBO/DE algorithm for these 4 randomly generated groups of main-2 relays are shown in Figure 13.4.

Table 13.3 Simulation results of the IEEE 6-bus test system when the extreme case of DPRS is applied – main-2 relays are a mixture of electromechanical and static.

Primary DOCR No	Simulation of group no. 1					Simulation of group no. 2					Simulation of group no. 3					Simulation of group no. 4				
	Main-1		Main-2			Main-1		Main-2			Main-1		Main-2			Main-1		Main-2		
	PS	*TMS*	*PS*	*TMS*	Type	*PS*	*TMS*	*PS*	*TMS*	Type	*PS*	*TMS*	*PS*	*TMS*	Type	*PS*	*TMS*	*PS*	*TMS*	Type
R_{1x}	1.01	0.603	0.8	0.8	Elec[a)]	1.65	0.528	1	0.8	Elec	0.96	0.638	0.6	0.725	Stat	0.83	0.589	0.5	0.75	Elec
R_{2x}	0.91	0.457	1	0.6	Elec	1.55	0.322	0.7	0.425	Stat	0.56	0.512	2.5	0.15	Stat	0.55	0.468	2.3	0.3	Stat
R_{3x}	2.35	0.479	2.3	0.375	Stat[b)]	1.21	0.449	2.5	0.35	Elec	1.58	0.302	0.5	0.7	Elec	0.71	0.494	1	0.45	Elec
R_{4x}	2.43	0.154	1	0.2	Elec	0.94	0.189	0.8	0.325	Stat	2.02	0.152	1.5	0.15	Elec	2.21	0.239	2	0.125	Stat
R_{5x}	2.08	0.218	0.8	0.35	Elec	1.59	0.338	2.4	0.175	Stat	1.23	0.284	0.7	0.35	Stat	2.2	0.318	2	0.3	Elec
R_{6x}	2.36	0.235	1.6	0.325	Stat	2.03	0.123	2	0.2	Elec	1.62	0.379	1.5	0.275	Elec	2.08	0.179	0.9	0.4	Stat
R_{7x}	1.63	0.422	0.6	0.475	Stat	0.55	0.676	0.6	0.6	Stat	0.55	0.585	1.5	0.375	Stat	1.76	0.337	0.7	0.5	Stat
R_{8x}	2.12	0.136	0.6	0.35	Elec	0.97	0.223	0.6	0.25	Elec	1.52	0.229	1.8	0.25	Stat	2.22	0.165	0.7	0.35	Stat
R_{9x}	1.17	0.298	1.5	0.3	Elec	1.23	0.263	1	0.25	Elec	1.81	0.269	0.5	0.55	Elec	1.87	0.192	1	0.4	Elec
R_{10x}	0.83	0.459	2.5	0.25	Elec	1.59	0.253	1.7	0.325	Stat	0.54	0.707	0.9	0.45	Stat	1.52	0.251	0.5	0.45	Elec
R_{11x}	2.06	0.265	1.4	0.325	Stat	0.7	0.733	0.5	0.6	Elec	1.4	0.354	1	0.7	Elec	1.47	0.35	0.6	0.6	Elec
R_{12x}	1.13	0.851	1.5	0.55	Elec	2.39	0.601	0.6	0.85	Stat	0.77	0.675	0.8	0.675	Stat	2.28	0.509	0.5	0.8	Stat
R_{13x}	1.18	0.369	2	0.1	Elec	2.02	0.148	0.8	0.325	Stat	1.99	0.058	2.5	0.35	Elec	0.66	0.763	1.5	0.2	Elec
R_{14x}	1.72	0.237	0.6	0.475	Stat	0.55	0.461	1	0.35	Elec	1.04	0.365	0.9	0.5	Stat	1.67	0.294	2.5	0.325	Stat
OBJ (s)	19.7223					**19.4669**					20.3409					20.0743				
Mean (s)	22.4577					**22.0182**					22.6952					22.2022				
StDev (s)	**1.1870**					1.2929					1.7982					1.4053				

a) Electromechanical (or Electromagnetic) Relay.
b) Static (or Solid-State) Relay.

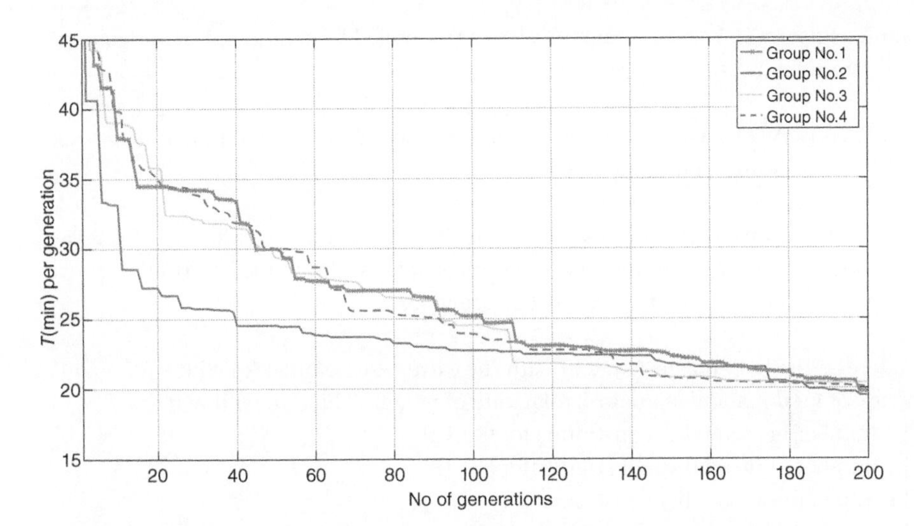

Figure 13.4 Curves of fitness functions of BBO and BBO/DE when the extreme case of DPRS is applied to the IEEE 6-bus test system – main-2 relays are a mixture of electromechanical and static.

Problems

Written Exercises

W13.1 Consider the IEEE 15-bus test system with DPRS. If all primary relays are of model A and all local backup relays are of model B, find the following:

1. The number of DOCRs and the dimension of the ORC problem if there is no any local backup relays.
2. The number of DOCRs and the dimension of the ORC problem if 50% of primary relays have local backup relays.
3. The number of DOCRs and the dimension of the ORC problem if the extreme case of DPRS is applied.

W13.2 Consider the IEEE 15-bus test system with the extreme case of DPRS where all main-1 relays are of model A and all main-2 relays are of model B. Find the following:

1. The number of inequality constraints for (13.12)
2. The number of inequality constraints for (13.15)
3. The number of inequality constraints for (13.16)
4. The number of side constraints for (13.17)
5. The number of side constraints for (13.19)

W13.3 Repeat Exercise W13.1 again when multiple standard TCCCs are taken into account. Use the one-scalar-variable approach.

W13.4 Repeat Exercise W13.1 again when user-defined TCCCs are taken into account. Use the three-scalar-variable approach.

W13.5 Repeat Exercise W13.1 again when the relays are supplied from different manufacturers with different features and capabilities.

W13.6 Repeat Exercise W13.3 again when the relays are supplied from different manufacturers with different features and capabilities.

W13.7 Repeat Exercise W13.4 again when the relays are supplied from different manufacturers with different features and capabilities.

W13.8 Consider the dimension formulas used in Exercise W13.1 as well as Exercises W13.3–W13.7 when the extreme case of DPRS is applied. From these formulas, prove that the IEEE 15-bus test system has just 21 branches.

W13.9 Consider the IEEE 42-bus test system with the extreme case of DPRS where all main-1 relays are of model X and all main-2 relays are of model Y. Find the following:

1. The number of inequality constraints for (13.12)
2. The number of inequality constraints for (13.15)
3. The number of inequality constraints for (13.16)
4. The number of side constraints for (13.17)
5. The number of side constraints for (13.19)

W13.10 The hybrid BBO/DE meta-heuristic optimization algorithm used in Examples 13.1 and 13.2 is based on the pseudocode given in Algorithm 5. It uses the simplest DE algorithm, which is called classic DE or DE/rand/1/bin. Re-write this pseudocode if the following variations are used:

1. DE/best/1/bin
2. DE/rand/2/bin
3. DE/best/2/bin

Computer Exercises

C13.1 Modify the BBO/DE program given in Example 13.1 for electromechanical main-2 relays by using the mutant vector of DE/best/1/bin.

C13.2 Modify the BBO/DE program given in Example 13.1 for static main-2 relays by using the mutant vector of DE/rand/2/bin.

C13.3 Modify the BBO/DE program given in Example 13.2 for electromechanical and static main-2 relays by using the mutant vector of DE/best/2/bin.

C13.4 Consider Exercise W13.2. Modify the BBO/DE program given in Example 13.2 to solve the ORC problem of the IEEE 15-bus test system when the extreme case of DPRS is applied. (**Hint:** It is normal to get violations on some P/B relay pairs. The classical ORC problem of the IEEE 15-bus test system (i.e. without DPRS) has been feasibly solved in Chapters 10, 11, and 12. Thus, the preceding violations do not disqualify the result. Instead, these violations mean that there are some pairs do not accept DPRS, which means that we cannot apply the extreme. Therefore, it is better to call them "incompatible pairs" instead of "infeasible pairs" (Al-Roomi and El-Hawary, 2019a).)

C13.5 Consider Exercise W13.9. Modify the BBO/DE program given in Example 13.2 to solve the ORC problem of the IEEE 42-bus test system when the extreme case of DPRS is applied. (**Hint:** Remember that this test system contains both directional and non-directional ITOCRs.)

14

Adaptive ORC Solver

After reading this chapter you should be able to:

- Understand the importance of adaptive coordination
- Know the main mechanism of adaptive ORC solvers
- Differentiate between operational and topological changes
- Solve adaptive ORC problems by using conventional LP, multi-start LP, meta-heuristic, and hybrid algorithms
- Estimate optimal settings by using artificial neural networks, support vector regression, and multiple linear regression

In Chapters 7, 8, 9, 10, 11, 12, and 13, we have seen how to optimally coordinate overcurrent protective devices when the network is at a steady-state. For real-world applications, this assumption is not correct because all the system parts are subject to many controllable and uncontrollable dynamic changes. Thus, it is required to update the optimal settings of protective devices frequently. Similar to contingency analysis, this means that the processing speed is very crucial in deciding whether the proposed ORC solver is effective or not. This chapter gives a brief introduction to the subject of adaptive-optimal coordination.

14.1 Introduction

Figure 6.1 shows the general flowchart to optimally coordinate overcurrent protective devices when the system is at a steady-state. If the study focuses only on one topology of the system, then the program can be terminated once the last stage of the preceding flowchart is reached. This means that the ORC problem is solved and thus optimal relay settings are reached. Practically speaking, there is no such system that can operate under a steady-state condition forever. The status of power components (generators, step-up transformers, IBTs, busbars, powerlines, uninterruptible power sources (UPS), loads, capacitor banks, etc.) is not constant. Thus, the mechanism shown in Figure 6.1 does not work. Therefore, if the researchers are interested to have adaptive optimizers for solving the issue of dynamic network configurations, an unterminated `while`-loop should be defined so the required stages of Figure 6.1 are repeated again and again. The practical way to tackle this serious technical problem is by modifying the preceding flowchart to be like the one shown in Figure 14.1. The user can program the algorithm to continuously scan the network for any dynamic change

Optimal Coordination of Power Protective Devices with Illustrative Examples, First Edition. Ali R. Al-Roomi.

Companion website: www.wiley.com/go/al-roomi/optimalcoordination

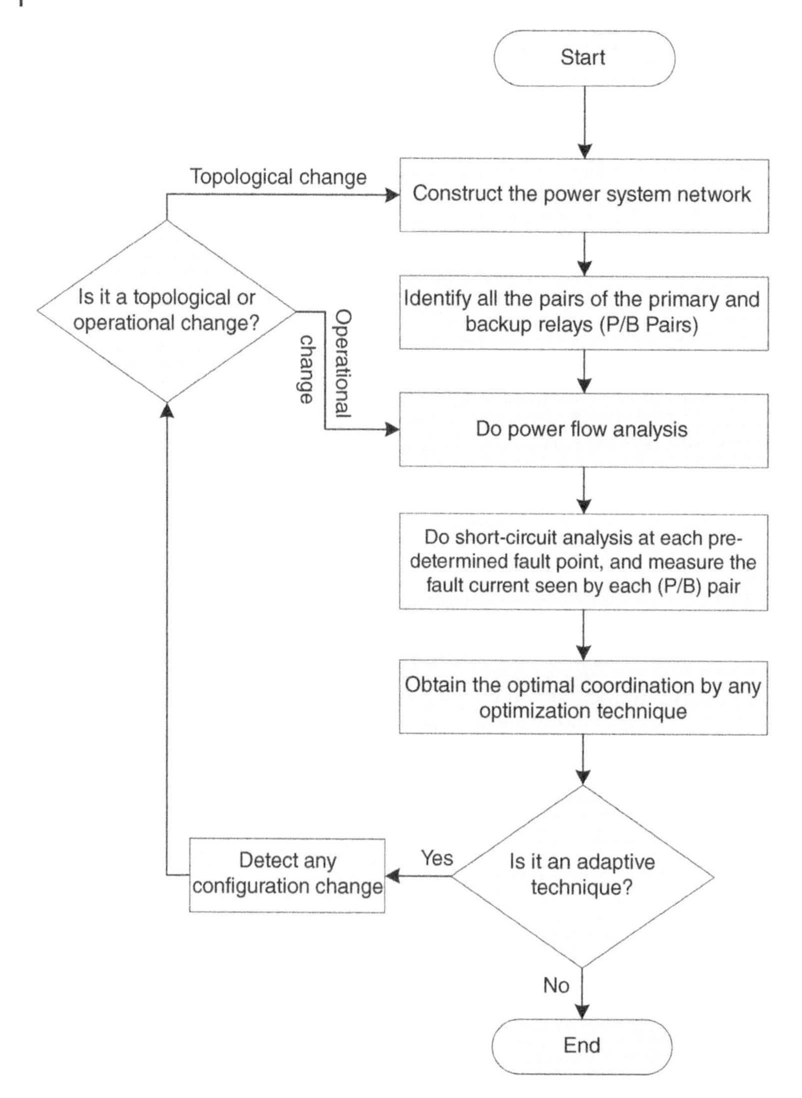

Figure 14.1 General flowchart to adaptively solve ORC problems.

without any delay between scans. This will consume lots of power due to continuous processing and might decrease the performance and lifespan of the computing machine due to *self-heating*. The other possible approach is to do this scanning process within fixed time-frames. Also, it can be linked with other power system studies, such as contingency analysis, optimal operation, and load forecasting. It is recommended to embed *watchdog*[1] so that an alarm is initiated when the current loop of Figure 14.1 exceeds a specified period.

14.2 Types of Network Changes

From the preceding flowchart, it can be seen that the ORC stage depends on two essential stages; power flow (PF) and fault analysis (FA). The dependency on these two stages can be illustrated

1 It is a function that is commonly used in PLCs.

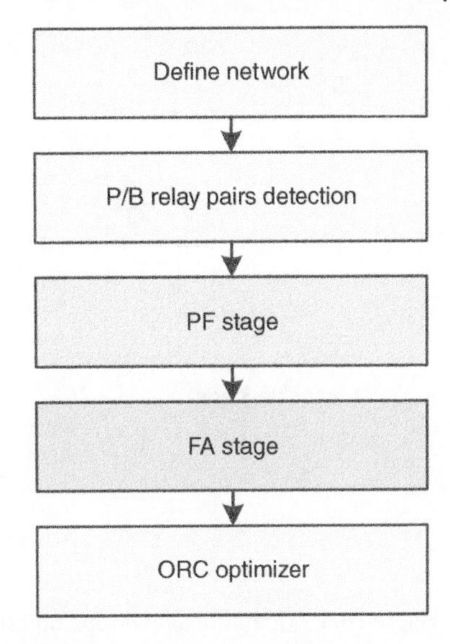

Figure 14.2 Flowchart of the basic optimal relay coordination process.

in Figure 14.2. We need to conduct PF and FA for any one of the dynamic changes listed in Section 14.1. For the first two stages,[2] it depends on the type of dynamic change. These changes are categorized under two groups called *operational changes* and *topological changes*. The first two stages are required just for the topological changes.

14.2.1 Operational Changes

These dynamic changes do not affect the topology of the network, which means that the structure and the P/B relay pairs remain without any changes. Some examples of operational changes are:

- The output power of a generating unit is increased/decreased
- The power consumed by end-users (i.e. load) is increased/decreased
- The tap of a transformer is changed
- Auto-adjustment of a unified power flow controller (UPFC)
- Variation in the system frequency
- Variation in the surrounding weather conditions (temperature, relative humidity, wind speed, wind direction, sunlight, etc.)
- Degraded efficiency and worn-out power components
- Upgraded/retrofitted/rehabilitated power components
- The effects of geomagnetically induced current (GIC)

14.2.2 Topological Changes

On the opposite side, there are some dynamic changes that affect the structure of the network and the list of P/B relay pairs. Some examples of topological changes are:

- A branch is taken out of/returned into service
- A node is energized/de-energized
- A generator is turned on/off
- The intermittent power of renewable resources
- Load shedding, partial service interruption, and power outages

2 That is, constructing power network and identifying P/B relay pairs.

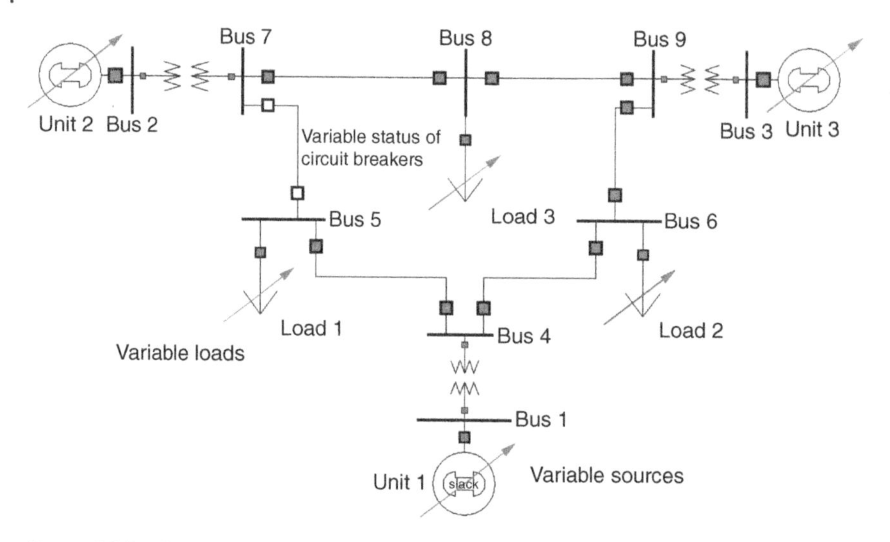

Figure 14.3 Operational and topological changes on the WSCC nine-bus test system.

Figure 14.3 shows how to implement different operational and topological changes on the WSCC nine-bus test system. In addition, we can consider the variations on system parameters due to the dynamic disturbances on the system frequency and the surrounding weather conditions (Al-Roomi and El-Hawary, 2017b,c). This means that even the lines and transformers are subject to operational changes.

Example 14.1 Consider the IEEE three-bus test system shown in Figure F.1. List the P/B relay pairs for the following cases:

1. **Case 1**: Steady-state condition
2. **Case 2**: Nodes 1–3 is taken out of service
3. **Case 3**: P_{G_1} is increased by 25%
4. **Case 4**: Node 1 is de-energized

Solution
The first case has the same P/B relay pairs listed in Appendix F. Because the third case is an operational change, so the steady-state list of the P/B relay pairs is not affected. On the opposite side, the dynamic changes of the second and fourth cases are topological. Thus, we have to re-identify the P/B relay pairs for each case. Table 14.1 lists the P/B relay pairs for each case.

14.3 AI-Based Adaptive ORC Solver

Although the modified flowchart shown in Figure 14.1 has the mechanism to track the optimality of relay settings based on the current state of the network, it does not guarantee to have a quick correction process. Thus, it is important to ensure that the ORC solver has the ability to compromise between the solution quality and the processing speed. We have discussed many ways to accelerate the processing speed, such as using multi-start LP and hybridizing LP with meta-heuristic optimization algorithms. However, this might not be enough for very large networks, which is the case with modern electric power systems. This can be clearly observed by looking to Figure 14.1!

Table 14.1 Lists of P/B relay pairs of Example 14.1.

Case 1		Case 2		Case 3		Case 4	
R^{pr}	R^{bc}	R^{pr}	R^{bc}	R^{pr}	R^{bc}	R^{pr}	R^{bc}
R_1	R_5	—	—	R_1	R_5	—	—
R_2	R_4	—	—	R_2	R_4	—	—
R_3	R_1	R_3	—	R_3	R_1	R_3	—
R_4	R_6	R_4	R_6	R_4	R_6	R_4	—
R_5	R_3	R_5	R_3	R_5	R_3	—	—
R_6	R_2	R_6	—	R_6	R_2	—	—

For each operational change, the ORC solver has to do PF and then short-circuit analysis before being able to optimize the relay settings. For topological changes, two additional stages are required to model the current state of the network and then identify its all P/B relay pairs. Furthermore, we might need to establish an interface between numerical programming languages and external engineering software, which is not an easy task.[3]

We can get rid of all these weary calculations and complicated steps by involving AI. For example, we can use an artificial neural network (ANN) to have a direct optimal or near optimal solution for any dynamic condition. The weights associated to the neurons can be obtained by training ANN based on a dataset that maps between various dynamic conditions and optimal relay settings. Thus, the ORC problem can be solved directly without going through all the stages shown in Figure 14.1. This means that both the solution quality and processing speed can be achieved with the help of AI. However, it is not clear how to generate this input/output dataset.

14.3.1 Generating Datasets

To make ANNs applicable in any numerical problem, it is important to feed these networks with a matrix of input values (called predictors or features) and a matrix of output values (called responses or targets). For this particular problem, it is important to say that some wise steps should be considered during creating the dataset. Some of these essential steps that should be considered are:

- Different settings of generating units and loads
- Reasonable predictors for the input matrix
- The status of all the branches and nodes
- The output matrix should be accurate

Three approaches could be used here:

- Using highly accurate solvers if the dataset is simulated based on realized models
- Using real power system readings and measurements through SCADA and EMS
- Combining both to achieve two important points:
  1. Exploring the entire search space of the problem by covering many scenarios[4]
  2. Reducing the uncertainty due to the gap between the actual power system and its model

3 This point has been covered in Chapter 6.

4 In Chapter 6, we said that we cannot do some dangerous scenarios, like opening live busbars or switching main thermal units, especially during peak season.

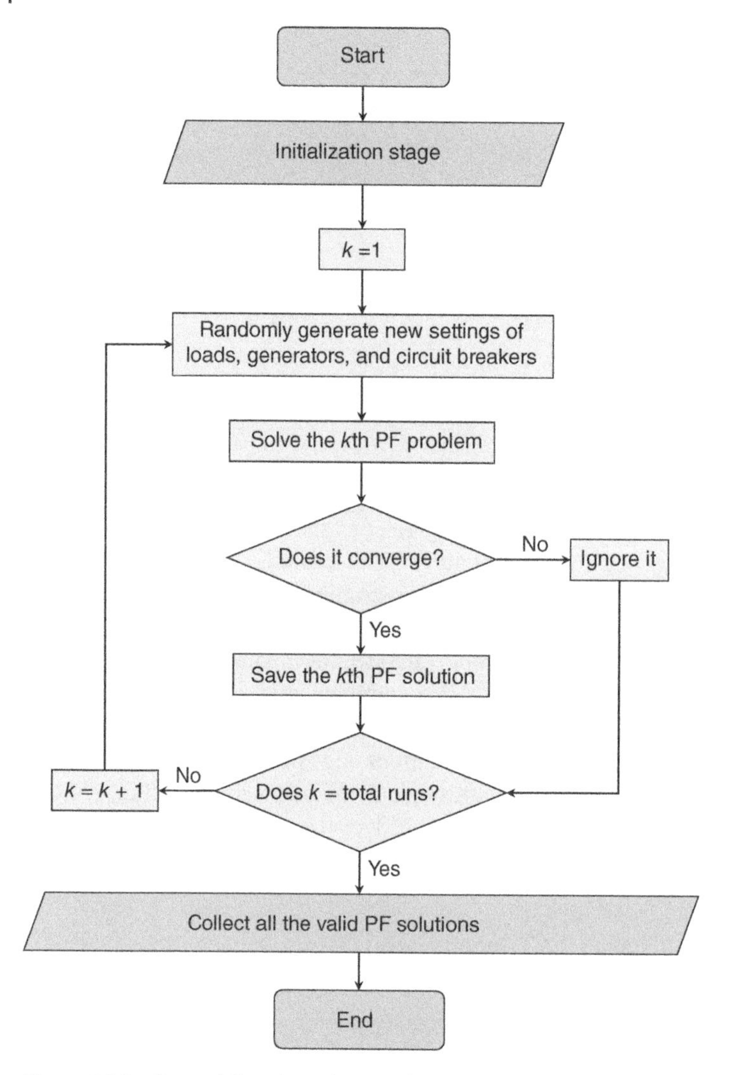

Figure 14.4 General flowchart that can be used to generate many PF solutions.

Figure 14.4 represents a systematic process that can be used to generate many PF solutions based on different dynamic configurations.[5]

14.3.2 Applying ANN to Solve ORC Problems

The P/B relay pairs can be identified by checking the status of all the branches as follows:

$$\mathfrak{L}_{ij} = \begin{cases} 1 & \text{if } y_{ij} \neq 0 \\ 0 & \text{otherwise} \end{cases} \tag{14.1}$$

where $\mathfrak{L}_{ij}$ is the status of the branch located between the ith and jth busbars. The branch is considered under outage (i.e. taken out of service) if it shows 0, and energized (i.e. returned back to service) if it shows 1.

5 That is, considering both the operational and topological changes.

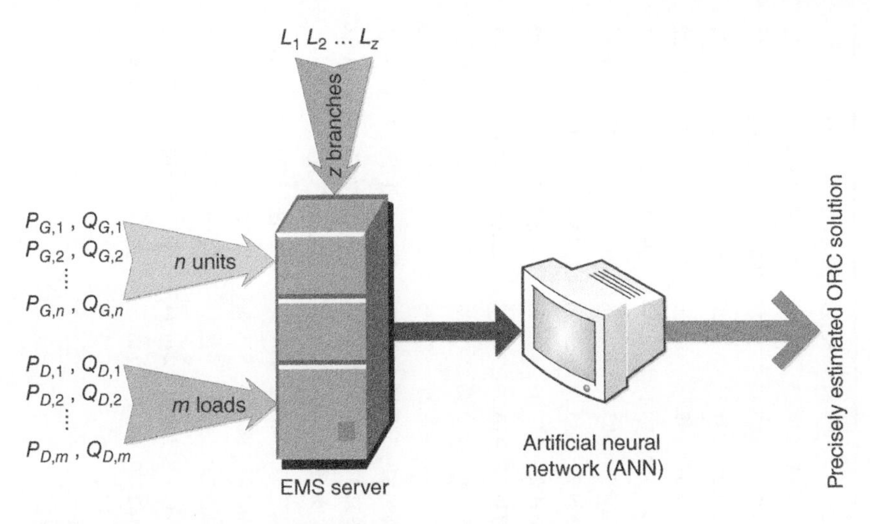

Figure 14.5 Mechanism of the proposed AI-based ORC solver.

After that, we can conduct FA for each dynamic configuration, which is the last stage before optimally coordinating the protective devices. Here, we have different approaches to map optimal relay settings with the dynamic changes of the system. Suppose that the physical properties of branches are not affected. The regression features could be the conventional PF variables tabulated in Table 6.1. The other more flexible approach is to use only active and reactive power of both generators and loads. That is, instead of using $\{V_{G_1}, \theta_{G_1}, P_{G_2}, V_{G_2}, \ldots, P_{G_n}, V_{G_n}\}$ for generators, we can use $\{P_{G_1}, Q_{G_1}, P_{G_2}, Q_{G_2}, \ldots, P_{G_n}, Q_{G_n}\}$. The architecture of such AI-based ORC solver is shown Figure 14.5. As can be seen, this technique does not depend on any power or instrument devices to measure voltage magnitudes and phase angles. It requires just to know the present status of branches and power settings[6] of generators and loads to give its precise estimation of settings of protective devices. This approach is highly practical and feasible because of the following three safety and economic facts (Al-Roomi and El-Hawary, 2019c):

- **Generation**: There is no any power plant that can operate its units without knowing their actual power settings.
- **Transmission/Sub-transmission**: The status of all overhead, underground, and underwater lines/cables are continuously updated and monitored.
- **Distribution/Utilization**: All power consumptions by customers and end-users are accurately measured and billed.

Example 14.2 Consider the simple three-bus test system given in Example 7.2. Table 14.2 shows the close-in 3ϕ faults of the six P/B relay pairs for 30 different dynamic conditions. Assume that all the relays are identical with $CTI = 0.2$ s. The IEC standard inverse TCCC is linearized for all the six relays by taking the following:

$$CTR = [300/5, 200/5, 200/5, 300/5, 200/5, 400/5]$$

$$PS = [3.0, 4.5, 3.0, 2.0, 4.5, 3.5]$$

Modify the LP file given in `Linearized_ORC_3Bus_LP.m` to find the optimal relay settings for each dynamic condition.

6 That is, the real and reactive power → P, Q.

Table 14.2 Short-circuit analysis of 30 different dynamic conditions for Example 14.2.

Condition No.	Active power (MW)			Reactive power (Mvar)			Primary fault currents (A)						Backup fault currents (A)					
	P_1	P_2	P_3	Q_1	Q_2	Q_3	$I^{pr}_{f,1}$	$I^{pr}_{f,2}$	$I^{pr}_{f,3}$	$I^{pr}_{f,4}$	$I^{pr}_{f,5}$	$I^{pr}_{f,6}$	$I^{bc}_{f,5}$	$I^{bc}_{f,4}$	$I^{bc}_{f,1}$	$I^{bc}_{f,6}$	$I^{bc}_{f,3}$	$I^{bc}_{f,2}$
1	200	l450	250	13	−260	290	1544.39	2181.48	2393.22	1201.93	1472	2004.68	654.5	323.81	558.39	1202.86	1473.09	1116.91
2	110	−360	250	5	−255	283	1271.31	2068.96	2222.26	1007.2	1390.44	1729.64	648.91	214.05	370.81	1007.97	1391.38	1105.32
3	190	−360	170	6	−255	281	1404.52	1965.06	2151.99	1088.45	1323.95	1809.51	593.69	308.15	529.7	1089.31	1324.83	1011.35
4	200	−600	400	108	−405	383	2039.1	2919.06	3221.89	1604.61	1988.96	2683.26	851.2	391.35	693.26	1606.67	1991.24	1490.51
5	30	−100	70	−117	−45	168	415.85	1132.1	1129.04	352.46	743.78	679.44	417.89	49.62	81.26	352.39	743.75	688.22
6	120	−150	30	−89	−95	193	719.65	1263.44	1307.07	559.24	835.54	961.4	439.38	178.13	296.64	559.32	835.76	728.97
7	200	−500	300	97	−400	369	1902.25	2633.32	2920.16	1489.58	1793.73	2475.3	760.5	383.18	676.78	1491.48	1795.79	1328.86
8	255	−470	215	−18	−200	259	1577.2	2022.14	2240.25	1206.78	1361.82	1973.2	599.71	383.26	652.56	1207.56	1362.45	1013.95
9	125	−400	275	32	−300	309	1441.1	2251.56	2438.26	1138.59	1518.87	1938.77	691.2	254.03	442.66	1139.63	1520.12	1186.33
10	240	−540	300	83	−370	355	1952.79	2610.94	2908.25	1520.89	1777.56	2510.62	748.77	414.39	727.73	1522.65	1779.42	1302.48
11	315	−415	100	32	−290	301	1778.9	2030.73	2283.39	1341.76	1372.65	2140.21	585.01	482.49	829.98	1342.96	1373.79	1001.8
12	210	−310	100	13	−270	286	1412.86	1884.2	2070.27	1084.93	1269.5	1782.62	567.65	338.21	580.95	1085.83	1270.37	968.62
13	385	−570	185	102	−390	365	2304.39	2541.82	2912.65	1751.47	1724.5	2786.54	686.81	597.36	1047.53	1753.64	1728.32	1198.41
14	218	−585	367	158	−490	428	2244.9	3043.25	3401.29	1763.11	2072.69	2916.12	856.37	450.28	808.29	1765.91	2078.49	1521.92
15	70	−110	40	−100	−76	183	553.36	1199.68	1219.19	447.01	791.18	811.02	428.48	103.4	171.93	447.01	791.27	708.9
16	135	−435	300	87	−390	360	1708.33	2536.81	2778.95	1347.19	1722.59	2269.76	753.37	315.17	556.47	1348.89	1724.41	1312.95
17	85	−510	425	128	−450	404	1881.44	2949.28	3201.9	1492.54	2004.66	2537.93	888.09	328.54	585.39	1494.73	2007.27	1564.46
18	10	−130	120	−93	−89	190	524.05	1304.88	1319.25	439.68	860.77	831.69	467.55	17.03	32.2	439.67	861	775.08
19	100	−130	30	−95	−85	188	642.08	1227.3	1258.81	505.65	810.35	887.06	432.73	148.2	246.44	505.67	810.58	716.86
20	98	−175	77	−79	−113	202	743.62	1357.89	1413.58	588.24	900.48	1028.15	462.48	146.68	246.33	588.33	900.73	769.66

21	41	−180	139	−76	−120	207	677.08	1438.04	1481.19	553.29	953.23	1006.94	496.6	68.62	118.42	553.4	953.19	827.32
22	40	−210	170	−43	−178	237	843.36	1639.7	1712.31	681.88	1091.98	1217.08	548.19	99.55	172.48	682.08	1092.51	921.48
23	68	−250	182	−36	−190	244	944.75	1717.48	1810.22	756.91	1146.45	1330.69	563.3	133.77	230.6	757.17	1146.99	948.93
24	108	−276	168	−21	−214	257	1084.88	1782.95	1905.79	857.49	1194.07	1474.19	567.55	190.46	327.15	857.78	1194.83	959.93
25	111	−305	194	6	−262	283	1229.7	1957.88	2108.84	971.14	1316.32	1659.77	610.22	217	375.61	971.95	1317.25	1040.16
26	147	−287	140	−7	−239	270	1213.85	1824.12	1973.87	947.75	1225.26	1597.9	565.93	247.8	425.36	948.21	1226.07	961.02
27	115	−325	210	21	−287	298	1315.04	2057.45	2224.81	1037.8	1385.96	1767.47	634.29	234.6	407.61	1038.72	1387.07	1085.7
28	160	−380	220	43	−320	317	1517.79	2204.6	2414.47	1188.84	1490.99	1994.03	659.6	299.06	521.08	1189.96	1492.33	1135.72
29	110	−420	310	71	−365	346	1597.22	2475.53	2689.78	1263.93	1675.92	2147.73	749.72	279.88	492.61	1265.36	1677.63	1300.74
30	185	−485	300	102	−410	374	1884.57	2640.04	2923.07	1478.45	1798.54	2462.38	764.4	372.86	659.8	1480.35	1800.63	1337.67

Solution

The modified MATLAB code and other files are available in a folder called `AdaptiveORC_3Bus`. By running `AdaptiveORC_3Bus_LP.m`, the following optimal settings are obtained in the command window:

```
Elapsed time is 0.344021 seconds.
Tab =
  30x8 table
```

Condition	TMS1	TMS2	TMS3	TMS4	TMS5	TMS6	Obj
1	0.10447	0.15033	0.19329	0.1	0.1	0.10499	2.1721
2	0.1	0.15248	0.19485	0.1	0.10227	0.1	2.274
3	0.10135	0.14295	0.19114	0.1	0.1	0.10043	2.2214
4	0.13346	0.17642	0.23491	0.10783	0.12977	0.12284	2.3148
5	0.15237	0.16853	0.16723	0.15032	0.15721	0.16022	5.3883
6	0.10634	0.15279	0.18621	0.11844	0.10394	0.1	3.1585
7	0.12812	0.16555	0.22048	0.10411	0.11935	0.11715	2.2655
8	0.12085	0.14421	0.19309	0.10192	0.10117	0.10637	2.2559
9	0.1	0.15477	0.19774	0.1	0.10317	0.10253	2.1929
10	0.13679	0.16937	0.22378	0.11321	0.12232	0.12406	2.3564
11	0.14838	0.15896	0.2034	0.13065	0.10957	0.12963	2.5373
12	0.11086	0.13992	0.19032	0.1	0.1	0.10029	2.2655
13	0.17865	0.18792	0.23411	0.15933	0.13147	0.16151	2.7435
14	0.15197	0.18826	0.24592	0.125	0.13829	0.13854	2.4694
15	0.11548	0.13981	0.17757	0.11454	0.12726	0.12228	3.7326
16	0.1051	0.16235	0.20595	0.1	0.10781	0.1101	2.1522
17	0.1141	0.17545	0.22757	0.1	0.12336	0.11455	2.205
18	0.15689	0.1711	0.17071	0.15457	0.1625	0.16497	4.6833
19	0.10665	0.15723	0.19001	0.11805	0.10866	0.1	3.3591
20	0.10742	0.15264	0.18525	0.11108	0.10149	0.1	3.0004
21	0.1323	0.15552	0.16443	0.13106	0.14338	0.14188	3.5989
22	0.11445	0.14653	0.21909	0.11343	0.13831	0.12618	3.0955
23	0.1	0.15506	0.19014	0.1	0.10142	0.1	2.5546
24	0.1	0.14942	0.18916	0.1	0.1	0.1	2.4207
25	0.1	0.14955	0.19103	0.1	0.1	0.1	2.2979
26	0.1	0.14478	0.18964	0.1	0.1	0.1	2.333
27	0.1	0.14976	0.19206	0.1	0.1	0.1	2.2385
28	0.1	0.15155	0.19357	0.1	0.1	0.1045	2.1601
29	0.1	0.16171	0.20346	0.1	0.10621	0.10727	2.1555
30	0.12511	0.16419	0.21925	0.101	0.11826	0.11478	2.2338

Problems

Written Exercises

W14.1 Consider Example 14.2. Among these 30 dynamic conditions, how many of them are categorized as operational changes, and how many are topological changes?

W14.2 A network consisted of ß branches is assumed. The planning department wants to replace many aged transmission lines with new ones manufactured based on a different technology. The upgrading phase is assumed to take around three to four years till all the cables are replaced. Is the AI-based architecture, shown in Figure 14.5, applicable? Explain.

W14.3 A network is operated under a frequency of 50 Hz. Because of bad control, there is a significant fluctuation in the frequency. Does this phenomenon affect the mechanism shown in Figure 14.5? Explain.

W14.4 Consider Example 14.2. Assume that all the relays are identical and emulated using one unified TCCC. Is there any possibility to enhance the solution quality of the adaptive ORC solver? Explain.

W14.5 Assume that a branch is taken out of service for just two hours for quick planned maintenance. Does the adaptive ORC solver need to identify the P/B relay pairs if the delay between scans is four hours? Explain.

W14.6 A very big smelter requires around 1.5 GW to produce aluminum. The smelter is powered by five thermal power plants. Due to the abundant demand in the global market, the production is kept at its maximum capacity. That is, a fixed load is connected to the power plants. This load is simply multiple arrays of electrolytic cells called pots. Assume that the production of power plants and the consumption of pot rooms are not changed during the last 12 months. Also, assume that the variation in the consumption of auxiliary plants and other small loads is neglected. The power network of this smelter is protected by protective devices, which were optimally coordinated during the winter. Do we need to re-optimize the settings of these protective devices during the summer? Explain.

Computer Exercises

C14.1 Modify the program given in Example 14.2 to emulate the IEEE very inverse TCCC.

C14.2 From Example 14.2, we can see that the fitness increases in some dynamic conditions. Modify the MATLAB code to find the best set of PS where $PS_i \in \{1.5, 2, 2.5, \dots, 4.5, 5\}$ A. Test its performance for both the IEC standard inverse and the IEEE very inverse TCCCs. (**Hint:** Refer to the multi-start LP program given in Example 8.3.)

C14.3 Repeat Exercise C14.2 by using BBO-LP. (**Hint:** Refer to the multi-start LP program given in Example 8.4.)

C14.4 Consider the dataset and results given in Example 14.2. Design a neural network to work based on the architecture shown in Figure 14.5. For the initialization stage, use the following settings:

- Maximum number of epochs to train: 20
- Performance goal: 0
- Maximum validation failures: 8

- Minimum performance gradient: 1×10^{-10}
- Number of hidden layers: 1
- Number of neurons: 10
- Ratio of vectors for training = 80%
- Ratio of vectors for validation = 10%
- Ratio of vectors for testing = 10%
- Activation function of hidden layer: `softmax`
- Activation function of output layer: `purelin`
- Training algorithm: Levenberg–Marquardt (LM)

Plot the actual and estimated optimal settings.

C14.5 By referring to Exercise C14.4, the performance can be enhanced by finding the optimal settings of neural networks. For example, the number of hidden layers, neurons in each layer, activation functions of output and hidden layers, the type of training algorithm and its internal settings, the features selected as predictors, the network topology, the feedback stream, etc. This process is known as optimal hyperparameters. The literature has many techniques to optimize these hyperparameters, such as the one reported in Al-Roomi and El-Hawary (2018c). The mission here is to use a trial-and-error technique to find acceptable hyperparameters. Because the dataset is so small, just use $\{P_2, P_3, Q_2, Q_3\}$ as predictors and fit just TMS_2. Also, the maximum number of hidden layers should be 2 and the total number of neurons should not exceed 4. Use the following settings for the training algorithm:

- Maximum number of epochs to train: 2000
- Performance goal: 0
- Maximum validation failures: 6
- Minimum performance gradient: 1×10^{-7}
- Maximum number of hidden layers: 2
- Maximum total number of neurons: 4
- Ratio of vectors for training = 70%
- Ratio of vectors for validation = 15%
- Ratio of vectors for testing = 15%

Plot the actual and estimated optimal settings.

C14.6 Repeat Exercise C14.4 by using support vector regression (SVR) with any setting you want. (**Hint:** Each relay will need one separate SVR.)

C14.7 Repeat Exercise C14.4 by using multiple linear regression (MLR). Fit the following polynomial equation for each ith relay:

$$\theta_0 + \theta_1 P_1 + \theta_2 P_2 + \theta_3 P_3 + \theta_4 Q_1 + \theta_5 Q_2 + \theta_6 Q_3$$

C14.8 Consider Exercise C14.7. The explanation level of the variability of the dataset could be improved by covering many other terms. Try to fit TMS_i again by using the following expanded regression model:

$$\begin{aligned} TMS_i = {} & \theta_0 + \theta_1 P_1 + \theta_2 P_2 + \theta_3 P_3 + \theta_4 P_1^2 + \theta_5 P_2^2 + \theta_6 P_3^2 + \theta_7 P_1 P_2 + \theta_8 P_1 P_3 \\ & + \theta_9 P_2 P_3 + \theta_{10} Q_1 + \theta_{11} Q_2 + \theta_{12} Q_3 + \theta_{13} Q_1^2 + \theta_{14} Q_2^2 + \theta_{15} Q_3^2 \\ & + \theta_{16} Q_1 Q_2 + \theta_{17} Q_1 Q_3 + \theta_{18} Q_2 Q_3 + \theta_{19} P_1 Q_1 + \theta_{20} P_2 Q_2 + \theta_{21} P_3 Q_3 \end{aligned}$$

C14.9 One of the main weaknesses of ANNs is that they need big datasets.[7] Also, ANNs are black-boxes where no mathematical models are generated. To avoid this technical problem, some recent ML tools have been successfully developed. One of them is called "universal functions originator (UFO)." After implementing this new ML computing system, the following mathematical model is generated for TMS_2:

$$TMS_2 \approx \left| a_1 + a_2 \left[\text{versin}\left(a_3 - \frac{a_4 P_1^{a_5}}{a_6 P_2^{a_7}} - a_8 P_3^{a_9} - \frac{a_{10} Q_1^{a_{11}}}{\left(a_{12} Q_2^{a_{13}}\right)\left(a_{14} Q_6^{a_{15}}\right)} \right) \right]^{a_{16}} \right|$$

where

$$a_1 = 0.1388$$
$$a_2 = 0.0011$$
$$a_3 = 0.064\,056\,737\,686\,959\,41$$
$$a_4 = -2.123\,696\,055\,790\,976$$
$$a_5 = 1.261\,697\,321\,148\,284\,3$$
$$a_6 = -4.229\,018\,684\,562\,822$$
$$a_7 = 1.596\,701\,960\,324\,278\,2$$
$$a_8 = -4.431\,448\,629\,703\,789$$
$$a_9 = -1.0$$
$$a_{10} = -3.288\,023\,153\,572\,991\,6$$
$$a_{11} = -0.314\,021\,019\,510\,202\,67$$
$$a_{12} = 1.021\,975\,618\,100\,725\,6$$
$$a_{13} = -0.945\,209\,872\,035\,861\,5$$
$$a_{14} = 1.844\,767\,178\,124\,776\,3$$
$$a_{15} = 0.976\,948\,097\,523\,529\,8$$
$$a_{16} = -1.0$$

Use non-linear regression (NLR) to fine-tune these coefficients based on the dataset given in Example 14.2. Plot the actual and estimated optimal settings.

7 For example, a few hidden layers and neurons are used in Exercise C14.5 because the dataset contains only 30 observations.

15

Multi-objective Coordination

After reading this chapter you should be able to:

- Understand the meaning of multi-objective optimization.
- Know the basic concepts and mechanisms used in algorithms.
- Convert any multi-objective problem into a mathematical model.
- Solve different multi-objective optimal relay coordination (ORC) problems.

In this chapter, we will apply the technique of multi-objective optimization to solve the optimal relay coordination (ORC) problem. It will cover the basic principles of the multi-objective optimization, translating the ORC problem into **multi-objective problem (MOP)**, and solving this MOP by modifying the conventional biogeography-based optimization (BBO) algorithm.

15.1 Basic Principles

Practically speaking, sometimes, the problem that has to be optimized contains more than one objective. Some of these objectives contradict to each other. This means that finding the optimal solution to one objective might led to a very bad solution to other objective(s). This special algorithm is known as **multi-objective optimization**.[1]

Suppose that there is an n-dimensional optimization problem with a design vector $X = [x_1, x_2, \dots, x_n]$ in the search space, and has an objective vector $F(X) = [f_1(X), f_2(X), \dots, f_k(X)]$ in the objective space. If the aim is to minimize one single objective and the objective vector is a subset of the real number (i.e. $F(X) \subseteq \mathbb{R}$), then (Gandibleux et al., 2004; Simon, 2013):

- A solution X^* will ***dominate*** X if $f_i(X^*) \leqslant f_i(X^*)\ \forall\ i \in [1, k]$ and at least one objective function $f_j(X^*) < f_j(X)$. The following notation is used to indicate that X^* dominate X:

$$X^* \succ X \tag{15.1}$$

1 It is also known as **multi-objective programming**, **Pareto optimization**, **multi-criteria optimization**, **multi-attribute optimization**, **multi-performance optimization** and **vector optimization** (Rao, 2009; Venkataraman, 2009).

Optimal Coordination of Power Protective Devices with Illustrative Examples, First Edition. Ali R. Al-Roomi.

Companion website: www.wiley.com/go/al-roomi/optimalcoordination

- If the second condition is not satisfied (i.e. $f_j(X^*) \nless f_j(X)$), then X^* will have ***weak domination*** to X. Thus, (15.1) becomes:

$$X^* \succcurlyeq X \tag{15.2}$$

- X^* is a ***nondominated*** point if there is no other solution X can dominate X^*.
- This nondominated X^* is called **Pareto optimal point**; or just **Pareto point**.
- If all these nondominated points X^* are collected in one set, then this set will be called **Pareto optimal set**; or just **Pareto set**:

$$P_s = \{X_1^*, X_2^*, \dots, X_m^*\} \tag{15.3}$$

  where m is the length of the Pareto set, or the total number of the Pareto points.
- The set of all the functions of X^* is called **Pareto front**:

$$P_f = \{f_1^*(X), f_2^*(X), \dots, f_m^*(X)\} \tag{15.4}$$

15.1.1 Conventional Aggregation Method

For k-objectives, MOPs can be mathematically expressed as:

$$\min_x F(X) = \min_x \left[f_1(X), f_2(X), \dots, f_k(X)\right] \tag{15.5}$$

For simplicity, we will use the **conventional weight aggregation (CWA)** method[2] (Drzadzewski and Wineberg, 2006), as a basis for solving MOPs. It is considered as a one of *non-Pareto-based* **multi-objective evolutionary algorithms (MOEAs)**. The idea behind this technique is very simple. A vector of objective functions can be converted into a single objective function as follows (Simon, 2013; Augusto et al., 2012):

$$\min_X F(X) \Rightarrow \min_X \sum_{i=1}^{k} w_i f_i(X), \quad \text{where} \quad \sum_{i=1}^{k} w_i = 1 \tag{15.6}$$

If only two objectives are assigned to a given function, then it could be said **biobjective function**. Thus, (15.6) becomes:

$$\min_X F(X) \Rightarrow \min_X \left[w f_1(X) + (1 - w) f_2(X)\right] \tag{15.7}$$

Some of the limitations of this simple method are listed as follows:

1. It is easy to implement the weight w in (15.7), but the proper selection may become very difficult.
2. This method can obtain only one Pareto optimal solution per each simulation run.
3. It fails to find non-convex Pareto fronts; so instead of finding the true Pareto front, it will provide an aggregation approximation.

However, some advanced modifications have been applied to this technique to give it the ability to find all the Pareto optimal set within one simulation run as well as finding the true Pareto front, for instance, the **dynamic weight aggregation (DWA)** method given in (Drzadzewski and Wineberg, 2006). Again, for the sake of simplicity, we will use the basic form and then combine it with BBO to have CWA-BBO for the optimization of biobjective functions. Also, we will use the **random weight aggregation (RWA)** method for the optimization of three objective functions. Thus, the original BBO algorithm described earlier in Figure 1.14 is modified to be as the one shown in Figure 15.1.

---

2 It is also called the **weighted sum method (WSM)** (Augusto et al., 2012).

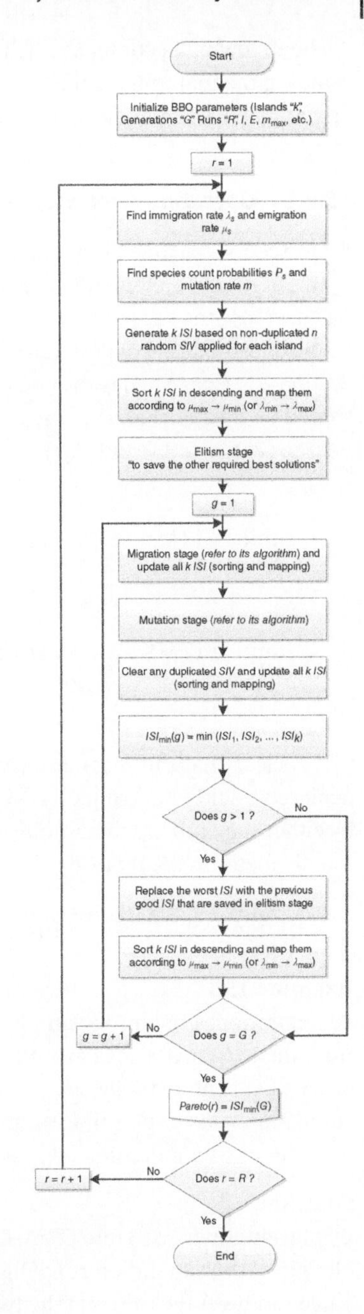

Figure 15.1 Flowchart of the proposed multi-objective BBO algorithm.

15.2 Multi-objective Formulation of ORC Problems

We can say that the multi-objective coordination is one of the latest research areas in power system protection. It is started just a few years ago. There are two goals that need to be achieved by employing this advanced optimization approach. The first one is to satisfy all the design constraints of classical ORC problems. The other goal is to compromise between different objectives, such as minimizing the total operating time of protective devices, minimizing the discrimination margin of primary and backup protective devices, system reliability, and system cost.

The following Sections 15.2.1, 15.2.2, and 15.2.3 give some possible objective functions. Understanding the mathematical formulation of these MOPs will let the reader to code and plug-in any other objective function(s), including those presented in the literature.

15.2.1 Operating Time vs. System Reliability

From Section 2.5, one of the main design criteria of any protective relay is its reliability. It consists of two conflicting elements; dependability and security. Increasing the dependability of the relay will decrease the security. There is a need to compromise between these two elements.

The dependability of protective relays can be adjusted by their response to in-zone faults; refer to Section 2.5. Hence, it depends on the value of plug setting (PS). As a role of thumb, the pickup value should be greater than the maximum overload current and less than the minimum fault current; refer to (7.20)–(7.23) and Figure 7.2. The two objective functions can be assumed as follows:

$$\min_{TMS,PS} f_1(TMS, PS), \quad f_1(TMS, PS) = \sum_{i=1}^{n} \left[\frac{\beta \times TMS_i}{\left(\frac{I_{R_i}}{PS_i}\right)^{\alpha} - 1} + \gamma \right] \tag{15.8}$$

$$\min_{TMS,PS} f_2(TMS, PS), \quad f_2(TMS, PS) = \frac{PS^{\max} - \sum_{i=1}^{n} PS_i}{\sum_{i=1}^{n} \left[\frac{\beta \times TMS_i}{\left(\frac{I_{R_i}}{PS_i}\right)^{\alpha} - 1} + \gamma \right]} \tag{15.9}$$

It is clear that the second objective function is also designed for minimization, which can be achieved by either maximizing $\sum_{i=1}^{n} PS_i$ or the denominator term. The latter one can be maximized by either maximizing time multiplier setting (TMS) or PS. We can fix TMS at predefined values to mainly focus on PS. By using the CWA method, (15.7) can be expressed as follows:

$$\min_{TMS,PS} F(TMS, PS) \Rightarrow \min_{TMS,PS} \left[w f_1 + (1 - w) f_2 \right] \tag{15.10}$$

Example 15.1 Consider the three-bus test system shown in Figure F.1 with just the normal configuration; similar to Example 5.6. Select and modify the proper single-objective BBO program, from the MATLAB codes given in Chapter 7, to deal with this multi-objective ORC problem as a CWA-BBO. The two objective functions of this problem are given in (15.8) and (15.9). Use 500 simulation runs to plot the Pareto front. In each simulation run, only the best individual should be archived as nondominated point (X^*). Show the final results and plots.

Solution

By plugging-in (15.10) into CWA-BBO, the graphical and numerical results are obtained after completing all the simulation runs. Figure 15.2 shows the Pareto front among all the individuals, and Table 15.1 gives the values of the two objectives for some selected simulation runs. All the MATLAB files are given in a folder called `TimeVsReliability_3BusSystem`.

15.2.2 Operating Time vs. System Cost

Again, from Section 2.5, the speed performance is one of the main design criteria. As it becomes faster, more money has to be spent for; it is a common fact for any protective design. We have discussed this point when different relay technologies have been used together. Electromechanical

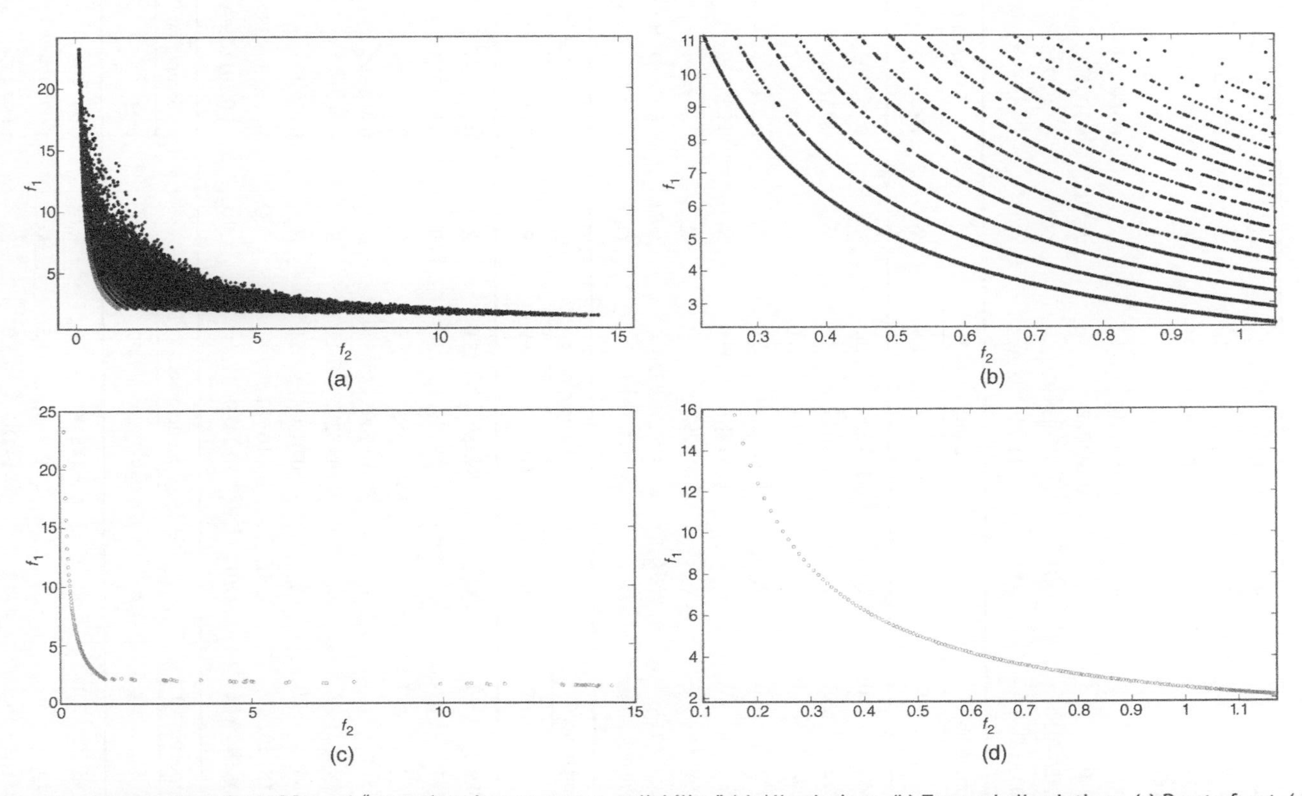

Figure 15.2 Archived Pareto front for MOP no. 1 “operating time vs. system reliability.” (a) All solutions. (b) Zoomed all solutions. (c) Pareto front. (d) Zoomed Pareto front.

Table 15.1 Optimal solutions of some selected simulation runs for MOP no. 1.

	Sample no. 1			Sample no. 2			Sample no. 3	
R_i	*TMS*	*PS*	R_i	*TMS*	*PS*	R_i	*TMS*	*PS*
R_1	0.119 400	1.5	R_1	0.105 768	2.5	R_1	0.102 943	4.5
R_2	0.101 198	1.5	R_2	0.105 584	1.5	R_2	0.103 754	1.5
R_3	0.101 331	2.0	R_3	0.106 095	4.0	R_3	0.106 795	5.0
R_4	0.101 151	2.0	R_4	0.117 264	1.5	R_4	0.109 559	1.5
R_5	0.101 494	1.5	R_5	0.109 570	1.5	R_5	0.100 264	4.0
R_6	0.101 909	1.5	R_6	0.100 322	1.5	R_6	0.100 876	1.5
OBJ_1	1.425 402 s		OBJ_1	1.558 916 s		OBJ_1	1.725 267 s	
OBJ_2	14.031 131 A/s		OBJ_2	11.225 748 A/s		OBJ_2	6.955 447 A/s	
	Sample no. 4			**Sample no. 5**			**Sample no. 6**	
R_i	*TMS*	*PS*	R_i	*TMS*	*PS*	R_i	*TMS*	*PS*
R_1	0.114 654	5.0	R_1	0.153 204	5.0	R_1	0.737 305	5.0
R_2	0.106 832	3.5	R_2	0.328 046	3.5	R_2	0.152 369	3.5
R_3	0.109 098	5.0	R_3	0.191 415	5.0	R_3	0.300 057	5.0
R_4	0.105 027	5.0	R_4	0.268 260	5.0	R_4	0.271 011	5.0
R_5	0.100 231	4.0	R_5	0.333 481	4.0	R_5	0.206 479	4.0
R_6	0.119 199	1.5	R_6	0.244 293	5.0	R_6	0.109 086	5.0
OBJ_1	2.083 621 s		OBJ_1	5.290 793 s		OBJ_1	6.258 461 s	
OBJ_2	2.879 603 A/s		OBJ_2	0.472 519 A/s		OBJ_2	0.399 459 A/s	
	Sample no. 7			**Sample no. 8**			**Sample no. 9**	
R_i	*TMS*	*PS*	R_i	*TMS*	*PS*	R_i	*TMS*	*PS*
R_1	0.886 589	5.0	R_1	0.593 897	5.0	R_1	0.703 833	5.0
R_2	0.102 594	3.5	R_2	0.724 935	3.5	R_2	0.859 720	3.5
R_3	0.175 340	5.0	R_3	0.202 458	5.0	R_3	0.474 051	5.0
R_4	0.574 351	5.0	R_4	0.396 969	5.0	R_4	0.464 855	5.0
R_5	0.356 309	4.0	R_5	0.288 321	4.0	R_5	0.362 082	4.0
R_6	0.175 084	5.0	R_6	0.602 440	5.0	R_6	0.639 326	5.0
OBJ_1	8.181 592 s		OBJ_1	10.082 895 s		OBJ_1	12.398 389 s	
OBJ_2	0.305 564 A/s		OBJ_2	0.247 945 A/s		OBJ_2	0.201 639 A/s	
	Sample no. 10			**Sample no. 11**			**Sample no. 12**	
R_i	*TMS*	*PS*	R_i	*TMS*	*PS*	R_i	*TMS*	*PS*
R_1	0.475 195	5.0	R_1	0.674 818	5.0	R_1	1.092 463	5.0
R_2	0.305 718	3.5	R_2	0.748 912	3.5	R_2	1.090 779	3.5
R_3	0.585 231	5.0	R_3	0.419 181	5.0	R_3	1.094 433	5.0
R_4	0.725 076	5.0	R_4	1.093 047	5.0	R_4	1.096 749	5.0
R_5	0.975 963	4.0	R_5	0.779 020	4.0	R_5	1.091 124	4.0
R_6	0.884 814	4.0	R_6	1.098 726	5.0	R_6	1.099 481	5.0
OBJ_1	14.349 629 s		OBJ_1	17.606 626 s		OBJ_1	23.249 276 s	
OBJ_2	0.174 221 A/s		OBJ_2	0.141 992 A/s		OBJ_2	0.107 530 A/s	

relays are slower than solid-state relays, and the latter type is slower than the state-of-the-art numerical relays. By referring to (5.3) in Chapter 5, it considers that the operating time of the backup relay should be at least equal to:

$$T_{j,k}^{\text{bc}} = T_{i,k}^{\text{pr}} + CTI = T_{i,k}^{\text{pr}} + T_{CB} + T_{OS} + T_{SM} \tag{15.11}$$

Practically, $T_{i,k}^{pr}$ is affected by two terms of errors, relay timing error (E_R) and current transformer (CT)-ratio error (E_{CT}). If both of these errors are considered, then (15.11) is replaced by (15.12) as follows (ALSTOM, 2002):

$$T_{j,k}^{\text{bc}} = \left[\frac{2E_R + E_{CT}}{100}\right] T_{i,k}^{\text{pr}} + T_{CB} + T_{OS} + T_{SM} \tag{15.12}$$

The last equation will not be used here, but it gives a clear figure about how the total operating time of any protective system can be effectively enhanced by using high quality[3] current transformers, protective relays, and circuit breakers. In contrast, the total price will also increase. Thus, it gives two conflicting objectives; *operating speed* and *total cost*.

The problem here is that there is no any standard marketing function that gives a common trend of the growing price based on the quality of the protective set. Instead, each manufacturer has its own price list that depends on the market's supply and demand. However, the common thing is that the total price will definitely increase as the quality of the protective equipment increases. In this part of the study, the total price will be simulated based on the following equation[4]:

$$\min_{TMS,PS} f_3(TMS, PS), \quad f_3(TMS, PS) = \frac{TMS^{\max} - \sum_{i=1}^{n} TMS_i}{\sum_{i=1}^{n} \left[\frac{\beta \times TMS_i}{\left(\frac{I_{R_i}}{PS_i}\right)^{\alpha} - 1} + \gamma\right]} \tag{15.13}$$

Thus, as $TMS \uparrow$, $T \uparrow$, $f_3 \downarrow$ and price $\downarrow$; and vice versa. Based on this, (15.7) can be expressed as follows:

$$\min_{TMS,PS} F(TMS, PS) \Rightarrow \min_{TMS,PS} \left[w f_1 + (1 - w) f_3\right] \tag{15.14}$$

Similarly, we can fix *PS* at predefined values to mainly focus on *TMS*. This approach can simplify the problem and accelerate the solution as seen before with linear programming (LP), multi-start LP, and BBO-LP.

Example 15.2 Repeat Example 15.1, but now with considering (15.13) instead of (15.9) and continuous *PS* instead of discrete. Show the final results and plots.

Solution

Again, CWA-BBO is used to solve the new single objective function given in (15.14). All the dominated and nondominated sets of all the simulation runs as well as the separated Pareto front "nondominated set" are shown in different plots of Figure 15.3. The results of some selected simulation runs are tabulated in Table 15.2. All the MATLAB files are given in a folder called `TimeVsCost_3BusSystem`.

3 In terms of accuracy and/or speed.

4 As said before, we can use any other equation. It does not matter. The important thing is to know the concept behind using multi-objective optimization.

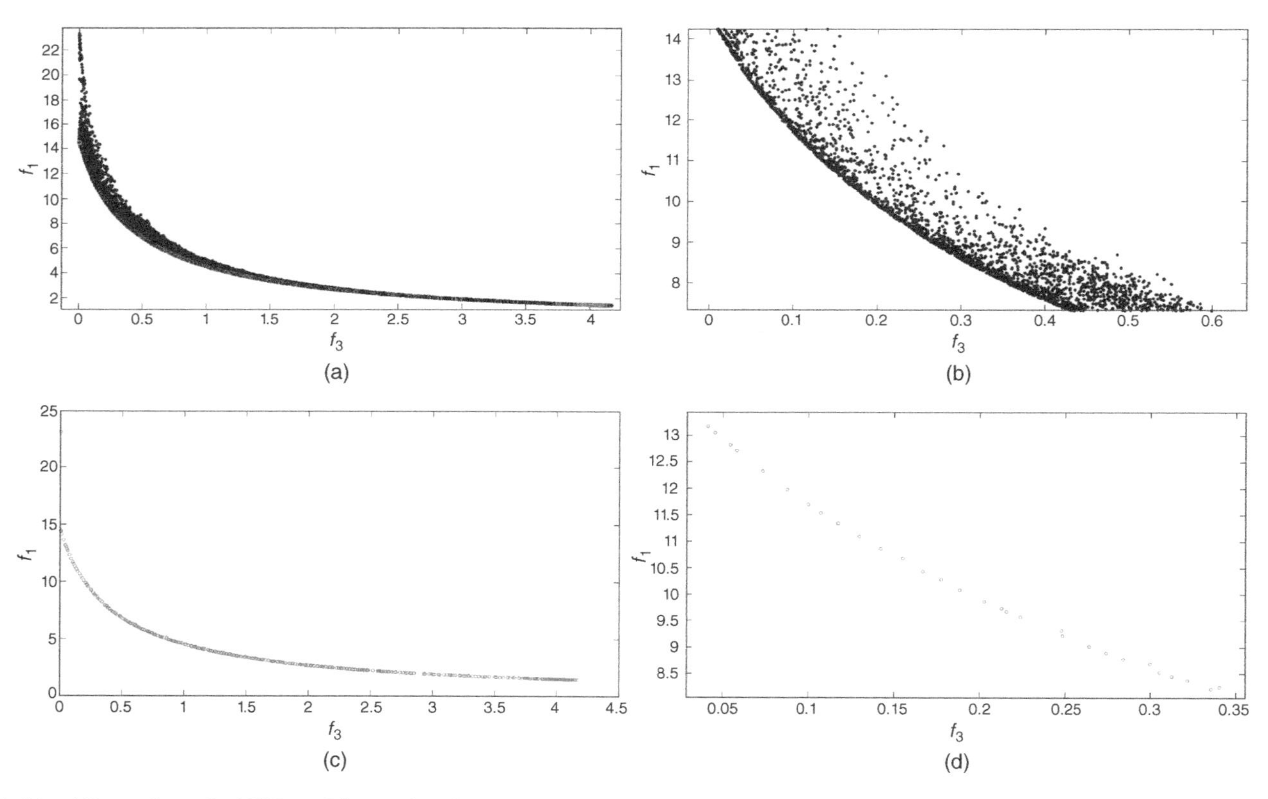

Figure 15.3 Archived Pareto front for MOP no. 2 "operating time vs. system cost." (a) All solutions. (b) Zoomed all solutions. (c) Pareto front. (d) Zoomed Pareto front.

Table 15.2 Optimal solutions of some selected simulation runs for MOP no. 2

	Sample no. 1			Sample no. 2			Sample no. 3	
R_i	*TMS*	*PS*	R_i	*TMS*	*PS*	R_i	*TMS*	*PS*
R_1	0.104 230	2.5	R_1	0.119 006	2.0	R_1	0.147 275	2.0
R_2	0.102 477	1.5	R_2	0.103 786	1.5	R_2	0.143 703	1.5
R_3	0.121 159	1.5	R_3	0.146 706	1.5	R_3	0.197 508	1.5
R_4	0.126 277	1.5	R_4	0.133 685	1.5	R_4	0.172 064	1.5
R_5	0.102 381	1.5	R_5	0.129 704	1.5	R_5	0.129 758	1.5
R_6	0.105 777	1.5	R_6	0.115 555	1.5	R_6	0.115 897	1.5
OBJ_1	1.505 199 s		OBJ_1	1.672 217 s		OBJ_1	2.015 320 s	
OBJ_2	3.944 791 s^{-1}		OBJ_2	3.499 283 s^{-1}		OBJ_3	2.825 257 s^{-1}	
	Sample no. 4			**Sample no. 5**			**Sample no. 6**	
R_i	*TMS*	*PS*	R_i	*TMS*	*PS*	R_i	*TMS*	*PS*
R_1	0.453 155	1.5	R_1	0.452 897	1.5	R_1	0.541 892	1.5
R_2	0.230 017	1.5	R_2	0.238 364	1.5	R_2	0.193 841	1.5
R_3	0.100 191	4.5	R_3	0.643 192	1.5	R_3	0.846 518	1.5
R_4	0.183 362	1.5	R_4	0.203 504	1.5	R_4	0.160 764	1.5
R_5	0.295 753	1.5	R_5	0.242 788	1.5	R_5	0.547 800	1.5
R_6	0.128 027	1.5	R_6	0.111 203	2.0	R_6	0.111 032	1.5
OBJ_1	3.145 016 s		OBJ_1	4.086 975 s		OBJ_1	5.112 410 s	
OBJ_2	1.656 429 s^{-1}		OBJ_2	1.151 965 s^{-1}		OBJ_2	0.821 169 s^{-1}	
	Sample no. 7			**Sample no. 8**			**Sample no. 9**	
R_i	*TMS*	*PS*	R_i	*TMS*	*PS*	R_i	*TMS*	*PS*
R_1	0.625 943	1.5	R_1	0.812 225	1.5	R_1	1.065 976	1.5
R_2	0.441 704	1.5	R_2	0.880 687	1.5	R_2	1.099 337	1.5
R_3	0.998 175	1.5	R_3	1.092 196	1.5	R_3	1.098 713	1.5
R_4	0.296 038	1.5	R_4	0.538 230	1.5	R_4	0.659 057	1.5
R_5	0.987 643	1.5	R_5	1.092 619	1.5	R_5	1.099 708	1.5
R_6	0.203 573	1.5	R_6	0.284 767	1.5	R_6	0.340 063	1.5
OBJ_1	7.590 269 s		OBJ_1	10.083 683 s		OBJ_1	11.540 127 s	
OBJ_2	0.401 425 s^{-1}		OBJ_2	0.188 351 s^{-1}		OBJ_2	0.107 204 s^{-1}	
	Sample no. 10			**Sample no. 11**			**Sample no. 12**	
R_i	*TMS*	*PS*	R_i	*TMS*	*PS*	R_i	*TMS*	*PS*
R_1	1.097 540	1.5	R_1	10.981 88	1.5	R_1	1.097 015	5.0
R_2	1.093 079	1.5	R_2	1.098 989	1.5	R_2	1.099 068	3.5
R_3	1.095 508	1.5	R_3	1.092 255	1.5	R_3	1.096 731	5.0
R_4	1.096 826	1.5	R_4	1.092 255	1.5	R_4	1.096 731	5.0
R_5	1.096 066	1.5	R_5	1.092 568	1.5	R_5	1.099 737	3.5
R_6	0.528 667	1.5	R_6	1.092 143	1.5	R_6	1.099 982	5.0
OBJ_1	13.049 695 s		OBJ_1	14.465 240 s		OBJ_1	23.139 654 s	
OBJ_2	0.045 389 s^{-1}		OBJ_2	0.002 372 s^{-1}		OBJ_2	0.000 376 s^{-1}	

15.2.3 Operating Time vs. System Reliability vs. System Cost

Based on the previous two MOPs, it is possible to combine these three objective functions into a one objective function. Equation (15.6) can be used, but it is very hard to tune the weights $\{w_1, w_2, w_3\}$ manually. Instead, based on (Jin, 2002), RWA-BBO is used here with some slight modifications. This approach can be briefly described as follows:

$$\begin{aligned} w_1 &= \text{rand} \\ w_2 &= (1 - w_1) \times \text{rand} \\ w_3 &= 1 - w_1 - w_2 \end{aligned} \tag{15.15}$$

Thus, the final expression of this MOP is:

$$\min_{TMS,PS} F(TMS, PS) \Rightarrow \min_{TMS,PS} \left[w_1 f_1 + w_2 f_2 + w_3 f_3\right] \tag{15.16}$$

Example 15.3 Consider the simple three-bus test system solved in the last two examples. Design a multi-objective optimization algorithm using RWA-BBO to solve the ORC problem of this test system when all the objective functions given in (15.8), (15.9), and (15.13) are taken into account. Use discrete *PS*. Show the final results and the surface plots. (**Hint:** Modify the MATLAB code given in Example 15.1 by embedding (15.13), or the MATLAB code given in Example 15.2 by embedding (15.9) and converting *PS* from being continuous to discrete. Use (15.16) to add the third objective.)

Solution

Because this MOP contains three objective functions, the optimizer should generate 3D Pareto front. All the modified MATLAB files are given in a folder called `TimeVsReliabilityVs-Cost_3BusSystem`. All the dominated and nondominated sets of all the simulation runs as well as the separated Pareto front "nondominated set" are shown in different plots of Figure 15.4. Table 15.3 contains the results of some selected simulation runs.

15.3 Further Discussions

The suggested objective functions (i.e. f_1, f_2, and f_3) can be modified by considering other aspects. For example, if the protective relays are very expensive, then the value of coordination time interval (*CTI*) will be smaller as the price reflects the technology embedded in the relays. This term can be included in $f_3(TMS, PS)$, so as the weight of f_3 increases, the value of *CTI* will decrease. One of the simplest methods can be presented as follow:

$$CTI(w) = \frac{CTI^{\max}}{1 + w} \tag{15.17}$$

Moreover, the expensive protective relays are those relays that come with the modern technology. They are numerical where *PS* can be set with semi-continuous values. This property can give more feasible solutions and enhance the points of the Pareto front.

Finally, the reader might find different objective functions in the literature. Also, the reader might see some highly advanced multi-objective approaches. This chapter explains the basic mechanism and the concepts behind this new world of research. The goal here is to know the meaning of multi-objective coordination and how to implement it practically. Understanding it gives the ability to apply any objective function the reader wants. Also, it can simplify the other more advanced MOEAs.

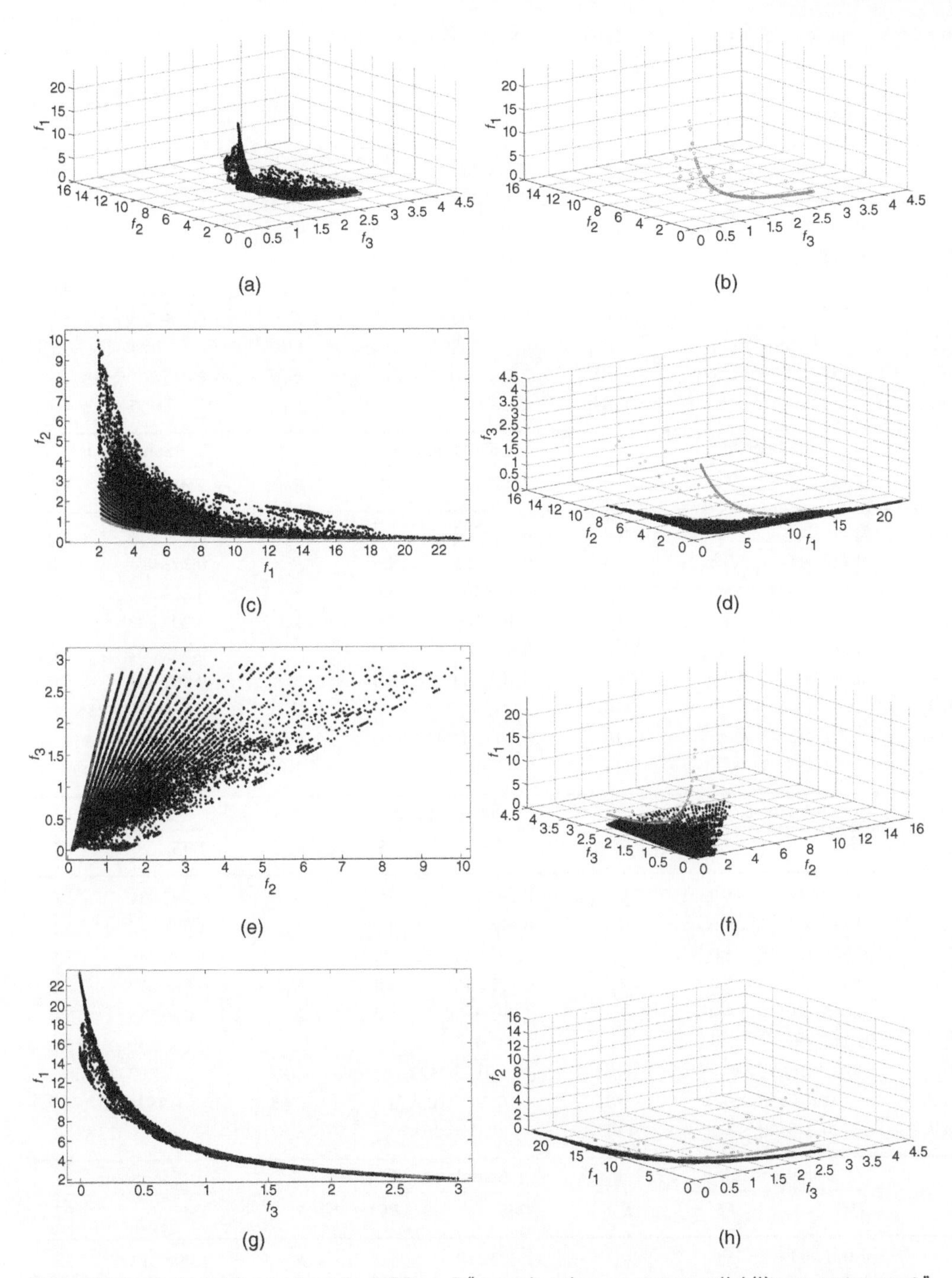

Figure 15.4 Archived Pareto front for MOP no. 3 "operating time vs. system reliability vs. system cost." (a) All solutions. (b) Pareto front. (c) $f_1 - f_2$ plane in 2D. (d) $f_1 - f_2$ plane in 3D. (e) $f_2 - f_3$ plane in 2D. (f) $f_2 - f_3$ plane in 3D. (g) $f_3 - f_1$ plane in 2D. (h) $f_3 - f_1$ plane in 3D.

Table 15.3 Optimal solutions of some selected simulation runs for MOP

	Sample no. 1			Sample no. 2			Sample no. 3	
R_i	*TMS*	*PS*	R_i	*TMS*	*PS*	R_i	*TMS*	*PS*
R_1	0.123 881	1.5	R_1	0.100 403	5.0	R_1	0.257 124	5.0
R_2	0.108 470	1.5	R_2	0.104 885	3.5	R_2	0.205 712	3.5
R_3	0.103 502	2.0	R_3	0.108 967	5.0	R_3	0.125 527	5.0
R_4	0.111 952	1.5	R_4	0.100 611	5.0	R_4	0.124 986	5.0
R_5	0.101 072	1.5	R_5	0.100 576	4.0	R_5	0.122 361	4.0
R_6	0.105 532	1.5	R_6	0.104 468	3.5	R_6	0.100 199	5.0
OBJ_1	1.463 057 s		OBJ_1	2.096 92 s		OBJ_1	3.245 278 s	
OBJ_2	14.011 760 A/s		OBJ_2	1.906 858 A/s		OBJ_2	0.770 350 A/s	
OBJ_3	4.063 815 s^{-1}		OBJ_3	2.850 795 s^{-1}		OBJ_3	1.745 333 s^{-1}	

	Sample no. 4			Sample no. 5			Sample no. 6	
R_i	*TMS*	*PS*	R_i	*TMS*	*PS*	R_i	*TMS*	*PS*
R_1	0.240 853	5.0	R_1	0.146 342	5.0	R_1	0.123 218	5.0
R_2	0.465 441	3.5	R_2	0.335 881	3.5	R_2	0.773 041	3.5
R_3	0.125 714	5.0	R_3	0.333 400	5.0	R_3	0.274 800	5.0
R_4	0.132 969	5.0	R_4	0.130 106	5.0	R_4	0.207 703	5.0
R_5	0.219 998	4.0	R_5	0.541 245	4.0	R_5	0.587 231	4.0
R_6	0.107 009	5.0	R_6	0.101 139	5.0	R_6	0.109 477	5.0
OBJ_1	4.290 510 s		OBJ_1	5.187 989 s		OBJ_1	6.641 823 s	
OBJ_2	0.582 681 A/s		OBJ_2	0.481 882 A/s		OBJ_2	0.376 403 A/s	
OBJ_3	1.237 153 s^{-1}		OBJ_3	0.966 056 s^{-1}		OBJ_3	0.681 218 s^{-1}	

	Sample no. 7			Sample no. 8			Sample no. 9	
R_i	*TMS*	*PS*	R_i	*TMS*	*PS*	R_i	*TMS*	*PS*
R_1	0.210 629	5.0	R_1	0.243 582	5.0	R_1	0.392 881	5.0
R_2	0.769 402	3.5	R_2	0.978 417	3.5	R_2	1.099 742	3.5
R_3	0.521 038	5.0	R_3	0.631 664	5.0	R_3	1.099 769	5.0
R_4	0.333 327	5.0	R_4	0.257 346	5.0	R_4	0.293 857	5.0
R_5	0.715 391	4.0	R_5	1.095 916	4.0	R_5	1.090 585	4.0
R_6	0.142 396	5.0	R_6	0.102 915	5.0	R_6	0.100 205	5.0
OBJ_1	8.766 152 s		OBJ_1	10.530 322 s		OBJ_1	13.037 029 s	
OBJ_2	0.285 188 A/s		OBJ_2	0.237 410 A/s		OBJ_2	0.191 761 A/s	
OBJ_3	0.445 785 s^{-1}		OBJ_3	0.312 446 s^{-1}		OBJ_3	0.193 523 s^{-1}	

	Sample no. 10			Sample no. 11			Sample no. 12	
R_i	*TMS*	*PS*	R_i	*TMS*	*PS*	R_i	*TMS*	*PS*
R_1	1.091 994	5.0	R_1	1.092 419	5.0	R_1	1.096 613	5.0
R_2	1.093 583	3.5	R_2	1.097 263	3.5	R_2	1.099 352	3.5
R_3	1.095 180	5.0	R_3	1.091 484	5.0	R_3	1.095 692	5.0
R_4	1.093 782	1.5	R_4	1.076 529	5.0	R_4	1.099 845	5.0
R_5	1.099 873	4.0	R_5	1.092 945	4.0	R_5	1.099 773	4.0
R_6	0.102 773	5.0	R_6	0.100 272	5.0	R_6	1.087 000	5.0
OBJ_1	16.941 034 s		OBJ_1	18.546 619 s		OBJ_1	23.273 284 s	
OBJ_2	0.354 170 A/s		OBJ_2	0.134 795 A/s		OBJ_2	0.107 419 A/s	
OBJ_3	0.060 375 s^{-1}		OBJ_3	0.056 565 s^{-1}		OBJ_3	0.000 933 s^{-1}	

Problems

Written Exercises

W15.1 What is the main drawback of the CWA method?

W15.2 A team of electrical engineers, data scientists, and computer programmers wants to design a simple MOEA to coordinate directional over-current relays (DOCRs) of a very-large distribution network. The MOP of this program will have two objective functions. The optimizer will use the CWA method where the weight (w) moves from 0 to 1 by a step-size of 0.01. If the average CPU time of the program is five minutes for each weight, estimate the total CPU time if the program is initiated with 20 simulation runs for each weight.

W15.3 Suppose that the preceding research team wants to sell the multi-objective relay coordinator designed in Exercise W15.2 to a big electric protection company as an adaptive program. If the company updates the relay settings every five days, can the research team convince the company to buy their software? Explain.

W15.4 Consider the results obtained in Exercises W15.2 and W15.3. What is the maximum acceptable trials per weight?

W15.5 Consider the results obtained in Exercises W15.2 and W15.3. What is the maximum acceptable CPU time per weight?

W15.6 Consider the results obtained in Exercises W15.2 and W15.3. What is the minimum acceptable step-size resolution? Round the answer to two decimal places.

W15.7 Suppose that you have been employed as a member of the team. State, at least, five possible modifications or adjustments to accelerate the processing speed of the multi-objective relay coordinator designed in Exercise W15.2.

Computer Exercises

C15.1 Modify Example 15.1 by using continuous *PS*.

C15.2 Modify Example 15.2 by using discrete *PS*.

C15.3 Modify Example 15.3 by using continuous *PS*.

C15.4 One way to accelerate the programs is by linearizing the optimization model. Fix the plug settings of all the relays mentioned in Example 15.2 and then use the same meta-heuristic algorithm to optimize *TMS*. (**Hint:** Do not use LP.)

C15.5 Modify the MATLAB code given in Example 15.1 by fixing *TMS* and replacing the second objective function by the following one:

$$\min_{PS} f_2(PS), \quad f_2(PS) = PS^{\max} - \sum_{i=1}^{n} PS_i$$

(**Hint:** Just fix *TMS*. Do not use the transformation-based approach to linearize the model.)

C15.6 Repeat Exercise C15.5, but by using continuous *PS*.

C15.7 Modify the MATLAB code given in Example 15.1 to solve the nine-bus test system shown in Figure F.13. Use the new objective function given in Exercise C15.5 for f_2.

C15.8 Repeat Exercise C15.7, but by fixing *TMS*.

C15.9 Repeat Exercise C15.7, but by fixing *PS* and replacing the second objective function by the following one:

$$\min_{TMS} f_2(TMS), \quad f_2(TMS) = \frac{TMS^{\max} - \sum_{i=1}^{n} TMS_i}{n \times \max\left\{TMS^{\max} - \left[TMS_1, TMS_2, \ldots, TMS_n\right]\right\}}$$

16

Optimal Coordination of Distance and Overcurrent Relays

After reading this chapter you should be able to:

- Understand the importance of coordinating distance and overcurrent relays
- Know the variation in operating time of distance relays at different fault zones
- Determine when do distance relays operate as primary or backup protective devices
- Express different mathematical models for this special optimal coordination problem

In Chapters 7, 8, 9, 10, 11, 12, 13, 14, and 15, we have seen many protective devices; including directional and nondirectional OCRs as well as fuses. We have seen how to deal with European and North American electromechanical, static, digital, and numerical OCRs. Also, we have seen the difference between definite-current/instantaneous, definite-time, and inverse-time characteristic curves. The reason of using OCRs instead of other relays is that OCRs can compromise between many performance criteria. They are fast, cheap, reliable, adequate, and simple and do not need lots of maintenance. However, for transmission and sub-transmission protection systems, the story is different. We could see only distance relays or a combination of distance and overcurrent relays (Perez and Urdaneta, 2001). For important transmission systems, two similar distance relays are used at each line.[1] For sub-transmission systems and some transmission systems, a combination of distance and overcurrent relays is implemented (Chabanloo et al., 2011b). Thus, the optimization models discussed in Chapters 7, 8, 9, 10, 11, 12, 13, 14, and 15 need to be modified to take into account distance relays.

16.1 Introduction

In distance protection, the voltage and current at the beginning of the line are utilized to determine the fault location where the impedance can be measured through a relay comparator. To prevent responding to faults behind the protective zone, some relay characteristics are employed. The main relay characteristics are:

- Impedance relay
- Reactance relay

1 They are called main-1 and main-2.

Optimal Coordination of Power Protective Devices with Illustrative Examples, First Edition. Ali R. Al-Roomi.

Companion website: www.wiley.com/go/al-roomi/optimalcoordination

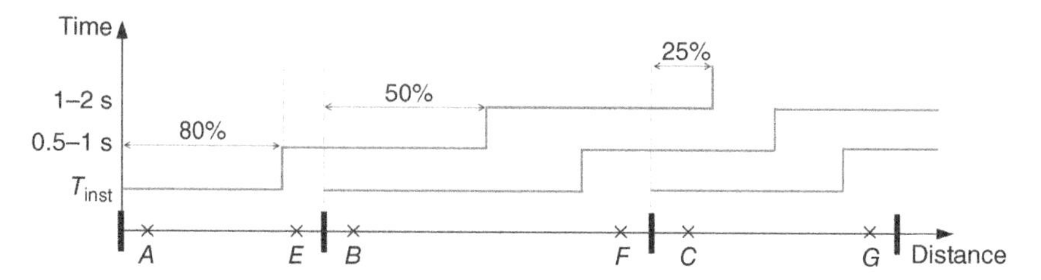

Figure 16.1 Illustration of TDCCs of distance relays.

- Ohm or admittance relay
- Mho or angle impedance relay
- Offset mho relay
- Quadrilateral relay

Figure 16.1 illustrates the time–distance characteristic curves (TDCCs) of distance relays. It shows that each distance relay operates instantaneously for any fault occurs in the first 80% of the branch. This length is known as the first zone of protection (or zone-1). Faults beyond this point and up to 50% of the next branch are cleared after awaiting a time delay of about 0.5–1 s. This length is known as the second zone of protection (or zone-2). A third zone can be established by a starting relay to protect the remaining 50% of the second branch and the first 25% of the third branch. The time delay of zone-3 is in the range of 1–2 s (El-Hawary, 2008). The detailed description about distance relays and the theory behind them can be found in many references. Our aim here is to optimally coordinate distance relays with overcurrent relays.

The literature contains many studies about optimal coordination of distance and overcurrent relays. Many mixed protection schemes have been suggested and solved by different optimization algorithms. Some of these studies are reported in Perez and Urdaneta (2001), Abyaneh et al. (2008), Chabanloo et al. (2011b), Singh et al. (2012a), Ahmadi et al. (2017), Roy et al. (2017), Singh et al. (2018), Yazdaninejadi et al. (2019), and Rivas et al. (2019). The following Sections 16.2, 16.3, and 16.4 cover three different mathematical models.

16.2 Basic Mathematical Modeling

Conventional OCRs are widely used to protect radial distribution systems and source nodes. For mesh distribution systems, DOCRs are employed to achieve the selectivity among primary and backup relays. The coordination between primary and backup relays is illustrated in Figure 16.2 when both R^{pr} and R^{bc} have the same inverse-time characteristic curve. We can see that for lower fault currents, which occur when the fault location is away from the near-end bus, the difference in operating times of R^{pr} and R^{bc} is bigger than that measured at the close-in fault. This is why satisfying the selectivity constraint at the near-end fault will ensure that the other constraints at less sever faults are also satisfied when overcurrent relays are equipped with the same TCCC.[2]

For transmission and sub-transmission systems, when both distance and overcurrent relays are imposed, the coordination task is more tricky. There are many ways to do that. One of the basic

2 This phenomenon has been covered in Chapter 11.

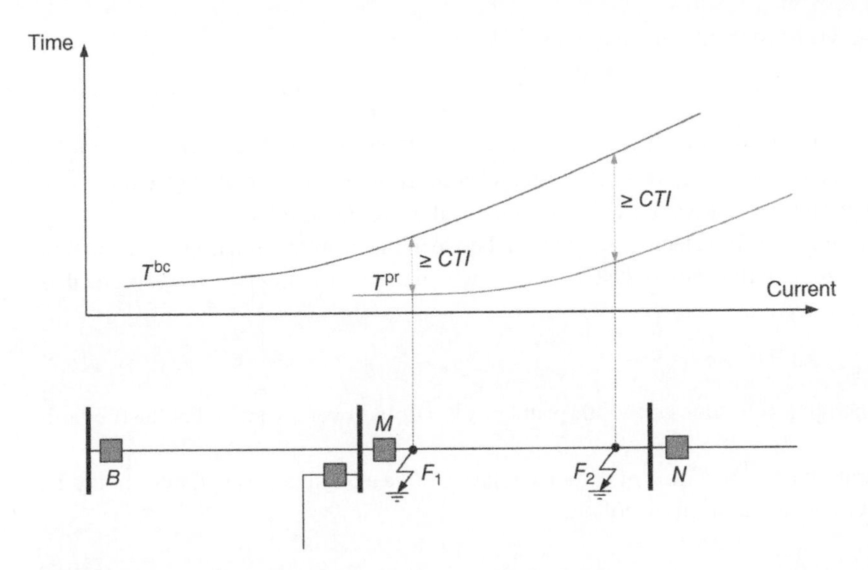

Figure 16.2 Basic coordination between overcurrent relays.

approaches is to satisfy two selectivity constraints for each distance–overcurrent pair. The first one is to ensure that the operating time of zone-2 of the primary distance relay is faster than that of the backup directional overcurrent relay and the difference is equal to or greater than the coordination time interval used for this mixed protection scheme. The other selectivity constraint is to ensure that the difference between the operating time of the primary directional overcurrent relay and that of the second zone of the backup distance relay is equal to or greater than the preceding time interval. The concept of this basic coordination scheme is illustrated in Figure 16.3. Thus, the conventional mathematical model discussed in Chapter 7 can be used here, but with some essential

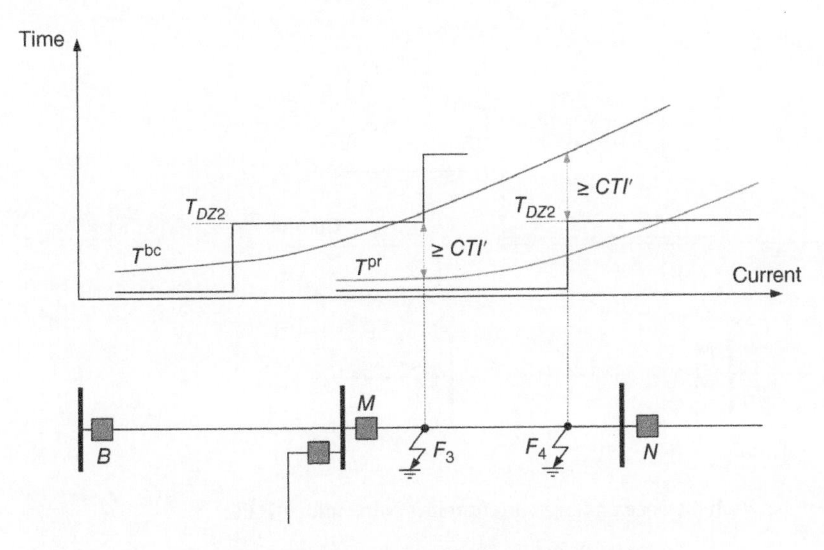

Figure 16.3 Coordination between distance and overcurrent relays.

modifications. First, we have to add the following selectivity constraint:

$$T^{\text{bc}}_{OCR_j,F_4} - T_{DZ2_i} \geqslant CTI' \tag{16.1}$$

where $T^{\text{bc}}_{OCR_j,F_4}$ is the operating time of the jth backup directional overcurrent relay for the fault occurred at F_4, T_{DZ2_i} is the operating time of the second zone of the ith primary distance relay, and CTI' is the coordination time interval between distance and overcurrent relays.[3]

The second step is to satisfy the selectivity constraint between the primary directional overcurrent relay and the second zone of the backup distance relay, which can be mathematically expressed as follows:

$$T_{DZ2_j} - T^{\text{pr}}_{OCR_i,F_3} \geqslant CTI' \tag{16.2}$$

where $T^{\text{pr}}_{OCR_i,F_3}$ is the operating time of the ith primary directional overcurrent relay for the fault occurred at F_3.

Because the operating time of zone-2 of distance relays can be adjusted, this variable should be bounded between two practical limits as follows:

$$T^{\min}_{DZ2} \leqslant T_{DZ2_i} \leqslant T^{\max}_{DZ2} \tag{16.3}$$

By fixing plug settings, we can easily find the optimal solution to this mixed protection ORC problem by using LP.

16.3 Mathematical Modeling with Considering Multiple TCCCs

The mathematical model presented in Section 16.2 has some limitations. For example, it is hard to apply multiple TCCCs,[4] minimize the discrimination times between P/B relays, or check the coordination for the fault located just close to the main circuit breaker. Figure 16.4 shows how

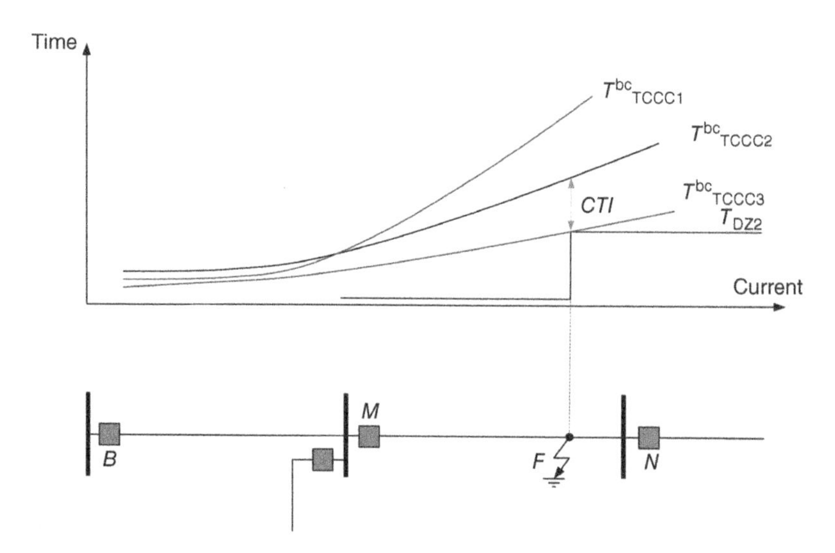

Figure 16.4 Coordination between distance and overcurrent relays with various TCCCs.

3 CTI' does not have to be the same CTI used in the selectivity constraints between overcurrent relays.
4 Please, refer to the inherent weaknesses of the multi-TCCCs approach highlighted in Chapter 10.

some TCCCs cause miscoordination. It is clear that the selectivity constraint at the critical fault point[5] is satisfied by the first and second characteristic curves of the backup DOCR, while this discrimination time is less than CTI' when the third characteristic curve is used. The study reported in Chabanloo et al. (2011b) tried to solve these infeasibility issues. We can summarize the mathematical model of this study as follows:

16.3.1 Inequality Constraints

By referring to Figure 16.2, where the coordination is done between two overcurrent relays, two selectivity constraints must be considered for each P/B pair; one for the near-end fault and the other for the far-end fault. Thus, we have:

$$T^{bc}_{OCR_j,F_1} - T^{pr}_{OCR_i,F_1} \geqslant CTI \tag{16.4}$$

$$T^{bc}_{OCR_j,F_2} - T^{pr}_{OCR_i,F_2} \geqslant CTI \tag{16.5}$$

which are derived from (7.25).

Now, let's consider Figure 16.5 as an example of combined distance–overcurrent relays coordination problems. The following two selectivity constraints must be included when the overcurrent relay acts as a backup protective device for the distance relay:

$$T^{bc}_{OCR_j,F_3} - T_{DZ1_i} \geqslant CTI' \tag{16.6}$$

$$T^{bc}_{OCR_j,F_4} - T_{DZ2_i} \geqslant CTI' \tag{16.7}$$

By referring to Figure 16.5, there is also one selectivity constraint that should be taken into account when the distance relay acts as a backup protective device for the primary overcurrent relay. This happens when the fault occurs at F_5. Therefore, the following inequality constraint should be included in the mathematical model of the optimization problem:

$$T_{DZ2_j} - T^{pr}_{OCR_i,F_5} \geqslant CTI'' \tag{16.8}$$

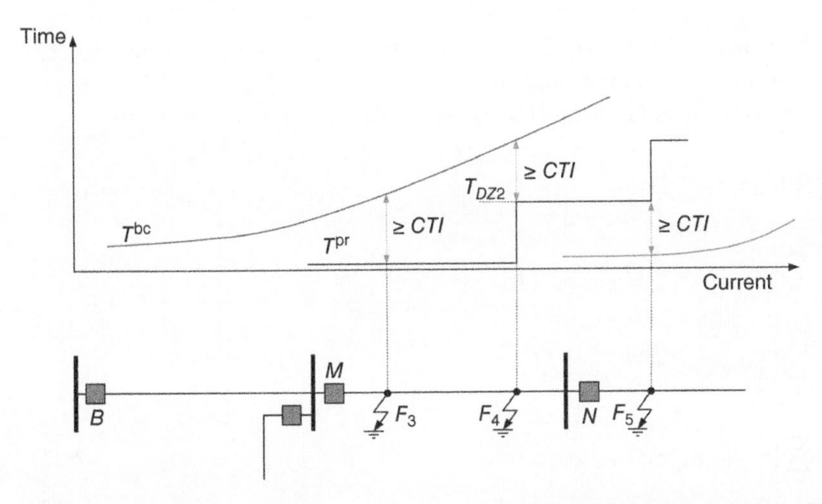

Figure 16.5 Coordination between distance and overcurrent relays at critical fault locations.

5 It is the changeover from zone-1 to zone-2.

where CTI'' is the coordination time interval when the overcurrent relay acts as a primary protective device.[6]

16.3.2 Objective Function

The cost function is based on the one given in (7.13), but with considering distance relays. It can be summarized as follows:

$$OBJ = \min\left[OBJ_{\text{oc-oc}} + OBJ_{\text{oc-dis}} + OBJ_{\text{dis-oc}}\right] \tag{16.9}$$

where these three terms are used as follows:

- $OBJ_{\text{oc-oc}}$ is used between the primary and backup overcurrent relays
- $OBJ_{\text{oc-dis}}$ is used when the overcurrent relay is the backup of the distance relay
- $OBJ_{\text{dis-oc}}$ is used when the distance relay is the backup of the overcurrent relay

The first term is used to consider (16.4) and (16.5) in the objective function. Based on (7.13), the mathematical expression can be derived from Figure 16.2 as follows:

$$\begin{aligned} OBJ_{\text{oc-oc}} = \kappa_1 \cdot \sum_{i=1}^{\varrho} T_i^2 + \lambda_1 \\ \times \sum_{j_1=1}^{\sigma_1}\left(T_{\text{oc-oc},ji,F_1} - \left|T_{\text{oc-oc},ji,F_1}\right|\right)^2 + \lambda_1 \\ \times \sum_{j_1=1}^{\sigma_1}\left(T_{\text{oc-oc},ji,F_2} - \left|T_{\text{oc-oc},ji,F_2}\right|\right)^2 + \lambda_2 \\ \times \sum_{j_1=1}^{\sigma_1}\left(T_{\text{oc-oc},ji,F_1} + \left|T_{\text{oc-oc},ji,F_1}\right|\right)^2 + \lambda_3 \\ \times \sum_{j_1=1}^{\sigma_1}\left(T_{\text{oc-oc},ji,F_2} + \left|T_{\text{oc-oc},ji,F_2}\right|\right)^2 \end{aligned} \tag{16.10}$$

where $T_{\text{oc-oc},ji,F_x}$ is the difference in the operating times of the jth backup and the ith primary overcurrent relays for a fault occurring at the xth location.

The second term is used to consider (16.6) and (16.7) in the objective function as follows:

$$\begin{aligned} OBJ_{\text{oc-dis}} = \lambda_1 \sum_{j_2=1}^{\sigma_2}\left(T_{\text{oc-dis},ji,F_3} - \left|T_{\text{oc-dis},ji,F_3}\right|\right)^2 \\ +\lambda_1 \sum_{j_2=1}^{\sigma_2}\left(T_{\text{oc-dis},ji,F_4} - \left|T_{\text{oc-dis},ji,F_4}\right|\right)^2 \\ +\lambda_4 \sum_{j_2=1}^{\sigma_2}\left(T_{\text{oc-dis},ji,F_3} + \left|T_{\text{oc-dis},ji,F_3}\right|\right)^2 \\ +\lambda_5 \sum_{j_2=1}^{\sigma_2}\left(T_{\text{oc-dis},ji,F_4} + \left|T_{\text{oc-dis},ji,F_4}\right|\right)^2 \end{aligned} \tag{16.11}$$

6 Note that the study reported in Perez and Urdaneta (2001) assumes $CTI'' = CTI'$; as can be seen in (16.2).

where $T_{\text{oc-dis},ji,F_x}$ is the difference in the operating times of the ith primary distance relay and the jth backup overcurrent relay for a fault occurring at the xth location.

Finally, (16.8) is included in the objective function by the third term as follows:

$$OBJ_{\text{dis-oc}} = \lambda_6 \sum_{j_3=1}^{\sigma_3} T_{\text{dis-oc},ji,F_5} \tag{16.12}$$

where $T_{\text{dis-oc},ji,F_x}$ is the difference in the operating times of the ith primary overcurrent relay and the jth backup distance relay for a fault occurring at the xth location.

From (7.12), $T_{\text{oc-oc},ji,F_x}$ can be calculated for the xth fault location as follows:

$$T_{\text{oc-oc},ji,F_x} = T^{\text{bc}}_{OCR_j,F_x} - T^{\text{pr}}_{OCR_i,F_x} - CTI \tag{16.13}$$

Also, by using the same concept, $T_{\text{oc-dis},ji,F_x}$ can be calculated for the xth fault location as follows:

$$T_{\text{oc-dis},ji,F_x} = T^{\text{bc}}_{OCR_j,F_x} - T_{DZ1_i} - CTI' \tag{16.14}$$

For $T_{\text{dis-oc},ji,F_x}$, the following expression can be used:

$$T_{\text{dis-oc},ji,F_x} = T_{DZ2_j} - T^{\text{pr}}_{OCR_i,F_x} - CTI'' \tag{16.15}$$

In the optimization algorithm, the second zone time delay (T_{DZ2_j}) can be updated at each iteration as follows:

$$T_{DZ2_j} = \max\left[T^{\text{pr}}_{OCR_i,F_5} + CTI'', T_z\right] \tag{16.16}$$

where T_z is the initial time delay for the second zone of the distance relay.

16.4 Mathematical Modeling with Considering Different Fault Locations

The mathematical model presented in Section 16.3 contains some selectivity constraints that are not covered in the basic one. However, this mixed coordination scheme requires to consider some additional fault points to ensure that the distance and overcurrent relays are fully coordinated. One of the studies that focus on this matter is reported in Labrador (2018) and Rivas et al. (2019). It uses only one TCCC for all overcurrent relays. The following Sections 16.4.1 and 16.4.2 highlight the objective function and the new selectivity constraints added to the mathematical model.

16.4.1 Objective Function

By taking into account the operating time of overcurrent relays at different fault locations and the operating time of zone-2 of distance relays, the cost function for this coordination approach can be mathematically derived from (7.5) as follows[7]:

$$OBJ = \min \sum_{i=1}^{\varrho} \left[\sum_{k=1}^{\ell} T^{\text{pr}}_{OCR_i,k} + T_{DZ2_i}\right] \tag{16.17}$$

7 Please note that the weight ($w_{i,k}$) that represents the fault probability in (7.5) is not considered here.

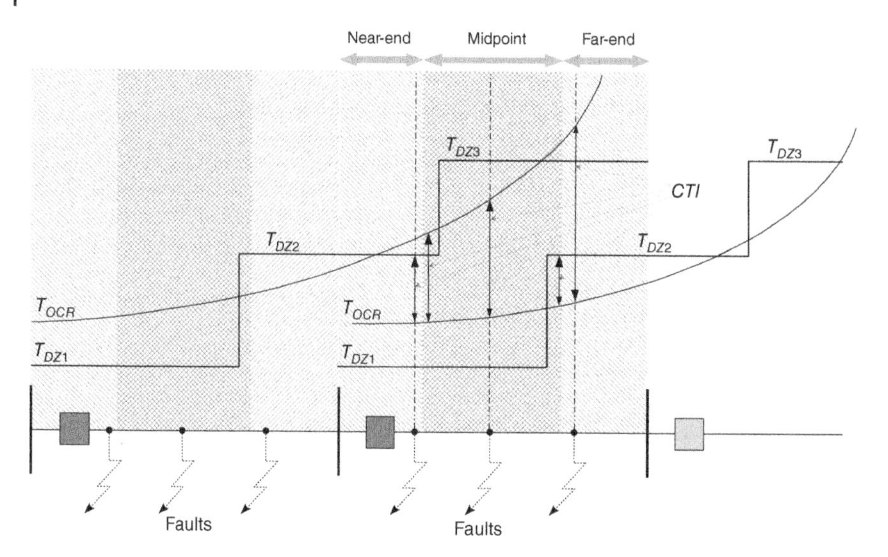

Figure 16.6 Coordination between overcurrent and distance relays at different fault locations.

16.4.2 Inequality Constraints

As said before, there are three zones of protection; zone-1, zone-2, and zone-3. Because the first zone provides an instantaneous trip, it is faster than DOCR. This means that DOCRs act as backup protective devices for zone-1 of distance relays. The problem appears with zone-2 where the selectivity between relays is required. Figure 16.6 shows the operating time of overcurrent and distance relays when they work together[8] (Labrador, 2018). It can be clearly seen that the operating time of distance relays are lower than that of overcurrent relays. This is valid for zone-1 where $T_{DZ1} < T_{OCR}$. Thus, by default, the near-end fault locations are cleared first by distance relays. The mathematical model presented in Rivas et al. (2019) tries to satisfy the selectivity criteria at the following fault points:

16.4.2.1 Near-End Faults

For the first zone, distance relays will always be the first. For the second zone, Figure 16.6 shows two selectivity constraints that need to be satisfied:

$$T_{DZ2_j} - T^{\text{pr}}_{OCR_i,\text{near}} \geqslant CTI \tag{16.18}$$

$$T^{\text{bc}}_{OCR_j,\text{near}} - T^{\text{pr}}_{OCR_i,\text{near}} \geqslant CTI \tag{16.19}$$

16.4.2.2 Middle-Point Faults

The third zone of distance relays shown in Figure 16.6 is not considered. The second zone is not under the protection scheme for faults located between 20% and 80% of the line. Therefore, there is only one selectivity constraint for primary and backup overcurrent relays:

$$T^{\text{bc}}_{OCR_j,\text{mid}} - T^{\text{pr}}_{OCR_i,\text{mid}} \geqslant CTI \tag{16.20}$$

8 Note that the figure assumes $CTI'' = CTI' = CTI$.

16.4.2.3 Far-End Faults

Similar to (16.18)–(16.19), the following selectivity constraints should be satisfied at the far-end faults:

$$T_{DZ2_i} - T^{\text{pr}}_{OCR_i,\text{far}} \geqslant CTI \tag{16.21}$$

$$T^{\text{bc}}_{OCR_j,\text{far}} - T^{\text{pr}}_{OCR_i,\text{far}} \geqslant CTI \tag{16.22}$$

Remember that the operating time of zone-2 of distance relays is bounded between two bounds. The mathematical expression of this side constraint is given in (16.3) where T^{min}_{DZ2} and T^{max}_{DZ2} are set to 18 and 36 cycles, respectively (Rivas et al., 2019).

17

Trending Topics and Existing Issues

This chapter covers the following points:

- Some new time–current characteristic curves that have been recently proposed for inverse-time overcurrent relays
- The technical issues that could be faced in smart grids during coordinating protective devices
- Adaptive coordination with taking into account the economic operation of power systems
- Improving the accuracy of mathematical models by realizing power system components
- Giving numerical DOCRs the ability to locate faults in mesh networks

The preceding sixteen chapters are devoted to cover the essential information about how to achieve optimal coordination. Many protection designs, protective devices, and mathematical models have been discussed through different illustrative and computer examples. Understanding this basis will nominate the reader to go further in this highly active research area. This chapter is intended to highlight some of the latest topics presented in the literature. It is very beneficial for researchers who want to proceed in advanced topics. It is also a good starter to know where did this research area reach and what are the possible challenges and the perspective techniques that could be seen in the future.

17.1 New Inverse-Time Characteristics

In Chapter 3, many models have been presented to mimic the characteristics of ITOCRs. However, due to some special needs, a few of recent studies suggested to use new formulas.

17.1.1 Scaled Standard TCCC Models

In some relay manuals, we could see that the models given in Chapter 3 for the North American and European relays are slightly modified. For example, for $M > 1$, the IEEE standard model given in (3.19) is modified to:

$$T = \left(\frac{A}{M^p - 1} + B\right) \times \frac{T_m}{\beta} \tag{17.1}$$

where T_m is the relay operating time at $10 \times I_p$.

Optimal Coordination of Power Protective Devices with Illustrative Examples, First Edition. Ali R. Al-Roomi.

Companion website: www.wiley.com/go/al-roomi/optimalcoordination

By comparing (17.1) with the original one given in (3.19), the following relation is obtained:

$$TDS = \frac{T_m}{\beta} \tag{17.2}$$

where the new coefficient can be determined if both TDS and T_m are known.

The same thing for the IEC standard model given in (3.25), which is modified to:

$$T = \frac{k}{PSM^{\alpha} - 1} \times \frac{T_m}{\beta} \tag{17.3}$$

where T_m is the relay operating time at $10 \times PS$.

Again, by comparing (17.3) with the original one given in (3.25), the following relation is obtained:

$$TMS = \frac{T_m}{\beta} \tag{17.4}$$

where the new coefficient can be determined if both TMS and T_m are known.

Applying the same concept to the special model of ANSI and IAC relays, which is given in (3.34), the following new model can be obtained:

$$T = \left[A + \frac{B}{M - C} + \frac{D}{M^2 - C} + \frac{E}{M^3 - C}\right] \times \frac{T_m}{\beta} \tag{17.5}$$

Also, the RI-type model given in (3.28) can be modified to:

$$T = \frac{1}{0.339 - 0.236PSM^{-1}} \times \frac{T_m}{\beta} \tag{17.6}$$

and (3.29) can be modified to:

$$T = \frac{-4.2373}{PSM^{-1} - 1.436\,44} \times \frac{T_m}{\beta} \tag{17.7}$$

More information about these models are given in Schneider Electric (2017a). The study reported in Mahindara et al. (2021) is based on (17.3).

17.1.2 Stepwise TCCCs

A combination of time–current characteristic curves (TCCCs) can be made in one relay. We have seen this in Section 2.6 of Chapter 2. However, the mixed characteristic curves are made between DCOCR, DTOCR, and/or ITOCR. For the last type of OCRs, we know that there are different standard TCCCs; such as standard inverse, very inverse, extremely inverse, etc. Some studies use stepwise functions to jump from one standard TCCC to another (Mahindara et al., 2021). For example, we can use the following three-mode of operation characteristic equation:

$$T_i = \begin{cases} TCCC_1, & ;\text{for } PSM_i^{\text{min}} \leqslant PSM_i < PSM_i^{\text{mode A}} \\ TCCC_2, & ;\text{for } PSM_i^{\text{mode A}} \leqslant PSM_i < PSM_i^{\text{mode B}} \\ TCCC_3, & ;\text{for } PSM_i^{\text{mode B}} \leqslant PSM_i \leqslant PSM_i^{\text{max}} \end{cases} \tag{17.8}$$

Beder et al. (2020) proposed a stepwise function for DOCRs when they work as primary or backup protective devices. The study combines the IEC standard inverse and very inverse TCCCs into a one equation. The first curve is activated when the ith relay acts as a backup relay, while the second one is for the primary operation mode. This stepwise TCCC can be mathematically expressed for the ith relay as follows:

$$T_i = \begin{cases} \dfrac{0.14\ TMS_i^{bc}}{PSM_i^{0.02} - 1}, & ;\text{for } 1.1 \leqslant PSM_i < H_i \quad \text{“backup setting”} \\ \dfrac{13.5\ TMS_i^{pr}}{PSM_i^{1} - 1}, & ;\text{for } H_i \leqslant PSM_i \leqslant 100 \quad \text{“primary setting”} \end{cases} \tag{17.9}$$

where $T_i^{pr} \leqslant 0.15$ s, $T_i^{bc} \leqslant 0.75$ s, and H_i is the value of the fault current in case of minimum phase fault in the ith primary relay over the pickup current.

17.1.3 New Customized TCCCs

Recently, it has been observed that many studies proposed new TCCCs. These non-standard formulas can be categorized under two groups (Abeid et al., 2020):

1. TCCCs based on current
2. TCCCs based on voltage

For more information about these models, the reader could refer to (Enríquez et al., 2003; Conde and Vázquez, 2007; Soria et al., 2014; Agrawal et al., 2016; Bayati et al., 2017; Jamali and Borhani-Bahabadi, 2017; Zhang et al., 2017; Tejeswini and Sujatha, 2017; Kılıçkıran et al., 2018).

17.2 Smart Grid

There is no doubt that most of the recent studies focus on smart grids. They are the future where many techniques and technologies are paving the way toward this untapped market. A significant portion of this market covers the next-generation protection systems, which includes the latest works done in the optimal coordination subject. The next Sections 17.2.1, 17.2.2, and 17.2.3 highlight some of the technical issues and problems that could be faced in smart girds.

17.2.1 Distributed Generation

Distributed generation (DG) is an electrical generation performed by a variety of small, grid-connected, or distribution system-connected devices referred to as distributed energy resources (DER). Typically, they are renewable energy sources, such as biomass/biogas, wind power, solar power, geothermal power, and small hydro. They could also be diesel or gas generators, which are commonly used in microgrids. These localized small-scale grids can be connected to the conventional large-scale grids[1] when there is a need for that, and they can be disconnected to operate autonomously. DER can be monitored, controlled, managed, and coordinated by the mean of advanced two-way communication technologies available in smart grids. These decentralized energy sources play an important role in modern distribution systems. In the past, the contribution of these devices was small. With the increase in the capital investment in renewable energy and storage elements, the impact of these devices is significant and thus it must be well-analyzed to avoid any consequences that might jeopardize the entire system.

---

1 That is, the centralized grids, which are powered by big power stations where coal, gas, nuclear, large-scale solar/wind farms, and hydroelectric dams are used.

In Chapter 14, we have seen that any variation in the load or/and generating units the power flow will change, and thus there is a need to re-initialize the ORC solver to update the settings of protective devices. The conventional way to do that is shown in Figure 14.1.

Because of the nature of DER systems, such as the intermittence of renewable energy sources and the pattern of consumers and end-users, so it is preferred to use an adaptive period between each two scans of the loop shown in the previous flowchart. The other fact is that these variations in power can be predicted, which means that the possibility to embed meteorology and forecasting science. This unfixed scanning scheme should be applied to the distributed gas and diesel generators, renewable energy sources, energy storage elements, and capacitor banks.

There are many attempts to study these effects, such as the studies reported in Baghaee et al. (2018), El-Naily et al. (2019), Elsadd et al. (2021), Bisheh et al. (2021), Draz et al. (2021), and ElSayed and Elattar (2021).

17.2.2 Series Compensation and Flexible Alternating Current Transmission System

The conventional mechanism of ORC solvers is described in Chapter 6, while the one given in Chapter 14 is used by adaptive solvers to consider the dynamic changes. These flowcharts are valid if there is normal power flow without any restriction or controlling unit across the branch. Unfortunately, this is not the case when a series compensator (SC) or a flexible alternating current transmission system (FACTS) is placed between the ith and jth busbars. This is very common in modern power systems where many SC and FACTS devices are employed to improve the system performance. Thus, the conventional adaptive ORC solver presented in Chapter 14 should be modified to take into account the variation in the flow of power during coordinating the protective devices.

There are some attempts to study the effects of these devices based on the feasibility and optimality of different relay coordination problems. Such studies are reported in Moravej et al. (2012), Benabid et al. (2013), Mancer et al. (2015), and Bougouffa and Chaghi (2019).

17.2.3 Fault Current Limiters

The other electric power device that is frequently used in modern power systems is called fault current limiter (FCL).[2] The goal of using this device is to reduce or limit the prospective short-circuit current when a fault occurs. Thus, the conventional short-circuit analysis done in the preceding chapters is not valid anymore for both primary and backup protective devices (Elmitwally et al., 2015; Asgharigovar et al., 2018).

17.3 Economic Operation

For dynamic electric power systems, adaptive ORC solvers have to be used. These changes are categorized as operational and topological changes. These two types are described in Chapter 14. Most of the operational changes come from the variation in the load connected to the grid and the output settings of the generating units committed to the grid. That is, the power injected into the grid is proportional to the power consumed from the grid. The question that should be asked here is: *How to meet the existing demand at the lowest possible cost? What about the losses in the network?*

2 It is also known as fault current controller (FCC).

The answer to these two questions is part of optimal economic operation. It is one of the hot subjects in the literature. This vital analysis should be solved to satisfy the load required by the end-users at the lowest possible production cost so that the net profit can be maximized. The other goal is to minimize the emission rates like oxides of nitrogen (NOx), oxides of sulfur (SOx), oxides of carbon (COx), soot, and unburned hydrocarbons (UHCs). Two main strategies can be involved here to achieve economic operation. The first one is based on scheduling the output of generating units to meet the required load demand at the lowest possible fuel consumption. The second strategy is based on minimizing the losses in the network by controlling the flow of power in each branch. The first strategy is called the economic load dispatch (ELD) problem, while the second strategy is called the minimum-loss problem, and both strategies can be optimized through what is called optimal power-flow (OPF) (El-Hawary, 1983; Grainger and Stevenson, 1994; Saadat, 1999; Husain, 2007; Wood and Wollenberg, 2013).

Thus, again, the delay between scans of adaptive ORC solvers will depend on the next schedule of the generating units. This optimal scheduling of generating units depend on different variables. For example, the existing load, fuel market, emissions, planned maintenance, and some design constraints.[3] More details can be found in Farag et al. (1995), Abido (2003), Mahor et al. (2009), Karthikeyan et al. (2013), and Al-Roomi and El-Hawary (2016c).

The practical way to solve all these issues is by applying forecasting science to three prediction levels; load pattern, fuel market, and power produced by renewable energy sources. This forces us to implement meteorology as well. Each one of these predictions will have a direct effect on the coordination and thus new optimal settings should be obtained for protective devices.

17.4 Power System Realization

Power system modeling aims to convert real-world power system problems into mathematical equations that can be emulated in computing machines. The gap between the actual problem and its soft mirror represents the accuracy of the model. In the past, the technology was insufficient. Thus, for some applications that require fast action,[4] more weight is given to the processing speed at the expense of solution quality. This simplification is applied in most modern power system studies. The following two Sections 17.4.1 and 17.4.2 are just two examples:

17.4.1 Power Lines

If someone opens any popular electric power engineering textbook, he/she will realize that the most important stage is PF (El-Hawary, 2008; Glover et al., 2012; Grainger and Stevenson, 1994; Saadat, 1999). The reason is that all the other stages[5] mainly depend on data received from PF. Therefore, if that data is incorrect or inaccurate, then the whole process will be affected! By looking at Figure 17.1, it is obvious that the PF analysis itself depends on the quality and accuracy of the model used to represent real transmission lines. Thus, even going with highly precise PF solvers, weak representation of transmission lines leads to significant errors in all power system analysis.

3 Such as spinning reserve constraint, line flow constraint, hydro-water discharge limits, reservoir storage limits, water balance equation, network security constraint, etc.

4 For example, DC and AC-DC power flow analysis are commonly used for **contingency analysis (CA)** (Al-Roomi and El-Hawary, 2020b).

5 Such as ELD, unit commitment, OPF, fault analysis, contingency analysis, stability and control, state estimation, relay coordination, etc.

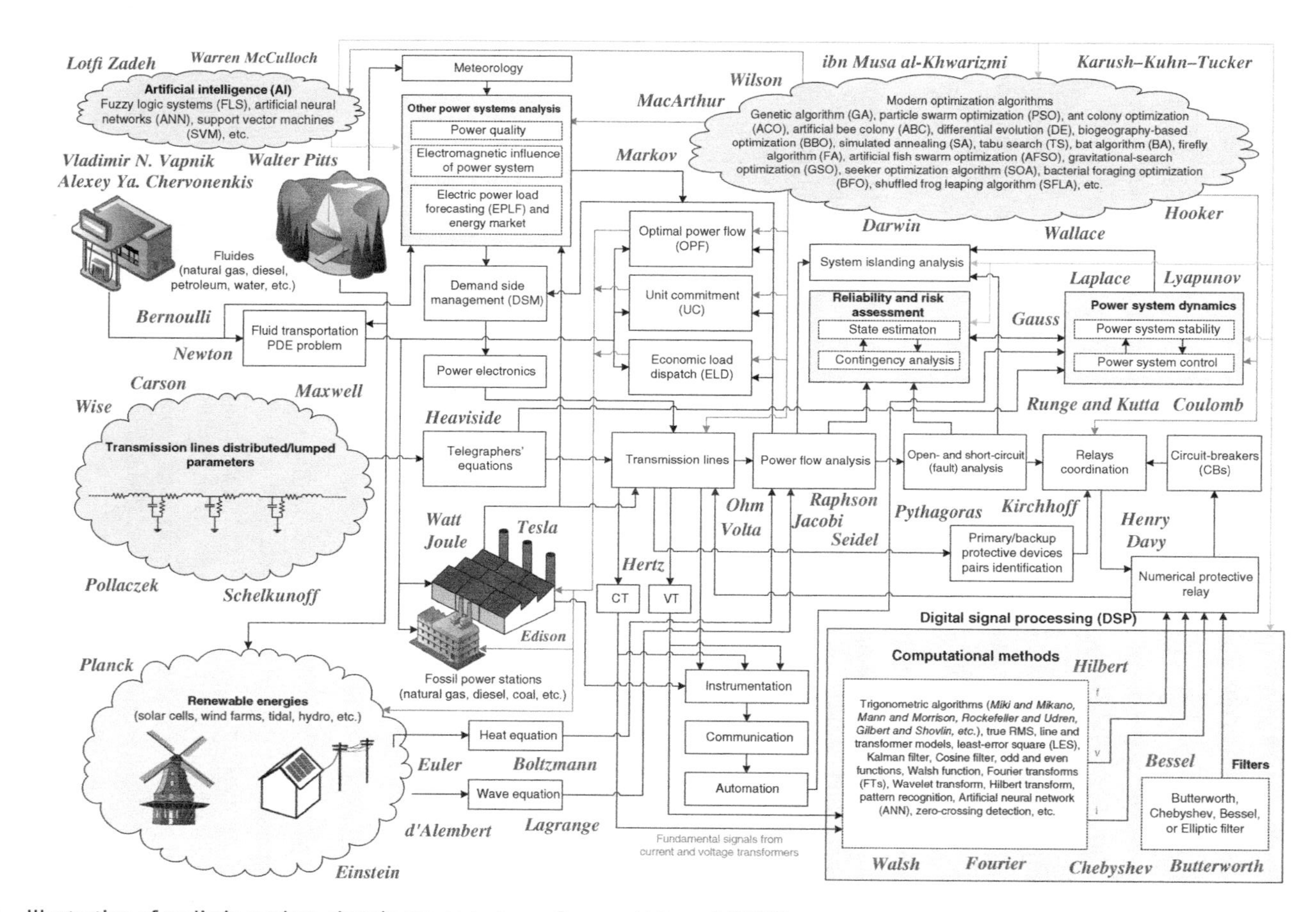

Figure 17.1 Illustration of realistic modern electric power systems. Source: Al-Roomi (2020).

In the **circuit theory**, some approximations are applied to simplify **Maxwell's equations** of electromagnetism.[6] As mentioned in Chapter 6, to represent real transmission lines as mathematical equations, there are two possible representations:

1. The distributed parameter model.
2. The lumped parameter models.

For simplicity and processing speed, the lumped parameter models are used for short- and medium-length transmission lines. This is the case for distribution networks, sub-transmission lines, and many transmission lines. Of course, the problem with these lumped parameter models is the accuracy. The study reported in Al-Roomi and El-Hawary (2020d) suggests using a new highly accurate medium-length transmission line model called the M-model.[7]

The other critical error associated with transmission line models is the dynamic changes of the distributed series and shunt parameters[8] due to operating conditions, weather conditions, surrounding environment, and cable design/status/age. Many attempts have been introduced in the literature to incorporate these variations in the mathematical models, such as the studies reported in Zarco and Exposito (2000), Bočkarjova and Andersson (2007), Santos et al. (2007), Leger and Nwankpa (2010), Cecchi et al. (2011), Frank et al. (2013), and Al-Roomi and El-Hawary (2017b). The effect on ORC problems was addressed in Al-Roomi and El-Hawary (2017c) for some significant variations. Most of these studies are built based on some temperature coefficients to estimate the variation in the preceding distributed series and shunt parameters. The scholars assumed that the temperature coefficients are known and available, which is – unfortunately – not a correct statement. It was the mission of (Al-Roomi and El-Hawary, 2020a) to govern these temperature coefficients by some mathematical equations. On the opposite side, the study reported in Al-Roomi and El-Hawary (2020c) offers two solutions that can account for these variations without referring to any temperature coefficient; directly through utilizing one of the properties of the M-model mentioned earlier. Actually, there are many factors that can affect the accuracy of line models. Many of these abandoned and hidden phenomena are addressed in Al-Roomi (2020).

17.4.2 Economic Operation

The classical economic operation has been covered in Section 17.3. Here, we are trying to highlight the realization that could be applied here. This discipline also suffers from model accuracy. Some of the design flaws are addressed in the following text: Because the distributed parameters of transmission lines are not constant, as mentioned in Section 17.4.1, so the active and reactive power losses also are not constant. Thus, the optimal power scheduled for generating units will drift away from the steady-state solution. This means that the power flow will change, which means that the fault analysis will also change, and thus the optimal settings of protective relays will change. Many assumptions are taken to calculate the power losses. For instance, Kron's loss formula is not accurate because it is built based on many assumptions, such as constant bus voltages, constant bus angles, fixed load, and steady-state network topology (El-Hawary, 1983; Kothari and Dhillon, 2010). The operating cost of thermal units is calculated based on the generated power. For spinning reserve units, the operating cost should be formulated as a function of another more significant variable, such as the angular speed of the prime-mover (Al-Roomi and El-Hawary, 2018a). Fuel assigned

6 In the period between 1884 and 1885, Heaviside successfully re-wrote Maxwell's 20 fundamental equations to obtain a new set of four compact equations, which were standardized by the late 1890s (Hunt, 2012).

7 Refer to Chapter 6.

8 That is, resistance, reactance, capacitance, and conductance.

to each machine could be gas, distillate, or a mixture of both (Al-Roomi and El-Hawary, 2017d). Even if only fuel gas is used, it could be a floating ratio of different fuel gases available in the market. Thus, if fuel mixtures are applied to some units, then the classical *one*-dimensional fuel-cost functions cannot be used anymore. Some units could be just commissioned from a **major overhaul** (or **C-inspection**), while some others could be weary or need some maintenance jobs. Some units could not be operated to their full capacity due to defective burners, unaligned prime-mover, unwell coated blades, vibration, the combustor behavior monitored by the **pulsation probe**, etc. Maximum power output is suppressed due to winding resistance or external problems with other equipment, like its **cooling system**, **excitation system**, or **step-up transformer**. Some units are forced to be stopped or partially operated based on strict orders received from the planning department and/or the **system control**.

Do the aforementioned points represent all the realization concerns? No! If we go deep into the subject from the real perspective, many other realizations will be found! Some further examples are given in the following text.

17.4.2.1 Combined-Cycle Power Plants

The second challenge that almost exists in all real power networks is about the busbars and whether they are energized by one individual unit or a group of synchronized units. Classical ELD optimizers work based on an assumption that each *i*th unit is connected to one specific busbar. Practically speaking, many, or even most of, units are connected to some common busbars as a group of synchronized units. For instance, a power station built based on two combined-cycle blocks, each powered by two gas turbines (160 MW GT – 13E2) and one steam turbine (160 MW ST – DKZ), is offered by ALSTOM.[9] Such a power station can supply national grids by 960 MW through one common busbar (i.e. from both blocks) or two common busbars (i.e. from each block). The other example from the same company is a power station built based on two combined cycle blocks where each block contains three gas turbines (97 MW GT – 13D2) and one steam turbine (117 MW ST – DK50) as shown in Figure 17.2. Thus, a total power of 816 MW can be supplied via one common busbar or two busbars with 408 MW for each. Figure 17.3 gives an additional illustration of this configuration. Some busbars could be energized by one or multiple units synchronized within an acceptable tolerance of frequency Δf. The other thing is that the production of STs of combined-cycle power plants (CCPPs) is a function of the production of GTs, and thus the fuel-cost function used in most of the studies of economic operation is not valid anymore.

17.4.2.2 Degraded Efficiency Phenomenon

This is one of the unrevealed constraints and it still does not have any attention in the literature. All the preceding information is based on the fact that the power output of units has a fixed fuel-power curve. In fact, the power output deviates with time and it can be predicted and measured based on the recorded past and current status of the unit. For example, if some burners of GT are not working or if there are some disturbances in the **air mixing valve**, then the machine efficiency will decrease. A similar thing can be applied here when the machine is just returned back from its major overhaul or when it is operating for a long time without any proper maintenance. That is, as **equivalent operating hours** (**EOH**) increases or as the machine status degrades due to some worn-out components, the machine efficiency decreases. This means that the unit produces less power with the same fuel rate consumption. Therefore, this phenomenon directly affects the fuel-power curve of the unit as illustrated in Figure 17.4.

9 French multinational company acquired by GE in November 2015.

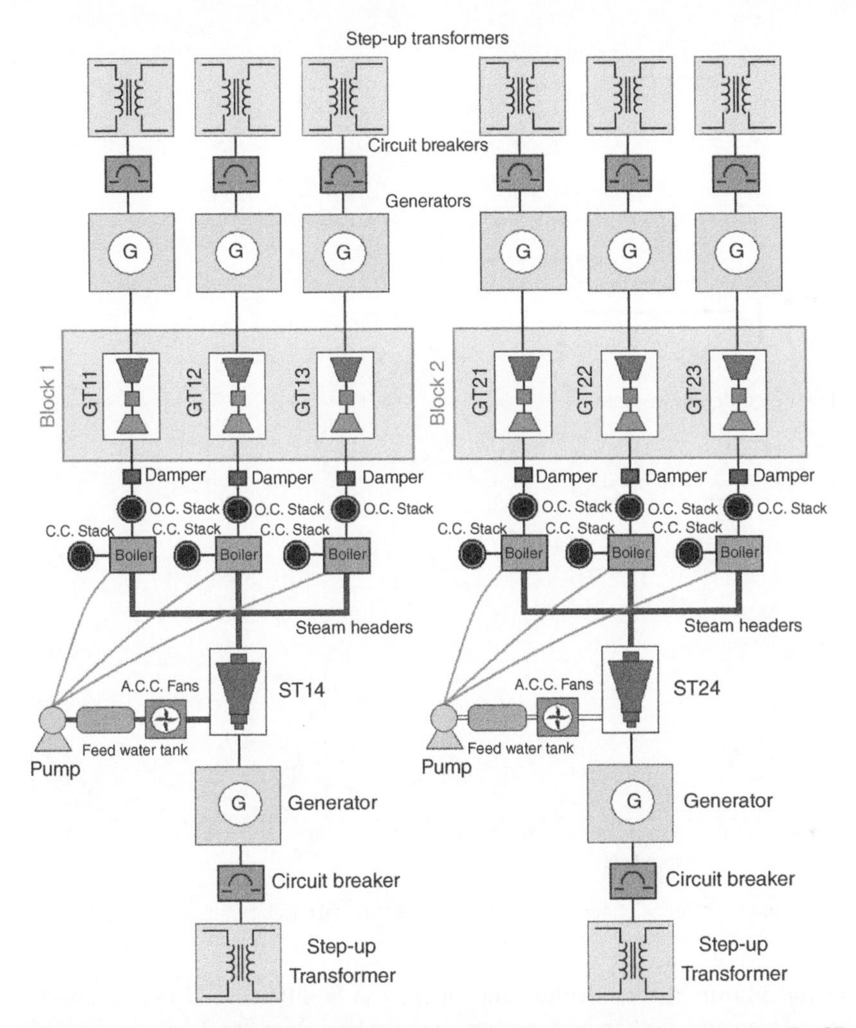

Figure 17.2 One of ALSTOM's CCPPs designed based on two blocks containing 3× GTs and 1× ST each.

17.4.2.3 Unaccounted Losses in Power Stations

By referring to the power losses mentioned earlier, that quantity represents the losses in the network. However, part of the total power produced by power stations is consumed by the station auxiliary plants such as boiler **feedwater pumps**, **air compressors**, gas stations – including **gas compressors** and **gas heaters** – water and lube oil cooling systems, **utility transformers**, **air intake systems**, **reverse osmosis** (**RO**) plants, etc. Thus, the net power should be used instead of the total power generated by power stations. These station and network losses represent just the losses of electrical energy. Realistically speaking, these electrical losses reflect just a small part of the real losses across the entire power system.

To explain these losses, first, let's look into the problem from the energy[10] side. To have electrical energy from thermal power stations, multiple energy conversions are needed. For CCPPs, the natural gas[11] is sent from its well to **gas stations**. That dirty fuel gas is cleaned through **scrubbers**.

10 Power = Energy/Time.
11 It can also be diesel fuel.

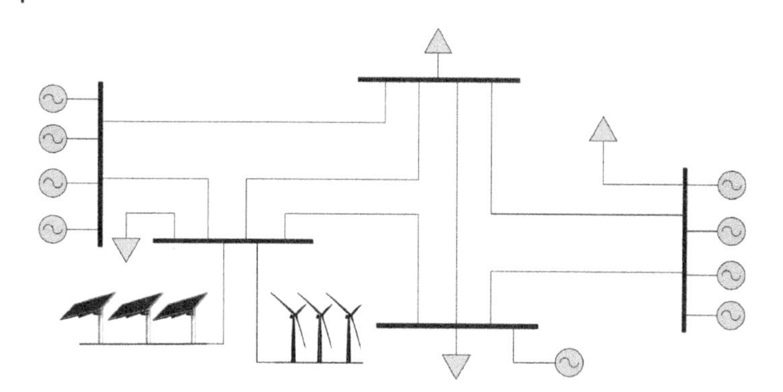

Figure 17.3 Illustrated real electric power system.

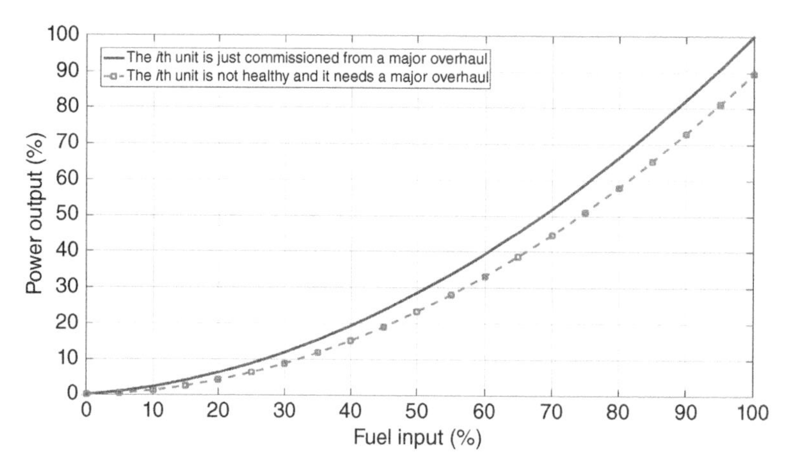

Figure 17.4 Illustrated degraded efficiency phenomenon due to the total EOH accumulated for the *i*th generating unit.

Also, by other **separation plants**, the condensate and sludge can be filtered out. Further, part of the fuel gas could be burned and discharged to the atmosphere by a **flare stack** to prevent the process components from being pressurized. Moreover, a small part of the fuel gas is utilized in some **heating plants** to heat the gas before being sent to GTs. Add to that, a very small part of unburned fuel gas is exhausted out.[12]

Once the fuel gas is burnt, a chemical to thermal energy conversion will take a place. This happens in the turbine stage of each GT. As a result, the prime-mover will start rotating, which is a thermal to mechanical energy conversion. Because the prime-mover is connected with a generator, so mechanical to electrical energy will take a place. Also, the total heat exhausted from all the GTs, installed in the same block,[13] is utilized to generate superheated steam via heat recovery steam generators (HRSGs), which is a **heat exchange process**. Then, the superheated steam is used to rotate the prime-mover of ST, which is a thermal to mechanical energy conversion. Again, that prime-mover will rotate a generator, which is a mechanical to electrical energy conversion. A part of electrical energy is used in the excitation system of each generator. After that, step-up transformers are used to increase the voltage and reduce the current so that the power losses in the

12 The amount depends on many factors, such as the combustor type, its status, the fuel gas temperature, the fuel/air ratio, etc.
13 Please, refer to Figure 17.2.

network can be minimized. Also, inter-bus transformers are used to connect branches of different voltage levels. Add to that, a significant part of the total power is consumed by the station auxiliary plants. Finally, a small part of electricity is used in the station buildings, air conditioning systems, main and local control rooms, lightings, etc.

As can be seen, many losses need to be considered in our realized ELD model. The operating cost is sensitive to many of these factors. The **Sankey diagram** shown in Figure 17.5 depicts most of these losses. Including them means a major improvement in the realized ELD model. This will improve the accuracy of ORC solvers, especially for adaptive schemes where some predictions could be required. For detailed information with many realization aspects, please refer to the comprehensive study reported in Al-Roomi (2020).

17.5 Locating Faults in Mesh Networks by DOCRs

One of the main advantages of using distance relays is that the location of faults can be found. On the opposite side, the relation between the fault location and the operating time of inverse-time overcurrent relays is nonlinear, and thus it is hard to locate faults by ITOCRs. The studies reported in Al-Roomi and El-Hawary (2017e, 2020f) are attempts to achieve this mission. The following lines briefly describe its mechanism.

Modern electric power systems are equipped with advanced tools to process all the preceding faults through several stages to maintain reliable and stable operation. These stages are summarized as: (i) **fault detection**, (ii) **fault classification**, (iii) **fault location**, (iv) **fault containment**, and (v) **fault recovery** (Chen et al., 2016; Verma et al., 2011). This study mainly focuses on the first three stages where the last two stages are parts of power system stability and control (Kundur, 1994; ALSTOM, 2002; Gers and Holmes, 2004). The relation between the first three stages is graphically described in Figure 17.6 (Chen et al., 2016).

The third stage, i.e. fault location, is the core of this study. It can be defined as a process to precisely locate faults in electric power systems. This process can be achieved by implementing many algorithms where the fault location function itself can be implemented as stand-alone fault locators, **digital fault recorders** (**DFRs**), or even just as post-fault analysis programs inside the state-of-the-art numerical protective relays (Saha et al., 2010). Nowadays, the market contains different types of fault locators where each one of them has its own capabilities and limitations. Such devices are: **Murray loop bridge**, **A-Frame**, **time domain reflectometer** (**TDR**), **Wheatstone bridge**, **high voltage "Thumpers,"** **integrated test sets**, etc. (Thue, 2011). Although modern protective relays can be upgraded to act as fault locators,[14] it has to be remembered that the main purpose of protective relays is to protect electric power components against any fault as fast and selective as possible. That is, protective relays are responsible to carry-out the first stage of Figure 17.6; i.e. fault detection. To clarify this essential point, Table 17.1 lists the main differences between fault locators and protective relays.

Powerlines (transmission, sub-transmission, and distribution lines) come with two or more terminals. Based on the number of terminals, powerlines can be protected by: (i) **one-end**, (ii) **two-end**, or (iii) **multi-end algorithms** (Saha et al., 2010; Chen et al., 2016; Monadi et al., 2016). The algorithms built based on the first approach are very simple and they do not require to use any communication link since the information received at the far-end terminal is not utilized. This is the reason why the fault location function can be embedded in modern numerical relays if the one-end approach is used. If the two-end algorithms are used to locate faults on faulty lines,

14 Because both devices have the same hardware.

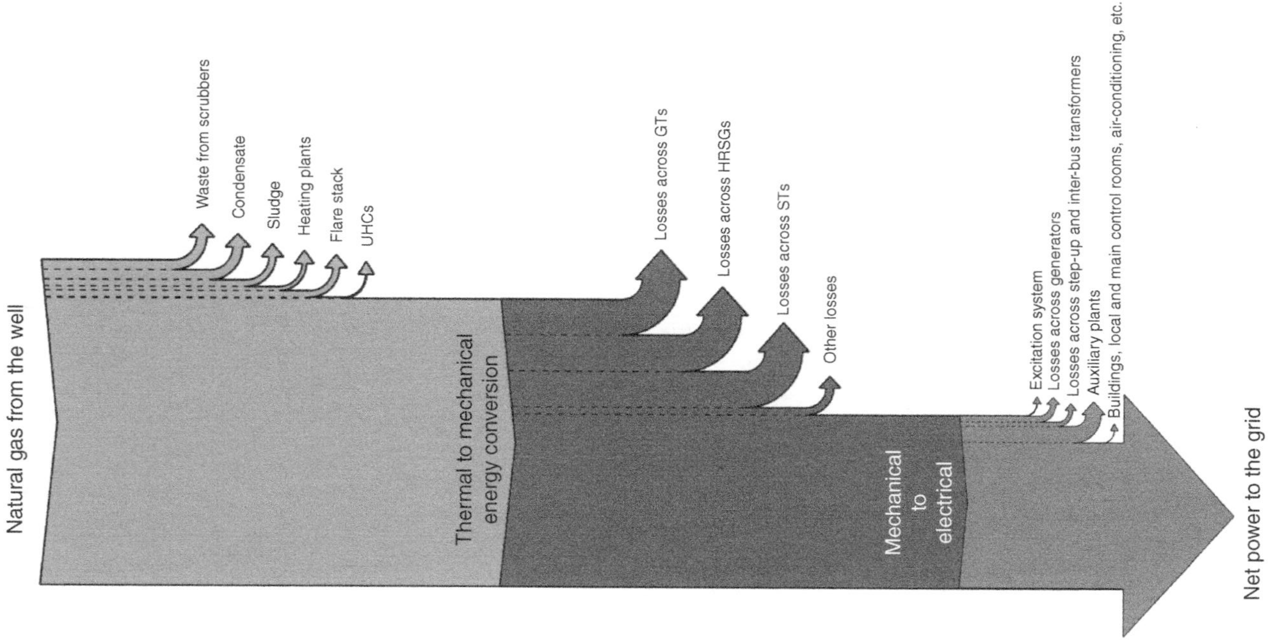

Figure 17.5 Sankey diagram of many energy losses that need to be considered in the realized ELD model.

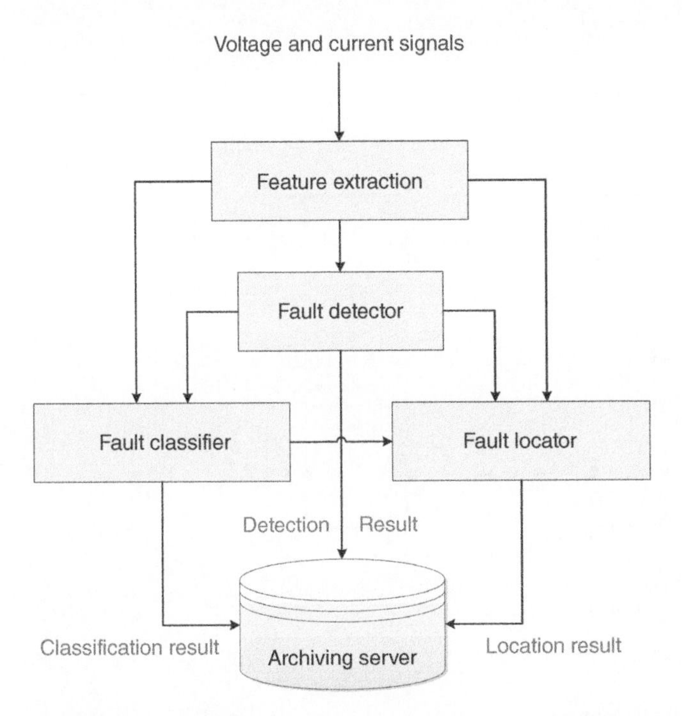

Figure 17.6 Simplified framework for fault detection, classification, and location.

Table 17.1 Main differences between protective relays and fault locators.

Fault locators	Protective relays
Larger data window	Smaller data window
More accurate	Less accurate
Slower	Faster

Source: Based on Saha et al. (2010).

then the information received from both ends (i.e. near-end and far-end terminals) should be processed. This job can be accomplished by providing a communication link between the two terminals. A global positioning system (GPS) antenna can also be installed to provide a time reference pulse to synchronize both relays, which is very useful in some applications. Therefore, the algorithms of the two-end approach consume more computation time than those of the first approach. With multi-terminal lines, the multi-end algorithms must be used to locate faults, which are more complicated than the preceding two approaches (Dinh et al., 2013).

The main goal of this study is to design a new two-end algorithm to locate faults in mesh networks based on the capabilities of modern numerical protective relays. This can be accomplished by utilizing the features that can be provided by modern EMS. For example, protection engineers and power operators can perform different operations, such as changing relays' settings and retrieving their online data remotely through some common communication protocols used today in numerical relays. Such protocols are **IEC 608750-5**, Modbus, **MMS /UCA2**, **Courier**, and **DNP** (Gers and Holmes, 2004). Therefore, this feature can be activated to import some important data from the two-end relays of a faulty line. Such data are the operating times, short-circuit currents and fault type, which can be effectively utilized to estimate the location of that fault.

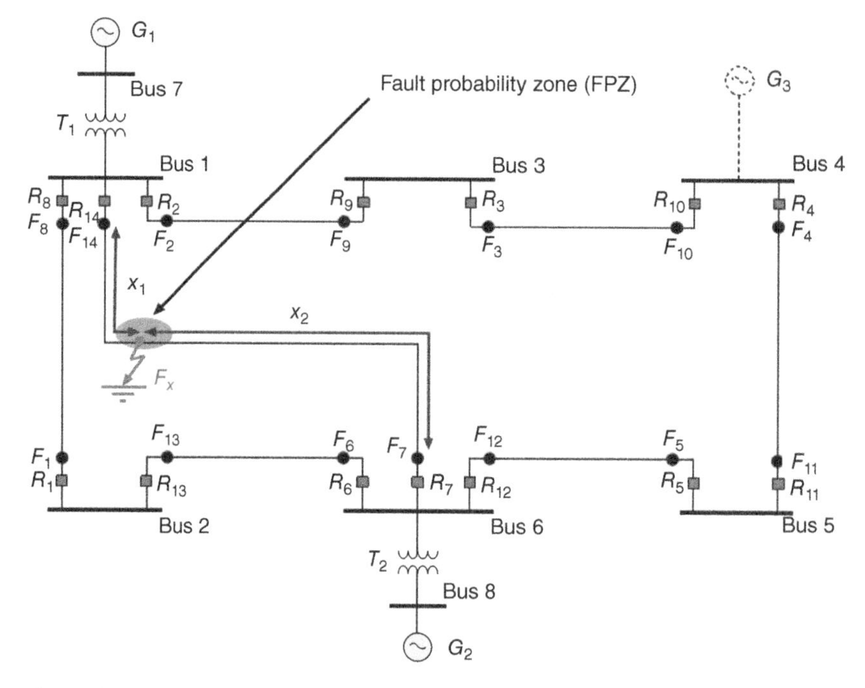

Figure 17.7 Single-line diagram of the IEEE eight-bus test system.

17.5.1 Mechanism of the Proposed Fault Location Algorithm

Figure 17.7 shows the IEEE eight-bus test system, which is one of the most popular ORC test systems presented in the literature (Al-Roomi, 2015b). Although it consists of eight buses and nine branches, only the buses and branches located after the two transformers are taken into account during solving its ORC problem (Noghabi et al., 2009; Amraee, 2012; Albasri et al., 2015). The network contains seven branches. Because each branch is protected by two DOCRs, so the total is 14 DOCRs. Based on (7.1), the dimension of this ORC problem is:

$$n = 2\varrho = 4ß = 2(14) = 4(7) = 28$$

The bold dot denoted beside each relay represents the near-end fault point, which is considered as the far-end point of the relay installed on the other terminal of the line. To clarify this point on the IEEE eight-bus test system, let's take bus 1 and bus 6 as bus x and bus y of Figure 7.1. By analogy, R_x and R_y are respectively R_{14} and R_7. Also, F_x and F_y are respectively F_{14} and F_7. Thus, R_{14} sees F_{14} as the near-end fault and R_7 sees it as the far-end fault, and vice versa for F_7.

Now, assume that a shunt fault F_x occurs on the line between bus 1 and bus 6 (i.e. L_{16}). Thus, the protective relays R_7 and R_{14} see F_x as an in-zone fault, so they act as primary relays for that fault. The backup relays R_1, R_5, R_9, and R_{13} see F_x as an out-zone fault, where R_5 and R_{13} operate when R_7 fails and R_1 and R_9 operate when R_{14} fails. From L_{16}, the length x_1 represents the distance between the near-end fault F_{14} and the actual fault F_x when it is seen by R_{14}, and the length x_2 represents the distance between the near-end fault F_7 and the actual fault F_x when it is seen by R_7. Also, it can be said that the length x_1 is the distance between F_x and the far-end fault F_{14} when it is seen by R_7, and the length x_2 is the distance between F_x and the far-end fault F_7 when it is seen

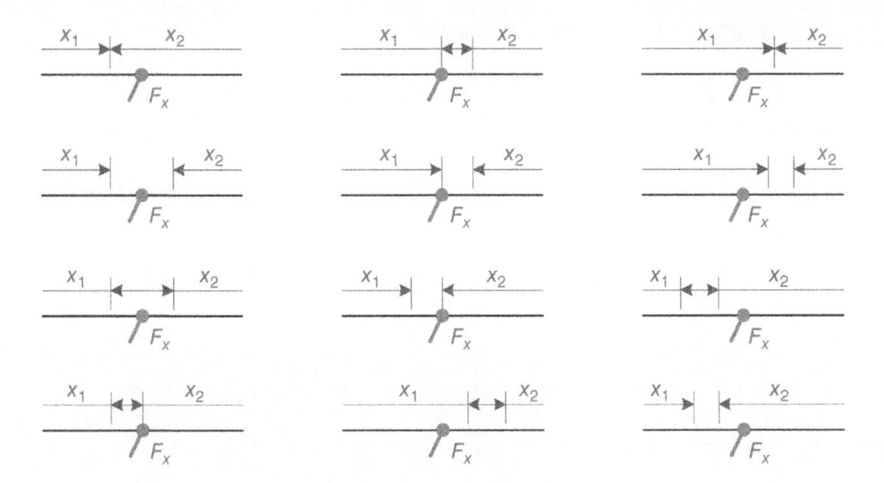

Figure 17.8 Twelve possible fault probability zones.

by R_{14}. However, we will stick with the first definition. Thus, the total length of the line L_{16} can be approximated as follows (Al-Roomi and El-Hawary, 2017e):

$$\text{length}(L_{16}) = l_{L_{16}} \approx x_1 + x_2 \tag{17.10}$$

From (17.10), it is clear that the sum of x_1 and x_2 measured by the two-end relays does not necessarily be the same length of L_{16}. There are many sources of uncertainty behind this phenomenon, such as the inherent errors of CTs, PTs, and relays (Short, 2007; ALSTOM, 2002). Twelve possible **fault probability zones (FPZs)** can be faced during calculating the distance between the actual fault location and its estimate where the destination of x_2 will not meet the destination of x_1 at F_x. These zones are graphically shown in Figure 17.8. By returning to Figure 17.7, these zones are created because of the errors associated with R_7 and R_{14}. Therefore, it is useful to take the average value of these two estimates to reduce the overall error as follows:

$$\overline{x}_{\text{bus 1}} = \frac{x_1 + \left(l_{L_{16}} - x_2\right)}{2} \tag{17.11}$$

$$\overline{x}_{\text{bus 6}} = \frac{x_2 + \left(l_{L_{16}} - x_1\right)}{2} \tag{17.12}$$

where $\overline{x}_{\text{bus 1}}$ is the distance estimated from bus 1 to F_x, and $\overline{x}_{\text{bus 6}}$ is the distance estimated from bus 6 to F_x. The estimates x_1 and x_2 are calculated from R_{14} and R_7, respectively. In this study, bus 1 is selected as a reference, so (17.11) should be used. For the sake of simplicity, the notation $\overline{x}_{\text{bus 1}}$ is replaced with $\overline{x}$ and the term $\left(l_{L_{16}} - x_2\right)$ is replaced with $\tilde{x}_2$, so the average estimate given in (17.11) becomes:

$$\overline{x} = \frac{x_1 + \tilde{x}_2}{2} \tag{17.13}$$

The goal is to find a direct relationship between the location of F_x and the operating times of the two-end relays, which are here R_7 and R_{14}. The operating time of the two-end relays can be calculated using any standard TCCC. These equations can determine the operating time of the ith relay based on the fault type and its value. Figure 17.9 shows how the IEC standard inverse TCCC behaves at different fault magnitudes. Thus, the operating times of these two relays, i.e. R_7 and R_{14},

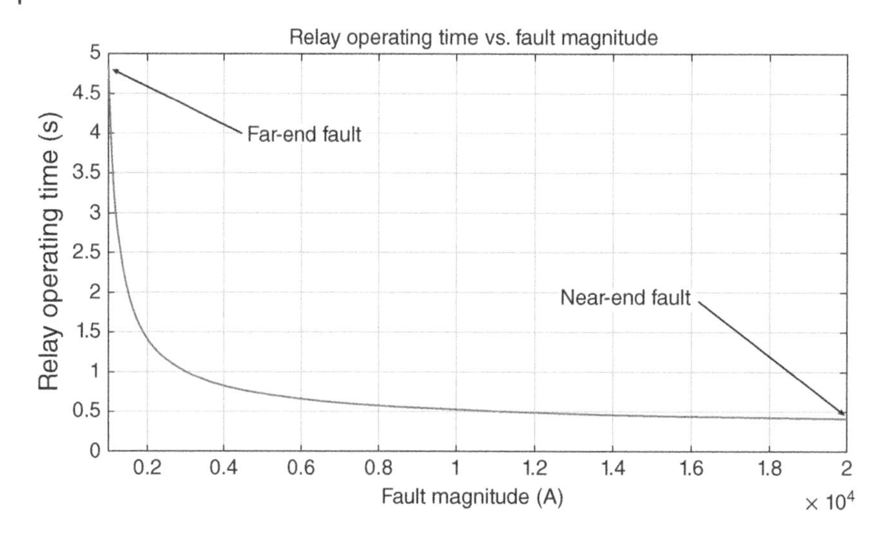

Figure 17.9 IEC IDMT-based TCCC for different fault magnitude values.

are inversely proportional to the fault magnitude. This magnitude depends on the fault type, and the operating time depends on the values of $\{TMS, PS, \alpha, \beta, \gamma\}$.

By focusing on the bold dots shown in Figure 17.7, let's say that the lowest and highest possible operating times of R_7 are $T_{R_7}^{\min} = T_{R_7}^{F_7}$ and $T_{R_7}^{\max} = T_{R_{14}}^{F_{14}}$, respectively. For the other end relay, which is here R_{14}, its lowest and highest possible operating times are then $T_{R_{14}}^{\min} = T_{R_{14}}^{F_{14}}$ and $T_{R_{14}}^{\max} = T_{R_{14}}^{F_7}$, respectively. Modern numerical protective relays can detect the time, value, and type of any fault. After that, they can store and send all these data automatically through the preceding protocols. For example, **AREVA MiCOM P12x OCRs** can detect all the five possible fault types and identify them by the code F80 (AREVA, 2010). Thus, the operating time T_{R_i} is a function of the fault magnitude I_{R_i,F_x} seen by the ith relay, where the value of I_{R_i,F_x} and the type of F_x can be provided too. Also, it has been seen that the fault magnitude I_{R_i,F_x} is a function of the fault location x. This relation can be mathematically expressed as follows:

$$I_{R_i,F_x} = I_{R_i}(x) \tag{17.14}$$

By substituting (17.14) in any formula mentioned in Chapter 3, the operating time of the ith relay can then be indirectly represented as follows:

$$T_{R_i}\left(I_{R_i}(x)\right) \equiv T_{R_i}(x) \tag{17.15}$$

The final step is to find the inverse function, so the location can be estimated as follows:

$$T_{R_i}(x) \Rightarrow x\left(T_{R_i}\right) \tag{17.16}$$

Based on this, a relationship can be made between the operating times of the two-end DOCRs and the location of the fault F_x. However, this method is not simple, because it is nonlinear and there are many sources of error due to the calculation, communication, and final data processing of the first three stages shown in Figure 17.6. Also, there are other sources of error due to some uncertainties that could be faced on actual electric networks. The next paragraphs describe the five approaches used in this study to estimate fault locations.

17.5.1.1 Approach No. 1: Classical Linear Interpolation

The linear interpolation method estimates the new jth point x_j by constructing a linear curve between two predetermined points. This approach is the simplest one. However, there is a significant error in its estimation. The reason behind this error comes from the nonlinearity nature of the relationship between the fault location x and the operating times of the corresponding two-end relays. The classical linear interpolation can be carried out for the ith relay using the following equation:

$$x_j = x_j^{\min} + \left(x_j^{\max} - x_j^{\min}\right)\left[\frac{T_{R_i}^{F_x} - T_{R_i}^{\min}}{T_{R_i}^{\max} - T_{R_i}^{\min}}\right] \tag{17.17}$$

where $x_j^{\min}$ and $x_j^{\max}$ are, respectively, the minimum and maximum distances measured from the jth terminal.

The estimate x_j is computed based on an assumption that all the points between $x_j^{\min}$ and $x_j^{\max}$ form a straight line, which is not correct for any ITOCR. In (17.17), the reference point is $x_j^{\min}$ for estimating the new point x_j. Thus, if $x_j^{\max}$ is taken as the reference point, then the alternative equation that can also be used is:

$$x_j = x_j^{\max} - \left(x_j^{\max} - x_j^{\min}\right)\left[\frac{T_{R_i}^{\max} - T_{R_i}^{F_x}}{T_{R_i}^{\max} - T_{R_i}^{\min}}\right] \tag{17.18}$$

In this study, (17.17) is used to estimate x_j. For the operating time of the other end relay, there is one tricky point that should be considered. In Figure 7.1, the operating time of R_x increases from the minimum value at x to the maximum value at y, while the operating time of R_y decreases as the fault location moves from x to y. Thus, to avoid using (17.11), the estimate of the other end relay can be directly computed as follows:

$$\tilde{x}_j = x_j^{\min} + \left(x_j^{\max} - x_j^{\min}\right)\left[\frac{T_{R_i}^{F_x} - T_{R_i}^{\max}}{T_{R_i}^{\min} - T_{R_i}^{\max}}\right] \tag{17.19}$$

Thus, there is no need to subtract the estimate from the total length. That is, (17.13) is involved here for the other end relay.

It has to be said that (17.17) should be carried out for each two-end relays. The final estimate of the exact location is then computed by using either (17.11) or (17.12); based on which bus is selected as a reference. If (17.11) is selected, then (17.13) is preferred.

The main problem associated with (17.17) is the accuracy. The amount of error generated by this static approach can be significantly minimized by using a more advanced dynamic linear interpolation. Thus, the closest lower and upper predefined points should be used in the position of the global minimum and maximum points of the static linear interpolation. This can be achieved by using the following equation:

$$\tilde{x}_j = x_j^{\text{clb}} + \left(x_j^{\text{cub}} - x_j^{\text{clb}}\right)\left[\frac{T_{R_i}^{F_x} - T_{R_i}^{\text{clb}}}{T_{R_i}^{\text{cub}} - T_{R_i}^{\text{clb}}}\right] \tag{17.20}$$

where x_j^{clb} and x_j^{cub} are, respectively, the closest lower and upper bounds or points predefined near the location of the actual point x. These closest points can be determined by mapping their closest lower and upper operating times $T_{R_i}^{\text{clb}}$ and $T_{R_i}^{\text{cub}}$ measured from R_i.

17.5.1.2 Approach No. 2: Logarithmic/Nonlinear Interpolation

This approach can be implemented to minimize the errors produced by the preceding approach. The magic ingredient applied here is that the relationship between x and $T_{R_i}^{F_x}$ is supposed to follow a logarithmic shape. Because that assumption is not correct, so the static version of this approach also suffers from the preceding accuracy problem faced with the classical linear interpolation approach. To apply it, the expression given in (17.17) is modified to be as follows:

$$x_j = x_j^{\text{min}} + \left(x_j^{\text{max}} - x_j^{\text{min}}\right) \cdot \left[\frac{\log T_{R_i}^{F_x} - \log T_{R_i}^{\text{min}}}{\log T_{R_i}^{\text{max}} - \log T_{R_i}^{\text{min}}}\right] \tag{17.21}$$

Similar to the preceding approach, the static equation given in (17.21) can be replaced with the following dynamic version:

$$x_j = x_j^{\text{clb}} + \left(x_j^{\text{cub}} - x_j^{\text{clb}}\right) \cdot \left[\frac{\log T_{R_i}^{F_x} - \log T_{R_i}^{\text{clb}}}{\log T_{R_i}^{\text{cub}} - \log T_{R_i}^{\text{clb}}}\right] \tag{17.22}$$

Also, to apply (17.13), the estimate of the other end relay should be modified to be:

$$\tilde{x}_j = x_j^{\text{clb}} + \left(x_j^{\text{cub}} - x_j^{\text{clb}}\right) \cdot \left[\frac{\log T_{R_i}^{F_x} - \log T_{R_i}^{\text{cub}}}{\log T_{R_i}^{\text{clb}} - \log T_{R_i}^{\text{cub}}}\right] \tag{17.23}$$

All these versions of the interpolation-based approaches are further explained through the pseudocode given in Algorithm 1.

Algorithm 1 Pseudocode of the Proposed Interpolation-Based Fault Locators

---

Require: Vectors of predetermined fault locations and operating times for each fault type
Require: Actual operating times supplied from both relays of a faulty line during the fault occurrence

1: Solve the ORC problem using any n-dimensional optimization algorithm. This stage will include load flow analysis, P/B relay pairs identification, and short-circuit analysis
2: Determine the fault type based on the data supplied from the two-end relays of the preceding faulty line
3: **if** the classical linear interpolation approach is used **then**
4:     **if** the static version is preferred **then**
5:         Apply (17.17)
6:     **else**
7:         Apply (17.20)
8:     **end if**
9: **else**
10:     **if** the static version is preferred **then**
11:         Apply (17.21)
12:     **else**
13:         Apply (17.22)
14:     **end if**
15: **end if**
16: Apply (17.13) to estimate the fault location

17.5.1.3 Approach No. 3: Polynomial Regression

The good thing of the interpolation-based approaches is that they only need to know the operating time of both end relays when the fault F_x happens at the lower and upper bounds. On the opposite side, the proposed regression-based approaches need a vector of predefined operating times measured at different fault locations. This means that it is required to conduct a short-circuit analysis at different points along each line. This process should be accomplished before energizing any line. Therefore, when this vector of short-circuit currents is substituted in the equation of TCCC for calculating the corresponding operating times of the two-end relays, the response vector can be obtained. Based on (17.14)–(17.16), the predictor must be the vector of the operating times of the *i*th relay and not the vector of the short-circuit currents, and the response must be the vector of the preceding predefined fault locations. These input/output vectors can then be used to construct linear and nonlinear regression models. Because each branch has two relays, so two regression models are required. Based on this, any *x*th fault location can be predicted by just supplying the actual operating times measured from the two-end relays. Then, the average estimate $\overline{x}$ of the actual fault location x can be calculated using either (17.11) or (17.13).

The first regression-based approach is built by using two polynomial equations for both end relays. These equations could be set as linear, quadratic, cubic, or even higher-order polynomial equations. In this study, different orders are used. A general model, with order d, can be expressed as follows:

$$x_j = \theta_{0,i} + \theta_{1,i} T_{R_i} + \theta_{2,i} T_{R_i}^2 + \theta_{3,i} T_{R_i}^3 + \cdots + \theta_{d,i} T_{R_i}^d \tag{17.24}$$

These $(d+1)$ theta coefficients can be obtained by using any regression software or package, such as SAS, R, SPSS, MINITAB, MATLAB, etc.

17.5.1.4 Approach No. 4: Asymptotic Regression

The **asymptotic regression model** can be mathematically expressed in the following general form:

$$x_j = \theta_{0,i} + \theta_{1,i} \exp\left(\theta_{2,i} T_{R_i}\right) \tag{17.25}$$

As can be clearly seen, this is a nonlinear regression model. Therefore, the optimal values of these three theta coefficients can be obtained by solving (17.25) numerically via using optimization algorithms. Such algorithms are **Gauss–Newton (GN)**, **Gradient Descent (GD)**, and **Levenberg–Marquardt (LM)**.

17.5.1.5 Approach No. 5: DTCC-Based Regression

This is the most advanced regression-based approach proposed in this study. The main goal behind it is to design a new nonlinear model that can act as a transposed function of $f_i(x)$ shown in (17.16); where f_i is the operating time received from the *i*th relay. To clarify this point, let's recall the general model given in (3.36):

$$T_{R_i} = TMS_i \times \left[\frac{\beta_i}{\left(\frac{I_{R_i}}{PS_i}\right)^{\alpha_i} - 1} + \gamma_i \right] \tag{3.36}$$

The next step is to transpose the preceding TCCC to our proposed **current–time characteristic curve** (**CTCC**) as follows:

$$I_{R_i} = PS_i \times \left[\frac{\beta_i \ TMS_i}{T_{R_i} - \gamma_i \ TMS_i} + 1 \right]^{1/\alpha_i} \tag{17.26}$$

Now, to apply the relation shown in (17.14) between the fault location x and its short-circuit current I_{R_i}, let's assume that a function $g(x)$ is used:

$$I_{R_i} = g_i(x) \tag{17.27}$$

For both end relays, the relation given in (17.27) is proportional to the first end relay and inversely proportional to the second end relay. For example, in Figure 7.1, if the fault F_m approaches the fault F_x, the short-circuit current I_{R_x} will increase and the short-circuit current I_{R_y} will decrease, and vice versa if F_m approaches F_y. These two behaviors can be mathematically explained as follows:

$$F_m \to F_x : \begin{cases} I_{R_x} \propto F_m \\ I_{R_y} \propto \frac{1}{F_m} \end{cases} \tag{17.28}$$

$$F_m \to F_y : \begin{cases} I_{R_x} \propto \frac{1}{F_m} \\ I_{R_y} \propto F_m \end{cases} \tag{17.29}$$

Some graphs and tables will cover these two opposite relations in the next numerical experiment with more detailed information.

Finally, after agreeing with the relation given in (17.27), a regression model can be designed to predict the fault location x by supplying the operating time received from the ith relay. This step has been successfully done by proposing a new transposed curve from CTCC to a **distance–time characteristic curve** (**DTCC**). To do that, the terms $\{PS_i, \beta_i \ TMS_i, -\gamma_i \ TMS_i, 1, 1/\alpha_i\}$ of (17.26) are replaced with theta regression coefficients $\{\theta_{0,i}, \theta_{1,i}, \theta_{2,i}, \theta_{3,i}, \theta_{4,i}\}$ to have the following new nonlinear regression model:

$$x_j = \theta_{0,i} \times \left[\frac{\theta_{1,i}}{T_{R_i} + \theta_{2,i}} + \theta_{3,i} \right]^{\theta_{4,i}} \tag{17.30}$$

Since the whole process is done through a regression analysis, a less accurate version of (17.30) can be designed by taking $\theta_{4,i}$ equal to 1 and then removing the brackets as follows:

$$x_j = \theta_{0,i} + \frac{\theta_{1,i}}{T_{R_i} + \theta_{2,i}} \tag{17.31}$$

But, be careful, the theta coefficients of (17.31) are different than that of (17.30). Also, they are different than that of (17.24) and (17.25). That is, each regression model has its own coefficients. This is why DTCC can be simplified from (17.30) to (17.31) by selecting proper initial coefficients during fitting their nonlinear curves.

These four regression-based approaches are further explained through the pseudocode given in Algorithm 2.

Algorithm 2 Pseudocode of the Proposed Regression-Based Fault Locators

Require: Vector of gradient operating times calculated for each type of faults bounded between $T_{R_i}^{\min}$ and $T_{R_i}^{\max}$ of the ith relay
Require: Actual operating times supplied from both relays of a faulty line during the fault occurrence
1: Solve the ORC problem using any n-dimensional optimization algorithm. This stage will include load flow analysis, P/B relay pairs identification, and short-circuit analysis
2: Determine the fault type based on the data supplied from the two-end relays of the preceding faulty line
3: **if** the polynomial regression approach is used **then**
4: Regress (17.24) for both x_1 and $\tilde{x}_2$
5: **else**
6: Set the optimization parameters (`max iterations`, `convergence tolerance`, `algorithm type`, `starting point`, etc.)
7: **if** the asymptotic regression approach is used **then**
8: Regress (17.25) for both x_1 and $\tilde{x}_2$
9: **else**
10: Regress (17.30) or (17.31) for both x_1 and x_2
11: **end if**
12: **end if**
13: Apply (17.13) to estimate the fault location

17.5.2 Final Structure of the Proposed Fault Locator

Figure 17.10 shows the final structure of the proposed fault locator. As can be seen from that diagram, it has an automatic mechanism to detect any dynamic change. They could be topology changes (such as opening branches and isolating busbars) or operational changes (such as changing the set-points of generating units and varying the settings of capacitor banks). For the topological change, the network needs to be re-configured again. Also, all the P/B relay pairs of the new topology need to be identified. If it is an operational change, then the preceding two sub-stages are bypassed. However, the network PF sub-stage needs to be updated for any topological or operational type. Then, to solve the ORC problem of the given network, it is important to carry out a 3ϕ short-circuit analysis with considering the proper coordination criteria of those listed in Table 7.1. Finally, the optimal values of *TMS* and *PS* of all ϱ DOCRs connected in the network can be obtained by using any global n-dimensional optimization algorithm. This adaptive ORC stage continuously checks if there is any topological or operational change, so the preceding steps are repeated again and again. This stage is extensively studied in the literature, and it is known as an **adaptive coordination** strategy.

The responsibility of the second stage is to design all the required fault location models for each two-end relays of all ß branches. Thus, the latest network state updated by PF is sent to the second

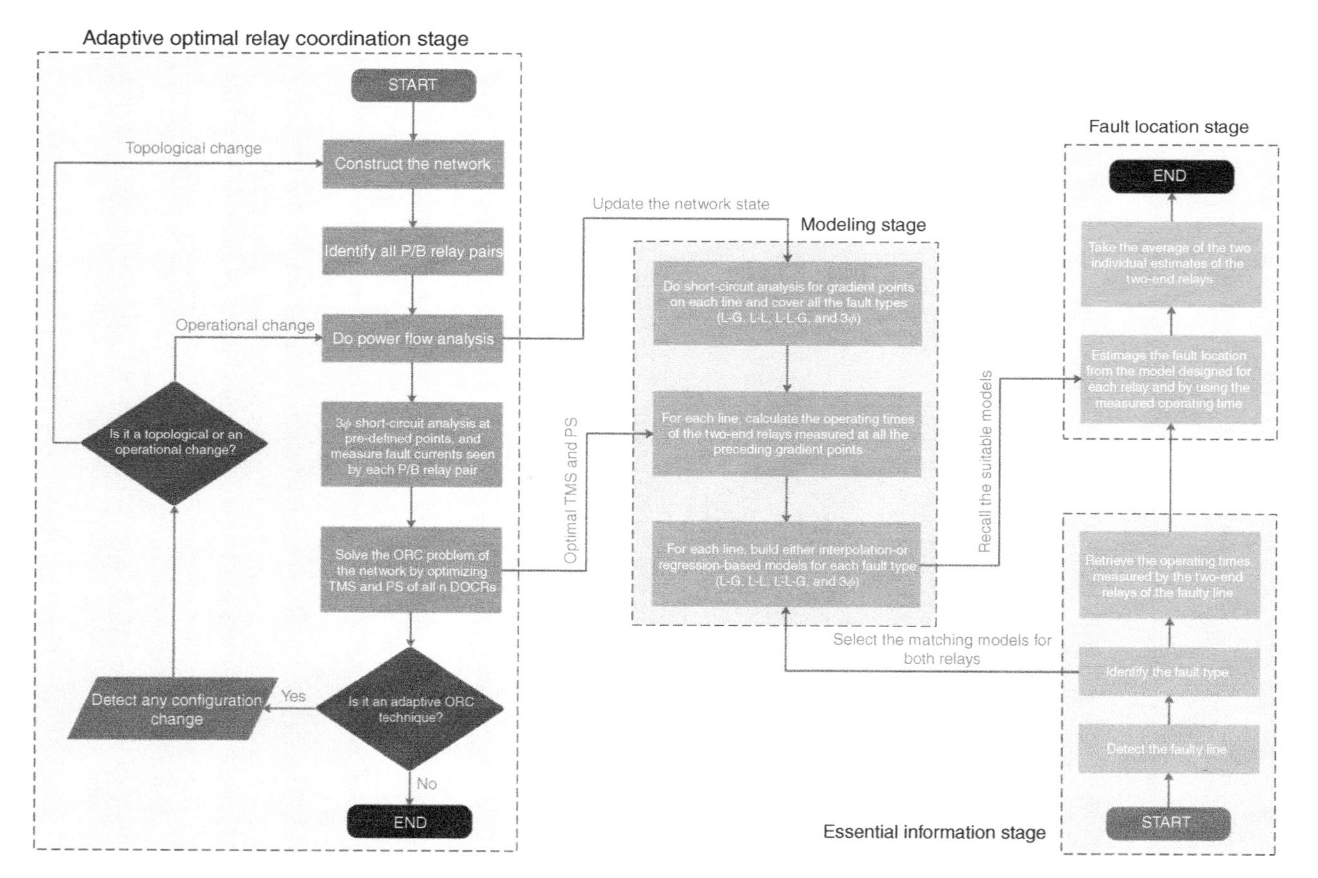

Figure 17.10 Final structure of the proposed adaptive fault locator including all the required stages.

stage. This information is utilized during doing another short-circuit analysis. This analysis is carried out for all the fault types, and it is repeated until covering all the branches. Furthermore, different predefined fault locations are considered for each branch. They should be gradually distributed from 1% (i.e. at the near-end point) to 99% (i.e. at the far-end point). Then, by knowing the optimal *TMS* and *PS* of each relay, the operating times of these relays can be calculated for any fault and at any point of any branch. These predefined operating times and fault locations can then be used to design all the required fault location models. These models could be designed based on linear interpolation, nonlinear interpolation, linear regression, or nonlinear regression.

The third and fourth stages contain optimization-free algorithms because all the optimization tasks are done in the previous two stages. The third stage is responsible to detect the faulty branch and the type of that fault. Also, the operating times recorded by the two-end relays of the faulty branch are retrieved again, which are sent to the fourth stage. These operating times act as the independent variables of the fault location models designed before in the second stage. The last step is to take the average of the two fault estimations calculated from the corresponding two-end relays.

17.5.3 Overall Accuracy vs. Uncertainty

It is obvious that, in most cases, the estimation errors of the two-end relays can be minimized if the average value is taken using (17.11) or (17.13). However, many sources of uncertainty could be faced in real-world applications. Based on this, the location of actual faults could not exactly match the DTCC curves of the two-end relays. Assume that $\wp$ sources of uncertainty are disturbing the operation of the network. Then, the deviation between the actual fault location x and its estimate $\overline{x}$ can be mathematically expressed as:

$$x = \overline{x} + \sum_{j=1}^{\wp} \varepsilon_j \tag{17.32}$$

That is, these $\wp$ disturbances are translated as the sum of errors $\sum_{j=1}^{\wp} \varepsilon_j$ associated with the final estimation model. With unbiased residuals that satisfy the normality test, the actual fault locations are supposed to be normally distributed above and below the fitted DTCC curves of the two-end relays. However, this claim has not been proven yet.

Regardless of the subject matter of uncertainty errors, the technique can still give a good estimation. For example, the impedance of branches can be affected by the surrounding temperature (Al-Roomi and El-Hawary, 2017b,c). It is known that the short-circuit current decreases as the impedance increases, and the later one increases as the surrounding temperature increases. Based on this, during the winter season, the surrounding temperature decreases and thus the impedance decreases as well. Therefore, the short-circuit current increases and the operating time of the ith relay decreases. Suppose, by chance, the operating time received from the ith relay intersects with one of the predefined observations. Thus, by taking into account the temperature effect, the correct predefined fault location that should be mapped with the measured operating time could become the closest lower or upper bound (i.e. x^{clb} or x^{cub}). However, interpolating these points will give an estimated location near that new bound. The same thing happens with linear and nonlinear regressions. This part of the study could be covered in future work where the most influent uncertainty sources can be highlighted and then trying either to eliminate or minimize their effects. One of the possible tools is to model the entire system with the temperature/frequency-based (TFB) technique reported in Al-Roomi and El-Hawary (2017b,c).

17.5.4 Further Discussion

Although the major parts of this technique are covered in Figure 17.10 there are many other enhancements could be made to improve its performance. For example:

- *How many predefined points need to be covered for each branch? In this study, gradient points of steps 5% are taken between the near-end and the far-end points of each ith relay. But, what is the best step-size? Is it 5%, 10%, or unequal steps and randomly selected points?*
- *In the modeling part, should the two-end relays be modeled using the same approach? What if the first model is designed using the dynamic nonlinear interpolation approach and the second model is designed using a dth order polynomial regression approach?*
- *What if one or more of "in-between" terms of polynomial regression models is removed instead of the last term[15]?*
- *Again, in the regression models, what if a piecewise linear or quadratic polynomial equations are used instead of increasing d up to 7?*
- *Could AI, such as ANNs and SVMs, be good add-ons to precisely estimate fault locations?*
- *In Figure* 17.10, *is it feasible to deactivate the adaptive link of the ORC stage?* This question is raised here because it is known that the fault location stage is a post-fault action, and thus there is no need to have a very fast fault locator. This is one of the main differences between protective relays and fault locators listed in Table 17.1.

Also, there are other pending points that need to be covered to ensure that the operation of Figure 17.10 is fully trusted. Such points are:

- Mal-operation case of primary DOCRs.
- Independent detection of in-zone and out-zone faults.
- Independent detection of fault types.
- Transient network topologies due to unequal operating speeds of CBs during solving the ORC problem.
- Networks with multi-terminal branches.
- Networks with FACTS devices.

All these points can be considered as open doors that need to be closed by conducting more future researches.

15 Such models are known as **non-hierarchical polynomial regression models**. That is, dealing with polynomial equations where some lower-order terms are missing. For example, using a septic polynomial equation without the quintic term.

Appendix A

Some Important Data Used in Power System Protection

This appendix contains the ANSI standard current transformer ratios and the device numbers and acronyms (IEEE, 2008a,b).

A.1 Standard Current Transformer Ratios

Table A.1 Standard ANSI CT-ratios.

No.	Primary	Secondary	Ratio
1	50	5	10
2	75	5	15
3	100	5	20
4	150	5	30
5	200	5	40
6	250	5	50
7	300	5	60
8	400	5	80
9	500	5	100
10	600	5	120
11	800	5	160
12	1000	5	200
13	1200	5	240
14	1500	5	300
15	2000	5	400
16	3000	5	600
17	4000	5	800

Optimal Coordination of Power Protective Devices with Illustrative Examples, First Edition. Ali R. Al-Roomi.

Companion website: www.wiley.com/go/al-roomi/optimalcoordination

A.2 Standard Device/Function Number and Function Acronym Descriptions

A.2.1 Standard Device/Function Numbers

1: Master element
2: Time-delay starting or closing relay
3: Checking or interlocking relay
4: Master contactor
5: Stopping device
6: Starting circuit breaker
7: Rate-of-change relay
8: Control power disconnecting device
9: Reversing device
10: Unit sequence switch
11: Multifunction device
12: Overspeed device
13: Synchronous-speed device
14: Underspeed device
15: Speed or frequency matching device
16: Data communications device
17: Shunting or discharge switch
18: Accelerating or decelerating device
19: Starting-to-running transition contactor
20: Electrically operated valve
21: Distance relay
22: Equalizer circuit breaker
23: Temperature control device
24: Volts per Hertz relay
25: Synchronizing or synchronism-check device
26: Apparatus thermal device
27: Undervoltage relay
28: Flame detector
29: Isolating contactor or switch
30: Annunciator relay
31: Separate excitation device
32: Directional power relay
33: Position switch
34: Master sequence device
35: Brush-operating or slip-ring short-circuiting device
36: Polarity or polarizing voltage device
37: Undercurrent or underpower relay
38: Bearing protective device
39: Mechanical condition monitor
40: Field (over/under excitation) relay
41: Field circuit breaker
42: Running circuit breaker
43: Manual transfer or selector device
44: Unit sequence starting relay
45: Abnormal atmospheric condition monitor
46: Reverse-phase or phase-balance current relay
47: Phase-sequence or phase-balance voltage relay
48: Incomplete sequence relay
49: Machine or transformer thermal relay
50: Instantaneous overcurrent relay
51: AC inverse time overcurrent relay
52: AC circuit breaker
53: Field excitation relay
54: Turning gear engaging device
55: Power factor relay
56: Field application relay
57: Short-circuiting or grounding device
58: Rectification failure relay
59: Overvoltage relay
60: Voltage or current balance relay
61: Density switch or sensor
62: Time-delay stopping or opening relay
63: Pressure switch
64: Ground detector relay
65: Governor
66: Notching or jogging device
67: AC directional overcurrent relay
68: Blocking or "out-of-step" relay
69: Permissive control device
70: Rheostat
71: Liquid level switch
72: DC circuit breaker
73: Load-resistor contactor
74: Alarm relay
75: Position changing mechanism
76: DC overcurrent relay
77: Telemetering device
78: Phase-angle measuring relay
79: AC reclosing relay
80: Flow switch
81: Frequency relay

82: DC load measuring reclosing relay
83: Automatic selective control or transfer relay
84: Operating mechanism
85: Pilot communications, carrier, or pilot-wire relay
86: Lockout relay
87: Differential protective relay
88: Auxiliary motor or motor generator
89: Line switch
90: Regulating device
91: Voltage directional relay
92: Voltage and power directional relay
93: Field-changing contactor
94: Tripping or trip-free relay
95–99: Used only for specific applications

A.2.2 Device/Function Acronyms

AFD: Arc flash detector
CLK: Clock or timing source
DDR: Dynamic disturbance recorder
DFR: Digital fault recorder
ENV: Environmental data
HIZ: High impedance fault detector
HMI: Human machine interface
HST: Historian
LGC: Scheme logic
MET: Substation metering
PDC: Phasor data concentrator
PMU: Phasor measurement unit
PQM: Power quality monitor
RIO: Remote input/output device
RTU: Remote terminal unit/data concentrator
SER: Sequence of events recorder
TCM: Trip circuit monitor

A.2.3 Suffix Letters

A.2.3.1 Auxiliary Devices

C: Closing relay/contactor
CL: Auxiliary relay, closed[1]
CS: Control switch
D: "Down" position switch relay
L: Lowering relay
O: Opening relay/contactor
OP: Auxiliary relay, open[2]
PB: Push button
R: Raising relay
U: "Up" position switch relay
X: Auxiliary relay
Y: Auxiliary relay
Z: Auxiliary relay

A.2.3.2 Actuating Quantities

A: Air/amperes/alternating
C: Current
D: Direct/discharge
E: Electrolyte
F: Frequency/flow/fault
GP: Gas pressure
H: Explosive/harmonics
I_0: Zero sequence current
I_-, I_2: Negative sequence current
I_+, I_1: Positive sequence current
J: Differential
L: Level/liquid
LR: Locked rotor
P: Power (real)/pressure
PF: Power factor
Q: Reactive power/oil
S: Speed/suction/smoke
T: Temperature
V: Voltage/volts/vacuum
VB: Vibration
W: Water/watts

1 It is energized when main device is in closed position.
2 It is energized when main device is in open position.

A.2.3.3 Main Device

A: Alarm/auxiliary power
AC: Alternating current
AN: Anode
B: Battery/blower/bus
BK: Brake
BL: Block (valve)
BP: Bypass
BT: Bus tie
C: Capacitor/condenser/compensator/carrier current/case/compressor
CA: Cathode
CH: Check (valve)
D: Discharge (valve)
DC: Direct current
E: Exciter
F: Feeder/field/filament/filter/fan
G: Generator/ground
H: Heater/housing
L: Line/logic
M: Metering/motor
MOC: Mechanism operated contact
N: Network/neutral
P: Phase comparison/pump
R: Reactor/rectifier/room/rotor
S: Secondary/stator/strainer/sump/suction (valve), synchronizing
T: Transformer/thyratron
TH: Transformer (high-voltage side)
TL: Transformer (low-voltage size)
TM: Telemeter
TOC: Truck-operated contact
TT: Transformer (tertiary-voltage side)
U: Unit

A.2.3.4 Main Device Parts

BK: Brake
C: Coil/condenser/capacitor
CC: Closing coil/closing contactor
HC: Holding coil
M: Operating motor
MF: Fly-ball motor
ML: Load-limit motor
MS: Speed adjusting or synchronizing motor
OC: Opening contactor
S: Solenoid
SI: Seal-in
T: Target
TC: Trip coil
V: Valve
W: Winding

A.2.3.5 Other Suffix Letters

A: Accelerating/automatic
B: Blocking
BU: Back up
BF: Breaker failure
C: Closed/cold
D: Decelerating/detonate/down/disengaged
DCB: Directional comparison blocking
DCUB: Directional comparison unblocking
DUTT: Direct underreaching transfer trip
E: Emergency/engaged
F: Failure/forward
GC: Ground check
GP: General Purpose
H: Hot/high
HIZ: High impedance fault
HR: Hand reset
HS: High speed
L: Left/local/low/lower/leading
M: Manual/master
O: Open/over
OFF: Off
ON: On
P: Polarizing
POTT: Permissive overreaching transfer trip
PUTT: Permissive underreaching transfer trip
R: Right/raise/reclosing /receiving/remote/reverse
S: Sending/swing

SHS: Semi-high speed
SOTF: Switch on to fault
T: Test/trip/trailing
TD: Time delay
TDC: Time-delay closing contact
TDDO: Time delayed relay coil drop-out
TDO: Time-delay opening contact
TDPU: Time delayed relay coil pickup
THD: Total harmonic distortion
U: Up/under
Z: Impedance

Appendix B

How to Install PowerWorld Simulator (Education Version)

Although PowerWorld Simulator is commercial software, a limited demo version is available for free. It is an interactive power system simulation package designed to simulate high voltage power system operation on a time frame ranging from several minutes to several days. The software contains a highly effective power flow (PF) analysis package capable of efficiently solving systems of up to 250,000 buses. It supports PF and Optimal PF, short-circuit analysis, transient stability, contingency analysis, and many others. This software is very useful to study optimal coordination of protective devices because it provides easy steps to do short-circuit analysis at busbars and any point on lines. It supports all the fault types and the calculations can be carried out using either actual quantities or per-unit quantities. To install the demo version of this software:

1. Go to https://www.powerworld.com/
2. From `Products` tab, select `Simulator`:

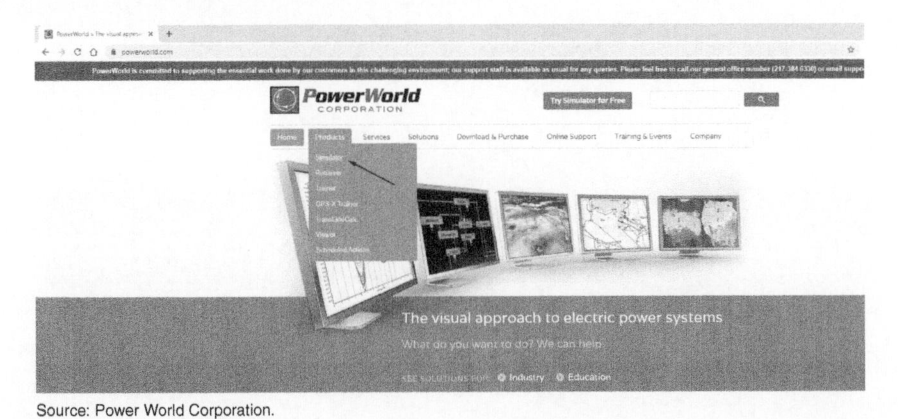

Source: Power World Corporation.

Optimal Coordination of Power Protective Devices with Illustrative Examples, First Edition. Ali R. Al-Roomi.

Companion website: www.wiley.com/go/al-roomi/optimalcoordination

3. Then, select `Download Simulator Demo`:

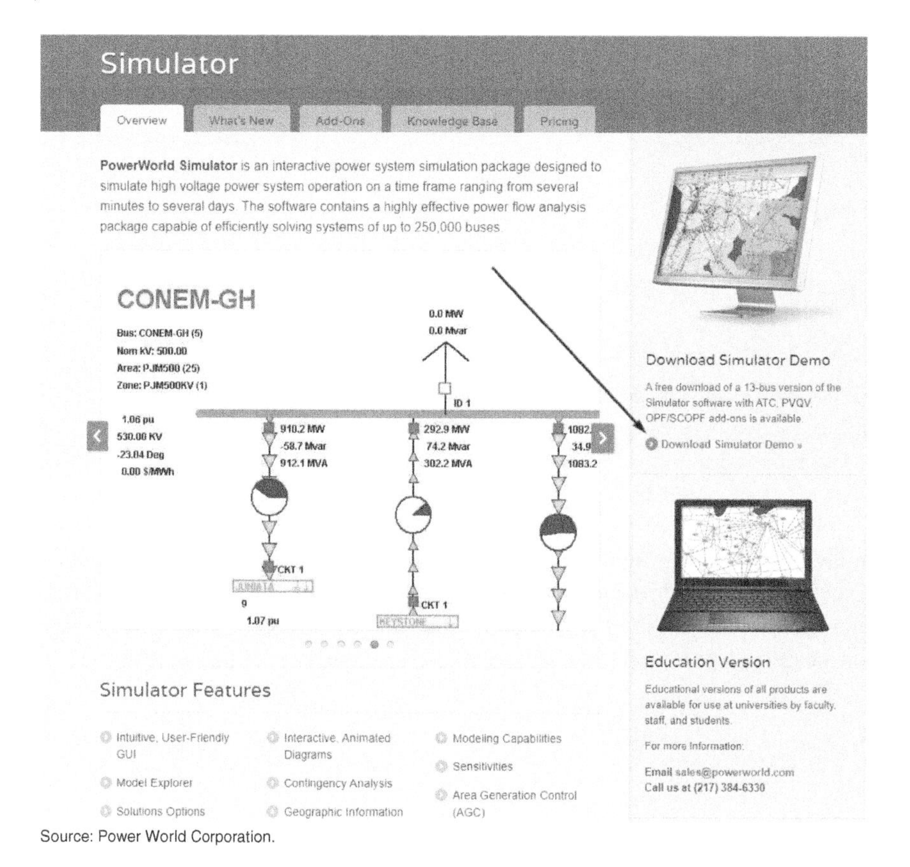

Source: Power World Corporation.

4. After that, select the latest version:

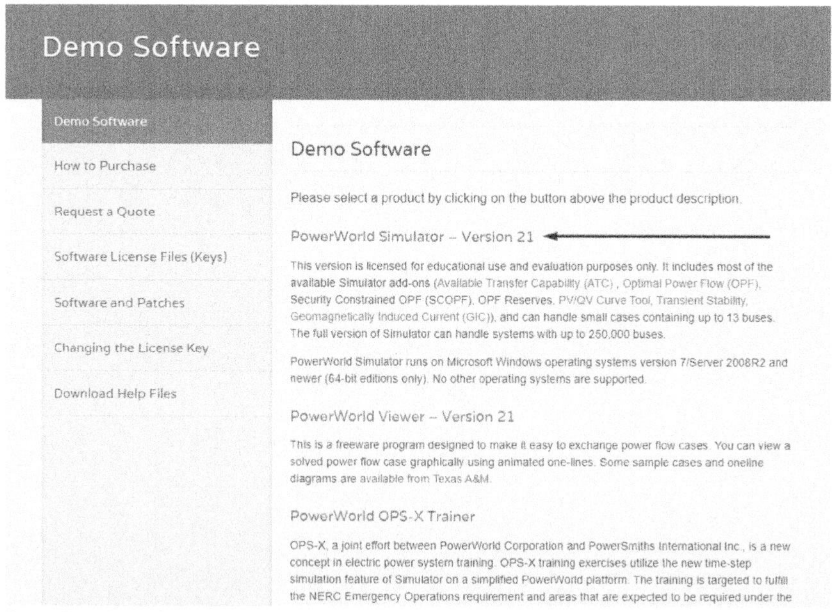

Source: Power World Corporation.

5. Then, fill the form. The mandatory fields are your first name, last name, email address, and company name. The optional fields are your title (Mr., Dr., Prof., etc.) and daytime phone number. Once completed, press on `Submit` button.

PowerWorld Simulator Version 21 Demo

Please check your email for the link to the download file. Thank you for your interest in PowerWorld software.

Submit another response

Source: Power World Corporation.

6. Finally, follow the link sent to your email address and download the software.

This demo version supports up to 13-bus. If you have any new version of this book Glover et al. (2012), then you can download a better version[1] through this link: https://www.powerworld.com/gloveroverbyesarma.

1 It supports more than 13-bus. This version is used as a supporting tool to cover the examples given in the book.

Appendix C

Single-Machine Infinite Bus

This appendix derives some equations used in the SMIB system. Suppose that the given SMIB system is similar to the one shown in Figure C.1.

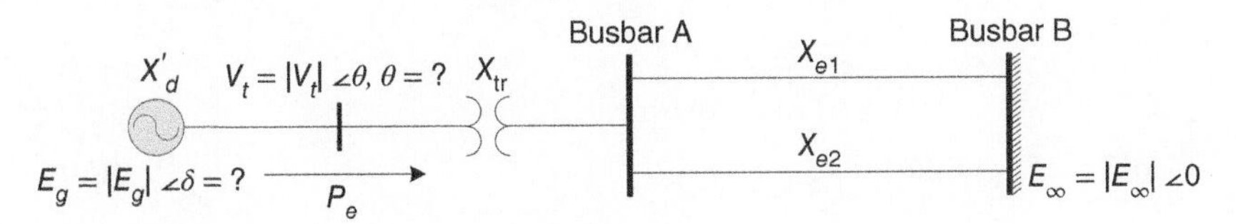

Figure C.1 Single-machine infinite bus system.

If all the per-unit impedances are approximated as inductances (because $X_L \gg R$), then the preceding network can be simplified to be as the one shown in Figure C.2.

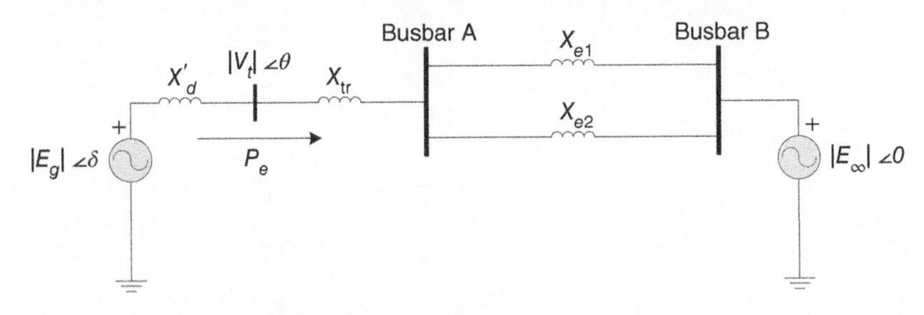

Figure C.2 Single-line diagram of the SMIB system.

From Figure C.2:

$$X_e = \frac{X_{e_1} \times X_{e_2}}{X_{e_1} + X_{e_2}} \tag{C.1}$$

$$X_l = X_{\text{tr}} + X_e \tag{C.2}$$

$$X_{\text{tot}} = X_d' + X_{\text{tr}} \tag{C.3}$$

$$S_g = E_g I^* = P_e + jQ_e \tag{C.4}$$

$$I = \frac{|E_g| \angle \delta^o - |E_\infty| \angle 0^o}{jX_d'} = \frac{|V_t| \angle \theta^o - |E_\infty| \angle 0^o}{jX_l} \tag{C.5}$$

Optimal Coordination of Power Protective Devices with Illustrative Examples, First Edition. Ali R. Al-Roomi.

Companion website: www.wiley.com/go/al-roomi/optimalcoordination

$$I^* = \frac{|E_g| \angle -\delta^o - |E_\infty| \angle 0^o}{-jX'_d} = \frac{|V_t| \angle -\theta^o - |E_\infty| \angle 0^o}{-jX_l} \tag{C.6}$$

$$P_e = \mathcal{R}\{S_g\} = \mathcal{R}\{E_g I^*\} \tag{C.7}$$

∵ All the impedances have only reactance component

∴ It is a lossless network, and thus P_e can be calculated as follows:

$$P_e = \mathcal{R}\{V_t I^*\} = \mathcal{R}\left\{\frac{V_t^2 - |V_t||E_\infty| \angle \theta^o}{-jX_l}\right\}$$

$$= \mathcal{R}\left\{j\frac{V_t^2}{X_l} - \frac{j}{X_l}\left[|V_t||E_\infty|\cos(\theta) + j|V_t||E_\infty|\sin(\theta)\right]\right\}$$

$$\therefore P_e = \frac{|V_t||E_\infty|}{X_l}\sin(\theta) \tag{C.8}$$

By analogy:

$$P_e = \frac{|E_g||E_\infty|}{X_{\text{tot}}}\sin(\delta) = P_{\max}\sin(\delta) \tag{C.9}$$

where E_g can be calculated from Figure C.3 as follows:

$$E_g = V_t + jX'_d I \tag{C.10}$$

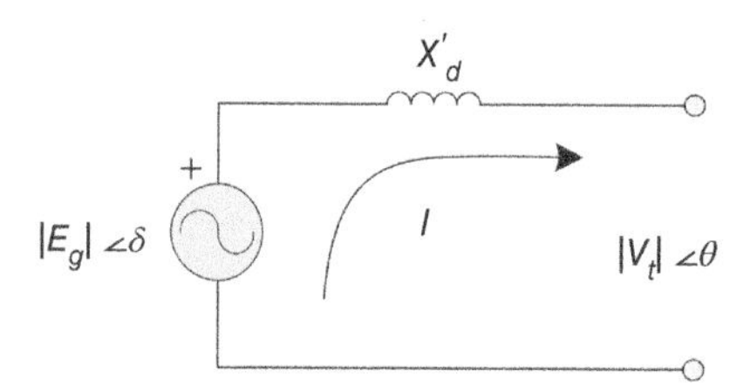

Figure C.3 Machine's internal voltage.

Appendix D

Linearizing Relay Operating Time Models

This appendix shows how to linearize the objective function of optimal relay coordination (ORC) problems using both the IEC/BS and ANSI/IEEE standard models.

D.1 Linearizing the IEC/BS Model of DOCRs by Fixing Time Multiplier Settings

The operating time of the ith directional over-current relay (DOCR) can be calculated at the kth location as follows:

$$T_{i,k} = TMS_i \times \frac{\beta}{\left(\frac{I_{i,k}}{PS_i}\right)^{\alpha} - 1} \tag{D.1}$$

By taking the reciprocal of both sides:

$$\frac{1}{T_{i,k}} = \frac{\left(\frac{I_{i,k}}{PS_i}\right)^{\alpha} - 1}{\beta\ TMS_i} \tag{D.2}$$

Distributing the exponent α and multiplying both sides by $\beta\ TMS_i$:

$$\frac{\beta\ TMS_i}{T_{i,k}} = \frac{I_{i,k}^{\alpha}}{PS_i^{\alpha}} - 1 = \frac{I_{i,k}^{\alpha} - PS_i^{\alpha}}{PS_i^{\alpha}} \tag{D.3}$$

By re-taking the reciprocal of both sides:

$$\frac{T_{i,k}}{\beta\ TMS_i} = \frac{PS_i^{\alpha}}{I_{i,k}^{\alpha} - PS_i^{\alpha}} \tag{D.4}$$

Multiplying both sides by $\beta\ TMS_i$:

$$\therefore T_{i,k} = \beta\ TMS_i \left(\frac{PS_i^{\alpha}}{I_{i,k}^{\alpha} - PS_i^{\alpha}}\right) = \boxed{\vartheta_i \widetilde{PS}_{i,k}} \tag{D.5}$$

where ϑ_i and $\widetilde{PS}_{i,k}$ are respectively equal to:

$$\boxed{\vartheta_i = \beta\ TMS_i} \tag{D.6}$$

$$\boxed{\widetilde{PS}_{i,k} = \frac{PS_i^{\alpha}}{I_{i,k}^{\alpha} - PS_i^{\alpha}}} \tag{D.7}$$

Optimal Coordination of Power Protective Devices with Illustrative Examples, First Edition. Ali R. Al-Roomi.

Companion website: www.wiley.com/go/al-roomi/optimalcoordination

To extract PS_i from (D.7), first, both sides should be multiplied by $\left(I_{i,k}^{\alpha} - PS_i^{\alpha}\right)$:

$$\widetilde{PS}_{i,k}\left(I_{i,k}^{\alpha} - PS_i^{\alpha}\right) = PS_i^{\alpha} \tag{D.8}$$

By applying the distributive property, the parentheses can be removed:

$$\widetilde{PS}_{i,k}I_{i,k}^{\alpha} - \widetilde{PS}_{i,k}PS_i^{\alpha} = PS_i^{\alpha} \tag{D.9}$$

Collecting the terms containing PS_i^{α} in one side:

$$\widetilde{PS}_{i,k}I_{i,k}^{\alpha} = PS_i^{\alpha} + \widetilde{PS}_{i,k}PS_i^{\alpha} = PS_i^{\alpha}\left(1 + \widetilde{PS}_{i,k}\right) \tag{D.10}$$

Therefore,

$$PS_i = \left[\frac{\widetilde{PS}_{i,k}I_{i,k}^{\alpha}}{1 + \widetilde{PS}_{i,k}}\right]^{1/\alpha} = \frac{\left(\widetilde{PS}_{i,k}I_{i,k}^{\alpha}\right)^{1/\alpha}}{\left(1 + \widetilde{PS}_{i,k}\right)^{1/\alpha}}$$

$$= \frac{\left(\widetilde{PS}_{i,k}\right)^{1/\alpha}\left(I_{i,k}^{\alpha}\right)^{1/\alpha}}{\left(1 + \widetilde{PS}_{i,k}\right)^{1/\alpha}} = \frac{\left(I_{i,k}\right)^{\alpha/\alpha}\sqrt[\alpha]{\widetilde{PS}_{i,k}}}{\sqrt[\alpha]{1 + \widetilde{PS}_{i,k}}}$$

$$\therefore \boxed{PS_i = I_{i,k} \times \sqrt[\alpha]{\frac{\widetilde{PS}_{i,k}}{1 + \widetilde{PS}_{i,k}}}} \tag{D.11}$$

D.2 Linearizing the ANSI/IEEE Model of DOCRs by Fixing Time Multiplier Settings

The operating time of the ith DOCR can be calculated at the kth location as follows:

$$T_{i,k} = TDS_i \times \left[\frac{\beta}{\left(\frac{I_{i,k}}{CTS_i}\right)^{\alpha} - 1} + \gamma\right] \tag{D.12}$$

Moving the gamma term to the left-side:

$$T_{i,k} - \gamma\ TDS_i = \frac{\beta\ TDS_i}{\left(\frac{I_{i,k}}{CTS_i}\right)^{\alpha} - 1} \tag{D.13}$$

By taking the reciprocal of both sides:

$$\frac{1}{T_{i,k} - \gamma\ TDS_i} = \frac{\left(\frac{I_{i,k}}{CTS_i}\right)^{\alpha} - 1}{\beta\ TDS_i} \tag{D.14}$$

Distributing the exponent α and multiplying both sides by $\beta\ TDS_i$:

$$\frac{\beta\ TDS_i}{T_{i,k} - \gamma\ TDS_i} = \frac{I_{i,k}^{\alpha}}{CTS_i^{\alpha}} - 1 = \frac{I_{i,k}^{\alpha} - CTS_i^{\alpha}}{CTS_i^{\alpha}} \tag{D.15}$$

By re-taking the reciprocal of both sides:

$$\frac{T_{i,k} - \gamma \ TDS_i}{\beta \ TDS_i} = \frac{CTS_i^{\alpha}}{I_{i,k}^{\alpha} - CTS_i^{\alpha}} \tag{D.16}$$

Multiplying both sides by $\beta \ TDS_i$:

$$T_{i,k} - \gamma \ TDS_i = \beta \ TDS_i \left(\frac{CTS_i^{\alpha}}{I_{i,k}^{\alpha} - CTS_i^{\alpha}} \right) \tag{D.17}$$

Moving the gamma term to the right-side:

$$\therefore \ T_{i,k} = \beta \ TDS_i \left(\frac{CTS_i^{\alpha}}{I_{i,k}^{\alpha} - CTS_i^{\alpha}} \right) + \gamma \ TDS_i = \boxed{\vartheta_i \widetilde{CTS}_{i,k} + \xi_i} \tag{D.18}$$

where ϑ_i, $\widetilde{CTS}_{i,k}$, and ξ_i are respectively equal to:

$$\boxed{\vartheta_i = \beta \ TDS_i} \tag{D.19}$$

$$\boxed{\widetilde{CTS}_{i,k} = \frac{CTS_i^{\alpha}}{I_{i,k}^{\alpha} - CTS_i^{\alpha}}} \tag{D.20}$$

$$\boxed{\xi_i = \gamma \ TDS_i} \tag{D.21}$$

Appendix E

Derivation of the First Order Thermal Differential Equation

This appendix contains the derivations of the models used in thermal overload protection for overhead lines and cables (Smith, 2015; Smith and Jain, 2017).

Suppose a current (I) is flowing in a conductor having a resistance (R). If the total heat (Q_T) generated by the root mean square (RMS) value of I equals the dissipated heat (Q_D) plus the heat imparted to the resistance (Q_R), then we will have the following relationship:

$$Q_T = Q_D + Q_R \tag{E.1}$$

which can also be calculated as follows:

$$Q_T = I^2 R dT \tag{E.2}$$

From heat transfer topic, we can calculate the heat dissipated to the surrounding:

$$Q_D = hA\theta dT \tag{E.3}$$

where

h heat transfer coefficient
A surface area of conductor
θ surface temperature (above ambient)
dT difference in time

Also, the heat absorbed by the conductor can be calculated as follows:

$$Q_R = cmd\theta \tag{E.4}$$

where

c specific heat capacity
m mass of conductor
$d\theta$ difference in temperature

Thus, by substituting (E.2)–(E.4) in (E.1) yields:

$$I^2 R dT = hA\theta dT + cmd\theta \tag{E.5}$$

Let's divide both sides by $hAdT$:

$$\frac{I^2 R}{hA} = \theta + \frac{cm}{hA}\frac{d\theta}{dT} \tag{E.6}$$

Optimal Coordination of Power Protective Devices with Illustrative Examples, First Edition. Ali R. Al-Roomi.

Companion website: www.wiley.com/go/al-roomi/optimalcoordination

Re-locating $\frac{d\theta}{dt}$ to the left side gives:

$$\frac{d\theta}{dT} = \frac{hA}{cm}\left(\frac{I^2R}{hA} - \theta\right) \tag{E.7}$$

The heat time constant is:

$$\tau = \frac{cm}{hA} \tag{E.8}$$

Therefore, by substituting (E.8) into (E.7), the heat balance equation becomes:

$$\frac{d\theta}{dT} = \frac{1}{\tau}\left(\frac{I^2R}{hA} - \theta\right) \tag{E.9}$$

Appendix F

List of ORC Test Systems

This appendix contains most of the classical optimal relay coordination (ORC) problems that are based only on directional over-current relays (DOCRs) with/without over-current relays (OCRs). It has to be noted that the problem dimension depends on the number of branches and not nodes. Thus, the problem dimension is proportional to the number of lines between busbars.

F.1 Three-Bus Test Systems

F.1.1 System No. 1

- This test system is shown in Figure F.1.
- It has 3-phase (3ϕ) fault at the mid-point of each line.
- It is considered as a one of the most popular test systems used in ORC.
- This system is used in the first paper published in the literature Urdaneta et al. (1988).
- The plug-setting (*PS*) is considered discrete with uniform steps of 0.5 A, while the time-multiplier setting (*TMS*) is considered continuous.
- The coordination time interval (*CTI*) is equal to 0.2 s.
- These settings should be changed to get fair comparisons with other studies conducted in the literature.
- In this test system, the transient selectivity constraints are considered too.
- All the current transformer-ratios (*CTR*s), primary/backup (P/B) relay pairs, 3ϕ fault at the mid-point of each line, generator and line data, and relay variable bounds are given in Tables F.1–F.5.
- This test system is used in many studies. Some of these studies are (Urdaneta et al., 1988; Mansour et al., 2007; Zeineldin, 2008; Amraee, 2012; El-Mesallamy et al., 2013; Ralhan and Ray, 2013; Albasri et al., 2015; Papaspiliotopoulos et al., 2017a).
- The online information about this system is available in: https://al-roomi.org/coordination/3-bus-systems/system-i.

F.1.2 System No. 2

- This test system is shown in Figure F.2.
- The near-end and far-end bolted 3ϕ faults are both considered for each line.
- The IEC standard inverse time–current characteristic curve (TCCC) is commonly used in this test system as per the cited references given in the following text.

Optimal Coordination of Power Protective Devices with Illustrative Examples, First Edition. Ali R. Al-Roomi.

Companion website: www.wiley.com/go/al-roomi/optimalcoordination

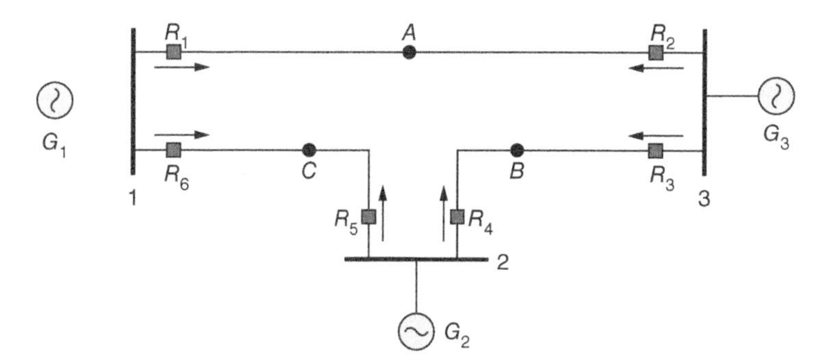

Figure F.1 Single-line diagram of the three-bus test system no. 1.

Table F.1 *CTR*s of the relays used in the three-bus system no. 1.

Relay	R_1	R_2	R_3	R_4	R_5	R_6
CTR	300 : 5	200 : 5	200 : 5	300 : 5	200 : 5	400 : 5

Table F.2 3ϕ faults of the three-bus system no. 1.

Primary R_i	Fault current (A)		Backup R_j	Fault current (A)	
	Normal	Transient		Normal	Transient
R_1	1978.90	2075.00	R_5	175.00	400.70
R_2	1525.70	1621.70	R_4	545.00	700.64
R_3	1683.90	1779.60	R_1	617.22	760.17
R_4	1815.40	1911.50	R_6	466.17	622.65
R_5	1499.66	1588.50	R_3	384.00	558.13
R_6	1766.30	1855.40	R_2	145.34	380.70

Table F.3 Generator data of the three-bus system no. 1.

Gen.	S_n (MVA)	V_p (kV)	x (%)
G_1	100	69	20
G_2	25	69	12
G_3	50	69	18

- Similar to the previous one, this network consists of three buses, three branches, and six DOCRs located at both ends of each line.
- As stated before in (7.6), the objective function of this system is mathematically expressed as follows:

$$\min \ Z = \sum_{p=1}^{\varrho^{\text{near}}} T_p^{\text{pr,near}} + \sum_{q=1}^{\text{far}} T_q^{\text{pr,far}}$$

Table F.4 Line data of the three-bus system no. 1.

Nodes	R (Ω)	X (Ω)	Length (km)
1–2	5.5	22.85	50
2–3	4.4	18.00	40
1–3	7.6	27.00	60

Table F.5 Optimization data of the three-bus system no. 1.

TMS^{min}	TMS^{max}	PS^{min} (A)	PS^{max} (A)	PS mode	CTI (s)
0.1	1.1	1.5	5.0	Discrete	0.2

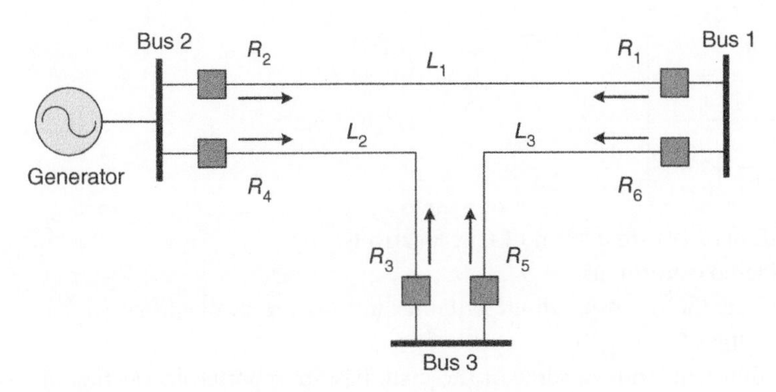

Figure F.2 Single-line diagram of the three-bus test system no. 2.

where $T_p^{pr,near}$ and $T_q^{pr,far}$ are, respectively, the operating time of the primary relay at the near-end 3ϕ fault (i.e. at the pth location) and the far-end 3ϕ fault (i.e. at the qth location), which can be calculated as follows:

$$T_p^{pr,near} = TMS_p \left[\frac{\beta}{\left(\dfrac{a_p}{PS_p \times b_p} \right)^{\alpha} - 1} \right] \tag{F.1}$$

$$T_q^{pr,far} = TMS_q \left[\frac{\beta}{\left(\dfrac{c_p}{PS_q \times d_p} \right)^{\alpha} - 1} \right] \tag{F.2}$$

- Note that the constants of $T_q^{pr,far}$ in (F.2) are calculated in a sequence of p instead of q.
- Also, these p and q notations are i and j in the cited references. This replacement is essential to prevent any confusion with other test systems presented in this book because the notations i and j are assigned for primary and backup relays, respectively.

- The number of selectivity constraints depends on the network where only 8 selectivity constraints out of 12 are considered. This results in a total number of 26 constraints.
- The mathematical expression of these selectivity constraints is already given in (7.25); and for the sake of some important modifications in the next two equations, (7.25) is repeated here:

$$T_{jk} - T_{ik} \geq CTI$$

where T_{jk} and T_{ik} are, respectively, the operating times of the jth backup and ith primary relays for a 3ϕ fault occurring at the kth location. They can be computed via the following two equations:

$$T_{jk} = TMS_j \left[\frac{\beta}{\left(\frac{e_p}{PS_j \times f_p} \right)^{\alpha} - 1} \right] \tag{F.3}$$

$$T_{ik} = TMS_i \left[\frac{\beta}{\left(\frac{g_p}{PS_i \times h_p} \right)^{\alpha} - 1} \right] \tag{F.4}$$

- The lower and upper bounds of (7.19) are 0.05 and 1.0, respectively.
- Both PS and TMS are considered continuous.
- All CTRs, P/B relay pairs, and the corresponding 3ϕ fault currents are available as $\{a, b, c, d, e, f, g, h\}$ constants in the Tables F.6–F.8.
- This test system is used in different studies. Some of these studies are reported in Deep et al. (2006), Bansal and Deep (2008), Deep and Bansal (2009), Thangaraj et al. (2010), Moirangthem et al. (2013), Singh et al. (2013), Chelliah et al. (2014), and Thakur and Kumar (2016).
- The online information about this system is available in: https://al-roomi.org/coordination/3-bus-systems/system-ii.

Table F.6 Values of a_p, b_p, c_p, and d_p of the three-bus system no. 2.

	$T^{near}_{pr,p}$			$T^{far}_{pr,p}$	
p	a_p	b_p	q	c_p	d_p
1	9.46	2.06	2	100.63	2.06
2	26.91	2.06	1	14.08	2.06
3	8.81	2.23	4	136.23	2.23
4	37.68	2.23	3	12.07	2.23
5	17.93	0.8	6	19.2	0.8
6	14.35	0.8	5	25.9	0.8

Table F.7 Values of e_p, f_p, g_p, and h_p of the three-bus system no. 2.

Tjk			T_{ik}		
j	e_p	f_p	i	g_p	h_p
5	14.08	0.8	1	14.08	2.06
6	12.07	0.8	3	12.07	2.23
4	25.9	2.23	5	25.9	0.8
2	14.35	0.8	6	14.35	2.06
5	9.46	0.8	1	9.46	2.06
6	8.81	0.8	3	8.81	2.23
2	19.2	2.06	6	19.2	0.8
4	17.93	2.23	5	17.93	0.8

Table F.8 Optimization data of the three-bus system no. 2.

TMS^{min}	TMS^{max}	PS^{min} (A)	PS^{max} (A)	PS mode	CTI (s)
0.05	1.1	1.25	1.5	Continuous	0.3

F.2 Four-Bus Test Systems

F.2.1 System No. 1

- This test system is shown in Figure F.3.
- The modeling of this test system is exactly similar to that given in the second three-bus test system where the rest of the information required to simulate this ORC problem is presented in Tables F.9–F.11.
- This test system is used in different studies. Some of these studies are reported in Deep et al. (2006), Bansal and Deep (2008), Deep and Bansal (2009), Thangaraj et al. (2010), Moirangthem et al. (2013), Singh et al. (2013), Chelliah et al. (2014), Albasri et al. (2015), and Thakur and Kumar (2016).
- The online information about this system is available in: https://al-roomi.org/coordination/4-bus-system/system-i.

F.2.2 System No. 2

- This test system is shown in Figure F.4.
- This test system is proposed as an adaptive ORC problem to study the impacts of various system operating conditions on protection coordination of DOCRs.
- It consists of four buses, four lines, and supplied by two sources.
- The four lines (i.e. $\{L_1, L_2, L_3, L_4\}$) are protected by 8 DOCRs; one at the end of each branch.

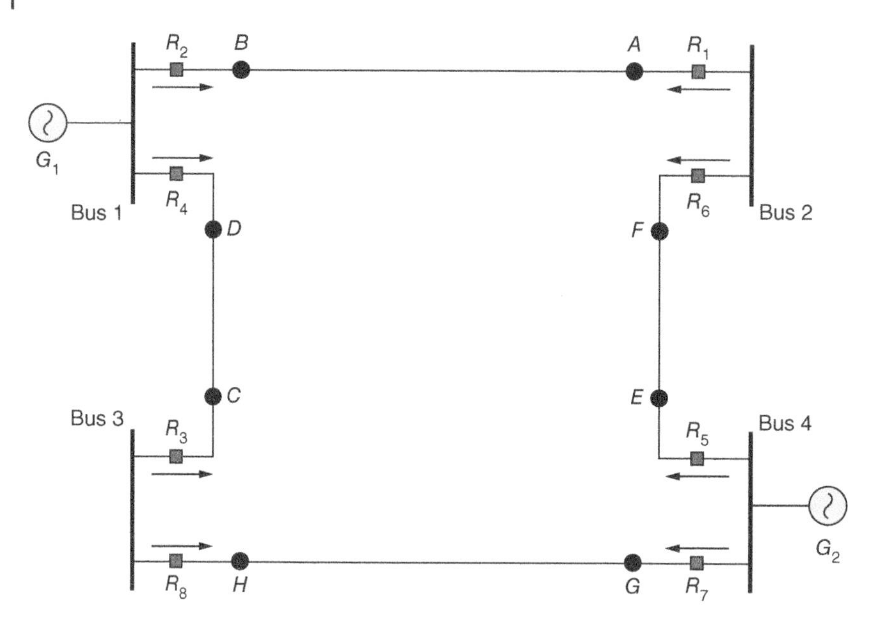

Figure F.3 Single-line diagram of the four-bus test system no. 1.

Table F.9 Values of a_p, b_p, c_p, and d_p of the four-bus system no. 1.

	$T^{near}_{pr,p}$			$T^{far}_{pr,p}$	
p	a_p	b_p	q	c_p	d_p
1	20.32	0.48	2	23.75	0.48
2	88.85	0.48	1	12.48	0.48
3	13.61	1.1789	4	31.92	1.1789
4	116.81	1.1789	3	10.38	1.1789
5	116.7	1.5259	6	12.07	1.5259
6	16.67	1.5259	5	31.92	1.5259
7	71.7	1.2018	8	11.00	1.2018
8	19.27	1.2018	7	18.91	1.2018

- The main power is supplied to the network through buses 1 and 2 where bus 1 is connected to a substation and bus 2 is connected to a 5 MVA generator.
- The short-circuit rating of the substation is 250 MVA, which is ten times that of the generator.
- Moreover, a DG is scheduled to be integrated at bus 4 with a short-circuit rating of 25 MVA; i.e. similar to the generator.
- The base quantities are 10.5 kV and 100 MVA.
- All the lines have the same impedance of $(0.02 + j0.05)$ pu.
- The active loads connected to buses 2, 3, and 4 are, respectively, 3, 4, and 5 MW.
- The reactive loads connected to the previous three buses are 2, 0.8, and 1.2 MVAR, respectively.
- The objective function used in this test system is the one given in (7.14).

Table F.10 Values of e_p, f_p, g_p, and h_p of the four-bus system no. 1.

Tjk			T_{ik}		
j	e_p	f_p	i	g_p	h_p
5	20.32	1.5259	1	20.32	0.48
5	12.48	1.5259	1	12.48	0.48
7	13.61	1.2018	3	13.61	1.1789
7	10.38	1.2018	3	10.38	1.1789
1	1.16	0.48	4	116.81	1.1789
2	12.07	0.48	6	12.07	1.1789
2	16.67	0.48	6	16.67	1.5259
4	11	1.1789	8	11	1.2018
4	19.27	1.1789	8	19.27	1.2018

Table F.11 Optimization data of the four-bus system no. 1.

TMS^{min}	TMS^{max}	PS^{min} (A)	PS^{max} (A)	PS mode	CTI (s)
0.05	1.1	1.25	1.5	Continuous	0.3

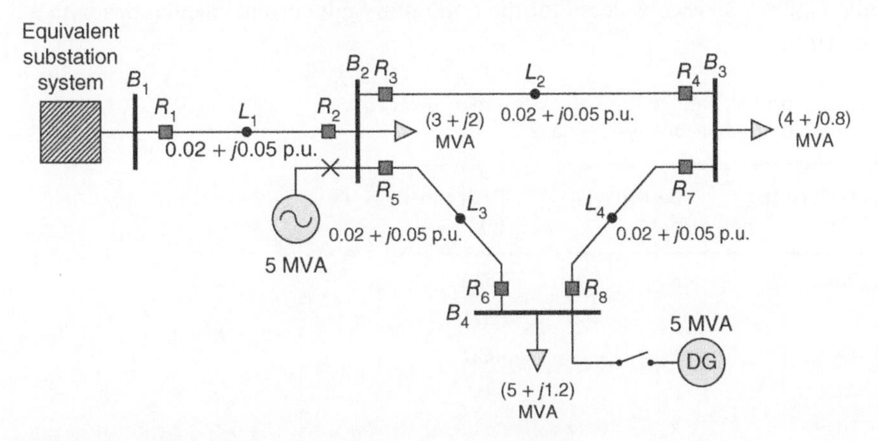

Figure F.4 Single-line diagram of the four-bus test system no. 2.

- The weights λ_1 and λ_2 give a compromise between the minimum operating times of the primary and backup relays (Alam et al., 2016).
- The optimal values of these two weighting factors can be determined by minimizing the average sum of the operating times of primary and backup relays (ASOTPBR) corresponding to different values of λ_1 and λ_2 (Alam, 2019):

$$\text{ASOTPBR} = \frac{1}{\varrho} \cdot \sum_{i=1}^{\varrho} T_i + \frac{1}{\sigma} \cdot \sum_{j=1}^{\sigma} T_j \tag{F.5}$$

Table F.12 Maximum load currents, minimum and maximum fault currents, and *CTR* at peak load operation of the four-bus system no. 2.

R_i	$I_{L,i}^{max}$ (A)	$I_{f,i}^{min}$ (A)	$I_{f,i}^{max}$ (A)	CTR_i
R_1	597	6543	8261	3000 : 5
R_2	—	1701	2148	3000 : 5
R_3	117	3412	4187	1200 : 5
R_4	—	2271	2788	800 : 5
R_5	178	3875	4735	1200 : 5
R_6	—	1426	1742	1000 : 5
R_7	62	3129	3810	1000 : 5
R_8	—	1827	2225	600 : 5

- All DOCRs are assumed to be numerical with continuous *TMS* and *PS* where $TMS \in [0.1, 1.1]$ and $PS \in [0.5, 2.0] \times$ current transformer (CT) secondary rating of each relay (in steps of 0.25).
- The CTI is set to 0.2 s, and the lower and upper bounds of T_i are 0.1 and 4 s, respectively.
- All DOCRs are emulated using the IEC standard inverse TCCC.
- The maximum load currents (I_L^{max}), minimum fault currents (I_f^{min}), and the maximum fault currents (I_f^{max}) of all DOCRs are tabulated in Table F.12 for the peak load condition of the system. Table F.12 also shows the *CTR* of all the relays used in the network.
- For the maximum fault current calculations, Figure 2.3e is considered with zero fault impedance, while Figure 2.3b is considered for the minimum fault current calculations with a fault impedance of 0.1 pu.

Table F.13 Short-circuit currents passing through all P/B relay pairs at the peak load condition of the four-bus system no. 2.

Primary relay R_i	Fault current (A)	Backup relay R_j	Fault current (A)
R_2	2527	R_4	1542
R_2	2527	R_6	821
R_3	4466	R_1	4627
R_3	4466	R_6	465
R_4	3048	R_8	436
R_5	4987	R_1	4481
R_5	4987	R_4	384
R_6	1989	R_7	1874
R_7	4066	R_3	1587
R_8	2491	R_5	2369

- Table F.13 lists the P/B relay pairs and the corresponding short-circuit currents at the peak load condition.
- If I_f^{bc} satisfies (F.6), then the corresponding P/B relay pair is ignored because $T_j - T_i \geqslant CTI$ will always be satisfied:

$$I_f^{bc} < \max\left[2 \times I_L^{max}, I_f^{min}\right] \tag{F.6}$$

- Table F.14 lists the P/B relay pairs and the corresponding short-circuit currents under generator outage at bus 2.
- Table F.15 lists the P/B relay pairs and the corresponding short-circuit currents under line 4 is taken out of service.
- Table F.16 lists the P/B relay pairs and the corresponding short-circuit currents under integration of DG at bus 4.
- More information about this test system can be found in Alam (2019).
- The online information about this system is available in: https://al-roomi.org/coordination/4-bus-systems/system-ii.

Table F.14 Short-circuit currents passing through some P/B relay pairs under generator outage at bus 2 in the four-bus system no. 2.

Primary relay R_i	Fault current (A)	Backup relay R_j	Fault current (A)
R_3	4155	R_1	4714
R_4	1120	R_8	834
R_5	4179	R_1	4707
R_6	1072	R_7	899
R_7	2309	R_3	2074
R_8	2241	R_5	2106

Table F.15 Short-circuit currents passing through some P/B relay pairs under line 4 is taken out of service in the four-bus system no. 2.

Primary relay R_i	Fault current (A)	Backup relay R_j	Fault current (A)
R_2	2396	R_4	2091
R_3	4770	R_1	4537
R_5	5978	R_1	4195
R_5	5978	R_4	1658

Table F.16 Short-circuit currents passing through some P/B relay pairs under integration of DG at bus 4 in the four-bus system no. 2.

Primary relay R_i	Fault current (A)	Backup relay R_j	Fault current (A)
R_2	3207	R_4	1693
R_2	3207	R_6	1373
R_3	4840	R_1	4523
R_3	4840	R_6	465
R_4	3382	R_8	919
R_5	5074	R_1	4457
R_5	5074	R_4	505
R_6	2866	R_7	1595
R_7	4091	R_3	1741
R_8	3342	R_5	2111

F.3 Five-Bus Test System

- This test system is shown in Figure F.5.
- This test system is proposed as an adaptive ORC problem to study the impacts of various system operating conditions on protection coordination of DOCRs.
- The 20-kV distribution network consists of five buses, five lines, four loads, two DGs, and an external HV grid supplied through a 150/20 kV step-down transformer having a power rating of 40 MVA with an impedance value of 12.8%.
- It represents a simplified two-feeder portion of the Hellenic distribution system.
- The short-circuit capacity of the external grid is 3000 MVA.
- DG1 and DG2 stations are composed of nine and two synchronous generators, respectively, interconnected with their own unit transformer.

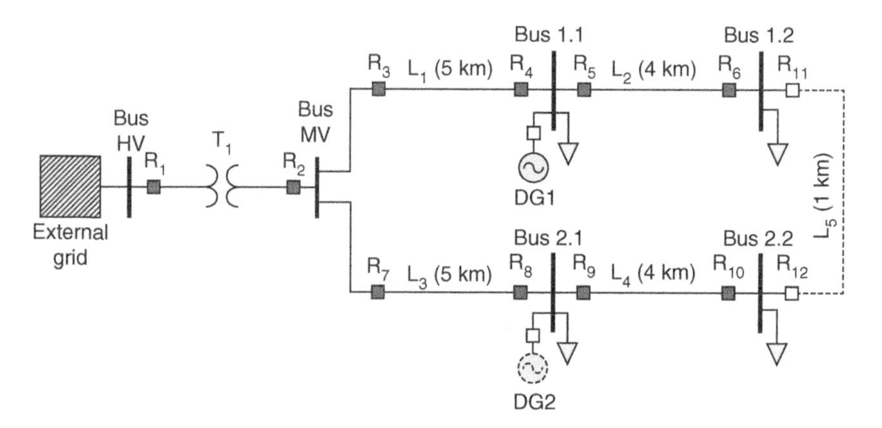

Figure F.5 Single-line diagram of the five-bus test system (config. no. 1).

Table F.17 Short-circuit currents passing through P/B relay pairs for three configurations applied to the five-bus system.

Primary DOCRs				Backup DOCRs			
Relay No.	Fault current (A)			Relay No.	Fault current (A)		
	C1	C2	C3		C1	C2	C3
R_1	11 547	11 547	11 547	—	—	—	—
R_2	4212	9145	9145	R_4	4212	4212	4048
R_2	4212	9145	9145	R_8	—	1866	2212
R_3	8614	10 203	10 404	R_1	1148	1134	1131
R_3	8614	10 203	10 404	R_8	—	1735	2010
R_4	10 642	10 642	11 486	R_6	—	—	1076
R_5	13 630	13 793	13 602	R_3	3470	3692	3417
R_6	—	—	2347	R_{12}	—	—	2347
R_7	12 012	12 012	11 883	R_1	1114	1114	1116
R_7	12 012	12 012	11 883	R_4	3848	3848	3664
R_8	—	2512	3845	R_{10}	—	—	1595
R_9	4213	6469	6379	R_7	4213	4137	4022
R_{10}	—	—	2894	R_{11}	—	—	2894
R_{11}	—	—	3580	R_5	—	—	3580
R_{12}	—	—	2782	R_9	—	—	2782

C1, C2, and C3 stand for configurations 1, 2, and 3, respectively.

- Each generator has 18.6% transient reactance with 0.4 kV, 2.1 MVA, and 0.85 pf ratings, while each unit transformer has 0.4/20 kV and 2.5 MVA ratings with a 6% impedance value.
- For lines L_1 and L_3, the conductor type is ACSR 95 mm^2 with a per-unit impedance of $0.215 + j0.334\ \Omega$/km, while the conductor type of L_2, L_4, L_5 is ACSR 35 mm^2 with a per-unit impedance of $0.576 + j0.397\ \Omega$/km.
- Table F.17 lists the P/B relay pairs and the corresponding short-circuit currents for the three configurations considered in this test system.
- Although the main reference uses the notations *TDS* and I_p, the IEC standard inverse TCCC is used with a range compatible with *TMS*.
- All the relays are assumed to be numerical with continuous *TMS* and *PS* where $TMS \in [0.1, 1.0]$ and $PS \in [1, 5]$ A.
- The CTI is set to 0.3 s, and the lower and upper bounds of T_i are 0.1 and 1.5 s, respectively.
- More information about this test system can be found in Papaspiliotopoulos et al. (2017b).
- The online information about this system is available in: https://al-roomi.org/coordination/5-bus-system.

F.4 Six-Bus Test Systems

F.4.1 System No. 1

- This test system is shown in Figure F.6.
- The modeling of this test system is exactly similar to that given in the second three-bus test system where the rest of the information required to simulate this ORC problem is presented in Tables F.18–F.20.
- This test system is used in different studies. Some of these studies are reported in Birla et al. (2006a,b), Thangaraj et al. (2010), Moirangthem et al. (2013), Singh et al. (2013), Chelliah et al. (2014), Corrêa et al. (2015), and Thakur and Kumar (2016).
- The online information about this system is available in:
https://al-roomi.org/coordination/6-bus-systems/system-i.

F.4.2 System No. 2

- This test system is shown in Figure F.7.
- In this system, the near-end bolted 3ϕ fault is considered.
- The generating units are connected to all busbars except buses 2 and 5, while the loads are not shown in the diagram.
- This system has the same number of DOCRs implemented in the first eight-bus test system.[1]

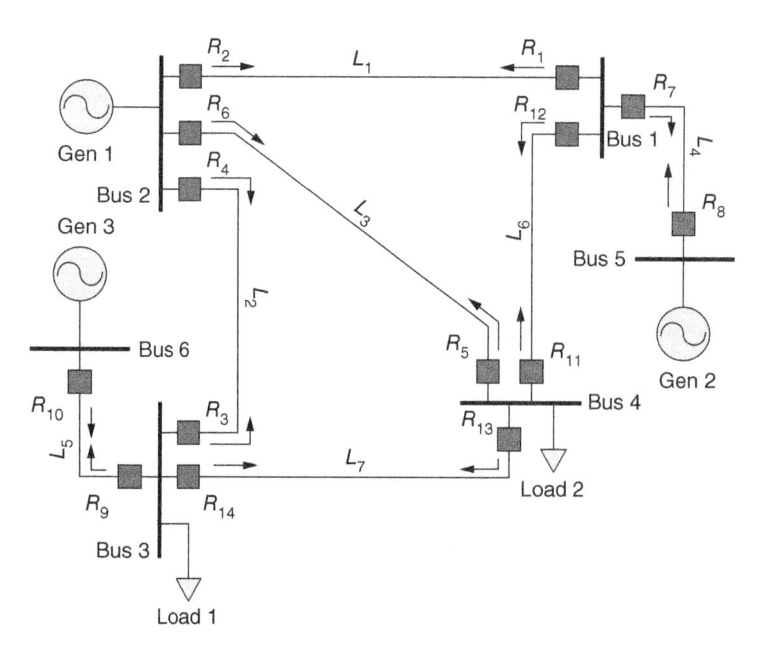

Figure F.6 Single-line diagram of the six-bus test system no. 1.

1 The reason is that the two additional buses (of the first eight-bus test system) are connected to the generating units and then feed the network through two transformers. Thus, there is no need to use any DOCRs on these two source busbars. The generators are supposed to be protected by their own protection systems; including non-directional OCRs.

Table F.18 Values of a_p, b_p, c_p, and d_p of the six-bus system no. 1.

	$T^{\text{near}}_{\text{pr},p}$			$T^{\text{far}}_{\text{pr},p}$	
p	a_p	b_p	q	c_p	d_p
1	2.5311	0.2585	2	5.9495	0.2585
2	2.7376	0.2585	1	5.3752	0.2585
3	2.9723	0.4863	4	6.6641	0.4863
4	4.1477	0.4863	3	4.5897	0.4863
5	1.9545	0.7138	6	6.2345	0.7138
6	2.7678	0.7138	5	4.2573	0.7138
7	3.8423	1.7460	8	6.3694	1.7460
8	5.6180	1.7460	7	4.1783	1.7460
9	4.6538	1.0424	10	3.8700	1.0424
10	3.5261	1.0424	9	5.2696	1.0424
11	2.5840	0.7729	12	6.1144	0.7729
12	3.8006	0.7729	11	3.9005	0.7729
13	2.4143	0.5879	14	2.9011	0.5879
14	5.3541	0.5879	13	4.3350	0.5879

- In Mansour et al. (2007), the relay operating time is modeled by using Sachdev's formula,[2] presented in Sachdev and Fleming (1978), instead of using the IEC standard (for European TCCCs) or the IEEE standard (for North American TCCCs).
- Also, ($TDS^{\text{max}} = 11$) is considered in Mansour et al. (2007) instead of ($TMS^{\text{max}} = 1.1$) as in Vijayakumar and Nema (2008) and Singh et al. (2012a).
- All the *CTR*s, P/B relay pairs, 3ϕ fault at the near-end point of each line, variable bounds of all relay settings, and *CTI* are given in Tables F.21 and F.22.
- The modeling of this test system is exactly similar to that given in the second three-bus test system where the rest of the information required to simulate this ORC problem is presented in Tables F.18–F.20.
- This test system is used in different studies. Some of these studies are reported in Mansour et al. (2007), Vijayakumar and Nema (2008), Singh et al. (2012a), El-Mesallamy et al. (2013), and Ralhan and Ray (2013).
- The online information about this system is available in: https://al-roomi.org/coordination/6-bus-systems/system-ii.

F.4.3 System No. 3

- This test system is shown in Figure F.8.
- In this system, the near-end bolted 3ϕ fault is considered.
- The generating units are connected to buses 1 and 6 through two power transformers, while the loads are not shown in the diagram.

2 Please, refer to Chapter 3.

Table F.19 Values of e_p, f_p, g_p, and h_p of the six-bus system no. 1.

T_{jk}			T_{ik}		
j	e_p	f_p	i	g_p	h_p
1	4.0909	1.7460	8	5.3752	0.2585
1	1.2886	0.7729	11	5.3752	0.2585
1	2.9323	1.7460	8	2.5311	0.2585
2	0.6213	0.4863	3	2.7376	0.2585
2	1.6658	0.4863	3	5.9495	0.2585
3	0.0923	1.0424	10	4.5897	0.4863
3	2.5610	1.0424	10	2.9723	0.4863
3	1.4995	0.5879	13	4.5897	0.4863
4	0.8869	0.2585	1	4.1477	0.4863
4	1.5243	0.2585	1	6.6641	0.4863
5	2.5444	0.7729	12	4.2573	0.7138
5	1.4549	0.7729	12	1.9545	0.7138
5	1.7142	0.5879	14	4.2573	0.7138
6	1.4658	0.4863	3	6.2345	0.7138
6	1.1231	0.2585	3	6.2345	0.7138
7	2.1436	0.7729	11	4.1783	1.7460
7	2.0355	0.2585	2	4.1783	1.7460
7	1.9712	0.7729	11	3.8423	1.7460
7	1.8718	0.2585	2	3.8423	1.7460
9	1.8321	0.5879	13	5.2696	1.0424
9	3.4386	0.4863	4	5.2696	1.0424
9	1.6180	0.5879	13	4.6538	1.0424
9	3.0368	0.4863	4	4.6538	1.0424
11	2.0871	0.5879	14	3.9005	0.7729
11	1.8138	0.7138	6	3.9005	0.7729
11	1.4744	0.5879	14	2.5840	0.7729
11	1.1099	0.7138	6	2.5840	0.7729
12	3.3286	1.7460	8	3.8006	0.7729
12	0.4734	0.2585	2	3.8006	0.7729
12	4.5736	1.7460	8	6.1144	0.7729
12	1.5432	0.2585	2	6.1144	0.7729
13	2.7269	0.7729	12	4.3350	0.5879
13	1.6085	0.7138	6	4.3350	0.5879
13	1.8360	0.7729	12	2.4143	0.5879
14	2.0260	1.0424	10	2.9011	0.5879
14	0.8757	0.4863	4	2.9011	0.5879
14	2.7784	1.0424	10	5.3541	0.5879
14	2.5823	0.4863	4	5.3541	0.5879

Table F.20 Optimization data of the six-bus system no. 1.

TMS^{min}	TMS^{max}	PS^{min} (A)	PS^{max} (A)	PS mode	CTI (s)
0.05	1.1	1.25	1.5	Continuous	0.2

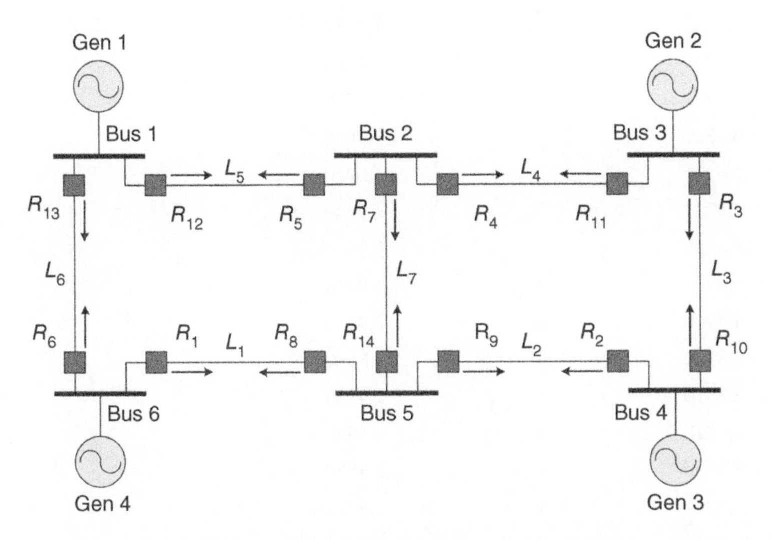

Figure F.7 Single-line diagram of the six-bus test system no. 2.

- For the same reason stated in the second six-bus test system, the number of DOCRs that need to be optimally coordinated for this system is equal to the first eight-bus test system. This is why Razavi et al. (2008) called it "8-bus" instead of "6-bus" test system.
- It can be observed from the given P/B relay pairs that the pair (R_7/R_{13}) is relaxed.
- As per the reference shown in the following text, the *CTR* are not given, but they are supposed to be similar to that used for the first eight-bus test system and the second six-bus test system provided in this appendix.
- Also, the specifications of power elements are different; as can be seen in Table F.23.
- The step-size resolutions of *PS* and *TMS* are 0.25 A and 0.05, respectively.
- The 3ϕ short-circuit currents for all primary and backup relays, variable bounds of all relay settings, and *CTI* are given in Tables F.24 and F.25.
- This test system is taken from Alkaran et al. (2015).
- The online information about this system is available in: https://al-roomi.org/coordination/6-bus-systems/system-iii.

F.4.4 System No. 4

- This test system is shown in Figure F.9.
- This test system consists of 6 buses, 11 lines, 3 loads, and 3 generating sources.
- The network is protected by 22 DOCRs having 64 P/B relay pairs.
- The relays are emulated using the IEC standard inverse TCCC.

Table F.21 3ϕ faults of the six-bus system no. 1.

Primary		Fault current (kA)	Backup R_j	Fault current (kA)
R_i	*CTR*			
R_1	1200 : 5	18.1720	R_{13}	0.6010
R_2	800 : 5	4.8030	R_3	1.3650
R_3	800 : 5	30.5470	R_4	0.5528
R_4	800 : 5	5.1860	R_{12}	3.4220
R_4	800 : 5	5.1860	R_{14}	1.7640
R_5	800 : 5	2.8380	R_{11}	1.0740
R_5	800 : 5	2.8380	R_{14}	1.7640
R_6	1200 : 5	18.3380	R_8	0.7670
R_7	800 : 5	4.4960	R_{11}	1.0740
R_7	800 : 5	4.4960	R_{12}	3.4220
R_8	800 : 5	2.3510	R_2	0.8680
R_8	800 : 5	2.3510	R_7	1.4830
R_9	800 : 5	6.0720	R_1	4.5890
R_9	800 : 5	6.0720	R_7	1.4830
R_{10}	600 : 5	4.0770	R_9	0.6390
R_{11}	800 : 5	30.9390	R_{10}	0.9455
R_{12}	800 : 5	17.7050	R_6	0.8610
R_{13}	1200 : 5	17.8210	R_5	0.9770
R_{14}	800 : 5	5.4570	R_1	4.5890
R_{14}	800 : 5	5.4570	R_2	0.8680

Table F.22 Optimization data of the six-bus system no. 2.

TMS^{min}	TMS^{max}	PS^{min} (A)	PS^{max} (A)	PS mode	*CTI* (s)
0.5	1.1[a)]	0.5	1.5	Continuous	0.2

a) Please, see the preceding in-text point for the upper limit used in Mansour et al. (2007).

- Table F.26 shows the maximum load current (I_L^{max}), minimum fault current (I_f^{min}), and maximum fault current (I_f^{max}) passing through all the relays.
- For I_f^{max}, Figure 2.3e is considered with zero fault impedance, while Figure 2.3b is considered for I_f^{min} with a fault impedance of 0.1 pu (Bedekar and Bhide, 2011a; Mathur et al., 2015).
- Table F.27 lists the *CTR*s of all the relays implemented in the network.

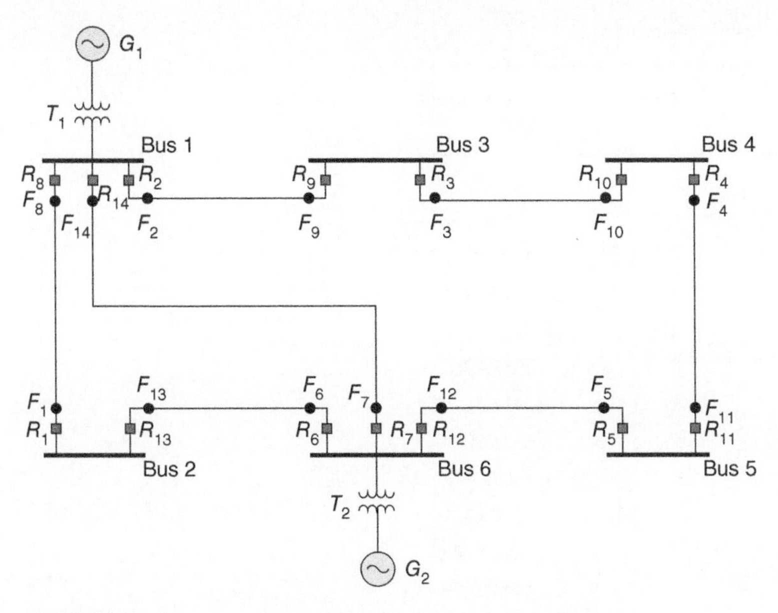

Figure F.8 Single-line diagram of the six-bus test system no. 3.

Table F.23 Network data of the six-bus system no. 3.

Element	***R* (pu)**	***X* (pu)**
Line 1	0.0018	0.0222
Line 2	0.0018	0.0222
Line 3	0.0018	0.02
Line 4	0.0022	0.02
Line 5	0.0022	0.02
Line 6	0.0018	0.02
Line 7	0.0022	0.0222
Generators	0.000 001	0.1
Transformers	0.000 001	0.026 666

- Based on these currents, the *CTR* of the *i*th relay can be calculated as follows (Gers and Holmes, 2004):

$$CTR_i = \max\left[I_{L,i}^{\max}, \frac{I_{f,i}^{\max}}{20}\right] \tag{F.7}$$

- In this test system, electromechanical and static DOCRs are used, and thus the ORC problem is modeled as a MINLP optimization problem where *TMS* is continuous and *PS* is discrete.

Table F.24 3ϕ faults of the six-bus system no. 3.

Primary R_i	Fault current (A)	Backup R_j	Fault current (A)
R_1	2682.4959	R_6	2682.4959
R_2	5362.2983	R_1	804.8782
R_2	5362.2983	R_7	1528.066
R_3	3334.5191	R_2	3334.5191
R_4	2234.3308	R_3	2234.3308
R_5	1352.8751	R_4	1352.8751
R_6	4965.0442	R_5	411.3675
R_6	4965.0442	R_{14}	1522.9084
R_7	4232.634	R_5	407.2472
R_8	4961.7704	R_7	1520.8911
R_8	4961.7704	R_9	410.8226
R_9	1443.6699	R_{10}	1443.6699
R_{10}	2334.6515	R_{11}	2334.6515
R_{11}	3480.7511	R_{12}	3480.7511
R_{12}	5365.0609	R_{13}	805.5618
R_{12}	5365.0609	R_{14}	1529.3638
R_{13}	2490.7454	R_8	2490.7454
R_{14}	4232.7243	R_1	794.092
R_{14}	4232.7243	R_9	407.2292

Table F.25 Optimization data of the six-bus system no. 3.

TMS^{min}	TMS^{max}	PS^{min} (A)	PS^{max} (A)	PS and TMS mode	CTI (s)
0.1	1.1	0.5	2	Discrete	0.4

- The side constraints of these two settings are: $TMS \in [0.1, 1.1]$ and $PS \in [0.5, 2.0] \times$ CT secondary rating of each relay (in steps of 0.25).
- The CTI is set to 0.2 s, and the lower and upper bounds of T_i are 0.1 and 4 s, respectively.
- All DOCRs are emulated using the IEC standard inverse TCCC.
- The optimization algorithm used in the main reference, i.e. Alam et al. (2016), consists of two phases. The first phase uses continuous PS where the lower and upper bounds of the ith relay are defined as follows:

$$PS_i^{\min} = \max\left[0.5, \min\left[1.25\frac{I_{L,i}^{\max}}{CTR_i}, \frac{I_{f,i}^{\min}}{3CTR_i}\right]\right] \tag{F.8}$$

$$PS_i^{\max} = \min\left[2, \frac{2I_{f,i}^{\min}}{3CTR_i}\right] \tag{F.9}$$

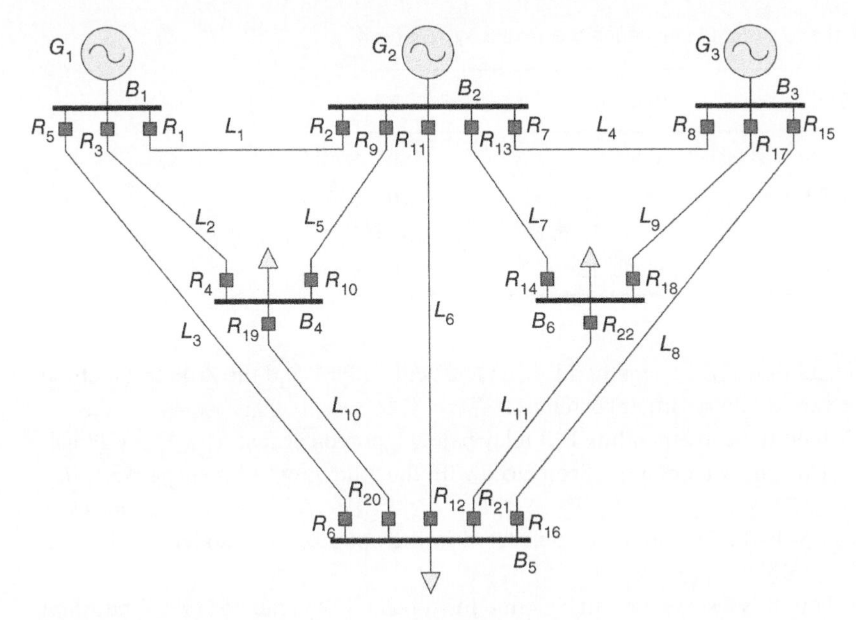

Figure F.9 Single-line diagram of the six-bus test system no. 4.

Table F.26 Maximum load currents, minimum and maximum fault currents of the six-bus system no. 4.

R_i	$I_{L,i}^{\max}$	$I_{f,i}^{\min}$	$I_{f,i}^{\max}$	R_i	$I_{L,i}^{\max}$	$I_{f,i}^{\min}$	$I_{f,i}^{\max}$
1	76	1795	4310	12	—	854	1605
2	—	1857	4461	13	72	1435	2741
3	117	1679	4374	14	—	741	1416
4	—	1305	3398	15	76	650	1061
5	92	1303	2924	16	—	931	1521
6	—	1047	2349	17	178	778	1295
7	23	1306	2415	18	—	1003	1669
8	—	699	1293	19	11	730	1177
9	138	2150	5471	20	—	704	1134
10	—	1343	3417	21	18	846	1416
11	56	1070	2010	22	—	691	1156

and the second phase uses discrete *PS* by applying the following side constraints:

$$PS_i^{\min} = \frac{\left\lfloor 4 \times PS_i^{\text{phase\_I}} \right\rfloor}{4} \tag{F.10}$$

$$PS_i^{\max} = \frac{\left\lceil 4 \times PS_i^{\text{phase\_I}} \right\rceil}{4} \tag{F.11}$$

- The short-circuit currents seen by all the P/B relay pairs are tabulated in Table F.28.

Table F.27 *CTR* of all the relays implemented in the six-bus system no. 4.

Relay no.	*CTR*	Relay no.	*CTR*
9	1600 : 5	6, 11	600 : 5
1, 2, 3	1200 : 5	12, 18	500 : 5
4, 10, 17	1000 : 5	8, 14, 15, 16, 21	400 : 5
5, 7, 13	800 : 5	19, 20, 22	300 : 5

- Newton–Raphson load flow (NRLF) method is used to calculate $I_L^{\max}$ and the Z-matrix method is used to calculate various short-circuit currents.
- If I_f^{bc} satisfies (F.6), then the corresponding P/B relay pair is ignored because $T_j - T_i \geqslant CTI$ will always be satisfied. This phenomenon has been faced with the following P/B relay pairs: R_1/R_4, R_2/R_{10}, R_5/R_2, R_5/R_4, R_7/R_{12}, R_{11}/R_8, R_{11}/R_{14}, R_{13}/R_{12}, R_{16}/R_{22}, R_{19}/R_9, R_{20}/R_{11}, and R_{21}/R_{15}. Thus, only 52 out of 64 P/B relay pairs are considered in the optimization model of this ORC problem.
- More information about this test system can be found in Wood and Wollenberg (1996) and Alam et al. (2016).
- The online information about this system is available in: https://al-roomi.org/coordination/6-bus-systems/system-iv.

F.5 Eight-Bus Test Systems

F.5.1 System No. 1

- This test system is shown in Figure F.10.
- In this system, the near-end bolted 3ϕ fault is considered.
- Bus 4 is connected to an external grid that is modeled by 400 MVA short-circuit capacity.
- Some studies named this system as a six-bus test system because buses 7 and 8 are connected to generating units and do not have any DOCRs involved in this ORC problem.
- The available set of the discretized *PS* is $\{0.5, 0.6, 0.8, 1.0, 1.5, 2, 2.5\}$.
- Although this test system is considered as a *low*-dimensional ORC problem, it has been noticed that getting both feasible and optimal solution at the same time is a very hard task; please refer to our study given in Albasri et al. (2015).
- The rest of the information required to simulate this ORC problem is presented in Tables F.29–F.34.
- This test system is very popular. Some of the studies that implemented it are reported in Braga and Saraiva (1996), Zeineldin et al. (2005,2006), Zeineldin (2008), Noghabi et al. (2009), Amraee (2012), Albasri et al. (2015), and Papaspiliotopoulos et al. (2017a).
- The online information about this system is available in: https://al-roomi.org/coordination/8-bus-systems/system-i.

Table F.28 Short-circuit currents passing through all P/B relay pairs of the six-bus system no. 6.

Faulty Line	Primary relay		Backup relay		Faulty Line	Primary relay		Backup relay	
	R_i	I_i^{pr} (A)	R_j	I_j^{bc} (A)		R_i	I_i^{pr} (A)	R_j	I_j^{bc} (A)
L_1	1	4310	4	184	L_6	12	1605	5	888
	1	4310	6	416		12	1605	15	316
	2	4461	8	1230		12	1605	19	260
	2	4461	10	186		12	1605	22	214
	2	4461	12	354	L_7	13	2741	1	1475
	2	4461	14	914		13	2741	8	308
L_2	3	4374	2	1079		13	2741	10	946
	3	4374	6	752		13	2741	12	125
	4	3398	9	2720		14	1416	17	890
	4	3398	20	679		14	1416	21	567
L_3	5	2924	2	528	L_8	15	1061	7	696
	5	2924	4	286		15	1061	18	589
	6	2349	11	608		16	1521	5	891
	6	2349	15	858		16	1521	11	356
	6	2349	19	304		16	1521	19	318
	6	2349	22	591		16	1521	22	101
L_4	7	2415	1	1417	L_9	17	1295	7	1016
	7	2415	10	898		17	1295	16	750
	7	2415	12	93		18	1669	13	1091
	7	2415	14	402		18	1669	21	582
	8	1293	16	659	L_{10}	19	1177	3	657
	8	1293	18	756		19	1177	9	536
L_5	9	5471	1	2068		20	1134	5	467
	9	5471	8	1001		20	1134	11	219
	9	5471	12	541		20	1134	15	265
	9	5471	14	813		20	1134	22	192
	10	3417	3	2800	L_{11}	21	1416	5	816
	10	3417	20	649		21	1416	11	334
L_6	11	2010	1	965		21	1416	15	135
	11	2010	8	119		21	1416	19	296
	11	2010	10	526		22	1156	13	694
	11	2010	14	44		22	1156	17	478

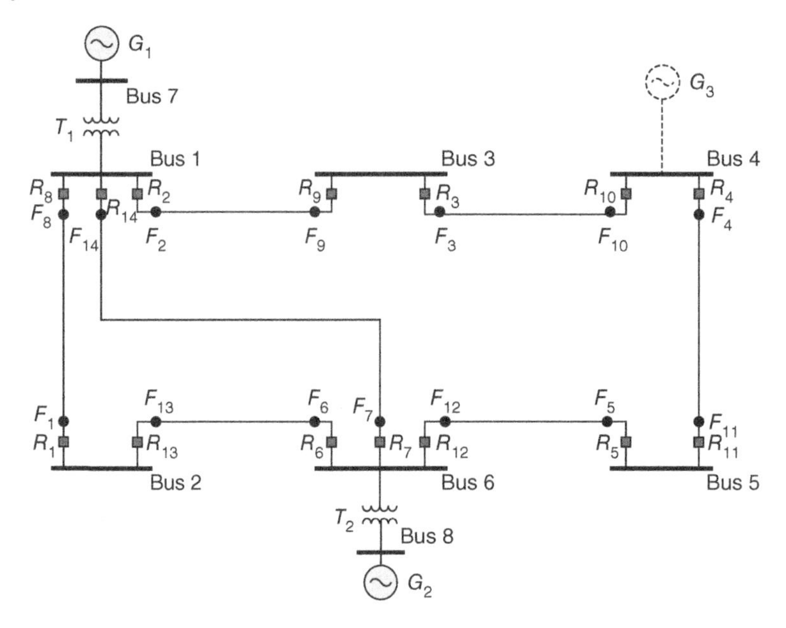

Figure F.10 Single-line diagram of the eight-bus test system no. 1.

Table F.29 Generator data of the eight-bus system no. 1.

Gen.	S_n (MVA)	V_p (kV)	x (%)
G_1	150	10	15
G_2	150	10	15

Table F.30 Transformer data of the eight-bus system no. 1.

Trans.	S_n (MVA)	V_p (kV)	V_s (kV)	x (%)
T_1	150	10	150	4
T_2	150	10	150	4

Table F.31 Line data of the eight-bus system no. 1.

Nodes	R (Ω/km)	X (Ω/km)	Y (S/km)	Length (km)
1–2	0.004	0.05	0.0	100
1–3	0.0057	0.0714	0.0	70
3–4	0.005	0.0563	0.0	80
4–5	0.005	0.045	0.0	100
5–6	0.0045	0.0409	0.0	110
2–6	0.0044	0.05	0.0	90
1–6	0.005	0.05	0.0	100

Table F.32 Load data of the eight-bus system no. 1.

Node	*P* (MW)	*Q* (MVAR)
2	40.0	20.0
3	60.0	40.0
4	70.0	40.0
5	70.0	50.0

Table F.33 3ϕ faults of the eight-bus system no. 1.

Primary		Fault current (A)	Backup	Fault current (A)
R_i	*CTR*		R_j	
R_1	1200 : 5	3232	R_6	3232
R_2	1200 : 5	5924	R_1	996
R_2	1200 : 5	5924	R_7	1890
R_3	800 : 5	3556	R_2	3556
R_4	1200 : 5	3783	R_3	2244
R_5	1200 : 5	2401	R_4	2401
R_6	1200 : 5	6109	R_5	1197
R_6	1200 : 5	6109	R_{14}	1874
R_7	800 : 5	5223	R_5	1197
R_7	800 : 5	5223	R_{13}	987
R_8	1200 : 5	6093	R_7	1890
R_8	1200 : 5	6093	R_9	1165
R_9	800 : 5	2484	R_{10}	2484
R_{10}	1200 : 5	3883	R_{11}	2344
R_{11}	1200 : 5	3707	R_{12}	3707
R_{12}	1200 : 5	5899	R_{13}	987
R_{12}	1200 : 5	5899	R_{14}	1874
R_{13}	1200 : 5	2991	R_8	2991
R_{14}	800 : 5	5199	R_1	996
R_{14}	800 : 5	5199	R_9	1165

Table F.34 Optimization data of the eight-bus system no. 1.

TMS^{min}	TMS^{max}	PS^{min} (*A*)	PS^{max} (*A*)	PS mode	*CTI* (s)
0.1	1.1	0.5	2.5	Discrete	0.3

F.5.2 System No. 2

- The same previous network is implemented in Liu and Yang (2012), but with a different fault analysis.
- The new short-circuit currents for all the P/B relay pairs are tabulated in Table F.35.
- The relays are emulated using the IEEE extremely inverse TCCC.
- Table F.36 lists the side constraints and the new value of *CTI*.
- A step-size resolution of 1 A is used for *CTS*.
- The online information about this system is available in: https://al-roomi.org/coordination/8-bus-systems/system-ii.

Table F.35 3ϕ faults of the eight-bus system no. 2.

Primary relay		Backup relay	
R_i	Fault current (A)	R_j	Fault current (A)
R_1	3230	R_6	3230
R_8	6080	R_9	1160
R_8	6080	R_7	1880
R_2	5910	R_1	993
R_9	2480	R_{10}	2480
R_2	5910	R_7	1880
R_3	3550	R_2	3550
R_{10}	3880	R_{11}	2340
R_6	6100	R_5	1200
R_6	6100	R_{14}	1870
R_{13}	2980	R_8	2980
R_{14}	5190	R_9	1160
R_7	5210	R_5	1200
R_{14}	5190	R_1	993
R_7	5210	R_{13}	985
R_4	3780	R_3	2240
R_{11}	3700	R_{12}	3700
R_5	2400	R_4	2400
R_{12}	5890	R_{13}	985
R_{12}	5890	R_{14}	1870

Table F.36 Optimization data of the eight-bus system no. 2.

TDS^{min}	TDS^{max}	CTS^{min} (A)	CTS^{max} (A)	CTS mode	CTI (s)
0.1	1.1[a)]	10	1000[b)]	Discrete	0.2

a) This range is used in European OCRs with *TMS* and *PS*.
b) The range is so wide compared with that of the preceding test system.

F.5.3 System No. 3

- This test system is reported in Ezzeddine et al. (2011).
- It is similar to the third six-bus test system, but by considering the two generator buses and using a different fault analysis.
- Table F.37 shows the new short-circuit currents used for all the P/B relay pairs.
- The relays are emulated using the IEC standard inverse TCCC.
- Table F.38 lists the side constraints and the new value of *CTI*.
- A step-size resolution of 0.25 A is used for *PS*.
- The online information about this system is available in:
  https://al-roomi.org/coordination/8-bus-systems/system-iii.

Table F.37 3ϕ faults of the eight-bus system no. 3.

Primary relay		Backup relay	
R_i	Fault current (A)	R_j	Fault current (A)
R_1	2666.3	R_6	2666.3
R_2	5374.8	R_1	804.7
R_2	5374.8	R_7	1531.5
R_3	3325.6	R_2	3325.6
R_4	2217.1	R_3	2217.1
R_5	1334.3	R_4	1334.3
R_6	4975	R_5	403.6
R_6	4975	R_{14}	1533
R_7	4247.6	R_5	403.6
R_7	4247.6	R_{13}	805.5
R_8	4973.2	R_7	1531.5
R_8	4973.2	R_9	403.2
R_9	1420.9	R_{10}	1420.9
R_{10}	2313.5	R_{11}	2313.5
R_{11}	3474.3	R_{12}	3474.3
R_{12}	5377	R_{13}	805.5
R_{12}	5377	R_{14}	1533
R_{13}	2475.7	R_8	2475.7
R_{14}	4246.4	R_1	804.7
R_{14}	4246.4	R_9	403.2

Table F.38 Optimization data of the eight-bus system no. 3.

TMS^{min}	TMS^{max}	PS^{min} (A)	PS^{max} (A)	PS mode	*CTI* (s)
0.05	1.1	0.5	2	Discrete	0.3

F.5.4 System No. 4

- This test system is shown in Figure F.11.
- The goal here is to set the phase and ground relays.
- This test system consists of 2 generators, 2 (Y-Y) earthed power transformers, 8 buses, and 7 transmission lines protected by 16 circuit breakers.
- The two generators have the following characteristics: 25 MVA, 12 kV, $X^+ = X^- = 15\%$, and $X^0 = 8\%$.
- The two transformers have the following characteristics: 30 MVA, 69/12 kV, $X = 9\%$, and $X/R = 10$.
- The per-unit-length impedances of all the lines are: $z_1 = 0.19 + j0.461\ \Omega/\text{km}$ and $z_0 = 0.65 + j1.254\ \Omega/\text{km}$.
- The lines have the following lengths: $L_1 = L_4 = L_6 = L_7 = 10$ km, $L_2 = L_5 = 20$ km, and $L_3 = 12$ km.
- In this system the transient configuration is neglected.
- The relays used are the Westinghouse CO9.[3]
- This system is equipped with both phase and ground relays.
- The setting of the phase relays is based on 3ϕ close-in faults, while the setting of the ground relays is based on 1ϕ close-in faults.
- The *TDS* of the DOCR associated with the circuit breakers $\{CB_7, CB_8\}$ are set for phase and ground relays as follows:
  - Phase relays: $TDS = 2.0$
  - Ground relays: $TDS = 1.0$
- The CTI is set to 0.3 s and the minimum and maximum limits of *TDS* are 0.5 and 11, respectively.

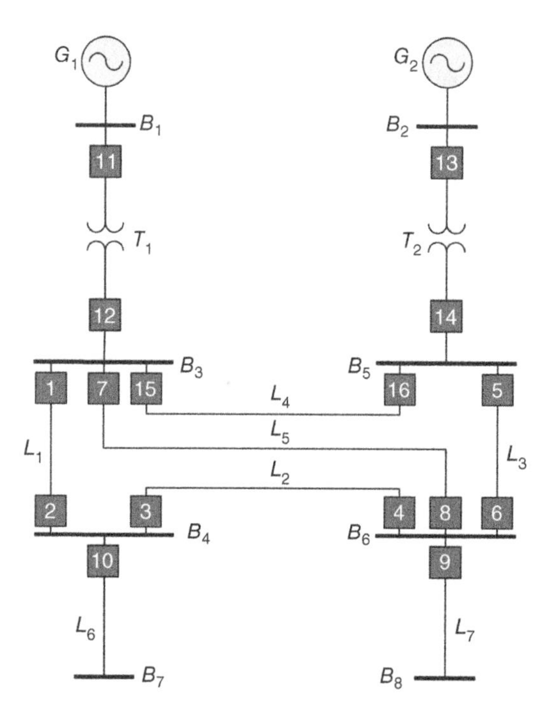

Figure F.11 Single-line diagram of the eight-bus test system no. 4.

3 That is the characteristic curve is modeled based on the CO very inverse time (CO9). Please, refer to Table 3.10.

Table F.39 Basic data of DOCRs used in the eight-bus system no. 4.

CB	*CTR*	Phase relays I_p (A)	Ground relays I_p (A)
CB_1	600 : 5	5.0	1.0
CB_2	400 : 5	5.0	1.0
CB_3	200 : 5	5.0	1.0
CB_4	200 : 5	5.0	1.0
CB_5	600 : 5	5.0	1.0
CB_6	400 : 5	5.0	1.0
CB_7	400 : 5	5.0	1.0
CB_8	400 : 5	5.0	1.0
CB_9	400 : 5	5.0	1.0
CB_{10}	300 : 5	5.0	1.0
CB_{11}	1600 : 5	5.0	1.0
CB_{12}	100 : 5	5.0	1.0
CB_{13}	1600 : 5	5.0	1.0
CB_{14}	100 : 5	5.0	1.0
CB_{15}	400 : 5	5.0	1.0
CB_{16}	400 : 5	5.0	1.0

- In Mansour et al. (2007), the relay operating time is modeled by using Sachdev's formula,[4] given in (3.14), instead of using the IEEE standard formula for CO9 relays.
- Urdaneta et al. (1999) linearized the problem by fixing the pick-up current (I_p). The settings of I_p and *CTR*s are given in Table F.39.
- Also, Urdaneta et al. (1999) analyzed three cases:
  1. DOCR with only time delay units (similar to IBC51 type);
  2. DOCR with time delay and instantaneous units ($T_{\text{ins}} = 0$); and
  3. DOCR with time delay and instantaneous units as the secondary protection, distance relays as the main protection, and **breaker failure relays** (**BFR**) as local backup.
- More information about this test system can be found in Urdaneta et al. (1999) and Mansour et al. (2007).
- The online information about this system is available in: https://al-roomi.org/coordination/8-bus-systems/system-iv.

F.5.5 System No. 5

- This test system is shown in Figure F.12.
- It consists of eight buses, eight lines, one generator, and one transformer.
- The remaining data of the network is given in Tables F.40–F.42.
- The short-circuit currents for all P/B relay pairs are tabulated in Table F.43.
- The first three relays (i.e. $\{R_1, R_2, R_3\}$), do not have backup protection.

4 Please, refer to Chapter 3.

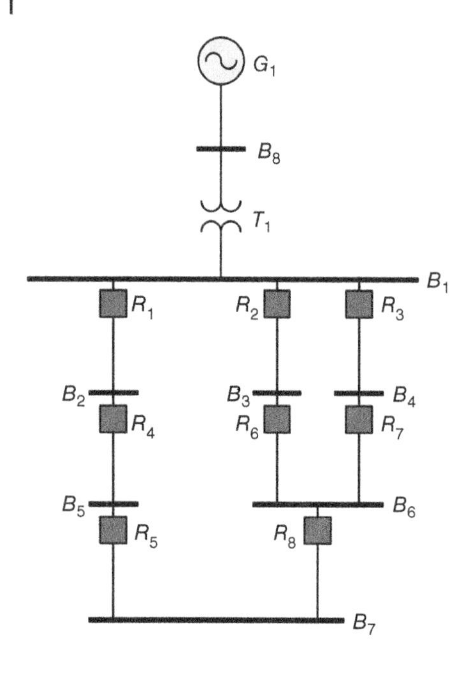

Figure F.12 Single-line diagram of the eight-bus test system no. 5.

- The fault currents seen by these three primary relays are not shown in the main reference. Rather, they are determined through a reverse calculation using the MATLAB file `ORC_8BusSystem5.m` given in the folder `Appendices`.
- The per-unit quantities of R and X are based on 100 MVA and 150 kV, respectively.
- All the lines are protected by normal inverse overcurrent relays.
- The model given in (3.8) is used to emulate these relays.
- Both *TMS* and *PS* are discrete where *TMS* varies from 0 to 1 in steps of 0.05 and *PS* varies from 0.5 to 2 A in steps of 0.25 A.
- The CTI is set to 0.4 s.
- The objective function given in (7.13) is used.

Table F.40 Line data of the eight-bus system no. 5.

Line	R (pu)	X (pu)	V (kV)
1	0.40	0.20	10
2	0.28	0.19	10
3	0.24	0.13	10
4	0.38	0.19	10
5	0.40	0.23	10
6	0.30	0.17	10
7	0.26	0.15	10
8	0.50	0.22	10

Table F.41 Generator data of the eight-bus system no. 5.

Generator	*R* (pu)	*X* (pu)	V (kV)
1	0.10	0.30	10

Table F.42 Transformer data of the eight-bus system no. 5.

Transformer	*R* (pu)	*X* (pu)
1	0.01	0.30

Table F.43 3ϕ faults of the eight-bus system no. 5.

Primary R_i	Fault current (kA)	Backup R_j	Fault current (A)
R_1	940.5	—	—
R_2	938.5	—	—
R_3	898.1	—	—
R_4	524.2	R_1	524.2
R_5	339.9	R_4	339.9
R_6	507.7	R_2	507.7
R_7	567.0	R_3	567.0
R_8	608.3	R_6	277.7
R_8	608.3	R_7	330.8

- The coefficients $\{\lambda_1, \lambda_2, \mu_2\}$ of (7.13) are defined for four cases as follows:
  - **Case 1**: $\lambda_1 = 1$, $\lambda_2 = 2$, and $\mu_2 = 100$
  - **Case 2**: $\lambda_1 = 20$, $\lambda_2 = 1$, and $\mu_2 = 10$
  - **Case 3**: $\lambda_1 = 1$, $\lambda_2 = 2$, and $\mu_2 = 0$
  - **Case 4**: $\lambda_1 = 1$, $\lambda_2 = 2$, and $\mu_2 = 100$
- More information about this test system can be found in Razavi et al. (2008) and Asadi et al. (2008).
- The online information about this system is available in: https://al-roomi.org/coordination/8-bus-systems/system-v.

F.6 Nine-Bus Test System

- This test system is shown in Figure F.13.
- It is a nine-bus interconnected distribution system with one single-end fed.
- All the lines have the same impedance of $(0 + j0.2)$ pu.

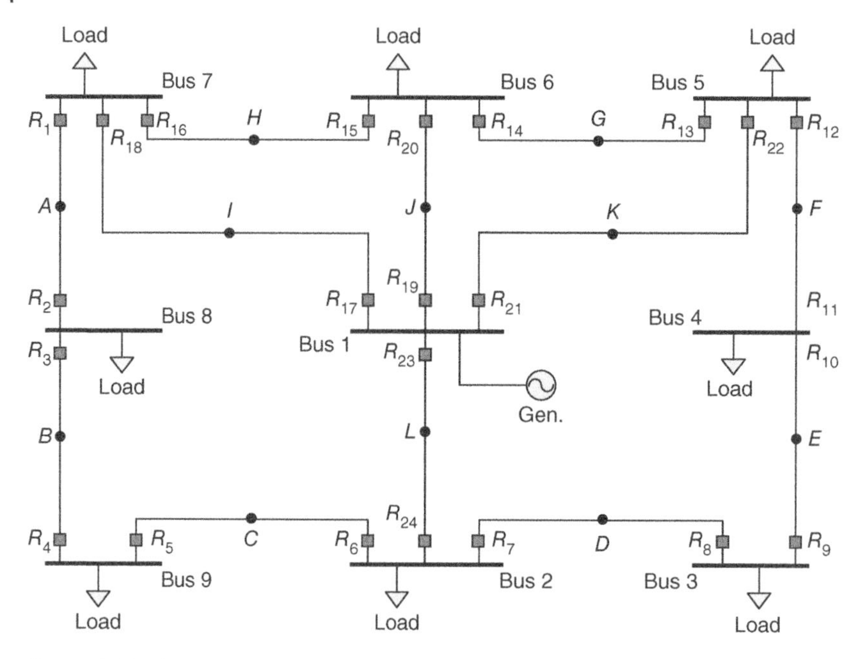

Figure F.13 Single-line diagram of the nine-bus test system.

- In this system, a mid-point bolted 3ϕ fault is taken for each line, as shown in Figure F.13, with no backup protection for relays $\{R_{17}, R_{19}, R_{21}, R_{23}\}$.
- Bus 1 is supplied by a power source of 100 MVA and 33 kV with a source impedance of $(0 + j0.1)$ pu.
- All DOCRs have the same *CTR* of 500 : 1.
- All these relays are considered to be numerical, in which both *PS* and *TMS* are continuous.
- In this ORC problem, the lowest acceptable operating time ($T^{\min}$) of relays should not be less than 0.2 s.
- The lower and upper limits (or side constraints) of *PS* of each DOCR are calculated based on the practical equations described by (7.22) and (7.23); where $OLF = 1.25$.
- The rest of the information required to simulate this ORC problem is presented in Tables F.44–F.46.
- Recently, this test system attracts the attention of some researchers. The following cited studies are just an example: Bedekar and Bhide (2011a), Adelnia et al. (2015), Alam et al. (2015), and Albasri et al. (2015).
- The online information about this system is available in: https://al-roomi.org/coordination/9-bus-system.

Table F.44 Maximum load and 3ϕ short-circuit currents of the nine-bus system.

Fault Point	Primary "R_i"	Max. load "$I_{L_{max}}$" (A)	Relay action mode		Backup(s) "R_j"
			$I_{pr,i}$ (A)	$I_{bc,j}$ (A)	
A	R_1	121.74	4863.6	1361.6	R_{15}, R_{17}
	R_2	212.74	1634.4	653.6	R_4
B	R_3	21.74	2811.4	1124.4	R_1
	R_4	21.74	2610.5	1044.2	R_6
C	R_5	78.26	1778.0	711.2	R_3
	R_6	78.26	4378.5	1226.0	R_8, R_{23}
D	R_7	78.26	4378.5	1226.0	R_5, R_{23}
	R_8	78.26	1778.0	711.2	R_{10}
E	R_9	21.74	2610.5	1044.2	R_7
	R_{10}	21.74	2811.4	1124.4	R_{12}
F	R_{11}	121.74	1634.4	653.6	R_9
	R_{12}	121.74	2811.4	787.2	R_{14}, R_{21}
G	R_{13}	30.44	3684.5	1031.7	R_{11}, R_{21}
	R_{14}	30.44	4172.5	1168.3	R_{16}, R_{19}
H	R_{15}	30.44	4172.5	1168.3	R_{13}, R_{19}
	R_{16}	30.44	3684.5	1031.7	R_2, R_{17}
I	R_{17}	441.3	7611.2	1293.9	—
	R_{18}	441.3	2271.7	1953.7	R_2, R_{15}
J	R_{19}	410.87	7435.8	1264.1	—
	R_{20}	410.87	2624.2	2256.8	R_{13}, R_{16}
K	R_{21}	441.3	7611.2	1293.9	—
	R_{22}	441.3	2271.7	1953.7	R_{11}, R_{14}
L	R_{23}	506.52	7914.7	1345.5	—
	R_{24}	506.52	1665.5	1432.3	R_5, R_8

Table F.45 Load currents of the nine-bus system.

Bus. no.	Load current (A)
Bus 1	—
Bus 2	350
Bus 3	100
Bus 4	100
Bus 5	350
Bus 6	350
Bus 7	350
Bus 8	100
Bus 9	100

Table F.46 Optimization data of the nine-bus system.

TMS^{min}	TMS^{max}	PS^{min} (A)	PS^{max} (A)	PS mode	CTI (s)
0.025	1.2	Equation (7.22)	Equation (7.23)	Continuous	0.2

F.7 14-Bus Test Systems

- This test system represents a portion of the American Electric Power System (in the Midwestern US) as of February, 1962.
- A hardcopy data was provided by Iraj Dabbagchi of AEP and entered in IEEE Common Data Format by Rich Christie at the University of Washington in August 1993.
- This test system consists of 14 buses, 2 generators, 3 synchronous condensers, 2 two-winding transformers, 1 three-winding transformer, 17 lines, and 11 loads.
- This test system does not have line limits.
- Compared to 1990s power systems, it has low base voltages and an overabundance of voltage control capability.
- The online information about this system is available in: https://al-roomi.org/power-flow/14-bus-system.

There were many attempts to use this test system as a real ORC problem. Most of these attempts show uncompleted data, such as the ones reported in Noghabi et al. (2010), Ezzeddine et al. (2011), Saleh et al. (2015), Huchel and Zeineldin (2016), Thakur and Kumar (2016), Shabani and Karimi (2018), Wadood et al. (2019), and Alam (2019). However, a few attempts are available with the necessary data to carry out optimal coordination experiments on this test system. Three of these attempts are covered in the following text:

The study reported in Liu and Yang (2012) covers all the busbars of the IEEE 14-bus test system during solving its ORC problem. On the opposite side, the study reported in Adelnia et al. (2015) covers just the 33 kV distribution section, which in turn makes the problem easier to solve.

F.7.1 System No. 1

- The single-line diagram of this test system is shown in Figure F.14.
- The power base and the voltage level of this system are 100 MVA and 138 kV, respectively.
- Because the test system consists of 20 branches (i.e. lines and transformers), so 40 DOCRs are used to protect this network.
- It is assumed that all the relays are North American, and thus the notations *TDS* and I_p have to be used instead of *TMS* and *PS*.
- The *TDS* values can range[5] continuously from 0.1 to 1.1, and the I_p values can range discretely between 10 and 1000 A with a step size of 1 A.
- The model has 92 inequality constraints for satisfying the selectivity criteria among P/B relay pairs.
- Table F.47 shows the P/B relay pairs and the corresponding close-in 3ϕ short-circuit currents passing through them.
- The objective function[6] used for optimizing the relay settings is the one given in (7.7).
- The CTI is set to 0.2 s.

Figure F.14 Single-line diagram of the 14-bus system no. 1.

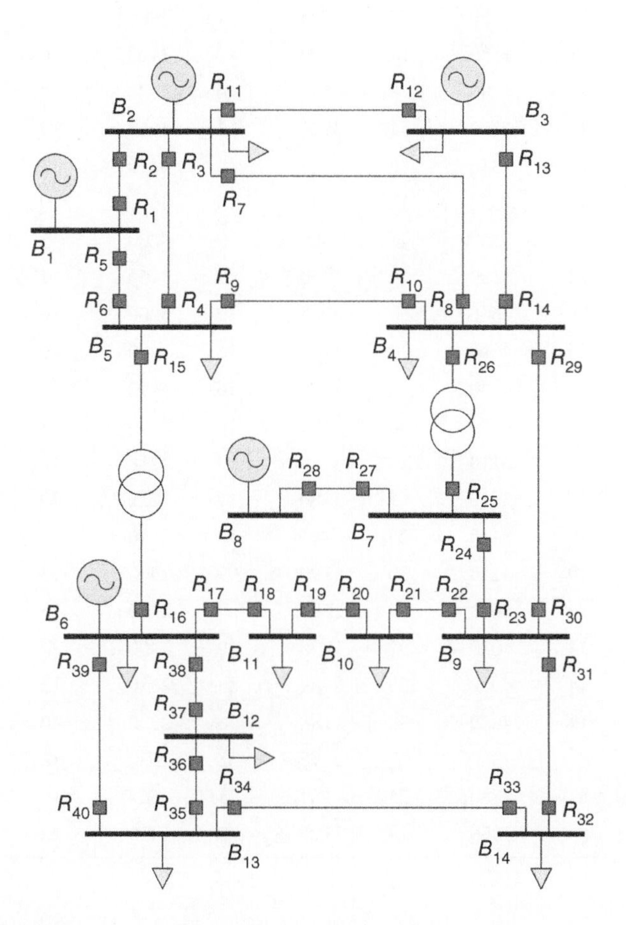

5 The range given here complies with *TMS* and not *TDS*.
6 Although the main reference assumes that all the relays have the same IEEE standard inverse TCCC, the optimization model uses the coefficients of the IEEE extremely inverse TCCC.

Table F.47 P/B relay pairs and the corresponding close-in 3ϕ short-circuit currents of the 14-bus system no. 1.

Primary relay		Backup relay		Primary relay		Backup relay		Primary relay		Backup relay	
R_i	$I_{f,i}$ (A)	R_j	$I_{f,j}$ (A)	R_i	$I_{f,i}$ (A)	R_j	$I_{f,j}$ (A)	R_i	$I_{f,i}$ (A)	R_j	$I_{f,j}$ (A)
1	11 650	6	654	8	3880	30	188	19	955	17	955
5	12 400	2	1980	29	4720	7	1220	18	725	20	725
2	4260	4	750	29	4720	9	1990	22	1930	29	499
2	4260	12	875	29	4720	13	1070	22	1930	24	1160
2	4260	8	723	29	4720	25	449	22	1930	32	280
3	7310	1	3920	6	3830	3	1280	23	1200	21	434
3	7310	12	848	6	3830	10	1990	23	1200	29	499
3	7310	8	689	6	3830	16	560	23	1200	32	281
11	7180	4	725	4	3920	5	1370	30	1810	21	424
11	7180	1	3920	4	3920	16	562	30	1810	24	1130
11	7180	8	695	4	3920	10	1990	30	1810	32	275
7	7330	1	3920	15	4610	5	1360	31	2060	21	428
7	7330	4	716	15	4610	3	1280	31	2060	24	1150
7	7330	12	845	15	4610	10	1970	31	2060	29	494
13	3280	11	1380	9	3260	5	1390	27	2030	26	1230
12	3130	14	1250	9	3260	3	1310	27	2030	23	808
26	4640	9	2080	9	3260	16	569	25	1430	28	633
26	4640	13	1120	16	1490	40	201	25	1430	23	806
26	4640	7	1270	16	1490	18	388	24	1870	26	1230
26	4640	30	179	16	1490	37	51	24	1870	28	634
10	3110	13	1140	17	2210	15	1110	37	572	35	572
10	3110	7	1290	17	2210	37	51	36	781	38	781
10	3110	25	495	17	2210	40	199	35	1480	33	368
10	3110	30	190	39	2400	15	1120	35	1480	39	1110
14	4030	9	2090	39	2400	18	389	34	1390	36	284
14	4030	7	1270	39	2400	37	47	34	1390	39	1110
14	4030	25	489	38	2530	15	1110	40	654	36	285
14	4030	30	188	38	2530	40	191	40	654	33	370
8	3880	9	2090	38	2530	18	386	32	547	34	547
8	3880	25	489	21	564	19	564	33	783	31	783
8	3880	13	1120	20	1310	22	1310	—	—	—	—

- More information about this ORC problem can be found in Liu and Yang (2012).
- The online information about this ORC problem is available in: https://al-roomi.org/coordination/14-bus-systems/system-i.

F.7.2 System No. 2

- The single-line diagram of the modified 33 kV distribution section of the IEEE 14-bus test system is shown in Figure F.15.
- This test system is implemented for an adaptive optimizer where the optimal relay settings are calculated off-line for each single contingency[7] and then applied according to the existing mode of the network. All the single contingencies are listed and the corresponding network operations[8] are identified. An event or lookup table is used for this set of profound network topologies.
- As can be clearly seen from the diagram, the 33 kV section is connected with a high voltage section through transformers via buses 6 and 9. The distribution section acts as an islanded micro-grid when these links are disconnected.
- There is only one synchronous generator connected to bus 6 in the original 33 kV section where the generator parameters are listed in Table F.48.
- Two similar generating units are added to buses 9 and 13 to increase the DG penetration of the system.
- Network line and load parameters are tabulated in Tables F.49 and F.50, respectively.
- The per-unit quantities are calculated based on a base load of 100 MVA.
- Table F.51 lists the single element contingencies coved in the main reference.
- The system is simulated using DigSILENT Powerfactory software.

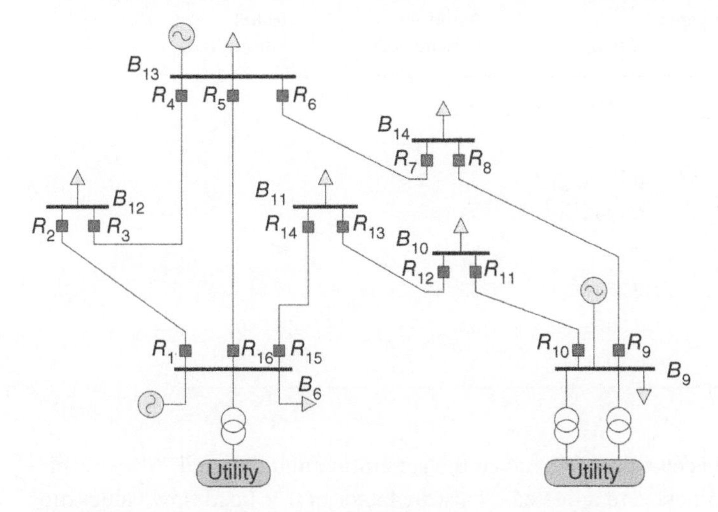

Figure F.15 Single-line diagram of the 14-bus system no. 2.

7 That is each line, transformer, generator, and utility can be disconnected one at a time.
8 That is the data of load flow and short-circuit current for each topology.

Table F.48 System generator parameters of the 14-bus system no. 2.

Parameter	Value
Connected bus type	PV
Active power	10 MW
Nominal voltage	33 kV
Power factor	0.90
Synchronous reactance (x_d)	2 p.u.
Synchronous reactance (x_q)	2 p.u.
Transient reactance (x_d')	0.3 p.u.
Sub-transient reactance (x_d'')	0.2 p.u.
Stator resistance (r_{str})	0.0 p.u.
Zero sequence reactance (x_0)	0.1 p.u.
Zero sequence resistance (r_0)	0.0 p.u.
Negative sequence reactance (x_2)	0.2 p.u.
Negative sequence resistance (r_2)	0.0 p.u.
Rotor type	Round rotor

Table F.49 System line parameters of the 14-bus system no. 2.

From bus	To bus	Line impedance *R* (p.u.)	*X* (p.u.)	Rated voltage (kV)	Rated current (kA)
6	12	0.1229	0.2558	33	1
6	13	0.0662	0.1303	33	1
6	11	0.095	0.1989	33	1
12	13	0.2209	0.1999	33	1
13	14	0.1709	0.348	33	1
9	14	0.1271	0.2704	33	1
9	10	0.0318	0.0845	33	1
10	11	0.0821	0.1921	33	1

- For the 13th topology, the load is curtailed to balance the generation and demand. Also, in this special topology, all the load values were adjusted by a scale factor of 0.3. Load flow values are used as lower limits when calculating the pickup settings of the relays.
- The load flow currents of all the preceding fourteen topologies and the *CTR*s of the relays are tabulated in Table F.52.

Table F.50 System load parameters of the 14-bus system no. 2.

Bus no.	Active power (MW)	Reactive power (MVAR)	Configuration[a)]
12	6.1	1.6	3ϕ-Δ R/X_L
13	13.5	5.8	3ϕ-Δ R/X_L
14	14.9	5	3ϕ-Δ R/X_L
9	29.5	16.6	3ϕ-Δ R/X_L
10	9	5.8	3ϕ-Δ R/X_L
11	3.5	1.8	3ϕ-Δ R/X_L
6	11.2	7.5	3ϕ-Δ R/X_L

a) 3ϕ-Δ R/X_L: three-phase, resistive-inductive load in a delta connection.

Table F.51 Formation of system topologies of the 14-bus system no. 2.

Topology no.	Disconnected element	Topology no.	Disconnected element
1	Line 6–12	8	Line 6–11
2	Line 12–13	9	None
3	Line 6–13	10	Gen – bus 13
4	Line 13–14	11	Gen – bus 9
5	Line 9–14	12	Gen – bus 6
6	Line 9–10	13	Utility
7	Line 10–11	14	All 3 DGs

- The fault analysis is done according to IEC-60909 where Table F.53 lists the 3ϕ short-circuit currents of all the P/B relay pairs for the first topology.[9]
- The near-end faults are taken as I_f^{max}, while the far-end faults are taken as I_f^{min}.
- It is assumed that all the relays are North American, and thus the notations *TDS* and I_p have to be used instead of *TMS* and *PS*.
- Both *TDS* and I_p are continuous variables.
- The side constraint of *TDS* is [0.05, 1.1], while I_p is bounded by (7.22) and (7.23) where the overload factor (*OLF*) is set to 1.5.
- Although the notations used are *TDS* and I_p, the study uses the IEC standard inverse TCCC and the range of *TDS* complies with *TMS*.
- The CTI is set to 0.2 s and the minimum and maximum relay operating times are 0.05 and 1 s, respectively.

9 The main reference does not show the maximum load and short-circuit currents calculated for the other topologies. These values can be replicated by using the provided network data. At present, Table F.53 can be used for non-adaptive protection scheme.

Table F.52 Load flow analysis for normal operation of the 14-bus system no. 2.

Relay no.	Load flow current (A)[a]														*CTR*
	T01	T02	T03	T04	T05	T06	T07	T08	T09	T10	T11	T12	T13	T14	
1	—	103	226	75	132	94	108	112	102	132	106	98	11	132	1500 : 5
2	—	−103	−226	−75	−132	−94	−108	−112	−102	−132	−106	−98	−11	−132	500 : 5
3	−105	—	122	−29	27	−11	6	11	−2	27	3	−7	−20	26	500 : 5
4	105	—	−122	29	−27	11	−6	−11	2	−27	−3	7	20	−26	1500 : 5
5	−285	−194	—	−87	−306	−166	−220	−240	−195	−306	−212	−177	21	−306	1600 : 5
6	120	134	63	—	273	97	166	191	134	91	155	110	70	88	1250 : 5
7	−120	−134	−63	—	−273	−97	−166	−191	−134	−91	−155	−110	−70	−88	600 : 5
8	−145	−130	−208	−266	—	−173	−106	−85	−130	−179	−114	−157	−13	−190	750 : 5
9	145	130	208	266	—	173	106	85	130	179	114	157	13	190	1500 : 5
10	106	110	89	−92	168	—	177	243	110	138	94	145	9	155	2000 : 5
11	−106	−110	−89	92	−168	—	−177	−243	−110	−138	−94	−145	−9	−155	1000 : 5
12	−94	−87	−124	−155	−26	−183	—	66	−87	−61	−118	−52	−44	−59	1500 : 5
13	94	87	124	155	26	183	—	−66	87	61	118	52	44	59	1000 : 5
14	−155	−148	−185	−215	−80	−249	−64	—	−149	−120	−180	−111	−63	−114	800 : 5
15	155	148	185	215	80	249	64	—	149	120	180	111	63	114	2000 : 5
16	285	194	—	87	306	166	220	240	195	306	212	177	−21	306	2000 : 5

a) T0*x* stands for the *x*th topology.

Table F.53 3ϕ short-circuit currents passing through all the P/B relay pairs for topology 1 of the 14-bus system no. 2.

Primary Relay no.	Backup Relay no.	3-Phase near-end fault		3-Phase far-end fault	
		Current seen by Primary relay (kA)	Current seen by Backup relay (kA)	Current seen by Primary relay (kA)	Current seen by Backup relay (kA)
1	5	—	—	—	—
1	14	—	—	—	—
2	4	—	—	—	—
3	1	0.000	0.000	0.000	0.000
4	16	17.707	7.116	4.999	2.009
4	7	17.707	2.200	4.999	0.621
5	3	10.811	0.000	5.864	0.000
5	7	10.811	2.228	5.864	0.873
6	16	15.415	7.051	3.752	1.479
6	3	15.415	0.000	3.752	0.000
7	9	4.805	4.805	2.268	2.268
8	6	3.699	3.699	2.123	2.123
9	11	19.591	2.655	4.891	0.411
10	8	19.378	2.085	10.219	0.983
11	13	3.442	3.442	2.749	2.749
12	10	10.057	10.057	4.753	4.753
13	15	5.923	5.923	3.466	3.466
14	12	4.699	4.699	2.764	2.764
15	5	19.177	5.706	6.006	1.660
15	2	19.177	0.000	6.006	0.000
16	14	16.311	2.710	7.319	0.965
16	2	16.311	0.000	7.319	0.000

- More information about this ORC problem can be found in Adelnia et al. (2015).
- The online information about this ORC problem is available in: https://al-roomi.org/coordination/14-bus-systems/system-ii.

F.8 15-Bus Test System

- This test system is shown in Figure F.16.
- This system consists of 15 buses and 21 branches, and hence it has 42 DOCRs and 82 P/B relay pairs with 84 variables.[10]

10 If only one TCCC is used for all DOCRs.

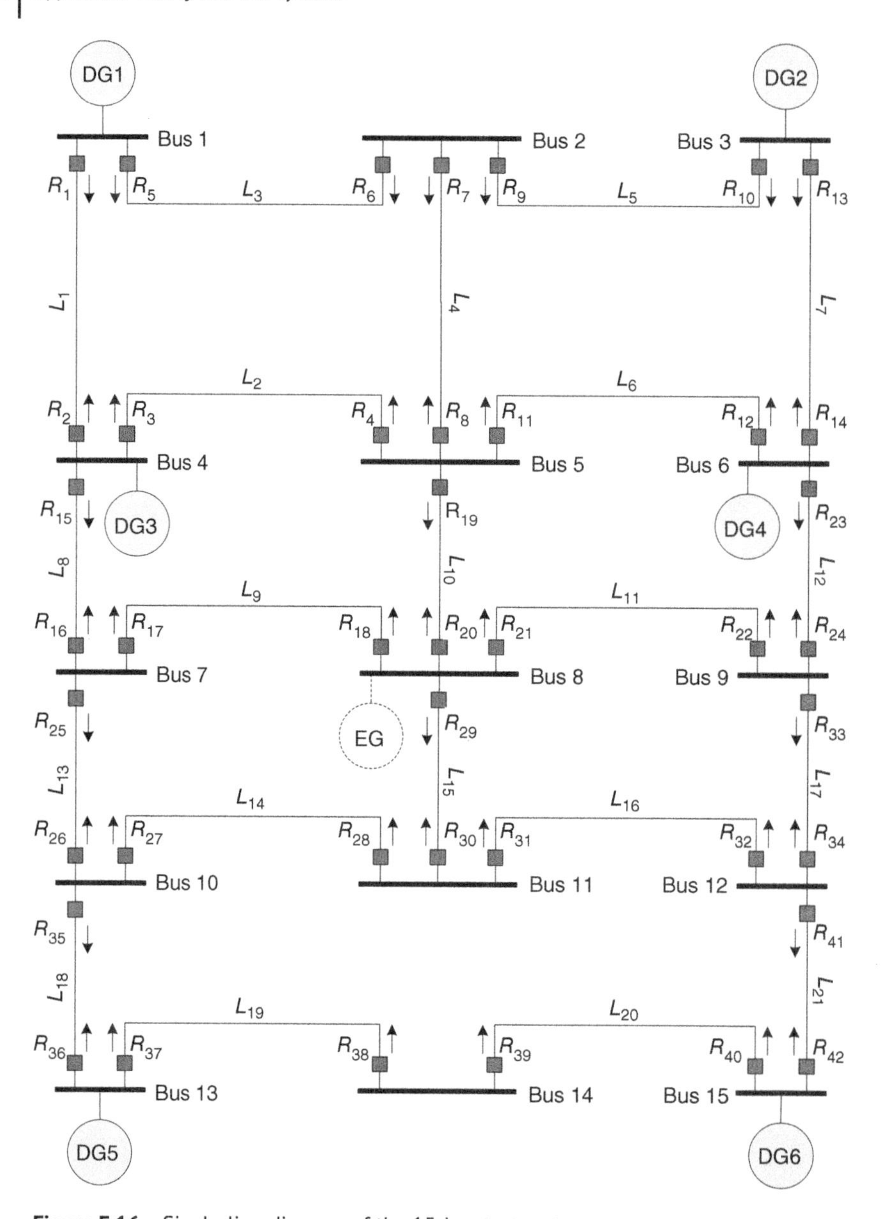

Figure F.16 Single-line diagram of the 15-bus test system.

- The total constraints are 250, and they are addressed as follows:
  - 82 inequality constraints for the P/B selectivity criteria
  - 42 inequality constraints for $T^{\min}$
  - 42 inequality constraints for $T^{\max}$

Table F.54 *CTR*s of the 15-bus system.

Relay no.	*CTR*
18, 20, 21, 29	1600 : 5
2, 4, 8, 11, 12, 14, 15, 23	1200 : 5
1, 3, 5, 10, 13, 19, 36, 37, 40, 42	800 : 5
6, 7, 9, 16, 24, 25, 26, 27, 28, 31, 32, 33, 35	600 : 5
17, 22, 30, 34, 38, 39, 41	400 : 5

  - 42 side constraints for *TMS*
  - 42 side constraints for *PS*

  where *TMS* is greater than 0.1 and *PS* is considered discrete in uniform steps of 0.5 A.
- All the generators have the same ratings of 15 MVA, 20 kV, and a synchronous reactance of $x = 15\%$.
- Also, all the lines have the same impedance of $Z = 0.19 + j0.46\ \Omega/\text{km}$.
- Bus 8 is connected to an external grid modeled by 200 MVA short-circuit capacity.
- The fault analysis is conducted based on the IEC standard.
- This network is presented as an example of highly penetrated/distributed generation (DG) distribution networks.
- In this test system, close-in bolted 3ϕ faults on all the lines are considered.
- The main article for this test system is Amraee (2012), but there are some typo errors on the short-circuit currents that have to be carefully addressed and corrected. They are highlighted in the following lines:
  - Based on the data given in Amraee (2012), the relay R_{13} sees 1503 A when it acts as a backup for R_{12}, and it sees 1053 A when it acts as a backup for R_{23}. Because all the near-end/close-in faults are simulated by a bolted 3ϕ short-circuit, so the backup relay should see the same I_f seen by the corresponding primary relays. Therefore, the question is: *which one should I select; 1503 A or 1053 A?* By testing the results reported in Amraee (2012), using the algorithms given in Chapter 7, the value 1053 A gives some violations when it is used. Based on that, 1503 A is selected instead.
  - A similar thing happens with R_{21} when it acts as a backup relay for R_{24} and R_{33}. With similar impedances (for all the branches), it is hard to select 175 A as the correct answer, especially if we see that all the other short-circuit currents are with high I_f values. Thus, 1326 A is selected as the correct value for the relay R_{21} when it acts as a backup relay.
- The rest of the information required to simulate this ORC problem is presented in Tables F.54–F.56.
- Recently, this test system attracts the attention of many researchers. The studies reported in Amraee (2012), Alipour et al. (2015), Alam et al. (2015), and Shah et al. (2016) are just an example.
- The online information about this system is available in:
  https://al-roomi.org/coordination/15-bus-system.

Table F.55 3ϕ short-circuit currents of the 15-bus system.

Primary	I (A)	Backup	I (A)	Primary	I (A)	Backup	I (A)
R_1	3621	R_6	1233	R_{20}	7662	R_{30}	681
R_2	4597	R_4	1477	R_{21}	8384	R_{17}	599
R_2	4597	R_{16}	743	R_{21}	8384	R_{19}	1372
R_3	3984	R_1	853	R_{21}	8384	R_{30}	681
R_3	3984	R_{16}	743	R_{22}	1950	R_{23}	979
R_4	4382	R_7	1111	R_{22}	1950	R_{34}	970
R_4	4382	R_{12}	1463	R_{23}	4910	R_{11}	1475
R_4	4382	R_{20}	1808	R_{23}	4910	R_{13}	1503
R_5	3319	R_2	922	R_{24}	2296	R_{21}	1326
R_6	2647	R_8	1548	R_{24}	2296	R_{34}	970
R_6	2647	R_{10}	1100	R_{25}	2289	R_{15}	969
R_7	2497	R_5	1397	R_{25}	2289	R_{18}	1320
R_7	2497	R_{10}	1100	R_{26}	2300	R_{28}	1192
R_8	4695	R_3	1424	R_{26}	2300	R_{36}	1109
R_8	4695	R_{12}	1463	R_{27}	2011	R_{25}	903
R_8	4695	R_{20}	1808	R_{27}	2011	R_{36}	1109
R_9	2943	R_5	1397	R_{28}	2525	R_{29}	1828
R_9	2943	R_8	1548	R_{28}	2525	R_{32}	697
R_{10}	3568	R_{14}	1175	R_{29}	8346	R_{17}	599
R_{11}	4342	R_3	1424	R_{29}	8346	R_{19}	1372
R_{11}	4342	R_7	1111	R_{29}	8346	R_{22}	642
R_{11}	4342	R_{20}	1808	R_{30}	1736	R_{27}	1039
R_{12}	4195	R_{13}	1503	R_{30}	1736	R_{32}	697
R_{12}	4195	R_{24}	753	R_{31}	2867	R_{27}	1039
R_{13}	3402	R_9	1009	R_{31}	2867	R_{29}	1828
R_{14}	4606	R_{11}	1475	R_{32}	2069	R_{33}	1162
R_{14}	4606	R_{24}	753	R_{32}	2069	R_{42}	907
R_{15}	4712	R_1	853	R_{33}	2305	R_{21}	1326
R_{15}	4712	R_4	1477	R_{33}	2305	R_{23}	979
R_{16}	2225	R_{18}	1320	R_{34}	1715	R_{31}	809
R_{16}	2225	R_{26}	905	R_{34}	1715	R_{42}	907
R_{17}	1875	R_{15}	969	R_{35}	2095	R_{25}	903
R_{17}	1875	R_{26}	905	R_{35}	2095	R_{28}	1192
R_{18}	8426	R_{19}	1372	R_{36}	3283	R_{38}	882
R_{18}	8426	R_{22}	642	R_{37}	3301	R_{35}	910
R_{18}	8426	R_{30}	681	R_{38}	1403	R_{40}	1403
R_{19}	3998	R_3	1424	R_{39}	1434	R_{37}	1434
R_{19}	3998	R_7	1111	R_{40}	3140	R_{41}	745
R_{19}	3998	R_{12}	1463	R_{41}	1971	R_{31}	809
R_{20}	7662	R_{17}	599	R_{41}	1971	R_{33}	1162
R_{20}	7662	R_{22}	642	R_{42}	3295	R_{39}	896

Table F.56 Optimization data of the 15-bus system.

TMS^{min}	TMS^{max}	PS^{min} (A)	PS^{max} (A)	PS mode	CTI (s)
> 0.1	1.1	0.5	2.5	Discrete	0.2

F.9 30-Bus Test Systems

- This test system represents a portion of the American Electric Power System (in the Midwestern US) as of December 1961.
- The base voltages of this test system are 11and 1.0 kV. These two quantities are guessed, which may not be the actual data.
- The system consists of 30 buses (132- and 33-kV buses), 37 lines, 4 transformers, and 6 generators.
- The network can be considered as a meshed sub-transmission/distribution system.
- This test system does not have line limits.
- The online information about this system is available in: https://al-roomi.org/power-flow/30-bus-system.

For ORC, there were many attempts to use this large test system as a real problem. Most of these attempts show uncompleted data, such as the ones reported in Elrefaie et al. (1994), El-Khattam and Sidhu (2008, 2009), Barzegari et al. (2010a,b), Chabanloo et al. (2011a), Ezzeddine et al. (2011), Singh et al. (2012b), Sharaf et al. (2014), Singh (2014), Saleh et al. (2015), and Saberi Noughabi et al. (2017). However, a few attempts are available with the necessary data to carry out optimal coordination experiments on this test system. Two of these attempts are covered in the following text:

The study reported in Mohammadi et al. (2011) solves the ORC problem of the 33 kV part of the IEEE 30-bus network, which in turn does not cover all the 30 buses. On the opposite side, the study reported in El-Fergany and Hasanien (2019) covers all the busbars, which represents a very large system. These two attempts are covered in the following text:

F.9.1 System No. 1

The first attempt is reported[11] in Mohammadi et al. (2011). The study solves the ORC problem of the 33 kV part of the IEEE 30-bus network, which in turn does not cover all the 30 buses. This test system has the following information:

- Figure F.17 shows the single-line diagram of the 33 kV part of the IEEE 30-bus network.
- The network is sustained by three 50 MVA, 132/33 kV transformers associated with buses 1, 6, and 13.
- This test system is modeled as an ORC problem with 86 directional and non-directional OCRs.
- Among these 86 relays, a total of 39 DOCRs is considered for the network, which are installed on each end of the lines.
- Although any TCCC can be used for conducting the simulation, the IEEE moderately inverse TCCC is used in the main reference.[12]

11 It has to be noted that the same test system is used in Khurshaid et al. (2019). However, the relays have different numbers than those reported in the main reference.
12 Please, refer to Table 3.10.

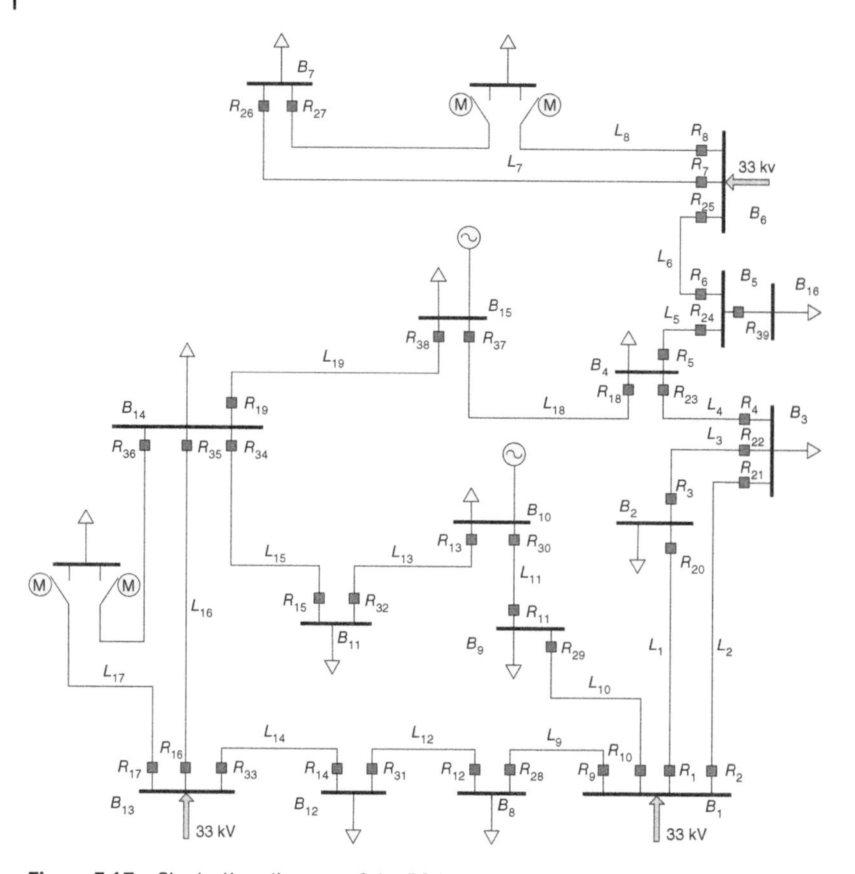

Figure F.17 Single-line diagram of the 30-bus test system no. 1.

- For L_{18}, pilot protection is considered.
- Two DGs are installed at buses 10 and 15, with transient reactance and capacity of 0.15 p.u. and 10 MVA, respectively.
- Two important loads have been located on buses 14 and 3.
- The relays $\{R_{13}, R_{30}\}$ and $\{R_{37}, R_{38}\}$ are, respectively, placed near generator buses (i.e. bus 10 and bus 15) with high set protections.
- The settings of the high set protections of the relays are selected to cover 60% of their lines.
- The *CTR* for each OCR is assumed as 500 : 1.
- The near-end bolted 3ϕ fault is considered where Table F.57 shows the short-circuit analysis.
- For the faults close to *CB* of relays $\{R_7, R_8, R_{25}\}$, the flowing fault currents of relays $\{R_{26}, R_{27}\}$ are very low, so these two relays cannot be used as backup protection for relays $\{R_7, R_8, R_{25}\}$.
- The optimization settings of *CTI* and the lower and upper bounds of *TDS* and *CTS* are tabulated in Table F.58.
- It has to be noted that some selectivity constraints are relaxed.
- More information about this ORC problem can be found in Mohammadi et al. (2011), Papaspiliotopoulos et al. (2017a), and Khurshaid et al. (2019).
- The online information about this ORC problem is available in: https://al-roomi.org/coordination/30-bus-systems/system-i.

Table F.57 3ϕ short-circuit currents of the 30-bus system no. 1.

Primary relay	Current (kA)	Backup relay	Current (kA)	Primary relay	Current (kA)	Backup relay	Current (kA)
3	4086.7	1	4086.7	9	7212.6	20	1103.5
4	5411.2	2	2138.8	10	7339.3	20	1095.8
22	4333.0	2	2147.0	1	7665.3	21	698.8
4	5411.2	3	3272.5	9	7212.6	21	721.2
21	5411.8	3	3243.6	10	7339.3	21	716.1
5	4960.8	4	3001.3	20	3481.5	22	3481.5
18	4719.4	4	3002.1	21	5411.8	23	2193.5
6	2416.0	5	2416.0	22	4333.0	23	2204.6
7	5669.0	6	1790.9	18	4719.4	24	1717.7
8	5607.7	6	1774.8	23	3689.7	24	1724.2
27	1472.3	7	1472.3	24	2695.0	25	2695.0
26	1026.8	8	1026.8	1	7665.3	28	1552.0
12	5034.9	9	5034.9	2	7985.7	28	1545.8
11	3457.1	10	3457.1	10	7339.3	28	1538.0
13	3727.3	11	2875.0	1	7665.3	29	1380.6
14	2906.5	12	2906.5	2	7985.7	29	1375.2
15	2660.5	13	2660.5	9	7212.6	29	1379.0
16	6185.6	14	1668.1	29	2518.9	30	2518.9
17	7492.9	14	1641.1	28	2036.8	31	2036.8
19	5445.2	15	1527.3	30	2998.8	32	2149.0
35	4222.0	15	1533.2	31	3263.6	33	3263.6
36	6420.2	15	1509.7	32	2930.4	34	2930.4
19	5445.2	16	3128.3	17	7492.9	35	1885.4
34	5796.6	16	3123.9	33	6456.2	35	1954.5
36	6420.2	16	3052.4	16	6185.6	36	490.9
19	5445.2	17	801.3	33	6456.2	36	500.6
34	5796.6	17	800.1	5	4960.8	37	1961.0
35	4222.0	17	794.0	23	3689.7	37	1968.5
38	3133.2	18	2292.2	34	5796.6	38	1886.8
37	3788.9	19	2940.9	35	4222.0	38	1896.7
2	7985.7	20	1053.9	36	6420.2	38	1867.7

Table F.58 Optimization data of the 30-bus system no. 1.

TDS^{min}	TDS^{max}	CTS^{min} (A)	CTS^{max} (A)	CTS mode	CTI (s)	References
0.1	1.1	1.5	6.0	Continuous	0.3	Papaspiliotopoulos et al. (2017a)
0.1	1.2	1.5	2.5	Continuous	0.3	Khurshaid et al. (2019)

F.9.2 System No. 2

The second attempt covered in this appendix is taken from El-Fergany and Hasanien (2019), which has the following information:

- The single-line diagram of this test system is shown in Figure F.18.
- As can be seen from the figure, this test system consists of 6 synchronous generating units having 6 non-directional OCRs, 41 lines having 82 DOCRs, two synchronous condensers with two non-directional OCRs, and 21 loads having 21 non-directional OCRs.
- Therefore, the total number of relays that need to be optimally coordinated in this large test system is 111, which represents a significant challenge to the proposed methodology and reveals a good realistic study.
- The maximum load currents and maximum near-end fault currents are determined via ETAP.
- Compared with the previous ORC problem, this one uses $\{TMS, PS\}$ instead of $\{TDS, CTS\}$, which means that it uses European relays instead of North American.[13]
- If the independent variables are only *TMS* and *PS*, then the problem dimension is 222 where half of them are continuous variables and the remaining half are discrete variables.[14] In El-Fergany and Hasanien (2019), the ORC problem is solved using multiple characteristic curves. Thus, the

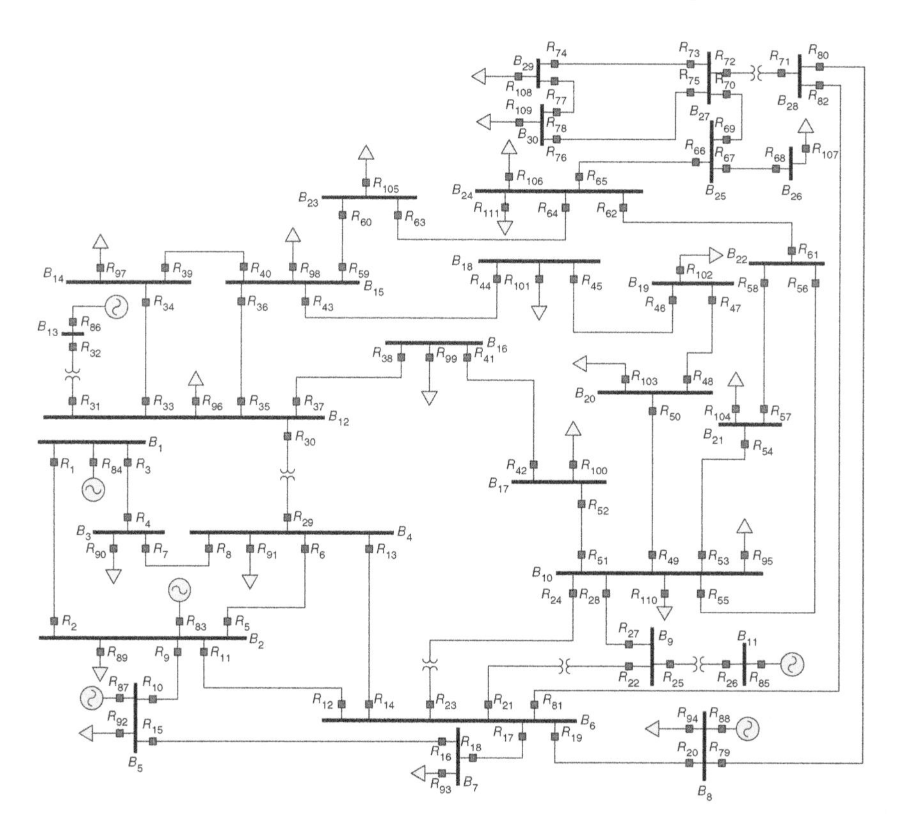

Figure F.18 Single-line diagram of the 30-bus system no. 2.

13 It can also be used for North American relays. However, it has to be noted that the results should not be compared with those obtained by European relays to have a fair performance comparison of optimization algorithms involved in both numerical experiments.

14 El-Fergany and Hasanien (2019) used discrete *TMS* and continuous *PS*.

problem dimension is 333 if only standard TCCCs are used, and 555 if user-defined TCCCs are implemented.
- The inequality constraints are 393 where 282 for *CTI* and 111 for $T^{\min}$.
- To realize the test system, a real industrial overcurrent relay produced by Schneider Electric, called Sepam-2000, is used.
- Although the relays used in this test system are European, the side constraint of *TDS* is used in the place of *TMS*. Thus, $TMS \in [0.1, 12.5]$, which might need to be transformed to proper limits by using either the fixed divisor or linear interpolation method.[15]
- For *PS*, (7.22) and (7.23) are used to determine the lower and upper bounds, which are taken for the *i*th relay as follows: $PS_i^{\min} = 120\% \times I_{L,i}^{\max}$ and $PS_i^{\max} = 200\% \times I_{L,i}^{\max}$.
- The range of *CTI* is from 150 to 600 ms.
- The maximum load currents are tabulated in Table F.59.
- Table F.60 shows the relay pairs and maximum near-end short-circuit currents.
- More information about this ORC problem can be found in El-Fergany and Hasanien (2019).
- The online information about this ORC problem is available in: https://al-roomi.org/coordination/30-bus-systems/system-ii.

Table F.59 Maximum load currents of the relays for the 30-bus system no. 2.

Relay no.	I_L (A)	Relay no.	I_L (A)	Relay no.	I_L (A)	Relay no.	I_L (A)	Relay no.	I_L (A)	Relay no.	I_L (A)
1	720.2	20	132.7	39	28.7	58	39.3	77	65.4	96	223.1
2	720.2	21	125.1	40	28.7	59	98.2	78	64.4	97	107.5
3	362.5	22	500.4	41	66.5	60	98.2	79	17.3	98	144.5
4	362.5	23	68.6	42	66.5	61	110.2	80	17.3	99	65.9
5	183.7	24	274.4	43	105	62	110.2	81	80.4	100	180.1
6	183.7	25	261.8	44	105	63	37.4	82	80.4	101	55.6
7	352.6	26	785.4	45	48.5	64	37.4	83	268.5	102	172.1
8	352.6	27	471.8	46	48.5	65	40.2	84	1079.0	103	39.2
9	345.5	28	471.8	47	124.2	66	40.2	85	785.3	104	352.0
10	345.5	29	200.8	48	124.2	67	73.3	86	521.7	105	60.9
11	253.0	30	803.2	49	163.3	68	73.3	87	159.6	106	188.1
12	253.0	31	173.9	50	163.3	69	82.1	88	160.8	107	73.3
13	319.8	32	521.7	51	115.9	70	82.1	89	76.9	108	44.7
14	319.8	33	136.0	52	115.9	71	81.5	90	11.5	109	190.0
15	87.3	34	136.0	53	312.9	72	20.4	91	33.6	110	347.4
16	87.3	35	316.9	54	312.9	73	109.6	92	416.2	111	76.9
17	165.5	36	316.9	55	149.0	74	109.6	93	110.3	—	—
18	165.5	37	132.2	56	149.0	75	124.6	94	183.7	—	—
19	132.7	38	132.2	57	39.3	76	124.6	95	102.7	—	—

15 We have to remember that the values of *TDS* and *TMS* are unitless.

Table F.60 3ϕ short-circuit currents of the 30-bus system no. 2.

No.	Relay pair		I_F (kA)		No.	Relay pair		I_F (kA)		No.	Relay pair		I_F (kA)	
	R_i	R_j	R_i	R_j		R_i	R_j	R_i	R_j		R_i	R_j	R_i	R_j
1	1	4	7.11	1.13	95	31	29	13.24	1.28	189	75	71	6.15	0.867
2	1	84	7.11	5.99	96	31	34	13.24	1.07	190	76	77	2.32	1.12
3	2	6	5.85	1.12	97	31	36	13.24	3.07	191	77	73	2.35	2.07
4	2	12	5.85	1.24	98	31	37	13.24	2.55	192	78	75	2.68	1.48
5	2	10	5.85	1.32	99	32	86	2.92	2.92	193	79	19	5.72	3.67
6	2	83	5.85	1.47	100	33	29	13.09	1.28	194	79	88	5.72	0.857
7	3	2	8.89	2.92	101	33	32	13.09	2.71	195	80	72	3.79	1.73
8	3	84	8.89	5.99	102	33	36	13.09	3.07	196	80	81	3.79	3.36
9	4	8	3.8	3.72	103	33	38	13.09	2.55	197	81	11	7.97	1.32
10	5	1	8.05	3.32	104	34	40	3.83	3.18	198	81	13	7.97	2.42
11	5	10	8.05	1.32	105	35	29	11.11	1.28	199	81	18	7.97	1.62
12	5	12	8.05	1.24	106	35	32	11.11	2.71	200	81	20	7.97	1.57
13	5	83	8.05	1.47	107	35	34	11.11	1.07	201	81	22	7.97	2.82
14	6	7	5.95	1.44	108	35	38	11.11	2.55	202	81	24	7.97	1.44
15	6	30	5.95	3.3	109	36	39	6.87	1.52	203	82	71	1.44	1.73
16	6	14	5.95	3.48	110	36	44	6.87	2.25	204	83	1	7.7	3.32
17	7	3	1.85	1.77	111	36	60	6.87	2.16	205	83	6	7.7	1.12
18	8	5	5.89	1.38	112	37	29	11.61	1.28	206	83	10	7.7	1.32
19	8	14	5.89	3.48	113	37	32	11.61	2.71	207	83	12	7.7	1.24
20	8	30	5.89	3.3	114	37	34	11.61	1.07	208	84	2	4.05	2.92
21	9	1	7.85	3.32	115	37	36	11.61	3.07	209	84	4	4.05	1.13
22	9	6	7.85	1.12	116	38	42	4.67	4.23	210	85	25	15.19	5.06
23	9	12	7.85	1.24	117	39	33	4.35	3.64	211	86	31	19.36	6.45
24	9	83	7.85	1.47	118	40	35	10.18	4.83	212	87	9	5.88	1.54
25	10	16	5.19	1.64	119	40	44	10.18	2.25	213	87	16	5.88	1.64
26	10	87	5.19	0.851	120	40	60	10.18	2.16	214	88	19	5.51	3.67
27	11	1	7.93	3.32	121	41	37	4.95	4.51	215	88	80	5.51	0.65
28	11	6	7.93	1.12	122	42	51	9.05	7.85	216	89	1	8.46	3.32
29	11	10	7.93	1.32	123	43	35	9.44	4.83	217	89	6	8.46	1.12
30	11	83	7.93	1.47	124	43	39	9.44	1.52	218	89	10	8.46	1.32
31	12	13	7.17	2.42	125	43	60	9.44	2.16	219	89	12	8.46	1.24
32	12	24	7.17	1.44	126	44	46	3.9	3.54	220	89	23	8.46	1.47
33	12	22	7.17	2.82	127	45	43	4.27	3.91	221	90	3	5.5	1.77
34	12	18	7.17	1.62	128	46	48	5.22	4.1	222	90	8	5.5	3.72
35	13	5	3.85	1.38	129	47	45	4.14	3.02	223	91	5	7.1	1.38
36	13	7	3.85	1.44	130	48	49	5.11	4.86	224	91	7	7.1	1.44
37	13	30	3.85	3.3	131	49	23	14.76	0.577	225	91	14	7.1	3.48
38	14	11	6.07	1.32	132	49	27	14.76	4.62	226	91	30	7.1	3.3

(continued)

Table F.60 (Continued)

No.	Relay pair		I_F (kA)		No.	Relay pair		I_F (kA)		No.	Relay pair		I_F (kA)	
	R_i	R_j	R_i	R_j		R_i	R_j	R_i	R_j		R_i	R_j	R_i	R_j
39	14	24	6.07	1.44	133	49	52	14.76	3.02	227	92	9	4.04	1.54
40	14	18	6.07	1.62	134	49	54	14.76	2.7	228	92	16	4.04	1.64
41	14	20	6.07	1.57	135	49	56	14.76	1.47	229	92	87	4.04	0.851
42	14	22	6.07	2.82	136	50	47	3.68	3.42	230	93	15	4.82	1.99
43	14	82	6.07	0.514	137	51	23	13.75	0.577	231	93	17	4.82	2.83
44	15	9	5.1	1.54	138	51	27	13.75	4.62	232	94	19	5.17	3.67
45	15	87	5.1	0.851	139	51	50	13.75	2.03	233	94	80	5.17	0.65
46	16	17	3.54	2.83	140	51	54	13.75	2.74	234	94	88	5.17	0.857
47	17	11	6.87	1.32	141	51	56	13.75	1.47	235	95	23	16.08	0.577
48	17	13	6.87	2.42	142	52	41	4.06	2.86	236	95	27	16.08	4.62
49	17	20	6.87	1.57	143	53	23	14.05	0.577	237	95	50	16.08	2.03
50	17	22	6.87	2.82	144	53	27	14.05	4.62	238	95	52	16.08	3.02
51	17	24	6.87	1.44	145	53	50	14.05	2.03	239	95	54	16.08	2.74
52	17	82	6.87	0.514	146	53	52	14.05	3.02	240	95	56	16.08	1.47
53	18	15	2.7	1.99	147	53	56	14.05	1.47	241	96	29	12.62	5.13
54	19	13	6.92	2.42	148	54	58	7.31	4.99	242	96	32	12.62	2.71
55	19	18	6.92	1.62	149	55	23	15.32	0.577	243	96	34	12.62	1.07
56	19	22	6.92	2.82	150	55	27	15.32	4.62	244	96	36	12.62	3.07
57	19	24	6.92	1.44	151	55	50	15.32	2.03	245	96	38	12.62	2.55
58	19	82	6.92	0.514	152	55	52	15.32	3.02	246	97	33	6.72	3.64
59	20	80	2.7	0.65	153	55	54	15.32	2.74	247	97	40	6.72	3.18
60	20	88	2.7	0.857	154	56	57	9.8	6.92	248	98	35	10.73	4.83
61	21	11	7.79	1.32	155	56	62	9.8	2.92	249	98	39	10.73	1.52
62	21	13	7.79	2.42	156	57	53	8.69	6.35	250	98	44	10.73	2.25
63	21	18	7.79	1.62	157	58	62	6.38	2.92	251	98	60	10.73	2.16
64	21	20	7.79	1.57	158	59	35	9.53	4.83	252	99	37	8.73	4.51
65	21	24	7.79	1.44	159	59	39	9.53	1.52	253	99	42	8.73	4.23
66	21	82	7.79	0.514	160	59	44	9.53	2.25	254	100	41	10.71	2.86
67	22	26	7.84	3.8	161	60	64	3.49	3.1	255	100	51	10.71	7.85
68	22	28	7.84	6.58	162	61	55	10.39	3.47	256	101	43	7.44	3.91
69	23	11	8.13	1.32	163	61	57	10.39	6.92	257	101	46	7.44	3.54
70	23	18	8.13	1.62	164	62	63	5.4	2.34	258	102	45	7.13	3.02
71	23	20	8.13	1.57	165	62	66	5.4	1.84	259	102	48	7.13	4.1
72	23	22	8.13	2.82	166	63	59	4.54	4.14	260	103	47	8.28	3.42
73	23	82	8.13	0.514	167	64	61	7.45	4.4	261	103	49	8.28	4.86
74	24	27	14.48	4.62	168	64	66	7.45	1.84	262	104	53	11.22	6.35
75	24	50	14.48	2.03	169	65	61	7.94	4.4	263	104	58	11.22	4.99
76	24	52	14.48	3.02	170	65	63	7.94	2.34	264	105	59	7.24	4.14

(continued)

Table F.60 (Continued)

No.	Relay pair		I_F (kA)		No.	Relay pair		I_F (kA)		No.	Relay pair		I_F (kA)	
	R_i	R_j	R_i	R_j		R_i	R_j	R_i	R_j		R_i	R_j	R_i	R_j
77	24	54	14.48	2.74	171	66	68	3.3	0.427	265	105	64	7.24	3.1
78	24	56	14.48	1.47	172	66	70	3.3	2.87	266	106	61	8.58	4.4
79	25	21	12.7	1.53	173	67	65	5.7	2.87	267	106	63	8.58	2.34
80	25	28	12.7	6.58	174	67	70	5.7	2.87	268	106	66	8.58	1.84
81	26	85	4.47	4.47	175	68	67	2.31	2.31	269	107	68	2.31	2.31
82	27	21	7.39	1.53	176	69	65	3.28	2.87	270	108	73	3.32	2.07
83	27	26	7.39	3.8	177	70	71	4.64	0.865	271	108	78	3.32	1.26
84	28	23	12.2	0.577	178	70	74	4.64	0.563	272	109	75	2.61	1.48
85	28	50	12.2	2.03	179	70	76	4.64	0.624	273	109	77	2.61	1.12
86	28	52	12.2	3.02	180	71	79	4.63	1.01	274	110	23	16.77	0.577
87	28	54	12.2	2.74	181	71	81	4.63	3.36	275	110	27	16.77	4.62
88	29	5	6.51	1.38	182	72	69	3.39	2.23	276	110	50	16.77	2.03
89	29	7	6.51	1.44	183	72	74	3.39	0.563	277	110	52	16.77	3.02
90	29	14	6.51	3.48	184	72	76	3.39	0.624	278	110	54	16.77	2.74
91	30	32	9.08	2.71	185	73	69	6.21	2.23	279	110	56	16.77	1.47
92	30	34	9.08	1.07	186	73	71	6.21	0.867	280	111	61	9.79	4.4
93	30	36	9.08	3.07	187	74	78	1.54	1.26	281	111	63	9.79	2.34
94	30	38	9.08	2.55	188	75	69	6.15	2.23	282	111	66	9.79	1.84

F.10 42-Bus Test System

This test system is considered one of the largest ORC test systems available in the literature. Some information is given in the following text:

- The 42-bus test system[16] is shown in Figure F.19.
- This test system is protected by a mixture of 97 directional and non-directional OCRs. That is, using (7.1), the dimension is 194 (i.e. 97 variables of type *TMS* and 97 variables of type *PS*) when only one unified TCCC is used for all the relays.
- If multiple TCCCs are used, then the dimension increases to 485 variables.
- If the double primary relay strategy (DPRS) is considered, then the dimension increases for both one unified TCCC and multiple TCCCs. For the extreme case (i.e. 194 relays: 97 main-1 relays and 97 main-2 relays), the dimension increases from being 194 to 388 variables if only one unified

16 This test system is known as the IEEE Std. 399-1997 system (IEEE, 1998).

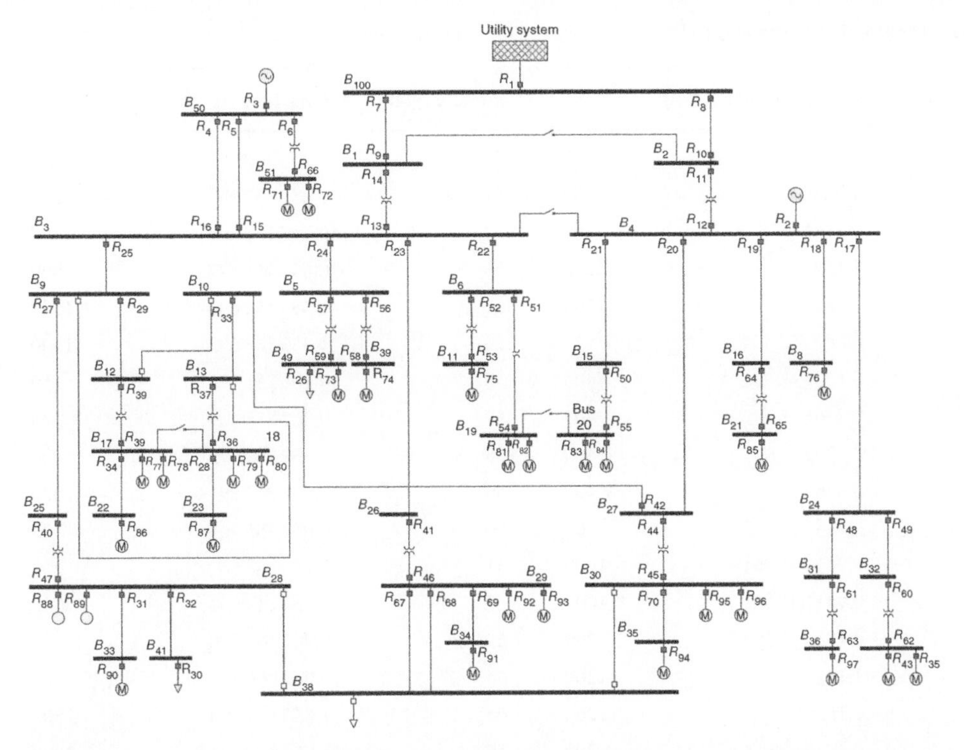

Figure F.19 Single-line diagram of the 42-bus test system (IEEE Std. 399-1997). Source: Based on IEEE (1998).

TCCC is used for all the relays. For multiple TCCCs, the dimension increases from being 485 to 970. Please, refer to Table 13.1.

- The same thing for the design constraints where the extreme case of this ORC problem (i.e. applying DPRS to all 194 relays of this test system) has the following constraints:
  - 114 inequality constraints for (13.12)
  - 194 inequality constraints for (13.15)
  - 194 inequality constraints for (13.16)
  - 194 side constraints for (13.17)
  - 194 side constraints for (13.19)
- The standard ANSI *CTR*s are listed in Table A.1.
- The overload currents and *CTR*s of all the primary relay sets are tabulated in Table F.61.
- The 3ϕ short-circuit analysis is carried out using DIgSILENT PowerFactory.
- The ANSI standard calculation is applied here with zero fault impedance.
- The short-circuit currents seen by all the P/B relay sets are tabulated in Table F.62.
- More details about this test system are given in IEEE (1998), Albasri et al. (2015), Al-Roomi (2015b), and Al-Roomi and El-Hawary (2019a).
- The online information about this system is available in: https://al-roomi.org/coordination/42-bus-system.

Table F.61 Overload currents and *CTR*s used for the 42-bus test system.

Primary relay	Over-load current (kA)	*CTR*	Primary relay	Over-load current (kA)	*CTR*
1	0.753	800	50	0.235 332 99	250
2	0.7844	800	51	0.235 332 99	250
3	0.7844	800	52	0.094 133 196	100
4	1.152	1200	53	0.541 265 877	600
5	1.152	1200	54	1.353 164 693	1500
6	0.062 755 464	75	55	1.353 164 693	1500
7	0.3765	400	56	0.108 253 175	150
8	0.3765	400	57	0.062 755 464	75
9	0.3765	400	58	0.359 109 092	400
10	0.3765	400	59	1.804 219 591	2000
11	0.188 266 392	200	60	0.062 755 464	75
12	0.941 331 961	1000	61	0.156 888 66	200
13	0.941 331 961	1000	62	1.804 219 591	2000
14	0.188 266 392	200	63	0.902 109 796	1000
15	1.152	1200	64	0.047 066 598	50
16	1.152	1200	65	1.353 164 693	1500
17	0.4725	500	66	1.804 219 591	2000
18	0.615	800	67	0.615	800
19	0.4725	500	68	0.615	800
20	0.4725	500	69	0.4725	500
21	0.4725	500	70	0.4725	500
22	0.4725	500	71	0.360 843 918	400
23	0.4725	500	72	0.085 700 431	100
24	0.4725	500	73	2.255 274 489	3000
25	0.4725	500	74	0.346 097 893	400
26	1.969 824 442	2000	75	0.017 140 086	50
27	0.4725	500	76	0.571 074 723	600
28	0.4725	500	77	0.087 472 806	100
29	0.4725	500	78	0.180 421 959	200
30	0.285 272 165	300	79	0.087 472 806	100
31	0.4725	500	80	0.180 421 959	200
32	1.5	1500	81	0.857 004 306	1000
33	0.4725	500	82	0.405 949 408	500
34	0.4725	500	83	0.649 519 053	800
35	0.090 210 98	100	84	0.599 903 014	600
36	2.706 329 387	3000	85	0.087 480 474	100
37	0.094 133 196	100	86	0.085 700 431	100

(continued)

Table F.61 (Continued)

Primary relay	Over-load current (kA)	*CTR*	Primary relay	Over-load current (kA)	*CTR*
38	0.094 133 196	100	87	0.085 700 431	100
39	2.706 329 387	3000	88	0.180 421 959	200
40	0.094 133 196	100	89	0.072 168 784	75
41	0.094 133 196	100	90	0.086 452 177	100
42	0.4725	500	91	0.198 464 155	200
43	0.087 478 85	100	92	0.225 527 449	250
44	0.094 133 196	100	93	0.838 962 11	1000
45	2.706 329 387	3000	94	0.086 452 177	100
46	2.706 329 387	3000	95	0.087 478 85	100
47	2.706 329 387	3000	96	0.090 210 98	100
48	0.4725	500	97	0.847 198 765	1000
49	0.4725	500	—	—	—

Table F.62 Three-phase fault currents of the 42-bus test system.

Primary relay	Current (kA)	Backup relay	Current (kA)	Primary relay	Current (kA)	Backup relay	Current (kA)
1	8.367	—	—	40	11.613	27	11.613
2	4.084	—	—	41	13.081	23	13.081
3	5.835	—	—	42	13.163	20	12.997
4	9.821	3	5.835	43	21.056	62	19.803
4	9.821	15	3.876	44	13.197	20	12.997
5	9.821	3	5.835	45	28.984	44	1.0081
5	9.821	16	3.876	46	28.963	41	1.0074
6	13.581	3	5.835	47	28.686	40	0.9978
7	9.115	1	8.367	48	13.07	17	12.923
7	9.115	10	0.748	49	13.321	17	12.923
8	9.113	1	8.367	50	12.526	21	12.526
8	9.113	9	0.746	51	12.812	22	12.722
9	0.767	13	3.835	52	13.391	22	12.722
10	0.768	12	3.84	53	6.046	52	1.0515
11	6.82	8	6.82	54	13.415	51	2.333
12	7.528	2	4.084	55	13.361	50	2.3237
13	7.495	4	2.873	56	13.281	24	13.142
13	7.495	5	2.873	57	13.451	24	13.142
14	6.822	7	6.822	58	3.659	56	1.103
15	11.004	4	2.873	59	30.761	57	1.0699
15	11.004	14	1.2764	60	12.892	49	12.892

(continued)

Table F.62 (Continued)

Primary relay	Current (kA)	Backup relay	Current (kA)	Primary relay	Current (kA)	Backup relay	Current (kA)
16	11.004	5	2.873	61	12.778	48	12.778
16	11.004	14	1.2764	62	19.803	60	0.6888
17	13.364	2	4.084	63	9.156	61	1.5923
17	13.364	11	1.2762	64	12.748	19	12.748
18	12.014	2	4.084	65	15.046	64	0.5233
18	12.014	11	1.2762	66	29.053	6	1.0105
19	13.818	2	4.084	67	35.125	46	28.963
19	13.818	11	1.2762	68	35.125	46	28.963
20	13.542	2	4.084	69	34.656	46	28.963
20	13.542	11	1.2762	70	33.409	45	28.984
21	13.359	2	4.084	71	31.497	66	29.053
21	13.359	11	1.2762	72	30.235	66	29.053
22	13.118	4	2.873	73	30.761	59	30.761
22	13.118	5	2.873	74	3.659	58	3.659
22	13.118	14	1.2764	75	6.046	53	6.046
23	13.698	4	2.873	76	11.69	18	11.69
23	13.698	5	2.873	77	28.487	39	24.941
23	13.698	14	1.2764	78	29.081	39	24.941
24	13.429	4	2.873	79	32.418	36	28.873
24	13.429	5	2.873	80	33.012	36	28.873
24	13.429	14	1.2764	81	15.033	54	13.415
25	13.511	4	2.873	82	16.828	54	13.415
25	13.511	5	2.873	83	15.749	55	13.361
25	13.511	14	1.2764	84	14.904	55	13.361
26	35.288	59	30.761	85	15.046	65	15.046
27	13.132	25	12.937	86	29.038	34	29.038
28	35.343	36	28.873	87	32.348	28	32.348
29	13.109	25	12.937	88	31.627	47	28.686
30	31.986	32	31.986	89	32.853	47	28.686
31	33.338	47	28.686	90	30.648	31	30.648
32	34.57	47	28.686	91	31.782	69	31.782
33	12.749	42	12.749	92	31.392	46	28.963
34	31.412	39	24.941	93	33.158	46	28.963
35	23.847	62	19.803	94	30.736	70	30.736
36	28.873	37	1.0043	95	32.321	45	28.984
37	12.579	33	12.579	96	32.526	45	28.984
38	12.963	29	12.964	97	9.156	63	9.156
39	24.941	38	0.8675	—	—	—	—

F.11 118-Bus Test System

- This test system represents a portion of the American Electric Power System (in the Midwestern US) as of December 1962.
- The base voltages of this test system are a very rough guess. The line MVA limits were not part of the original data and are made up.
- The original system consists of 118 buses, 177 lines, 9 transformers, 19 generators, 35 synchronous condensers, and 91 loads.
- It has a lot of voltage control devices and is quite robust, converging in 5 or so iterations with a fast decoupled load flow.
- The online information about this system is available in: https://al-roomi.org/power-flow/118-bus-system.

For ORC, there was only one attempt, which is reported in Alam et al. (2016). With 372 DOCRs of mixed characteristics and manufacturing technologies, it is the largest test system available in the literature. Some general information about the 118-bus test system is given in the following text:

- The ORC problem of this test system is shown in Figure F.20.
- There are 1184 P/B relay pairs among these 372 DOCRS.
- Only 829 P/B relay pairs are considered where the rest are ignored because $T_j - T_i \geqslant CTI$; please refer to (F.6).
- For $I_f^{\max}$, Figure 2.3e is considered with zero fault impedance, while Figure 2.3b is considered for $I_f^{\min}$ with a fault impedance of 0.1 pu (Bedekar and Bhide, 2011a; Mathur et al., 2015).
- Based on these currents, the *CTR* of the *i*th relay can be calculated as follows (Gers and Holmes, 2004):

$$CTR_i = \max\left[I_{L,i}^{\max}, \frac{I_{f,i}^{\max}}{20}\right] \tag{F.12}$$

- The first 300 relays are numerical or digital, while the rest are electromechanical. Thus, the ORC problem is modeled as a MINLP optimization problem where *TMS* is continuous for all the relay types and *PS* is continuous for the numerical/digital relays and discrete for the electromechanical.
- The side constraints of these two settings are: $TMS \in [0.1, 1.1]$ and $PS \in [0.5, 2.0] \times$ CT secondary rating of each relay (in steps of 0.25 in case *PS* is discrete).
- The CTI is set to 0.2 s, and the lower and upper bounds of T_i are 0.1 and 4 s, respectively.
- The characteristic curves of these 372 old and new DOCRs are listed as follows:
  1. $\boldsymbol{R_1}$–$\boldsymbol{R_{200}}$: The IEC standard inverse TCCC
  2. $\boldsymbol{R_{201}}$–$\boldsymbol{R_{260}}$: The IEC very inverse TCCC
  3. $\boldsymbol{R_{261}}$–$\boldsymbol{R_{300}}$: The IEC extremely inverse TCCC
  4. $\boldsymbol{R_{301}}$–$\boldsymbol{R_{372}}$: The IEC standard inverse TCCC
- The optimization algorithm used in the main reference, i.e. Alam et al. (2016), consists of two phases. The first phase uses continuous *PS* where the lower and upper bounds of the *i*th relay are defined by (F.8) and (F.9), respectively. The second phase uses discrete *PS* for the electromechanical relays by respectively applying the side constraints given in (F.10) and (F.11).

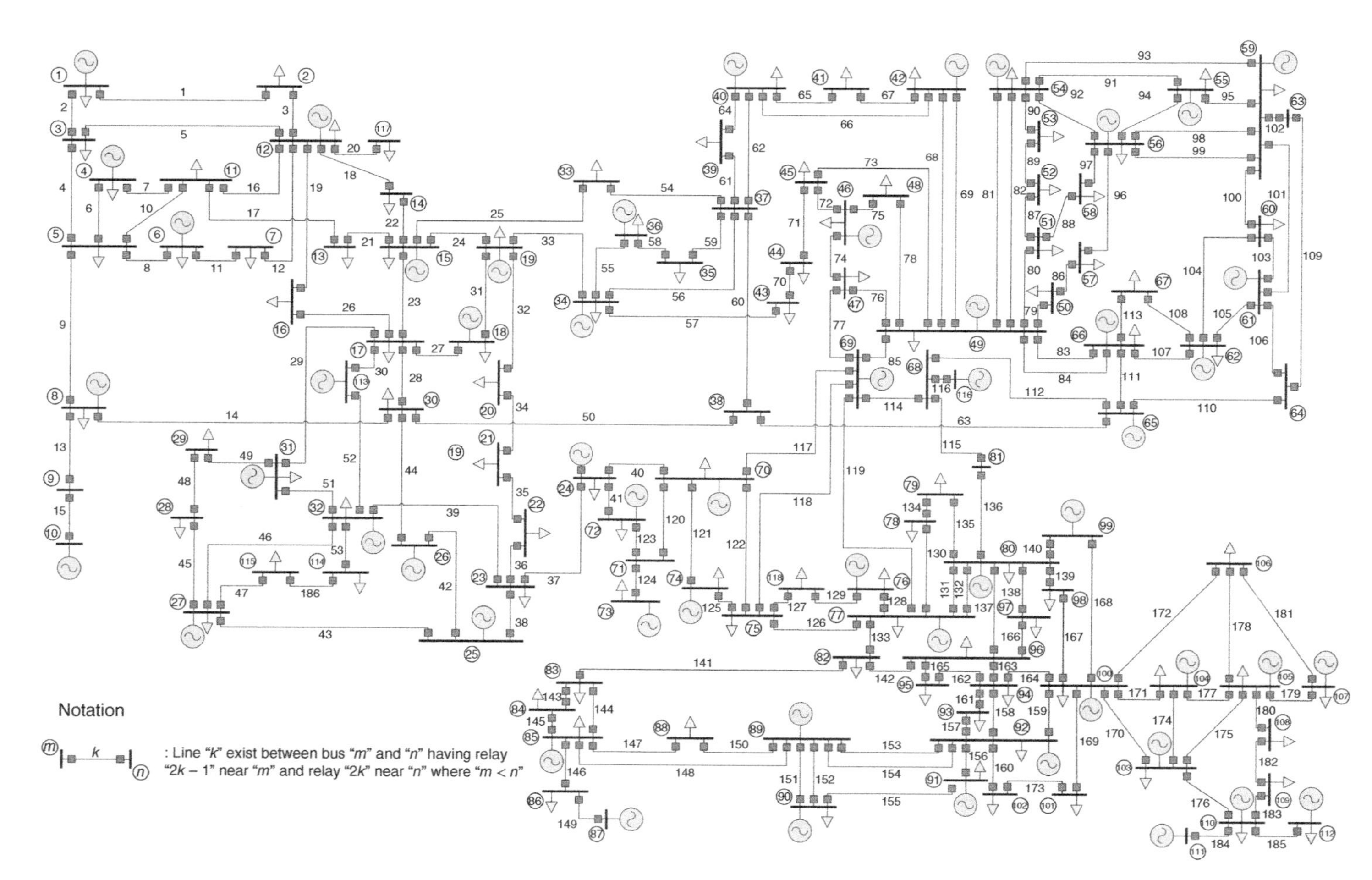

Figure F.20 Single-line diagram of the 118-bus test system.

- NRLF method is used to calculate $I_L^{\max}$ and the Z-matrix method is used to calculate various short-circuit currents.
- More information about this test system can be found in Alam et al. (2016).
- Because of the space limitation, the detailed numerical data (R_i/R_j, CTR_i, $I_{L,i}^{\max}$, $I_{f,i}^{\min}$, $I_{f,i}^{\max}$, $I_{f,i}^{\mathrm{pr}}$, and $I_{f,j}^{\mathrm{bc}}$) are not given in the preceding reference. Therefore, we have contacted the author to provide these essential information. We have transformed these numerical data, which can be found and downloaded from the following link:
  https://al-roomi.org/coordination/118-bus-system.

References

ABB (1992). Type CO Overcurrent Relay, January 1992. https://library.e.abb.com/public/7042f1d45630fca185256eac0053e7d4/41-101U.pdf (accessed on 09 October 2020).

ABB (1998). Overcurrent Relay (REJ 523), September issued: 1998, version: D/14.11.2005. https://library.e.abb.com/public/5dec034667bfc4acc12570eb0046bf02/REJ523tech750940ENd.pdf (accessed on 13 October 2020).

ABB (2018). Relion Protection and Control: 615 Series ANSI, December 2018. https://library.e.abb.com/public/70602692769a4ffa87ca027e6fb1af1d/RE_615_tech_756887_ENn.pdf (accessed on 22 October 2020).

Abeid, S., Hu, Y., and Hu, Y. (2020). Overcurrent relays coordination optimisation methods in distribution systems for microgrids: a review. *IET Conference Proceedings*, (1), pp. 042 (8 pp.). https://digital-library.theiet.org/content/conferences/10.1049/cp.2020.0019 (accessed 30 July 2021).

Abido, M.A. (2003). Environmental/economic power dispatch using multiobjective evolutionary algorithms. *IEEE Transactions on Power Systems* 18 (4): 1529–1537. https://doi.org/10.1109/TPWRS.2003.818693.

Abyane, H.A., Faez, K., and Karegar, H.K. (1997). A new method for overcurrent relay (O/C) using neural network and fuzzy logic. *TENCON '97 Brisbane - Australia. Proceedings of IEEE TENCON '97. IEEE Region 10 Annual Conference. Speech and Image Technologies for Computing and Telecommunications (Cat. No.97CH36162)*, Volume 1, pp. 407–410.

Abyaneh, H.A., Al-Dabbagh, M., Karegar, H.K. et al. (2003). A new optimal approach for coordination of overcurrent relays in interconnected power systems. *IEEE Transactions on Power Delivery* 18 (2): 430–435. https://doi.org/10.1109/TPWRD.2002.803754.

Abyaneh, H.A., Kamangar, S.S.H., Razavi, F., and Chabanloo, R.M. (2008). A new genetic algorithm method for optimal coordination of overcurrent relays in a mixed protection scheme with distance relays. *2008 43rd International Universities Power Engineering Conference*, pp. 1–5. https://doi.org/10.1109/UPEC.2008.4651442.

Alroomi, A.R., Albasri, F.A., and Talaq, J.H. (2013a). Solving the associated weakness of biogeography-based optimization algorithm. *International Journal of Soft Computing* 4 (4): 1–20. http://dx.doi.org/10.5121/ijsc.2013.4401.

Alroomi, A.R., Albasri, F.A., and Talaq, J.H. (2013b). A comprehensive comparison of the original forms of biogeography-based optimization algorithms. *International Journal on Soft Computing, Artificial Intelligence and Applications* 2 (5/6): 11–30. http://dx.doi.org/10.5121/ijscai.2013.2602.

Adelnia, F., Moravej, Z., and Farzinfar, M. (2015). A new formulation for coordination of directional overcurrent relays in interconnected networks. *International Transactions on Electrical Energy Systems* 25 (1): 120–137. https://doi.org/10.1002/etep.1828. ETEP-13-0348.R1.

Optimal Coordination of Power Protective Devices with Illustrative Examples, First Edition. Ali R. Al-Roomi.

Companion website: www.wiley.com/go/al-roomi/optimalcoordination

Agrawal, A., Singh, M., and Tejeswini, M.V. (2016). Voltage current based time inverse relay coordination for PV feed distribution systems. *2016 National Power Systems Conference (NPSC)*, pp. 1–6. https://doi.org/10.1109/NPSC.2016.7858866.

Ahmadi, S.-A., Karami, H., and Gharehpetian, B. (2017). Comprehensive coordination of combined directional overcurrent and distance relays considering miscoordination reduction. *International Journal of Electrical Power & Energy Systems* 92: 42–52. https://doi.org/https://doi.org/10.1016/j.ijepes.2017.04.008.

Al-Roomi, A.R. (2014). Optimal relays settings by using biogeography based optimization "BBO" technique. Master's thesis. Sakhir, Bahrain: University of Bahrain.

Al-Roomi, A.R. (2015a). Unconstrained single-objective benchmark functions repository. http://al-roomi.org/benchmarks/unconstrained (accessed 29 August 2020).

Al-Roomi, A.R. (2015b). Optimal relay coordination test systems repository. https://www.al-roomi.org/coordination (accessed 29 August 2020).

Al-Roomi, A.R. (2015c). Power flow test systems repository. https://al-roomi.org/power-flow (accessed 29 August 2020).

Al-Roomi, A.R. (2020). Improved power system realization and integration. PhD thesis. Halifax, NS, Canada: Dalhousie University. http://hdl.handle.net/10222/78503 (accessed 7 May 2020).

Al-Roomi, A.R. and El-Hawary, M.E. (2016a). Optimal coordination of directional overcurrent relays using hybrid BBO/DE algorithm and considering double primary relays strategy. *2016 IEEE Electrical Power and Energy Conference (EPEC)*, pp. 1–7, October 2016. https://doi.org/10.1109/EPEC.2016.7771762.

Al-Roomi, A.R. and El-Hawary, M.E. (2016b). Metropolis biogeography-based optimization. *Information Sciences* 360: 73–95. https://doi.org/https://doi.org/10.1016/j.ins.2016.03.051.

Al-Roomi, A.R. and El-Hawary, M.E. (2016c). Economic load dispatch using hybrid MpBBO-SQP algorithm. In: *Nature-Inspired Computation in Engineering, Studies in Computational Intelligence*, vol. 637, 1e (ed. X.-S. Yang), 217–250. Switzerland: Springer International Publishing. ISBN 978-3-319-30233-1. https://doi.org/10.1007/978-3-319-30235-5.

Al-Roomi, A.R. and El-Hawary, M.E. (2017a). Optimal coordination of directional overcurrent relays using hybrid BBO-LP algorithm with the best extracted time-current characteristic curve. *2017 IEEE 30th Canadian Conference on Electrical and Computer Engineering (CCECE)*, pp. 1–6, April 2017a. https://doi.org/10.1109/CCECE.2017.7946625.

Al-Roomi, A.R. and El-Hawary, M.E. (2017b). Effective weather/frequency-based transmission line models—Part I: Fundamental equations. *2017 IEEE Electrical Power and Energy Conference (EPEC)*, pp. 1–6, October 2017b. https://doi.org/10.1109/EPEC.2017.8286206.

Al-Roomi, A.R. and El-Hawary, M.E. (2017c). Effective weather/frequency-based transmission line models—Part II: Prospective applications. *2017 IEEE Electrical Power and Energy Conference (EPEC)*, pp. 1–8, October 2017c. https://doi.org/10.1109/EPEC.2017.8286207.

Al-Roomi, A.R. and El-Hawary, M.E. (2017d). A novel multiple fuels' cost function for realistic economic load dispatch needs. *2017 IEEE Electrical Power and Energy Conference (EPEC)*, pp. 1–6, October 2017d. https://doi.org/10.1109/EPEC.2017.8286205.

Al-Roomi, A.R. and El-Hawary, M.E. (2017e). A new technique to locate faults in distribution networks based on optimal coordination of numerical directional overcurrent relays. *2017 IEEE 30th Canadian Conference on Electrical and Computer Engineering (CCECE)*, pp. 1–6, April 2017e. https://doi.org/10.1109/CCECE.2017.7946624.

Al-Roomi, A.R. and El-Hawary, M.E. (2018a). Fuel cost modeling for spinning reserve thermal generating units. *2018 IEEE Canadian Conference on Electrical Computer Engineering (CCECE)*, pp. 1–6, May 2018. https://doi.org/10.1109/CCECE.2018.8447753.

Al-Roomi, A.R. and El-Hawary, M.E. (2018b). Can linear heat sensors be a good and practical replacement of traditional protective fuses? *2018 IEEE Electrical Power and Energy Conference (EPEC)*, pp. 1–6, October 2018a. https://doi.org/10.1109/EPEC.2018.8598382.

Al-Roomi, A.R. and El-Hawary, M.E. (2018c). Optimizing load forecasting configurations of computational neural networks. *2018 IEEE Canadian Conference on Electrical Computer Engineering (CCECE)*, pp. 1–6, May 2018b. https://doi.org/10.1109/CCECE.2018.8447739.

Al-Roomi, A.R. and El-Hawary, M.E. (2019a). Optimal coordination of double primary directional overcurrent relays using a new combinational BBO/DE algorithm. *IEEE Canadian Journal of Electrical and Computer Engineering* 42 (3): 135–147. https://doi.org/10.1109/CJECE.2018.2802461.

Al-Roomi, Ali R.A.R. and El-Hawary, Mohamed E.M.E. (2019b). Is it enough to just rely on near-end, middle, and far-end points to get feasible relay coordination? *2019 IEEE Canadian Conference of Electrical and Computer Engineering (CCECE) (CCECE 2019)*, Edmonton, Canada, May 2019.

Al-Roomi, Ali R.A.R. and El-Hawary, Mohamed E.M.E. (2019c). Estimating complex power magnitudes using a bank of pre-defined PFs embedded in ANNs. *2019 IEEE Electrical Power and Energy Conference (EPEC) (EPEC19)*, pp. 1–71–7, Montreal, Canada, October 2019.

Al-Roomi, A.R. and El-Hawary, M.E. (2020a). How to estimate temperature coefficients of series and shunt parameters of transmission lines with sag. *2020 IEEE Electric Power and Energy Conference (EPEC)*, pp. 1–6. https://doi.org/10.1109/EPEC48502.2020.9320034.

Al-Roomi, A.R. and El-Hawary, M.E. (2020b). Fast AI-based power flow analysis for high-dimensional electric networks. *2020 IEEE Electric Power and Energy Conference (EPEC)*, pp. 1–6. https://doi.org/10.1109/EPEC48502.2020.9320057.

Al-Roomi, A.R. and El-Hawary, M.E. (2020c). A novel approach to precisely calculate lumped parameters for transmission lines with sag using the M-model equivalent circuit. *2020 IEEE Electric Power and Energy Conference (EPEC)*, pp. 1–6. https://doi.org/10.1109/EPEC48502.2020.9320040.

Al-Roomi, A.R. and El-Hawary, M.E. (2020d). M-Model: a new precise medium-length transmission line model. *2020 IEEE Canadian Conference on Electrical and Computer Engineering (CCECE) (IEEE CCECE 2020)*, pp. 1–6, London, Canada, April 2020d.

Al-Roomi, A.R. and El-Hawary, M.E. (2020e). Mathematical schemes to linearize operating times of overcurrent relays by sequentially fixing plug settings and time multiplier settings. *2020 IEEE Canadian Conference on Electrical and Computer Engineering (CCECE) (IEEE CCECE 2020)*, pp. 1–6, London, Canada, April 2020e.

Al-Roomi, A.R. and El-Hawary, M.E. (2020f). Hybridizing UFO with other ML tools to locate faults by just knowing relay operating times. *2020 IEEE Canadian Conference on Electrical and Computer Engineering (CCECE) (IEEE CCECE 2020)*, pp. 1–6, London, Canada, April 2020f.

Al-Roomi, A.R. and El-Hawary, M.E. (2020g). Peak-load forecasting of Nova Scotia during winter using support vector machine with optimally configured hyperparameters. *2020 IEEE Electric Power and Energy Conference (EPEC) (EPEC 2020)*, Edmonton, Canada, November 2020a.

Al-Roomi, A.R. and El-Hawary, M.E. (2020h). Universal functions originator. *Applied Soft Computing* 94: 106417. https://doi.org/10.1016/j.asoc.2020.106417.

Alam, M.N. (2019). Adaptive protection coordination scheme using numerical directional overcurrent relays. *IEEE Transactions on Industrial Informatics* 15 (1): 64–73. https://doi.org/10.1109/TII.2018.2834474.

Alam, M.N., Das, B., and Pant, V. (2015). A comparative study of metaheuristic optimization approaches for directional overcurrent relays coordination. *Electric Power Systems Research* 128: 39–52. https://doi.org/http://dx.doi.org/10.1016/j.epsr.2015.06.018.

Alam, M.N., Das, B., and Pant, V. (2016). An interior point method based protection coordination scheme for directional overcurrent relays in meshed networks. *International Journal of Electrical Power & Energy Systems* 81: 153–164. https://doi.org/10.1016/j.ijepes.2016.02.012.

Alam, M.N., Sallem, A., and Masmoudi, N. (2020). Protection coordination using mixed characteristics of directional overcurrent relays in interconnected power distribution networks. *2020 6th IEEE International Energy Conference (ENERGYCon)*, pp. 761–766. https://doi.org/10.1109/ENERGYCon48941.2020.9236488.

Albasri, F.A., Alroomi, A.R., and Talaq, J.H. (2015). Optimal coordination of directional overcurrent relays using biogeography-based optimization algorithms. *IEEE Transactions on Power Delivery* 30 (4): 1810–1820. https://doi.org/10.1109/TPWRD.2015.2406114.

Albrecht, R.E., Nisja, M.J., Feero, W.E. et al. (1964). Digital computer protective device coordination program, Part I: General program description. *IEEE Transactions on Power Apparatus and Systems* PAS-83: 402–410.

Alipour, M., Teimourzadeh, S., and Seyedi, H. (2015). Improved group search optimization algorithm for coordination of directional overcurrent relays. *Swarm and Evolutionary Computation* 23: 40–49. https://doi.org/http://dx.doi.org/10.1016/j.swevo.2015.03.003.

Alkaran, D.S., Vatani, M.R., Sanjari, M.J. et al. (2015). Overcurrent relays coordination in interconnected networks using accurate analytical method and based on determination of fault critical point. *IEEE Transactions on Power Delivery* 30 (2): 870–877. https://doi.org/10.1109/TPWRD.2014.2330767.

Almas, M.S., Leelaruji, R., and Vanfretti, L. (2012). Over-current relay model implementation for real time simulation amp; hardware-in-the-loop (HIL) validation. *IECON 2012 - 38th Annual Conference on IEEE Industrial Electronics Society*, pp. 4789–4796, October 2012. https://doi.org/10.1109/IECON.2012.6389585.

ALSTOM (2002). *Network Protection & Automation Guide*. Levallois-Perret, France: ALSTOM. ISBN 2-9518589-0-6.

Amraee, T. (2012). Coordination of directional overcurrent relays using seeker algorithm. *IEEE Transactions on Power Delivery* 27 (3): 1415–1422. https://doi.org/10.1109/TPWRD.2012.2190107.

Anderson, P.M. (1998). *Power System Protection, IEEE Press Series on Power Engineering*. New York: Wiley-IEEE Press. ISBN 0-7803-3427-2.

Anthony, M.A. (1995). *Electric Power System Protection and Coordination: A Design Handbook for Overcurrent Protection*. New York: McGraw-Hill Inc. ISBN 0-07-002671-8.

Aravindhababu, P., Ganapathy, S., and Nayar, K.R. (2001). A novel technique for the analysis of radial distribution systems. *International Journal of Electrical Power & Energy Systems* 23 (3): 167–171. https://doi.org/http://dx.doi.org/10.1016/S0142-0615(00)00048-X.

AREVA (2010). *MiCOM P120/P121/P122/P123 Overcurrent Relays*. Headquarter: Palo Alto, California, United States: Schneider Electric, version 12 edition.

AREVA (2011). Directional / non Directional Overcurrent Protection. *Technical Data Sheet*, P12xy/EN TDS/H76.

Arrillaga, J. and Arnold, C.P. (1990). *Computer Analysis of Power Systems*. Chichester: Wiley. ISBN 0-471-92760-0.

Arrillaga, J. and Watson, N.R. (2001). *Computer Modelling of Electrical Power Systems*, 2e. Chichester: Wiley. ISBN 0-471-81249-0.

Asadi, M.R., Abyaneh, H.A., Mahmoodan, M. et al. (2008). Optimal overcurrent relays coordination using genetic algorithm. *2008 11th International Conference on Optimization of Electrical and Electronic Equipment*, pp. 197–202. https://doi.org/10.1109/OPTIM.2008.4602366.

Asgharigovar, S., Seyedi, H., and Dibazari, S.P. (2018). Optimal coordination of overcurrent protection in the presence of SFCL and distributed generation. *Turkish Journal of Electrical Engineering and Computer Sciences* 26: 2056–2065. https://doi.org/10.3906/elk-1709-81.

Askarian, H., Mohammadi, R., Razavi, F. et al. (2007). A New Genetic Algorithm Method for Overcurrent Relays and Fuses Coordination. https://citeseerx.ist.psu.edu/viewdoc/download?doi=10.1.1.728.5250&rep=rep1&type=pdf (accessed 28 November 2020).

Augusto, O.B., Bennis, F., and Caro, S. (2012). A new method for decision making in multi-objective optimization problems. *Pesquisa Operacional* 32 (2): 331–369. https://doi.org/10.1590/s0101-74382012005000014.

Babu, P.R., Kumar, M.P.V.V.R., Hemachandra, V.S., and Vanamali, M.P.R. (2010). A novel power flow solution methodology for radial distribution systems. *2010 IEEE Region 8 International Conference on Computational Technologies in Electrical and Electronics Engineering (SIBIRCON)*, pp. 507–512, July 2010. https://doi.org/10.1109/SIBIRCON.2010.5555377.

Baghaee, H.R., Mirsalim, M., Gharehpetian, G.B., and Talebi, H.A. (2018). MOPSO/FDMT-based pareto-optimal solution for coordination of overcurrent relays in interconnected networks and multi-DER microgrids. *IET Generation, Transmission and Distribution* 12 (12): 2871–2886. https://doi.org/10.1049/iet-gtd.2018.0079.

Bakshi, U.A. and Bakshi, V.U. (2010). *Protection of Power System*. Pune: Technical Publications. ISBN 9788184316063.

Bandyopadhyay, M.N. (2006). *Electrical Power Systems: Theory and Practice*. New Delhi: Prentice-Hall of India Pvt. Ltd. ISBN 81-203-2783-7.

Bansal, J.C. and Deep, K. (2008). Optimization of directional overcurrent relay times by particle swarm optimization. *2008 IEEE Swarm Intelligence Symposium*, pp. 1–7, September 2008. https://doi.org/10.1109/SIS.2008.4668290.

Bartz-Beielstein, T. (2006). *Experimental Research in Evolutionary Computation: The New Experimentalism, Natural Computing Series*. Berlin: Springer-Verlag. ISBN 978-3-540-32026-5.

Barzegari, M., Bathaee, S.M.T., and Alizadeh, M. (2010a). Optimal coordination of directional overcurrent relays using harmony search algorithm. *2010 9th International Conference on Environment and Electrical Engineering*, pp. 321–324. https://doi.org/10.1109/EEEIC.2010.5489935.

Barzegari, M., Fard, A.N., Hamidi, M.M., and Shahrood, A.J. (2010b). Optimal coordination of directional overcurrent relays in the presence of distributed generation using FCLs. *2010 IEEE International Energy Conference*, pp. 826–829. https://doi.org/10.1109/ENERGYCON.2010.5771796.

Basler Electric (2000). Instruction Manual for Overcurrent Relay BE1-50/51-B-218, December 2000. https://sertecrelays.net/wp-content/uploads/2019/02/2520_995_RevC.pdf (accessed 15 October 2020).

Baughman, D.R. and Liu, Y.A. (1995). *Neural Networks in Bioprocessing and Chemical Engineering*. San Diego, CA: Academic Press. ISBN 9781483295657.

Bayati, N., Dadkhah, A., Sadeghi, S.H.H. et al. (2017). Considering variations of network topology in optimal relay coordination using time-current-voltage characteristic. *2017 IEEE International Conference on Environment and Electrical Engineering and 2017 IEEE Industrial and Commercial Power Systems Europe (EEEIC / I CPS Europe)*, pp. 1–5. https://doi.org/10.1109/EEEIC.2017.7977810.

Bedekar, P.P. and Bhide, S.R. (2011a). Optimum coordination of directional overcurrent relays using the hybrid GA-NLP approach. *IEEE Transactions on Power Delivery* 26 (1): 109–119. https://doi.org/10.1109/TPWRD.2010.2080289.

Bedekar, P.P. and Bhide, S.R. (2011b). Optimum coordination of overcurrent relay timing using continuous genetic algorithm. *Expert Systems with Applications* 38 (9): 11286–11292. https://doi.org/http://dx.doi.org/10.1016/j.eswa.2011.02.177.

Beder, H., Mohandes, B., El Moursi, M.S. et al. (2020). A new communication-free dual setting protection coordination of microgrid. *IEEE Transactions on Power Delivery* 1. https://doi.org/10.1109/TPWRD.2020.3041753.

Benabid, R., Zellagui, M., Chaghi, A., and Boudour, M. (2013). Optimal coordination of IDMT directional overcurrent relays in the presence of series compensation using differential evolution algorithm. *3rd International Conference on Systems and Control*, pp. 1049–1054. https://doi.org/10.1109/ICoSC.2013.6750984.

Bergen, A.R. (1986). *Power Systems Analysis, Electrical and Computer Engineering*. Englewood Cliffs, NJ: Prentice-Hall Inc. ISBN 0-13-687864-4.

Birla, D. (2014). Use of LINKNET structure for relay pair identification in power networks. *National Conference NCETSDET-2014, Emerging Trends for Sustainable Development in Engineering and Technology*, SS College of Engineering, India, January 2014.

Birla, D., Maheshwari, R.P., and Gupta, H.O. (2004). Novel technique for relay pair identification: a relay coordination requirement. *2004 IIT 13th National Power System Conference*, Indian Institute of Technology, Chennai, December 2004.

Birla, D., Maheshwari, R.P., Gupta, H.O. et al. (2006a). Application of random search technique in directional overcurrent relay coordination. *International Journal of Emerging Electric Power Systems* 7 (1): 1–14. https://doi.org/10.2202/1553-779X.1271.

Birla, D., Maheshwari, R.P., and Gupta, H.O. (2006b). A new nonlinear directional overcurrent relay coordination technique, and banes and boons of near-end faults based approach. *IEEE Transactions on Power Delivery* 21 (3): 1176–1182. https://doi.org/10.1109/TPWRD.2005.861325.

Bisheh, H., Fani, B., and Shahgholian, G. (2021). A novel adaptive protection coordination scheme for radial distribution networks in the presence of distributed generation. *International Transactions on Electrical Energy Systems*. https://doi.org/10.1002/2050-7038.12779.

Blackburn, J.L. and Domin, T.J. (2006). *Protective Relaying: Principles and Applications, Power Engineering*, 3e. Boca Raton, FL: CRC Press. ISBN 1-57444-716-5.

Bougouffa, L. and Chaghi, A. (2019). Dual simplex method for optimal coordination of DOCR's in distribution system with D-FACTS. *The International Journal of Information Science & Technology* 3 (3): 3–9. ISSN 2550-5114.

Bočkarjova, M. and Andersson, G. (2007). Transmission line conductor temperature impact on state estimation accuracy. *2007 IEEE Lausanne Power Tech*, pp. 701–706, July 2007. https://doi.org/10.1109/PCT.2007.4538401.

Braga, A.S. and Saraiva, J.T. (1996). Coordination of overcurrent directional relays in meshed networks using the simplex method. *Proceedings of 8th Mediterranean Electrotechnical Conference on Industrial Applications in Power Systems, Computer Science and Telecommunications (MELECON 96)*, Volume 3, pp. 1535–1538, May 1996. https://doi.org/10.1109/MELCON.1996.551243.

Brown, H.E., Carter, G.K., Happ, H.H., and Person, C.E. (1963). Power flow solution by impedance matrix iterative method. *IEEE Transactions on Power Apparatus and Systems* 82 (65): 1–10. https://doi.org/10.1109/TPAS.1963.291392.

Can, B., Beham, A., and Heavey, C. (2008). A comparative study of genetic algorithm components in simulation-based optimisation. *Simulation Conference, 2008. WSC 2008. Winter*, pp. 1829–1837, December 2008. https://doi.org/10.1109/WSC.2008.4736272.

Cecchi, V., Leger, A.S., Miu, K., and Nwankpa, C.O. (2011). Incorporating temperature variations into transmission-line models. *IEEE Transactions on Power Delivery* 26 (4): 2189–2196. https://doi.org/10.1109/TPWRD.2011.2159520.

Chabanloo, R.M., Abyaneh, H.A., Agheli, A., and Rastegar, H. (2011a). Overcurrent relays coordination considering transient behaviour of fault current limiter and distributed generation in distribution power network. *IET Generation, Transmission and Distribution* 5: 903–911.

Chabanloo, R.M., Abyaneh, H.A., Kamangar, S.S.H., and Razavi, F. (2011b). Optimal combined overcurrent and distance relays coordination incorporating intelligent overcurrent relays characteristic selection. *IEEE Transactions on Power Delivery* 26 (3): 1381–1391. https://doi.org/10.1109/TPWRD.2010.2082574.

Chan, S. and Maurer, R. (1992). Modeling overcurrent relay characteristics. *IEEE Computer Applications in Power* 5 (1): 41–45. https://doi.org/10.1109/67.111471.

Chang, Y.C., Yang, W.T., and Liu, C.C. (1994). Improvement on GGDF for power system security evaluation. *IEE Proceedings - Generation, Transmission and Distribution* 141 (2): 85–88. https://doi.org/10.1049/ip-gtd:19949868.

Chelliah, T.R., Thangaraj, R., Allamsetty, S., and Pant, M. (2014). Coordination of directional overcurrent relays using opposition based chaotic differential evolution algorithm. *International Journal of Electrical Power & Energy Systems* 55: 341–350. https://doi.org/http://dx.doi.org/10.1016/j.ijepes.2013.09.032.

Chen, K., Huang, C., and He, J. (2016). Fault detection, classification and location for transmission lines and distribution systems: a review on the methods. *High Voltage* 1 (1): 25–33. https://doi.org/10.1049/hve.2016.0005.

Chong, E.K.P. and Zak, S.H. (2001). *An Introduction to Optimization, Wiley Series in Discrete Mathematics and Optimization*, 2e. New York: Wiley-Interscience. ISBN 0-471-39126-3.

Chung, C.Y., Tse, C.T., and David, A.K. (1997). New load flow technique based on load transfer and load buses elimination. *Advances in Power System Control, Operation and Management, 1997. APSCOM-97. 4th International Conference on (Conf. Publ. No. 450)*, Volume 2, pp. 614–619, November 1997. https://doi.org/10.1049/cp:19971905.

Cody, M.L. (2006). *Plants on Islands: Diversity and Dynamics on a Continental Archipelago*. Berkeley, CA: University of California Press. ISBN 978-0520247291.

Cole, S. and Belmans, R. (2011). MatDyn, a New MATLAB based toolbox for power system dynamic simulation. *2011 IEEE Power and Energy Society General Meeting*, p. 1. https://doi.org/10.1109/PES.2011.6038913.

Conde, A. and Vázquez, E. (2007). Functional structure for performance improvement of time overcurrent relays. *Electric Power Components & Systems* 35 (3): 261–278. https://doi.org/10.1080/15325000600978635.

Conejo, A.J. (2011). Load flow. https://eecs.wsu.edu/~ee521/Material/20121105%20midterm2/Problem4/01_LoadFlow_R1.pdf (accessed 23 August 2021).

Corrêa, R., Cardoso, G. Jr., de Araújo, O.C.B., and Mariotto, L. (2015). Online coordination of directional overcurrent relays using binary integer programming. *Electric Power Systems Research* 127: 118–125. https://doi.org/http://dx.doi.org/10.1016/j.epsr.2015.05.017.

Costa, M.H., Saldanha, R.R., Ravetti, M.G., and Carrano, E.G. (2017). Robust coordination of directional overcurrent relays using a matheuristic algorithm. *IET Generation, Transmission and Distribution* 11 (2): 464–474. https://doi.org/10.1049/iet-gtd.2016.1010.

Cox, C.B. and Moore, P.D. (1993). *Biogeography: An Ecological and Evolutionary Approach*. Oxford, England, Cambridge, MA: Blackwell Scientific Publications. ISBN 9780632029679.

Damborg, M.J., Ramaswami, R., Venkata, S.S., and Postforoosh, J.M. (1984). Computer aided transmission protection system design Part I: Algorithms. *IEEE Transactions on Power Apparatus and Systems* PAS-103 (1): 51–59. https://doi.org/10.1109/TPAS.1984.318576.

Dantzig, G.B. (1948). Programming in a Linear Structure. *Technical Report of the September 9, 1948 Meeting in Madison*. Econometrica.

Das, J.C. (2002). *Power System Analysis: Short-Circuit Load Flow and Harmonics, Power Engineering*. New York: Marcel Dekker Inc. ISBN 0-8247-0737-0.

Das, D. (2006). *Electrical Power Systems*. New Delhi: New Age International (P) Ltd.. ISBN 978-81-224-2515-4.

Das, J.C. (2012). *ARC Flash Hazard Analysis and Mitigation, IEEE Press Series on Power Engineering*. Hoboken, NJ: Wiley-IEEE Press. ISBN 978-1-118-16381-8.

Deb, K. (2010). *Optimization for Engineering Design: Algorithms and Examples*, 11e. Prentice-Hall of India. ISBN 9788120309432.

Debs, A.S. (1988). *Modern Power Systems Control and Operation*. Boston, FL: Kluwer Academic Publishers. ISBN 0-89838-265-3.

Deep, K. and Bansal, J.C. (2009). Optimization of directional overcurrent relay times using laplace crossover particle swarm optimization (LXPSO). *Nature Biologically Inspired Computing, 2009. NaBIC 2009. World Congress on*, pp. 288–293, December 2009. https://doi.org/10.1109/NABIC.2009.5393722.

Deep, K., Birla, D., Maheshwari, R.P. et al. (2006). A population based heuristic algorithm for optimal relay operating times. *World Journal of Modeling and Simulation* 2 (3): 167–176.

De Rubira, T.T. (2012). Alternative Methods for Solving Power Flow Problems. http://www.stanford.edu/class/cme334/docs/2013-10-29-rubira_power_flow.pdf (accessed 30 July 2021).

Dinh, M.T.N., Bahadornejad, M., Shahri, A.S.A., and Nair, N.K.C. (2013). Protection schemes and fault location methods for multi-terminal lines: a comprehensive review. *2013 IEEE Innovative Smart Grid Technologies-Asia (ISGT Asia)*, pp. 1–6, November 2013. https://doi.org/10.1109/ISGT-Asia.2013.6698781.

Draz, A., Elkholy, M.M., and El-Fergany, A.A. (2021). Soft computing methods for attaining the protective device coordination including renewable energies: review and prospective. *Archives of Computational Methods in Engineering*. https://doi.org/10.1007/s11831-021-09534-5.

Drzadzewski, G. and Wineberg, M. (2006). The importance of scalability when comparing dynamic weighted aggregation and pareto front techniques. In: *Artificial Evolution, Lecture Notes in Computer Science*, vol. 3871 (ed. E.-G. Talbi, P. Liardet, P. Collet et al.), 143–154. Berlin, Heidelberg: Springer-Verlag. ISBN 978-3-540-33589-4. https://doi.org/10.1007/11740698_13.

Dudor, J.S. and Padden, L.K. (1994). Protective relaying on medium and high voltage systems, some lessons to be learned. *Petroleum and Chemical Industry Conference, 1994. Record of Conference Papers., IEEE Incorporated Industry Applications Society 41st Annual*, pp. 53–61, September 1994. https://doi.org/10.1109/PCICON.1994.347632.

Dwarakanath, M.H. and Nowitz, L. (1980). An application of linear graph theory for coordination of directional overcurrent relays. *Proceedings of Electric Power Problems - The Mathematical Challenge, SIAM Meeting*, pp. 104–114.

Eiben, A.E. and Smith, J.E. (2003). *Introduction to Evolutionary Computing, Natural Computing Series*. Berlin, Heidelberg: Springer-Verlag. ISBN 978-3-662-05094-1.

El-Fergany, A.A. and Hasanien, H.M. (2019). Water cycle algorithm for optimal overcurrent relays coordination in electric power systems. *Soft Computing* 23 (23): 12761–12778. https://doi.org/10.1007/s00500-019-03826-6.

El-Hawary, M.E. (1983). *Electrical Power Systems: Design and Analysis*. Reston, VA: Prentice-Hall Inc. ISBN 0-8359-1627-8.

El-Hawary, M.E. (2008). *Introduction to Electrical Power Systems*. Hoboken, NJ: Wiley. ISBN 978-0470-40863-6.

El-Khattam, W. and Sidhu, T.S. (2008). Restoration of directional overcurrent relay coordination in distributed generation systems utilizing fault current limiter. *IEEE Transactions on Power Delivery* 23 (2): 576–585. https://doi.org/10.1109/TPWRD.2008.915778.

El-Khattam, W. and Sidhu, T.S. (2009). Resolving the impact of distributed renewable generation on directional overcurrent relay coordination: a case study. *IET Renewable Power Generation* 3: 415–425.

El-Mesallamy, M., El-Khattam, W., Hassan, A., and Talaat, H. (2013). Coordination of directional overcurrent relays using artificial bee colony. *22nd International Conference and Exhibition on Electricity Distribution (CIRED 2013)*, pp. 1–4, June 2013. https://doi.org/10.1049/cp.2013.0618.

El-Naily, N., Saad, S.M., Hussein, T., and Mohamed, F.A. (2019). Optimal PA novel constraint and non-standard characteristics for optimal over-current relays coordination to enhance microgrid protection scheme. *IET Generation, Transmission and Distribution* 13: 780–793.

Elmitwally, A., Gouda, E., and Eladawy, S. (2015). Optimal application of fault current limiters for assuring overcurrent relays coordination with distributed generations. *Arabian Journal for Science and Engineering* 41 (9): 3381–3397. https://doi.org/10.1007/s13369-015-1917-1.

Elmore, W.A. (1994). *Protective Relaying: Theory And Applications*. New York: ABB - Marcel Dekker Inc.. ISBN 0-8247-9152-5.

Elrafie, H.B. and Irving, M.R. (1993). Linear programming for directional overcurrent relay coordination in interconnected power systems with constraint relaxation. *Electric Power Systems Research* 27 (3): 209–216. https://doi.org/http://dx.doi.org/10.1016/0378-7796(93)90047-I.

Elrayyah, A., Sozer, Y., and Elbuluk, M. (2013). A novel load flow analysis for particle-swarm optimized microgrid power sharing. *Applied Power Electronics Conference and Exposition (APEC), 2013 28th Annual IEEE*, pp. 297–302, March 2013. https://doi.org/10.1109/APEC.2013.6520224.

Elrefaie, H.B., Irving, M.R., and Zitouni, S. (1994). A parallel processing algorithm for co-ordination of directional overcurrent relays in interconnected power systems. *IEE Proceedings - Generation, Transmission and Distribution* 141 (6): 514–520.

Elsadd, M.A., Kawady, T.A., Taalab, A.-M.I., and Elkalashy, N.I. (2021). Adaptive optimum coordination of overcurrent relays for deregulated distribution system considering parallel feeders. *Electrical Engineering*. https://doi.org/10.1007/s00202-020-01187-0.

ElSayed, S.K. and Elattar, E.E. (2021). Hybrid harris hawks optimization with sequential quadratic programming for optimal coordination of directional overcurrent relays incorporating distributed generation. *Alexandria Engineering Journal* 60 (2): 2421–2433. https://doi.org/10.1016/j.aej.2020.12.028.

Enríquez, A.C., Vázquez-Martínez, E., and Altuve-Ferrer, H.J. (2003). Time overcurrent adaptive relay. *International Journal of Electrical Power & Energy Systems* 25 (10): 841–847. https://doi.org/https://doi.org/10.1016/S0142-0615(03)00059-0.

ETAP (2020). Short circuit software. https://etap.com/product/short-circuit-software (accessed 7 November 2020).

ETAP Company (2011). Load Flow Analysis: One Program, One Database, One Solution. Brochure. https://etap.com/docs/default-source/brochures/fact-sheets/load-flow.pdf (accessed 23 August 2021).

Ezzeddine, M. and Kaczmarek, R. (2011a). A new algorithm to avoid the infeasibility of linear programming techniques used to solve the over-current relays coordination problem. *Electric Power Components & Systems* 39 (14): 1596–1607. https://doi.org/10.1080/15325008.2011.596500.

Ezzeddine, M. and Kaczmarek, R. (2011b). A novel method for optimal coordination of directional overcurrent relays considering their available discrete settings and several operation characteristics. *Electric Power Systems Research* 81 (7): 1475–1481. https://doi.org/http://dx.doi.org/10.1016/j.epsr.2011.02.014.

Ezzeddine, M., Kaczmarek, R., and Iftikhar, M.U. (2011). Coordination of directional overcurrent relays using a novel method to select their settings. *IET Generation, Transmission and Distribution* 5: 743–750.

Farag, A., Al-Baiyat, S., and Cheng, T.C. (1995). Economic load dispatch multiobjective optimization procedures using linear programming techniques. *IEEE Transactions on Power Systems* 10 (2): 731–738. https://doi.org/10.1109/59.387910.

Ferreira, L.A.F.M. (1990). Tellegen's theorem and power systems-new load flow equations, new solution methods. *IEEE Transactions on Circuits and Systems* 37 (4): 519–526. https://doi.org/10.1109/31.52753.

Fletcher, R. (1987). *Practical Methods of Optimization*, 2e. Chichester, New York: Wiley-Interscience Publication. Wiley. ISBN 0471915475.

Floreano, D. and Mattiussi, C. (2008). *Bio-Inspired Artificial Intelligence: Theories, Methods, and Technologies, Intelligent Robotics and Autonomous Agents Series*. Cambridge, MA: The MIT Press. ISBN 9780262062718.

Forst, W. and Hoffmann, D. (2010). *Optimization– Theory and Practice*. New York: Springer. ISBN 978-0-387-78977-4.

Frank, S., Sexauer, J., and Mohagheghi, S. (2013). Temperature-dependent power flow. *IEEE Transactions on Power Systems* 28 (4): 4007–4018. https://doi.org/10.1109/TPWRS.2013.2266409.

Gajbhiye, R.K., De, A., Helwade, R., and Soman, S.A. (2005). A simple and efficient approach to determination of minimum set of break point relays for transmission protection system coordination. *2005 International Conference on Future Power Systems*, 5 pp. https://doi.org/10.1109/FPS.2005.204245.

Gajbhiye, R.K., De, A., and Soman, S.A. (2007). Computation of optimal break point set of relays—an integer linear programming approach. *IEEE Transactions on Power Delivery* 22 (4): 2087–2098. https://doi.org/10.1109/TPWRD.2007.905539.

Gandibleux, X., Sevaux, M., Sorensen, K., and T'Kindt, V. (eds.) (2004). *Metaheuristics for Multiobjective Optimisation, Lecture Notes in Economics and Mathematical Systems*, 2004 edition. Springer. ISBN 3-540-20637-X.

Gendreau, M. and Potvin, J.-Y. (2010). *Handbook of Metaheuristics, International Series in Operations Research & Management Science*. Springer. ISBN 9781441916655.

Gers, J.M. and Holmes, E.J. (2004). *Protection of Electricity Distribution Networks, IEE Power & Energy Series*, vol. 47, 2e. London: Institution of Engineering and Technology. ISBN 0-86341-357-9.

Glover, J.D., Sarma, M.S., and Overbye, T.J. (2012). *Power System Analysis and Design*, 5e. Stamford, CT: Cengage Learning. ISBN 978-1-111-42577-7.

Gokhale, S.S. and Kale, V.S. (2014). Application of the firefly algorithm to optimal over-current relay coordination. *2014 International Conference on Optimization of Electrical and Electronic Equipment (OPTIM)*, pp. 150–154, May 2014. https://doi.org/10.1109/OPTIM.2014.6850887.

Gómez, J.C., Tourn, D.H., Nesci, S. et al. (2017). Why the operation failure of high breaking capacity fuses is so frequent? *CIRED - Open Access Proceedings Journal* (1): 1545–1549. https://doi.org/10.1049/oap-cired.2017.0278.

Gonen, T. (2016). *Modern Power System Analysis*. Boca Raton, FL: CRC Press. ISBN 9781466570825.

Gong, W., Cai, Z., Ling, C.X., and Li, H. (2010a). A real-coded biogeography-based optimization with mutation. *Applied Mathematics and Computation* 216 (9): 2749–2758. https://doi.org/http://dx.doi.org/10.1016/j.amc.2010.03.123.

Gong, W., Cai, Z., and Ling, C.X. (2010b). DE/BBO: A hybrid differential evolution with biogeography-based optimization for global numerical optimization. *Soft Computing* 15 (4): 645–665. https://doi.org/10.1007/s00500-010-0591-1.

Grainger, J.J. and Stevenson, W.D. (1994). *Power System Analysis*. New York: McGraw-Hill Education. ISBN 0-07-113338-0.

Grigsby, L.L. and Grigsby, L.L. (2007). *Power System Stability and Control*. Hoboken, NJ: CRC Press. ISBN 9781420009248.

Hadj-Alouane, A.B. and Bean, J.C. (1997). A genetic algorithm for the multiple-choice integer program. *Operations Research* 45 (1): 92–101.

Hase, Y., Khandelwal, T., and Kameda, K. (2020). *Power System Dynamics with Computer-Based Modeling and Analysis*. Hoboken, NJ: Wiley. ISBN 9781119487463.

Hegde, B., Midlam-Mohler, S., and Tulpule, P.J. (2015). Thermal model of fuse dynamics for simulation under intermittent DC faults. *2015 ASME Dynamic Systems and Control Conference (DSCC)*, Volume 2. Columbus, Ohio, USA, October 2015. ISBN 978-0-7918-5725-0. https://doi.org/10.1115/DSCC2015-9815.

Hewitson, L., Brown, M., and Balakrishnan, R. (2004). *Practical Power System Protection*. Newnes, Oxford Burlington, MA: Practical Professional Books. ISBN 0750663979.

Hieber, J.E. (1965). Empirical equations of overcurrent relay curves for computer application. *IEEE Winter Power Meeting*, pp. 65–91.

Hodgkiss, J.W. (1995). Overcurrent protection. In: *Power System Protection: Systems and Methods, Power & Energy Series 905*, vol. 2 (ed. The Electricity Training Association), 1–65. Cambridge: IEE-Cambride University Press. ISBN 978-0-85296-836-9.

Huchel, L. and Zeineldin, H.H. (2016). Planning the coordination of directional overcurrent relays for distribution systems considering DG. *IEEE Transactions on Smart Grid* 7 (3): 1642–1649. https://doi.org/10.1109/TSG.2015.2420711.

Hui, W.A.N. (2006). Protection coordination in power system with distributed generations. PhD thesis. Hung Hom, Kowloon: The Hong Kong Polytechnic University. http://hdl.handle.net/10397/84834 (accessed 17 October 2021).

Hunt, B.J. (2012). Oliver heaviside: a first-rate oddity. *Physics Today* 65 (11): 48. https://doi.org/10.1063/PT.3.1788.

Husain, A. (2007). *Electrical Power Systems*, 5e. New Delhi: CBS Publishers & Distributors. ISBN 8123914482.

IEC (1989). *Single Input Energizing Quality Measuring Relays With Dependent or Independent*. IEC Publication 255-3.

IEEE IEEE Std. 399-1997 (1998). *IEEE Recommended Practice for Industrial and Commercial Power Systems Analysis (The IEEE Brown Book)*. New York: Institute of Electrical & Electronics Engineers. ISBN 1-55937-968-5.

IEEE IEEE Std. C37.112-1996 (1999). *IEEE Standard Inverse-Time Characteristic Equations for Overcurrent Relays*. Volume 14, issue (3), pp. 868–872, July 1999. ISSN 0885-8977. https://doi.org/10.1109/61.772326.

IEEE IEEE Std C37.2-2008 (2008a). *IEEE Standard Electrical Power System Device Function Numbers, Acronyms, and Contact Designations*. IEEE Std C37.2-2008 (Revision of IEEE Std C37.2-1996), pp. 1–48. https://doi.org/10.1109/IEEESTD.2008.4639522.

IEEE IEEE Std C37.2-2008 (2008b). *IEEE Standard Electrical Power System Device Function Numbers, Acronyms, and Contact Designations*. (Revision of IEEE Std C37.2-1996), pp. 1–48. https://doi.org/10.1109/IEEESTD.2008.4639522.

Igelnik, B. and Simon, D. (2011). The eigenvalues of a tridiagonal matrix in biogeography. *Applied Mathematics and Computation* 218 (1): 195–201. https://doi.org/http://dx.doi.org/10.1016/j.amc.2011.05.054.

Ilic-Spong, M. and Zaborszky, J. (1982). A different approach to load flow. *IEEE Transactions on Power Apparatus and Systems* PAS-101 (1): 168–179. https://doi.org/10.1109/TPAS.1982.317334.

Illingworth, W.T. (1989). Beginner's guide to neural networks. *Proceedings of the IEEE National Aerospace and Electronics Conference*, pp. 1138–1144, Volume 3, May 1989. https://doi.org/10.1109/NAECON.1989.40352.

ISA (2011). User Manual: T1000 Plus Application Guide, September 2011. http://userequip.com/files/specs/5690/MIE91093-T1000%20PLUS_APPLICATION_GUIDE.pdf (accessed 15 October 2020).

Jalilzadeh Hamidi, R., Ahmadian, A., Patil, R., and Asadinejad, A. (2019). Optimal time-current graded coordination of multistage inverse-time overcurrent relays in distribution networks. *International Transactions on Electrical Energy Systems* 29 (5): e2841. https://doi.org/https://doi.org/10.1002/2050-7038.2841. ITEES-18-0637.R1.

Jamali, S. and Borhani-Bahabadi, H. (2017). Recloser time-current-voltage characteristic for fuse saving in distribution networks with DG. *IET Generation, Transmission and Distribution* 11 (7): 272–279. https://digital-library.theiet.org/content/journals/10.1049/iet-gtd.2016.0979.

Jenkins, L., Khincha, H.P., Shivakumar, S., and Dash, P.K. (1992). An application of functional dependencies to the topological analysis of protection schemes. *IEEE Transactions on Power Delivery* 7 (1): 77–83. https://doi.org/10.1109/61.108892.

Jin, Y. (2002). Effectiveness of Weighted Aggregation of Objectives for Evolutionary Multiobjective Optimization: Methods, Analysis and Applications. Unpublished Manuscript. http://www.soft-computing.de/edwa2002.pdf (Accessed 17 October 2021).

Johns, A.T. and Salman, S.K. (1995). *Digital Protection for Power Systems, IEE Power Series 15*. London: P. Peregrinus on behalf of the Institution of Electrical Engineers. ISBN 0-86341-303-X.

Johnson, R.A. (2000). *Miller and Freund's Probability and Statistics for Engineers*, 6e. Upper Saddle River, NJ: Prentice Hall Inc. ISBN 0-13-014158-5.

Jones, R.L. (1980). *Biogeography: Structure, Process, Pattern, and Change Within the Biosphere*. Hulton Educational Publications Limited. ISBN 9780717508723.

Kajee-Bagdadi, Z. (2007). Differential evolution algorithms for constrained global optimization. Master's thesis. Johannesburg, South Africa: University of the Witwatersrand. http://wiredspace.wits.ac.za/bitstream/handle/10539/4733/Thesis.pdf (accessed 17 October 2021).

Karegar, H.K., Abyaneh, H.A., Ohis, V., and Meshkin, M. (2005). Pre-processing of the optimal coordination of overcurrent relays. *Electric Power Systems Research* 75 (2–3): 134–141. https://doi.org/http://dx.doi.org/10.1016/j.epsr.2005.02.005.

Karthikeyan, V., Senthilkumar, S., and Vijayalakshmi, V.J. (2013). A new approach to the solution of economic dispatch using particle swarm optimization with simulated annealing. *International Journal on Computational Sciences & Applications (IJCSA)* 3 (3): 37–49. https://doi.org/10.5121/ijcsa.2013.3304 37.

Khodr, H.M., Ocque, L., Yusta, J.M., and Rosa, M.A. (2006). New load flow method S-E oriented for large radial distribution networks. *2006 IEEE/PES Transmission Distribution Conference and Exposition: Latin America*, pp. 1–6, August 2006. https://doi.org/10.1109/TDCLA.2006.311490.

Khurshaid, T., Wadood, A., Farkoush, S.G. et al. (2019). An improved optimal solution for the directional overcurrent relays coordination using hybridized whale optimization algorithm in complex power systems. *IEEE Access* 7: 90418–90435. https://doi.org/10.1109/ACCESS.2019.2925822.

Kılıçkıran, H.C., Akdemir, H., Şengör, I. et al. (2018). A non-standard characteristic based protection scheme for distribution networks. *Energies* 11 (5). https://doi.org/10.3390/en11051241.

Kimbark, E.W. (1995). *Power System Stability, IEEE Press Power Systems Engineering Series*. New York: IEEE Press. ISBN 9780780311350.

Kirtley, J.L. (2010). *Electric Power Principles: Sources, Conversion, Distribution and Use*. Wiley. ISBN 978-0-470-68636-2.

Knable, A.H. (1961). A computer approach to setting directional over-current relays on industrial network type systems. *IEEE CP-61-933*.

Knable, A.H. (1969). A standardized approach to relay coordination. *IEEE Winter Power Meeting, 69 CP 58*.

Knoll, D.A. and Keyes, D.E. (2004). Jacobian-free Newton–Krylov methods: a survey of approaches and applications. *Journal of Computational Physics* 193 (2): 357–397. https://doi.org/http://dx.doi.org/10.1016/j.jcp.2003.08.010.

Korashy, A., Kamel, S., Alquthami, T., and Jurado, F. (2020). Optimal coordination of standard and non-standard direction overcurrent relays using an improved moth-flame optimization. *IEEE Access* 8: 87378–87392. https://doi.org/10.1109/ACCESS.2020.2992566.

Kothari, D.P. and Dhillon, J.S. (2010). *Power System Optimization*, 2nd revised edition. New Delhi: PHI Learning. ISBN 978-8120340855.

Kothari, D.P. and Dhillon, J.S. (2011). *Power System Optimization*, 2e. New Delhi: PHI Learning Private Limited. ISBN 9788120340855.

Kothari, D.P. and Nagrath, I.J. (2003). *Modern Power System Analysis*. New Delhi: Tata McGraw-Hill Publishing Company. ISBN 9780070494893.

Kumar, R.J.R. and Rao, P.S.N. (2014). A new successive displacement type load flow algorithm and its application to radial systems. *2014 IEEE 2nd International Conference on Electrical Energy Systems (ICEES)*, pp. 15–19, January 2014. https://doi.org/10.1109/ICEES.2014.6924134.

Kundur, P. (1994). *Power System Stability and Control*, 1e. New York: McGraw-Hill Education. ISBN 9780070359581.

Kutner, M.H., Nachtsheim, C.J., Neter, J., and Li, W. (2004). *Applied Linear Statistical Models, McGraw-Hill/Irwin Series Operations and Decision Sciences*, 5e. McGraw-Hill/Irwin. ISBN 9780072386882.

Labrador, A. (2018). Coordination of distance and overcurrent relays using a mathematical optimization technique. Master's thesis. Londrina, Paraná, Brazil: Universidade Estadual de Londrina. http://www.uel.br/pos/meel/disserta/2018-Angel%20Labrador.pdf (Accessed 13 October 2020).

Lee, Lee and El-Sharkawi, El-Sharkawi (eds.) (2008). *Modern Heuristic Optimization Techniques: Theory and Applications to Power Systems, IEEE Press Series on Power Engineering*. Piscataway, NJ, Hoboken, NJ: IEEE Press Wiley-Interscience. ISBN 978-0471-45711-4.

Leger, A.St. and Nwankpa, C. (2010). OTA-based transmission line model with variable parameters for analog power flow computation. *International Journal of Circuit Theory and Applications* 38 (2): 199–220. https://doi.org/10.1002/cta.555.

Leondes, C.T. (1998). *Optimization Techniques, Neural Network Systems Techniques and Applications*, Pt. 2, vol. 2. San Diego, CA: Academic Press. ISBN 0-12-443862-8.

Liu, L. and Fu, L. (2017). Minimum breakpoint set determination for directional overcurrent relay coordination in large-scale power networks via matrix computations. *IEEE Transactions on Power Delivery* 32 (4): 1784–1789. https://doi.org/10.1109/TPWRD.2016.2583222.

Liu, A. and Yang, M.-T. (2012). A new hybrid nelder-mead particle swarm optimization for coordination optimization of directional overcurrent relays. *Mathematical Problems in Engineering* 1–18. https://doi.org/10.1155/2012/456047.

Lohokare, M.R., Pattnaik, S.S., Panigrahi, B.K., and Das, S. (2013). Accelerated biogeography-based optimization with neighborhood search for optimization. *Applied Soft Computing* 13 (5): 2318–2342. https://doi.org/http://dx.doi.org/10.1016/j.asoc.2013.01.020.

Lomolino, M.V., Riddle, B.R., and Brown, J.H. (2009). *Biogeography*, 3e. Sunderland, MA: Sinauer Associates, Incorporated. ISBN 0878934863.

Losos, J.B. and Ricklefs, R.E. (2010). *The Theory of Island Biogeography Revisited*. Princeton, NJ: Princeton University Press. ISBN 9781400831920.

Luke, S. (2009). *Essentials of Metaheuristics*. Lulu. Available for free at http://cs.gmu.edu/sean/book/metaheuristics/.

Powell, L. (2005). *Power System Load Flow Analysis*. New York: McGraw-Hill Inc. ISBN 978-0-07-178261-6.

Ma, H. and Simon, D. (2011). Blended biogeography-based optimization for constrained optimization. *Engineering Applications of Artificial Intelligence* 24: 517–525. https://doi.org/http://dx.doi.org/10.1016/j.engappai.2010.08.005.

MacArthur, R.H. (1972). *Geographical Ecology: Patterns in the Distribution of Species*. New York: Harper & Row Publishers Inc..

MacArthur, R.H. and Connell, J.H. (1966). *The Biology of Populations*. New York: Wiley.

MacArthur, R.H. and Wilson, E.O. (1963). An equilibrium theory of insular zoogeography. *Evolution* 17 (4): 373–387. https://doi.org/10.2307/2407089.

MacArthur, R.H. and Wilson, E.O. (1967). *The Theory of Island Biogeography, Landmarks in Biology Series*. Princeton University Press. ISBN 9780691088365.

MacDonald, G.M. (2003). *Biogeography: Introduction to Space, Time, and Life*. New York: Wiley. ISBN 9780471241935.

Madureira, V.S. and Vieira, T.C. (2018). Coordination of inverse-time overcurrent relays with fuses using genetic algorithm. *2018 Simposio Brasileiro de Sistemas Eletricos (SBSE)*, pp. 1–6. https://doi.org/10.1109/SBSE.2018.8395627.

Mahindara, V.R., Rodriguez, D.F.C., Pujiantara, M. et al. (2021). Practical challenges of inverse and definite-time overcurrent protection coordination in modern industrial and commercial power distribution system. *IEEE Transactions on Industry Applications* 57 (1): 187–197. https://doi.org/10.1109/TIA.2020.3030564.

Mahor, A., Prasad, V., and Rangnekar, S. (2009). Economic dispatch using particle swarm optimization: a review. *Renewable and Sustainable Energy Reviews* 13 (8): 2134–2141. https://doi.org/http://dx.doi.org/10.1016/j.rser.2009.03.007.

Mancer, N., Mahdad, B., and Srairi, K. (2015). Optimal coordination of directional overcurrent relays using PSO-TVAC considering series compensation. *Advances in Electrical and Electronic Engineering* 13 (2). https://doi.org/10.15598/aeee.v13i2.1178.

Mansour, M.M., Mekhamer, S.F., and El-Kharbawe, N.E.-S. (2007). A modified particle swarm optimizer for the coordination of directional overcurrent relays. *IEEE Transactions on Power Delivery* 22 (3): 1400–1410. https://doi.org/10.1109/TPWRD.2007.899259.

Marriott, K. and Stuckey, P. (1998). *Programming with Constraints: An Introduction*. The MIT Press. ISBN 0262133415.

Mason, C.R. (1956). *The Art and Science of Protective Relaying, General Electric Series*. New York: Wiley, Chapman & Hall. ISBN 0471575526.

Mathur, A., Pant, V., and Das, B. (2015). Unsymmetrical short-circuit analysis for distribution system considering loads. *International Journal of Electrical Power & Energy Systems* 70: 27–38. https://doi.org/10.1016/j.ijepes.2015.02.003.

Matthews, R.C., Reno, M.J., and Summers, A.K. (2019). A graph theory method for identification of a minimum breakpoint set for directional relay coordination. *Electronics* 8 (12): 1376. https://doi.org/10.3390/electronics8121376.

Mbamalu, G.A.N., El-Hawary, M.E., and El-Hawary, F.E.R.I.A.L. (1995). A pseudo-inverse-based probabilistic power flow approach. *Electric Machines & Power Systems* 23 (2): 107–118. https://doi.org/10.1080/07313569508955611.

McCleer, P.J. (1987). *The Theory and Practice of Overcurrent Protection*. Jackson, MI: Mechanical Products Inc. ISBN 0-9618814-0-2.

Meliopoulos, A.P.S., Kang, S.W., Cokkinides, G.J., and Dougal, R. (2003). Animation and visualization of spot prices via quadratized power flow analysis. *Proceedings of the 36th Annual Hawaii International Conference on System Sciences, 2003*, 7 pp. January 2003. https://doi.org/10.1109/HICSS.2003.1173844.

Michalewicz, Z. (1995). A survey of constraint handling techniques in evolutionary computation methods. *Proceedings of the 4th Annual Conference on Evolutionary Programming*, pp. 135–155. MIT Press.

Milano, F. (2005). An open source power system analysis toolbox. *IEEE Transactions on Power Systems* 20 (3): 1199–1206. https://doi.org/10.1109/TPWRS.2005.851911.

Mohammadi, R., Abyaneh, H., Razavi, F. et al. (2010). Optimal relays coordination efficient method in interconnected power systems. *Journal of Electrical Engineering* 61 (2): 75–83. https://doi.org/10.2478/v10187-010-0011-x.

Mohammadi, R., Abyaneh, H.A., Rudsari, H.M. et al. (2011). Overcurrent relays coordination considering the priority of constraints. *IEEE Transactions on Power Delivery* 26 (3): 1927–1938. https://doi.org/10.1109/TPWRD.2011.2123117.

Moirangthem, J., Krishnanand, K.R., Dash, S.S., and Ramaswami, R. (2013). Adaptive differential evolution algorithm for solving non-linear coordination problem of directional overcurrent relays. *IET Generation, Transmission and Distribution* 7 (4): 329–336. https://doi.org/10.1049/iet-gtd.2012.0110.

Momoh, J.A. (2001a). *Electric Power System Applications of Optimization, Power Engineering (Willis)*. New York: Marcel Dekker, Inc. ISBN 0-8247-9105-3.

Momoh, J.A. (2001b). *Electric Power System Applications of Optimization, Power Engineering*. New York: Marcel Dekker Inc. ISBN 0-8247-9105-3.

Monadi, M., Koch-Ciobotaru, C., Luna, A. et al. (2016). Multi-terminal medium voltage DC grids fault location and isolation. *IET Generation, Transmission and Distribution* 10 (14): 3517–3528. https://doi.org/10.1049/iet-gtd.2016.0183.

Moravej, Z., Jazaeri, M., and Gholamzadeh, M. (2012). Optimal coordination of distance and over-current relays in series compensated systems based on MAPSO. *Energy Conversion and Management* 56: 140–151. https://doi.org/10.1016/j.enconman.2011.11.024.

Murthy, P.S.R. (2014). *Power System Analysis*. Hyderabad: BS Pub. ISBN 978-81-7800-161-6.

Myers, A.A. and Giller, P.S. (eds.) (1990). *Analytical Biogeography: An Integrated Approach to the Study of Animal and Plant Distributions*, 1e. London: Chapman and Hall. ISBN 9780412400506.

Mysore, P. (2010). Power System Protection Lectures - Lecture 1 (Fundamentals of Protection), August 2010. https://cusp.umn.edu/power-system-protection-videos (accessed 17 October 2021).

Nasar, S.A. (1990). *Schaum's Outline of Electrical Power Systems*. New York: McGraw-Hill Inc. ISBN 0-07-045917-7.

NEOS (2012). Bound Constrained Optimization, June 2012. https://neos-guide.org/content/bound-constrained-optimization (accessed 17 October 2021).

Nierenberg, W.A. (ed.) (1995). *Encyclopedia of Environmental Biology, Three-Volume Set: 1–3*, 1e. Academic Press. ISBN 9780122267307.

Nocedal, J. and Wright, S. (2006). *Numerical Optimization, Operations Research and Financial Engineering*, 2e. New York: Springer. ISBN 9780387303031.

Noghabi, A.S., Sadeh, J., and Mashhadi, H.R. (2009). Considering different network topologies in optimal overcurrent relay coordination using a hybrid GA. *IEEE Transactions on Power Delivery* 24 (4): 1857–1863. https://doi.org/10.1109/TPWRD.2009.2029057.

Noghabi, A.S., Mashhadi, H.R., and Sadeh, J. (2010). Optimal coordination of directional overcurrent relays considering different network topologies using interval linear programming. *IEEE Transactions on Power Delivery* 25 (3): 1348–1354. https://doi.org/10.1109/TPWRD.2010.2041560.

Ou, T.C., Tsao, T.P., Lin, W.M. et al. (2013). A novel power flow analysis for microgrid distribution system. *2013 IEEE 8th Conference on Industrial Electronics and Applications (ICIEA)*, pp. 1550–1555, June 2013. https://doi.org/10.1109/ICIEA.2013.6566614.

Padiyar, K.R. (2008). *Power System Dynamics: Stability and Control*, 2e. Hyderabad: BSP, BS Publications. ISBN 81-7800-024-5.

Paithankar, Y.G. and Bhide, S.R. (2003). *Fundamentals of Power System Protection*. New Delhi: Prentice-Hall of India Pvt. Ltd. ISBN 81-203-2194-4.

Papaspiliotopoulos, V.A., Kurashvili, T.A., and Korres, G.N. (2014). Optimal coordination of directional overcurrent relays in distribution systems with distributed generation based on a hybrid PSO-LP algorithm. *MedPower 2014*, pp. 1–6, November 2014. https://doi.org/10.1049/cp.2014.1697.

Papaspiliotopoulos, V.A., Korres, G.N., and Maratos, N.G. (2017a). A novel quadratically constrained quadratic programming method for optimal coordination of directional overcurrent relays. *IEEE Transactions on Power Delivery* 32 (1): 3–10. https://doi.org/10.1109/TPWRD.2015.2455015.

Papaspiliotopoulos, V.A., Korres, G.N., Kleftakis, V.A., and Hatziargyriou, N.D. (2017b). Hardware-in-the-loop design and optimal setting of adaptive protection schemes for distribution systems with distributed generation. *IEEE Transactions on Power Delivery* 32 (1): 393–400. https://doi.org/10.1109/TPWRD.2015.2509784.

Pattnaik, S.S., Lohokare, M.R., and Devi, S. (2010). Enhanced biogeography-based optimization using modified clear duplicate operator. *2010 2nd World Congress on Nature and Biologically Inspired Computing (NaBIC)*, pp. 715–720, December 2010. https://doi.org/10.1109/NABIC.2010.5716322.

Perez, L.G. and Urdaneta, A.J. (2001). Optimal computation of distance relays second zone timing in a mixed protection scheme with directional overcurrent relays. *IEEE Transactions on Power Delivery* 16 (3): 385–388. https://doi.org/10.1109/61.924815.

Phadke, A.G., Horowitz, S.H., and Thorp, J.S. (1989). Integrated hierarchical computer systems for adaptive protective relaying and control of electric transmission power systems, November 1989. https://www.osti.gov/servlets/purl/5382017 (accessed 17 October 2021).

Pierre, C.R.S. and Wolny, T.E. (1986). Standardization of benchmarks for protective device time-current curves. *IEEE Transactions on Industry Applications* IA-22 (4): 623–633. https://doi.org/10.1109/TIA.1986.4504772.

Radke, G.E. (1963). A method for calculating time-overcurrent relay settings by digital computer. *IEEE Transactions on Power Apparatus and Systems* PAS-82 (Special Supplement): 189–205.

Ralhan, S. and Ray, S. (2013). Directional overcurrent relays coordination using linear programming intervals: a comparative analysis. *2013 Annual IEEE India Conference (INDICON)*, pp. 1–6, December 2013. https://doi.org/10.1109/INDCON.2013.6725883.

Ram, B. and Vishwakarma, D.N. (1995). *Power System Protection and Switchgear*. New Delhi: Tata McGraw-Hill. ISBN 0-07-462350-8.

Ramaswami, R., Venkata, S.S., Damborg, M.J., and Postforoosh, J.M. (1984). Computer aided transmission protection system design Part II: Implementation and results. *IEEE Transactions on Power Apparatus and Systems* PAS-103 (1): 60–65.

Ramaswami, R., Damborg, M.J., and Venkata, S.S. (1990). Coordination of directional overcurrent relays in transmission systems-a subsystem approach. *IEEE Transactions on Power Delivery* 5 (1): 64–71. https://doi.org/10.1109/61.107257.

Rao, S.S. (2008). *Switchgear Protection and Power Systems: Theory, Practice & Solved Problems*, 13e. Delhi: Khanna Publishers. ISBN 81-7409-232-3.

Rao, S.S. (2009). *Engineering Optimization: Theory and Practice*, 4e. Hoboken, NJ: Wiley. ISBN 978-0-470-18352-6.

Razavi, F., Abyaneh, H.A., Al-Dabbagh, M. et al. (2008). A new comprehensive genetic algorithm method for optimal overcurrent relays coordination. *Electric Power Systems Research* 78 (4): 713–720. https://doi.org/https://doi.org/10.1016/j.epsr.2007.05.013.

Rebizant, W., Szafran, J., and Wiszniewski, A. (2011). *Digital Signal Processing in Power System Protection and Control, Signals and Communication Technology*. London: Springer. ISBN 978-0-85729-802-7.

Rivas, A.E.L., Pareja, L.A.G., and Abr ao, T. (2019). Coordination of distance and directional overcurrent relays using an extended continuous domain ACO algorithm and an hybrid ACO algorithm. *Electric Power Systems Research* 170: 259–272.

Roy, S., Babu, P.S., and Babu, N.V.P. (2017). Optimal combined overcurrent and distance relays coordination using teaching learning based optimization. *2017 14th IEEE India Council International Conference (INDICON)*, pp. 1–6. https://doi.org/10.1109/INDICON.2017.8487876.

Rudin, C. (2019). Stop explaining black box machine learning models for high stakes decisions and use interpretable models instead. *Nature Machine Intelligence* 1 (5): 206–215. https://doi.org/10.1038/s42256-019-0048-x.

Saadat, H. (1999). *Power System Analysis, McGraw-Hill Series in Electrical and Computer Engineering*. Singapore: WCB/McGraw-Hill. ISBN 9780071167581.

Saberi Noughabi, A., Badrsimaei, H., and Farshad, M. (2017). A probabilistic method to determine the optimal setting of combined overcurrent relays considering uncertainties. *Tabriz Journal of Electrical Engineering* 47 (1): 141–153.

Sachdev, M.S., Fleming, R., and Singh, J. (1978). Mathematical models representing time-current characteristics of overcurrent relays for computer applications. *Proceedings of IEEE PES - Winter Meeting*, pp. 1–8, January 1978.

Saha, M.M., Izykowski, J.J., and Rosolowski, E. (2010). *Fault Location on Power Networks (Power Systems)*. London: Springer. ISBN 978-1-84882-886-5.

Saleh, K.A., Zeineldin, H.H., Al-Hinai, A., and El-Saadany, E.F. (2015). Optimal coordination of directional overcurrent relays using a new time-current-voltage characteristic. *IEEE Transactions on Power Delivery* 30 (2): 537–544. https://doi.org/10.1109/TPWRD.2014.2341666.

Santos, J.R., Expósito, A.G., and Sáanchez, F.P. (2007). Assessment of conductor thermal models for grid studies. *IET Generation, Transmission and Distribution* 1 (1): 155–161. https://doi.org/10.1049/iet-gtd:20050472.

Schneider Electric (2015). Overcurrent Relay (VAMP 11F), July 2015. https://m.vamp.fi/dmsdocument/312 (accessed 13 October 2020).

Schneider Electric (2017a). Sepam Series 20, February 2017a. https://download.schneider-electric.com/files?p_Doc_Ref=PCRED301005EN (accessed 13 October 2020).

Schneider Electric. (2017b). Easergy P3: Universal Relays P3U10, P3U20 and P3U30, October 2017b. http://www.rza.by/upload/iblock/058/Easergy%20P3Ux0_User%20Manual_P3U_EN_M_B001_26-10-2017.pdf (accessed 16 October 2020).

Schrijver, A. (1998). *Theory of Linear and Integer Programming*. Chichester, New York: Wiley. ISBN 978-0-471-98232-6.

SEL (2013). SEL-421 Relay: Protection and Automation System, June 2013. (accessed 15 October 2020).

Shabani, M. and Karimi, A. (2018). A robust approach for coordination of directional overcurrent relays in active radial and meshed distribution networks considering uncertainties. *International Transactions on Electrical Energy Systems* 28 (5): e2532. https://doi.org/10.1002/etep.2532.

Shah, P.A., Nanoty, A.S., and Rajput, V.N. (2016). Comparative analysis of different optimization methods for optimal coordination of directional overcurrent relays. *2016 International Conference on Electrical, Electronics, and Optimization Techniques (ICEEOT)*, pp. 3927–3931, March 2016. https://doi.org/10.1109/ICEEOT.2016.7755451.

Sharaf, H.M., Zeineldin, H.H., Ibrahim, D.K., and El Zahab, E.E.D.A. (2014). Protection coordination of directional overcurrent relays considering fault current direction. *IEEE PES Innovative Smart Grid Technologies, Europe*, pp. 1–5. https://doi.org/10.1109/ISGTEurope.2014.7028793.

Shirmohammadi, D., Hong, H.W., Semlyen, A., and Luo, G.X. (1988). A compensation-based power flow method for weakly meshed distribution and transmission networks. *IEEE Transactions on Power Systems* 3 (2): 753–762. https://doi.org/10.1109/59.192932.

Short, T. (2007). Fault Location on Distribution Systems: An Update on EPRI and DOE Research. *Technical Report P128.003*. Headquarter: Palo Alto, CA, United States: Electric Power Research Institute (EPRI). http://grouper.ieee.org/groups/td/dist/presentations/2007-01-short.pdf (accessed 30 July 2021).

SIEMENS (n.d.). LV HRC Fuses: DIN Type. https://assets.new.siemens.com/siemens/assets/api/uuid:e689f97d-a567-4e3d-b9b0-fedebf129454/3na-hrc-din-fuse.pdf (Accessed 28 November 2020).

Sierksma, G. (2002). *Linear and Integer Programming: Theory and Practice*, 2e. New York: CRC Press. ISBN 978-0-8247-0673-9.

Simon, D. (2008a). Biogeography-based optimization. *IEEE Transactions on Evolutionary Computation* 12 (6): 702–713. https://doi.org/10.1109/TEVC.2008.919004.

Simon, D. (2008b). The Matlab Code of Biogeography-Based Optimization, August 2008b. http://academic.csuohio.edu/simond/bbo/. (accessed 17 October 2021).

Simon, D. (2011). A probabilistic analysis of a simplified biogeography-based optimization algorithm. *Evolutionary Computation* 19 (2): 167–188. https://doi.org/10.1162/EVCO_a_00018.

Simon, D. (2013). *Evolutionary Optimization Algorithms: Biologically-Inspired and Population-Based Approaches to Computer Intelligence*. Hoboken, NJ: Wiley. ISBN 978-0-470-93741-9.

Simon, D., Ergezer, M., and Du, D. (2009). Population distributions in biogeography-based optimization algorithms with elitism. *IEEE International Conference on Systems, Man and Cybernetics, 2009. SMC 2009*, pp. 991–996, October 2009. https://doi.org/10.1109/ICSMC.2009.5346058.

Singh, M. (2014). Non-adaptive over current relay coordination in distribution systems. *2014 IEEE International Conference on Power Electronics, Drives and Energy Systems (PEDES)*, pp. 1–6, December 2014. https://doi.org/10.1109/PEDES.2014.7041952.

Singh, J., Sachdev, M.S., Fleming, R.J., and Krause, A. (1980). Digital IDMT directional overcurrent relays. *Developments in Power System Protection IEE Conference Publication No. 185*, pp. 152–157, April 1980.

Singh, M., Panigrahi, B.K., and Abhyankar, A.R. (2011). Optimal overcurrent relay coordination in distribution system. *2011 International Conference on Energy, Automation and Signal*, pp. 1–6, December 2011. https://doi.org/10.1109/ICEAS.2011.6147214.

Singh, M., Panigrahi, B.K., and Abhyankar, A.R. (2012a). Combined optimal distance to overcurrent relay coordination. *2012 IEEE International Conference on Power Electronics, Drives and Energy Systems (PEDES)*, pp. 1–6. https://doi.org/10.1109/PEDES.2012.6484300.

Singh, M., Panigrahi, B.K., and Mukherjee, R. (2012b). Optimum coordination of overcurrent relays using CMA-ES algorithm. *2012 IEEE International Conference on Power Electronics, Drives and Energy Systems (PEDES)*, pp. 1–6. https://doi.org/10.1109/PEDES.2012.6484275.

Singh, M., Panigrahi, B.K., and Abhyankar, A.R. (2013). Optimal coordination of directional over-current relays using teaching learning-based optimization (TLBO) algorithm. *International Journal of Electrical Power & Energy Systems* 50: 33–41. https://doi.org/http://dx.doi.org/10.1016/j.ijepes.2013.02.011.

Singh, M., Telukunta, V., and Srivani, S.G. (2018). Enhanced real time coordination of distance and user defined over current relays. *International Journal of Electrical Power & Energy Systems* 98: 430–441. https://doi.org/https://doi.org/10.1016/j.ijepes.2017.12.018.

Sivanagaraju, S. and Sreenivasan, G. (2010). *Power System Operation and Control*. Chennai: Pearson Education. ISBN 978-81-317-2662-4.

Sivananaithaperumal, S., Amali, S.M.J., Baskar, S., and Suganthan, P.N. (2011). Constrained self-adaptive differential evolution based design of robust optimal fixed structure controller. *Engineering Applications of Artificial Intelligence* 24 (6): 1084–1093. https://doi.org/http://dx.doi.org/10.1016/j.engappai.2011.05.003.

Smith, K. (2015). Modeling Relion Thermal Overload Curves in Engineering Software, October 2015. https://search.abb.com/library/Download.aspx?DocumentID=1MAC004548&LanguageCode=en&DocumentPartId=&Action=Launch (accessed 22 October 2020).

Smith, K. and Jain, S. (2017). The necessity and challenges of modeling and coordinating microprocessor based thermal overload functions for device protection. *2017 70th Annual Conference for Protective Relay Engineers (CPRE)*, pp. 1–22. https://doi.org/10.1109/CPRE.2017.8090053.

So, C.W. and Li, K.K. (2000). Time coordination method for power system protection by evolutionary algorithm. *IEEE Transactions on Industry Applications* 36 (5): 1235–1240. https://doi.org/10.1109/28.871269.

Soares, A.H.M. (2009). Metodologia Computacional Para Coordenaç ao Automática de Dispositivos de Proteç ao Contra Sobrecorrente em Sistemas Elétricos Industriais. Master's thesis. S ao Carlos, SP, Brasil: Escola de Engenharia de S ao Carlos. https://www.teses.usp.br/teses/disponiveis/18/18154/tde-17112009-090033/publico/Mineiro.pdf (accessed 28 November 2020).

Soman, S.A. (2010). Power System Protection, January 2010. http://rkgitw.ac.in/nptel/108101039/Power%20System%20Protection/Course%20Objective.html (accessed 28 October 2020).

Soria, O.A., Enríquez, A.C., and Guajardo, L.A.T. (2014). Overcurrent relay with unconventional curves and its application in industrial power systems. *Electric Power Systems Research* 110: 113–121. https://doi.org/https://doi.org/10.1016/j.epsr.2013.12.012.

Stagg, G.W. and El-Abiad, A.H. (1968). *Computer Methods in Power System Analysis*. Tokyo: McGraw-Hill Inc. ISBN 0-07-085764-4.

Storn, R. (1996). On the usage of differential evolution for function optimization. *Fuzzy Information Processing Society, 1996. NAFIPS., Biennial Conference of the North American*, pp. 519–523, June 1996. https://doi.org/10.1109/NAFIPS.1996.534789.

Storn, R. and Price, K. (1947). Differential Evolution - A Simple and Efficient Adaptive Scheme for Global Optimization over Continuous Spaces. *Technical Report TR-95-012*. Center Street, Berkeley, CA 94704-1198, Suite 600, Fax: 510-643-7684: International Computer Science Institute, March 1995. http://www1.icsi.berkeley.edu/storn/litera.html (accessed 31 July 2021).

Sun, Y., Liu, D., He, G., and Mei, S. (2009). A power flow algorithm with three-order convergence rate. *2009 Asia-Pacific Power and Energy Engineering Conference*, pp. 1–4, March 2009. https://doi.org/10.1109/APPEEC.2009.4918107.

Talbi, E.-G. (2009). *Metaheuristics: From Design to Implementation, Wiley Series on Parallel and Distributed Computing*. Hoboken, NJ: Wiley. ISBN 9780470278581.

Tejeswini, M.V. and Sujatha, B.C. (2017). Optimal protection coordination of voltage-current time based inverse relay for PV based distribution system. *2017 Second International Conference on Electrical, Computer and Communication Technologies (ICECCT)*, pp. 1–7. https://doi.org/10.1109/ICECCT.2017.8118006.

Thakur, M. and Kumar, A. (2016). Optimal coordination of directional over current relays using a modified real coded genetic algorithm: a comparative study. *International Journal of Electrical Power & Energy Systems* 82: 484–495. https://doi.org/http://dx.doi.org/10.1016/j.ijepes.2016.03.036.

Thangaraj, R., Pant, M., and Deep, K. (2010). Optimal coordination of over-current relays using modified differential evolution algorithms. *Engineering Applications of Artificial Intelligence* 23 (5): 820–829. https://doi.org/http://dx.doi.org/10.1016/j.engappai.2010.01.024. Advances in metaheuristics for hard optimization: new trends and case studies.

Thue, W.A. (2011). *Electrical Power Cable Engineering, Power Engineering (Willis)*, 3e. Taylor & Francis. ISBN 9781439856437.

Tizhoosh, H.R. (2005). Opposition-based learning: a new scheme for machine intelligence. *Proceedings of the International Conference on Computational Intelligence for Modelling, Control and Automation and International Conference on Intelligent Agents, Web Technologies and Internet Commerce Vol-1 (CIMCA-IAWTIC'06) - Volume 01, CIMCA '05*, pp. 695–701. Washington, DC, USA: IEEE Computer Society. ISBN 0-7695-2504-0-01. http://dl.acm.org/citation.cfm?id=1134823.1135200.

Tostado-Véliz, M., Kamel, S., and Jurado, F. (2019). A robust power flow algorithm based on Bulirsch—Stoer method. *IEEE Transactions on Power Systems* 34 (4): 3081–3089. https://doi.org/10.1109/TPWRS.2019.2900513.

Tostado-Véliz, M., Kamel, S., and Jurado, F. (2020a). Power flow approach based on the S-iteration process. *IEEE Transactions on Power Systems* 35 (6): 4148–4158. https://doi.org/10.1109/TPWRS.2020.2989270.

Tostado-Véliz, M., Kamel, S., and Jurado, F. (2020b). An efficient and reliable power flow solution method for large scale Ill-conditioned cases based on the Romberg's integration scheme. *International Journal of Electrical Power & Energy Systems* 123: 106264. https://doi.org/10.1016/j.ijepes.2020.106264.

Trias, A. (2012). The holomorphic embedding load flow method. *Power and Energy Society General Meeting, 2012 IEEE*, pp. 1–8, July 2012. https://doi.org/10.1109/PESGM.2012.6344759.

Tsien, H.Y. (1964). An automatic digital computer program for setting transmission line directional overcurrent relays. *IEEE Transactions on Power Apparatus and Systems* PAS-83 (10): 1048–1053.

Ungrad, H. (1995). *Protection Techniques in Electrical Energy Systems*, 1e. New York: CRC Press. ISBN 9780824796600.

Urdaneta, A.J., Nadira, R., and Perez Jimenez, L.G. (1988). Optimal coordination of directional overcurrent relays in interconnected power systems. *IEEE Transactions on Power Delivery* 3 (3): 903–911. https://doi.org/10.1109/61.193867.

Urdaneta, A.J., Perez, L.G., and Restrepo, H. (1997). Optimal coordination of directional overcurrent relays considering dynamic changes in the network topology. *IEEE Transactions on Power Delivery* 12 (4): 1458–1464. https://doi.org/10.1109/61.634161.

Urdaneta, A.J., Nadira, R., and Perez, L.G. (1999). Optimal coordination of directional overcurrent relays considering definite time backup relays. *IEEE Transactions on Power Delivery* 14 (4): 1276–1284.

Uy, N.Q., Hoai, N.X., O'Neill, M. et al. (2011). Semantically-based crossover in genetic programming: application to real-valued symbolic regression. *Genetic Programming and Evolvable Machines* 12 (2): 91–119. https://doi.org/10.1007/s10710-010-9121-2.

Venkata, S.S., Damborg, M.J., and Jampala, A.K. (1991). Power system protection: software issues. In: *Control and Dynamic Systems*, (ed. C.T. Leondes), pp. 57–109. Elsevier. https://doi.org/10.1016/b978-0-12-012742-9.50007-0.

Venkataraman, P. (2009). *Applied Optimization: with MATLAB Programming*, 2e. Hoboken, NJ: Wiley. ISBN 978-0-470-08488-5.

Verma, T.S.M. and Rao, H.K. (1976). Inverse time overcurrent relays using linear components. *IEEE Transactions on Power Apparatus and Systems* 95: 1738–1743. https://doi.org/10.1109/t-pas.1976.32273.

Verma, A.K., Ajit, S., and Kumar, M. (2011). *Dependability of Networked Computer-based Systems, Springer Series in Reliability Engineering*. London: Springer. ISBN 9780857293183.

Vijayakumar, D. and Nema, R.K. (2008). Superiority of PSO relay coordination algorithm over non-linear programming: a comparison, review and verification. *2008 Joint International Conference on Power System Technology and IEEE Power India Conference*, pp. 1–6, October 2008. https://doi.org/10.1109/ICPST.2008.4745385.

Wadood, A., Khurshaid, T., Farkoush, S.G. et al. (2019). Nature-inspired whale optimization algorithm for optimal coordination of directional overcurrent relays in power systems. *Energies* 12 (12): 2297. https://doi.org/10.3390/en12122297.

Wang, C., Liu, L., and Jiang, C. (2005). A new power flow arithmetic based on hybrid analysis. *2005 IEEE/PES Transmission Distribution Conference Exposition: Asia and Pacific*, pp. 1–6. https://doi.org/10.1109/TDC.2005.1547026.

Wang, X.-F., Song, Y., and Irving, M. (2008). *Modern Power Systems Analysis*. New York: Springer US. ISBN 978-0-387-72853-7. https://doi.org/10.1007/978-0-387-72853-7.

Wang, Y., Wagner, N., and Rondinelli, J.M. (2019). Symbolic regression in materials science. *MRS Communications* 9 (3): 793–805. https://doi.org/10.1557/mrc.2019.85.

Warrington, A.R.C. (1962). *Protective Relays, Their Theory and Practice*, Chapter 4, vol. I. New York: Wiley.

Wong, K.P., Yuryevich, J., and Li, A. (2003). Evolutionary-programming-based load flow algorithm for systems containing unified power flow controllers. *IEE Proceedings - Generation, Transmission and Distribution* 150 (4): 441–446. https://doi.org/10.1049/ip-gtd:20030405.

Wood, A.J. and Wollenberg, B.F. (1996). *Power Generation, Operation, and Control*, 2e. New York: Wiley-Interscience. ISBN 978-0471586999.

Wood, A.J. and Wollenberg, B.F. (2013). *Power Generation, Operation, and Control*, 3e. New York: Wiley-Interscience. ISBN 978-0-471-79055-6.

Wood, H.C., Sidhu, T.S., Sachdev, M.S., and Nagpal, M. (1989). A general purpose hardware for microprocessor based relays. *International Conference on Power System Protection '89*, pp. 43–59, September 1989. http://www.eng.uwo.ca/people/tsidhu/Documents/General%20purpose%20hardware.pdf.

Wu, F.-J., Chou, C.-J., Lu, Y., and Chung, J.-L. (2012). Modeling electromechanical overcurrent relays using singular value decomposition. *Journal of Applied Mathematics* 1–18. https://doi.org/10.1155/2012/104952.

Yazdaninejadi, A., Nazarpour, D., and Talavat, V. (2019). Coordination of mixed distance and directional overcurrent relays: miscoordination elimination by utilizing dual characteristics for DOCRs. *International Transactions on Electrical Energy Systems* 29 (3): e2762. https://doi.org/https://doi.org/10.1002/etep.2762. ITEES-18-0348.R1.

Yeniay, O. (2005). Penalty function methods for constrained optimization with genetic algorithms. *Mathematical and Computational Applications* 10: 45–56.

Yu, X. and Gen, M. (2010). *Introduction to Evolutionary Algorithms, Decision Engineering*. London: Springer Science+Business Media. ISBN 9781849961295.

Zarco, P. and Exposito, A.G. (2000). Power system parameter estimation: a survey. *IEEE Transactions on Power Systems* 15 (1): 216–222. https://doi.org/10.1109/59.852124.

Zeineldin, H.H. (2008). Optimal coordination of microprocessor based directional overcurrent relays. *2008 Canadian Conference on Electrical and Computer Engineering*, pp. 000289–000294, May 2008. https://doi.org/10.1109/CCECE.2008.4564541.

Zeineldin, H., El-Saadany, E.F., and Salama, M.A. (2005). Optimal coordination of directional overcurrent relay coordination. *IEEE Power Engineering Society General Meeting, 2005*, pp. 1101–1106, Volume 2, June 2005. https://doi.org/10.1109/PES.2005.1489233.

Zeineldin, H.H., Sharaf, H.M., Ibrahim, D.K., and El-Zahab, E.E.A. (2015). Optimal protection coordination for meshed distribution systems with DG using dual setting directional over-current relays. *IEEE Transactions on Smart Grid* 6 (1): 115–123. https://doi.org/10.1109/TSG.2014.2357813.

Zeineldin, H.H., El-Saadany, E.F., and Salama, M.M.A. (2006). Optimal coordination of overcurrent relays using a modified particle swarm optimization. *Electric Power Systems Research* 76 (11): 988–995. https://doi.org/http://dx.doi.org/10.1016/j.epsr.2005.12.001.

Zellagui, M. and Abdelaziz, A.Y. (2015). Optimal coordination of directional overcurrent relays using hybrid PSO-DE algorithm. *International Electrical Engineering Journal (IEEJ)* 6 (4): 1841–1849.

Zhang, Z., Crossley, P., and Li, L. (2017). A positive-sequence-fault-component-based improved reverse power protection for spot network with PV. *Electric Power Systems Research* 149: 102–110. https://doi.org/https://doi.org/10.1016/j.epsr.2017.04.022.

Zhu, J. (2009). *Optimization of Power System Operation*. Piscataway, NJ; Chichester: Wiley-IEEE Press. ISBN 978-0-470-29888-6.

Zocholl, S.E., Akamine, J.K., Hughes, A.E. et al. (1989). Computer representation of overcurrent relay characteristics. *IEEE Transactions of Power Delivery* 4 (3): 1659–1667.

Index

high-dimensional problem 23
2-dimensional problem 31, 52
n-dimensional problem 4, 9, 40, 45, 53, 169, 217, 294, 307, 333, 374, 377
one-dimensional problem 4, 24, 43, 364

a

A-frame 367
adaptive coordination xxiv–xxvii, 169, 185, 188, 190, 202, 242, 245, 249, 271, 279, 280, 319, 320, 329, 345, 360, 361, 367, 377, 380
admittance matrix 129, 175, 179, 181–184, 202–204
AI xxvi, 3, 174, 280, 290, 322, 323, 325, 329, 380
ambient temperature 62, 68, 118, 119, 124, 397
ant colony optimization xxv, 33
anti-clockwise loop 150, 151, 156, 161–166
Appendix 251, 252, 264, 276, 278, 285, 288, 299, 301, 310, 322
applied mathematics and optimization 1
artificial intelligence xxvi, 3, 93, 290
artificial neural network xxvii, 3, 94, 319, 323
asymptotic regression 375, 377
attracted armature relay 62
auxiliary CT 65, 68
auxiliary PT 65, 68
auxiliary transformer 62

b

backup-1/backup-2 relay 306, 308, 311
bee colony optimization xxv
benchmark function
  constrained 21
  continuous 21
  convex 21
  differentiable 21
  dynamic 21
  multi-objective 21
  multimodal 21
  noisy 21
  non-convex 21
  non-noisy 21
  non-rotated 21
  non-shifted 21
  non-smooth 21
  rotated 21
  scalable 21
  separable 21
  shifted 21
  single-objective 21
  smooth 21
  static 21
  unconstrained 21
  unimodal 21
bimetallic relay 61, 67, 69, 77
biobjective function 334
biogeography 34
biogeography-based optimization xxvii, 33, 34, 47, 297
  algorithm
    clear duplication stage 43, 297
    elitism stage 44
    flowchart 46
    migration stage 40
    mutation range 43
    mutation stage 41
    overall structure 45
  habitat suitability index 40
  island suitability index 40
  species count probability 41

Optimal Coordination of Power Protective Devices with Illustrative Examples, First Edition. Ali R. Al-Roomi.

Companion website: www.wiley.com/go/al-roomi/optimalcoordination

biogeography-based optimization (*contd.*)
  suitability index variable 40
black-box optimization algorithm 33
blackout 58, 185
Branin's benchmark function no.1 55
breaker failure relay 425
brute-force search 242
Buchholz relay 61, 67
Bulirsch-Stoer method 174

c

calibration 62
candidate solution 51
capacitor bank 319, 360, 377
Carpentier 181
case studies
  118-bus system 453, 454
  14-bus system 206, 430–437
  15-bus system 232, 233, 252, 276, 277, 285, 291, 294, 299, 300, 306, 315, 316, 437–441
  2-bus system 254–257
  3-bus system 157, 177, 191, 202, 203, 209, 213, 218, 231, 232, 250, 258–260, 265–268, 272, 294, 305, 306, 322, 325, 336, 342, 400–403, 410, 411
  30-bus system 294, 306, 441–446
  39-bus system 206
  4-bus system 149, 150, 162, 232, 252, 263, 403–408
  42-bus system 206, 264, 278, 291, 294, 301, 305, 306, 316, 317, 448–451
  5-bus system 161, 165, 166, 186, 230, 408, 409
  6-bus system 165, 205, 232, 265, 294, 301, 306, 310, 312–315, 410–419
  8-bus system 182–184, 192, 193, 205, 224, 228, 232, 252, 253, 262, 288, 289, 370, 410, 413, 418, 420–427
  9-bus system 195, 196, 209, 232, 233, 252, 322, 346, 427–430
Cauchy mutation 41
cause-and-effect 185
CB 68, 130, 131, 140, 146–149, 169, 185, 208, 214–216, 380
CCPP 364, 365
characteristic curves
  definite current 70
  definite-current 139, 263, 272, 347
  definite-time 71, 139, 263, 272, 273, 347
  instantaneous 70, 139, 272, 273, 347
  inverse-time xxvi, xxvii, 72, 79, 98, 99, 104, 139, 149, 157, 207, 261, 263, 272–274, 276, 280, 284, 294, 347, 348, 357, 367
  mixed 73, 263, 284
    definite-time plus instantaneous 73
    inverse definite minimum time 76, 139, 276
    inverse-time plus definite-time 75
    inverse-time plus definite-time plus instantaneous 74
    inverse-time plus instantaneous 74
circuit breaker 68, 131, 136, 140, 143, 146, 148, 149, 161, 164, 185, 194, 202, 214, 305–307, 309–311, 339, 350
circuit theory 363
classic DE 45, 56, 316
classical algorithms
  GD 375
  GN 375
  GS xxv, xxvi, 157, 173, 174, 182, 183
  ILP xxv, 29, 31, 163, 250, 297
  INLP xxv, 29, 31
  JM 173, 174
  LM 330, 375
  LP xxv–xxvii, 25, 26, 28, 29, 31, 32, 48, 49, 51, 218, 235–239, 242, 244–247, 249–252, 255, 257, 260, 265–269, 273, 281, 287, 291, 297, 319, 322, 325, 339, 345, 350
  MILP 29, 31
  MINLP xxiv, xxv, 29, 31, 148, 217, 233
  NLP xxv, 28, 31, 48, 49, 235
  NR 173–175, 177, 179, 182, 183, 185, 202, 205
classical optimization algorithm 3, 23, 46
classical random search method 15
clockwise loop 150, 161–166
coasting time 62, 148
combinatorial optimization 29, 43, 53, 163, 217, 243, 245, 251, 305
combined-cycle power plant 364
combined-cycle power plants 364
complexity xxvi, 10, 67, 69, 160, 222, 223, 244, 247, 249, 253, 261, 263–265, 271, 272, 289, 290, 305
constraint-handling technique
  adaptive multiplication penalty function 19, 232

binary static penalty function 18, 224, 226, 246, 310
constant
infinite barrier penalty function 221
constant penalty function 17
eclectic evolutionary algorithm 18, 231
exponential dynamic penalty function 19, 232
infinite barrier penalty function 17
Karush-Kuhn-Tucker 218
random search method 17
self-adaptive penalty function 20, 232
superiority of feasible points
type I (SFP-I) 18, 231
type II (SFP-II) 18, 231
typical dynamic penalty function 19, 232
constraint-handling techniques 13, 16, 218
consumption point 171
contingency analysis 174, 319, 320, 361, 387
conventional weight aggregation 334
coordination criteria xxvi, 207, 209, 210, 288, 289, 377
desired design 209, 230
enhanced 209
mid-point 209, 210
minimum 209, 288
coordination delay time 148
coordination time interval xxiii, 148, 164, 212, 214, 266, 295, 308, 349, 350, 352, 399
cost function 5, 31, 210, 352, 353
criterion 5
critical clearing time 117, 126, 211, 296, 299
crossover rate 48, 50, 56, 288, 310
CT 61, 62, 68–70, 76, 142, 144, 150, 151, 157, 254, 256, 266, 307, 310, 371
CT saturation 62, 68
CTR 77, 144, 145, 148, 151, 212, 218, 230, 231, 254, 256–259, 266, 267, 272, 273, 325, 339, 381, 399, 400, 402, 406, 411, 413–416, 418, 421, 425, 428, 439, 449–451, 453
current transformer 62, 69, 77, 144, 339
current-time characteristic curve 376
curve fitting 82, 86, 89, 113, 261
CWA 334, 336, 345

d

DC grid 193
DCS 185
DE/rand/1/bin 45, 316
death penalty function 17, 52, 218, 245
decaying DC 62, 68
decision variables 4
dependability 66, 336
dependent variable 1, 40, 51, 211, 296, 308
binary 17
discrete 14, 17
mixed-integer 17
DER 359, 360
derivation 118, 136, 176, 238, 239, 397
design constant 4
design constraint
functional constraint
equality constraint 7–9
inequality constraint 7–10, 15, 25, 211–213, 216, 223, 228, 254, 295, 296, 308, 311, 315, 316, 351, 354, 438, 445, 449
side constraint xxiv, 7, 9, 10, 13, 44, 101, 113, 117, 118, 189, 211, 212, 250, 254, 274, 280, 281, 296, 308, 309, 311, 315, 316, 355, 428, 439, 449
design parameter 4
design variable 1
design vector 4
DFR 367
DG 359, 439, 442
differential evolution xxv, 45
differential relay 61, 66
digital fault recorder 367
digital signal processors 65
direct search method 15
discrimination margin xxiii, 140, 145, 146, 148, 149, 157, 212, 288, 290, 335
distance-time characteristic curve 376
distributed control system 185
distributed energy resources 359
distributed generation 359, 439
distributed parameter model 173, 363
double primary relay strategy xxvii, 304, 306, 448
DPRS xxvii, 304–307, 309–317, 448, 449
extreme case xxvii, 303–307, 309–317, 448, 449
DSP 62, 65
duality principle 6
DWA 334

dynamic change xxvi, 284, 319, 321, 322, 325, 360, 363, 377
dynamic disturbances 322
dynamic network 188, 202, 249, 319
dynamic weight aggregation 334

e

EA xxv, 15, 18, 33, 48, 49, 51, 217, 218, 223, 235, 236, 238, 239, 245, 249–251, 273
early stopping criterion 5
earth science 34
ecology 34
economic load dispatch 294, 361
Edward O. Wilson 34
eigenvalue 38
eigenvector 38
ELD 294, 305, 306, 311, 361, 364, 367, 368
electrical software
  ASPEN OneLiner 182
  ATP-EMTP 182
  CAPE 182
  CYME 182, 190
  DIgSILENT PowerFactory 182, 183, 190, 193, 195, 196, 204, 206, 449
  DSATools 182, 190
  EasyPower 182, 190
  ERACS 182
  ETAP 170, 182, 183, 190, 193, 204, 444
  EuroStag 182, 190
  GridView 182
  Kalkitech PowerApps 182
  MilSoft Power 182, 190
  NEPLAN 182, 190
  Nexant 182
  OpenDSS 182
  Paladin 182
  PowerWorld Simulator 182–184, 190, 192–194, 205, 215, 216, 387
  PSCAD 182
  PSLF 182
  PSS/E 182, 190
  SIMPOW 182, 190
  SKM Systems
    CAPTOR 182
    DAPPER 182, 190
  SPARD mp 182, 190
  SPICE 182
  Transmission 2000, 182
EMS 171, 323, 369
energy management system 171
EPF 15
epochs 329, 330
equal-area criterion 126, 127, 131–133
equivalent operating hours 364
evolutionary algorithm xxv, xxvii, 18, 34, 45, 280
evolutionary algorithms 33
evolutionary biology 34
evolutionary computation 1
excessive heat 77, 117, 118, 125
excitation system 364, 366
exhaustive search 242
explainability 94
exterior penalty function 15, 17, 218
  additive approach 15
  multiplicative approach 15

f

FACTS 360, 380
fast decoupled load flow 453
fast-decoupled load flow 174, 182
fault 57
fault analysis 189, 190, 192, 320, 325, 361, 363, 367, 439
fault classification 367, 369
fault containment 367
fault current controller 360
fault current limiter 360
fault detection 367, 369
fault location 126, 127, 133, 135, 136, 140, 141, 143, 193, 208–210, 213, 214, 217, 238, 251, 255, 257, 266, 279, 287, 288, 290, 295, 307, 347, 348, 351, 353, 354, 357, 367, 369–373, 374–377, 379, 380
fault probability zone 371
fault recovery 367
faults classification 57
  active 58, 60
  asymmetrical 61
  balanced 60
  open-circuit 57, 58
  passive 58
  permanent 58
  persistent 58, 76, 77
  series 58, 76

short-circuit 57, 58, 60, 68, 72, 73, 118, 137, 138, 141, 145–147, 149, 150, 157, 188–190, 192–197, 201, 202, 204, 206, 208, 211, 213–216, 218, 230, 231, 251, 254, 256, 258, 259, 266, 283, 287, 288, 294–296, 307, 309, 323, 326, 360, 369, 375–377, 379, 387, 413, 418, 429, 432, 439, 440, 443, 446, 449
short-circuit analysis 189
shunt 58, 68, 76, 370
symmetrical 60
temporary 58, 76, 77
transient 58, 76
unbalanced 61
FC 223, 224, 226, 228, 265, 268, 276, 279, 287, 297, 305, 311
FCC 360
FCL 360
FDLF 174, 182, 183
feasibility checker 223, 224, 245–247, 249, 265, 281, 287, 297
feasible solution xxiv, 7–9, 217, 222–224, 287, 297, 311, 342
final temperature 119, 120
fine-tuning xxvi, 23, 25, 51, 235, 238, 239, 244, 245, 250–252, 281, 297, 331
fitness 5, 7, 21, 28, 49, 51, 221, 222, 226, 242–244, 248, 249, 264, 265, 278, 279, 311, 313, 315, 329
fixed divisor method 106–108, 114, 165, 213, 274, 275
fixed topology 186, 215, 218, 231, 363
FL 94
flexible alternating current transmission system 360
flowchart 3, 45, 46, 49, 112, 169, 170, 224, 245–247, 249, 250, 275, 280, 298, 319–322, 324, 335, 360
flywheel 171
FME 25
forecasting 320, 360, 361
Fourier 25
Fourier-Motzkin elimination 25
FPZ 371
Frank Lauren Hitchcock 25
fuel-cost function 364
fuse xxiii–xxvi, 62, 67–69, 77, 99, 195, 253, 261–264, 271, 290, 309, 347
fuzzy logic 94
fuzzy optimization 4, 7

g

gas turbine 126, 127, 171, 364
Gauss elimination 173
Gauss–Seidel 173
Gauss-Newton 375
Gauss-Seidel xxv, xxvi, 157, 159, 163, 173
Gaussian mutation 41
general purpose relay 65, 66, 77
generate and test 242, 243
generations 5, 19, 21, 34, 45, 50, 55, 56, 221, 231, 245, 246, 265, 278, 279, 286, 287, 299, 310, 311
generator bus 172, 255
generators 61
genetic algorithm xxv, 33
geomagnetically induced current 321
George Bernard Dantzig 25
GIC 321
global optimizer 33
global optimum solution 11, 12, 33, 43, 245
global-local optimization strategy 28
GPS 369
Gradient Descent 375
gradient process 32
gradient-based method 23
grading 139–141, 143
current xxvi, 139–141, 143
inverse-time 75, 139, 143
time xxvi, 139, 140
grading margin 148
graph theory xxiv, 163, 186–188, 204
gravitational search algorithm 33

h

harmonic 62, 68
heuristic-based algorithm
heuristic 33
meta-heuristic 33
HMI 185
holomorphic embedding method 174
human-machine interface 185
hybrid algorithms
BBO-LP 48, 49, 51, 243, 245–247, 249, 251, 281, 284–288, 291, 329, 339

hybrid algorithms (*contd.*)
  BBO/DE 50, 51, 229, 305, 310, 311, 313, 315–317
  DE/LP 251
  GA-LP 228, 229
  GA-NLP 233
hybrid optimization algorithm 3, 236, 238, 239, 245, 249, 250, 273, 319
hyperparameters 94, 330

i
IBT xxiii, 319
IDMT 76, 139, 276, 286, 288, 306, 372
IED 65, 113, 122, 124, 136
independent variable 1, 4, 7, 12, 13, 40, 45, 50, 51, 172, 173, 307, 379
individual 51
induction disk 89–93
induction relay 62
inertia 62
infeasible solution 7, 17, 223, 245
initial temperature 119, 120
initialization parameters 23, 221, 232, 246, 247, 252, 310, 313
injected source 171
inspection method 186, 205
instantaneous trip 354
integer linear programming 29
integer nonlinear programming 29
integrated test set 367
intelligent electronic devices 65, 122
inter-bus transformer xxiii, 367
interior penalty function 15
interpolation 82, 374, 375
  linear 79–82, 108, 109, 114, 115, 165, 213, 274, 275, 280, 373, 374, 379
  nonlinear 79–82, 374, 379, 380
interpretability 94
IPF 15
island biography
  dynamic equilibrium 34
  emigration rate 34
  fauna 34
  flora 34
  immigration rate 34
  island 34
  simplified linear migration model 34
  species
    emigrated species 35
    endemic species 34
    extinct species 34
    immigrated species 34
iteration 5
iterations xxv, xxvi, 8, 19, 23, 33, 36, 39, 40, 51, 151–154, 156, 173, 174, 176, 177, 179, 222, 223, 235, 281, 301, 310, 353

j
Jacobi method 173
Jacobian matrix 176, 179, 181, 202
jumping rate 245

k
K. V. Price 45
KCL 144, 175
Kirchhoff's current law 144, 175
KKT 218
Kron's loss formula 363

l
Lèvy mutation 41
LCR 185
least squares 82, 86
Leonid Kantorovich 25
Levenberg-Marquardt 330, 375
line impedance 203, 204, 214, 258
linear optimization 25
linear programming 25, 253, 256, 258, 297
  algorithm
    dual-simplex 28
    revised simplex 26
  conditions 25
  histrocial time-line 25
  interior-point 26
  method
    graphical approach 25
    non-simplex 25
    simplex 25
  solvers 26
linear programming problem 26
linear regression xxvii, 5, 82, 90, 93, 237, 379
linearization 31, 235–237, 242, 245, 250
LINKNET 186
load bus 172, 174

load flow 170, 174, 175, 177, 182, 183, 205, 418, 455
load shedding 57, 321
load starting current 146
local backup relay xxiii, xxv, xxvii, 66, 163, 303, 304, 309, 310, 315, 425
local control room 185, 367
local optimum solution 11
loops 5, 8, 44, 150, 243, 310
lower-upper 173
  LU 173
LPP 26
lumped component model 173
lumped element model 173
lumped parameter model 173, 363
  I-model 173
  O-model 173
  T-model 173
  ∞-model 173
  ⊂-model 173
  ⊃-model 173
  M-model 173, 363
  Γ-model 173
  Π-model 173
  Ꞁ-model 173
  equals-sign-model 173

m

M.J. Damborg 84, 85, 87, 99, 114, 209
M.S. Sachdev 86, 88, 99, 411, 425
machine learning xxvii, 94
  ANN xxvii, 3, 94, 323, 324, 331, 380
  LR xxvii, 5, 93, 94
  MLR 330
  NLR xxvii, 5, 93, 94, 331
  SVD 94
  SVM xxvii, 94, 380
  SVR 94, 330
  UFO 5, 331
main control room 185, 367
main-1/main-2 relay xxvii, 303–316, 347, 448
mathematical formulation 26, 79, 207, 294, 306, 348, 350, 353
  multi-objective problem 335
mathematical model 8, 25, 30
mathematical optimization 1, 309
mathematical programming 25
mathematical software
  LINGO 26
  Maple 26
  Mathematica 26
  MS Excel 26, 221
maxima 1
maximum temperature 126, 137
Maxwell's equations 363
MBPS xxv, xxvi, 139, 160–164
MCR 185
merit 5
meta-heuristic algorithms
  ACO xxv, 33
  BBO xxvii, xxviii, 33, 34, 36, 40, 41, 43–46, 49, 51, 53, 55, 106, 111, 218, 221, 223, 245, 249, 250, 264, 265, 268, 273, 276–278, 281, 284–287, 291, 297–299, 310, 311, 313, 315, 333–336
  BCO xxv
  DE xxv, xxvii, 45, 51, 56, 104, 111, 218, 232, 233, 310, 316
  GA xxv, xxviii, 33, 106, 111, 228, 229, 233, 265
    binary 33
    micro 33
    real 33
    stud 33
  GSA 33
  OBL 106, 111, 245
  PSO xxv, xxviii, 33
  RSA 17, 223, 281
  SA xxv, 33, 41
  TS xxv, 33
meta-heuristic optimization algorithm xxvi, xxvii, 1, 25, 29, 32–34, 46, 52, 53, 89, 104, 106, 163, 207, 217, 222, 235, 242–246, 253, 265, 280, 284, 285, 297, 316, 319, 322, 345
meteorology 360, 361
microcontroller 65
microgrid 359
microprocessor 65
minima 1
minimum acceptable error 5
minimum break-point set xxv, xxvi, 139, 160, 161
miscoordination 351
mixed-integer linear programming 29
mixed-integer nonlinear programming 29

mixed-integer optimization xxiv, 43, 217, 251, 297, 311
ML xxvii, 94, 331
modeling-free technique 3
modern optimization algorithm 3, 33
modified algorithms
  BBO + FC 286, 311, 312
  BBO-LP + FC 286
  BBO/DE + FC 312
MOEA 334, 342, 345
MOEAs
  CWA-BBO 334, 336, 339
  RWA-BBO 342
MOP xxvii, 333, 334, 336–338, 340, 342, 343, 345
motor operated relay 62
moving coil relay 62
multi-attribute optimization 333
multi-criteria optimization 333
multi-function relay 65, 77
multi-objective evolutionary algorithm 334
multi-objective optimization xxvii, 333, 339, 342
multi-objective problem xxvii, 333
multi-objective programming 333
multi-optimum points 6
multi-performance optimization 333
multi-point algorithm xxv, 48
multi-start linear programming xxvi, 28, 29, 51, 235, 242, 243, 245, 249, 319, 322, 329, 339
multi-start strategy 31, 89, 242–244, 251
multiple linear regression 319, 330
Murray loop bridge 367
mutant vector 48, 50

n
nature-inspired algorithm 33, 41
near feasibility threshold 19
near-global solution 33
Newton-Krylov 174
Newton-Raphson 25, 173, 175, 179, 418, 455
NFE 8
NFT 19
nominal current 119
nominal temperature 119, 136, 137
non-classical optimization algorithm 33
non-death penalty function 299
non-traditional optimization algorithm 3, 33
nonlinear programming 28
nonlinear regression xxvii, 5, 90, 93, 263, 375, 376, 379
number of function evaluations 8
numerical analysis 1, 94

o
objective function 1, 5, 12, 208, 210, 294, 307, 352, 353
  dual setting 210
  multiple 336, 342
OCR type
  DCOCR 70, 77, 99, 139, 143, 146, 163, 207, 273, 358
  DTOCR 71, 72, 77, 99, 139, 143, 146, 163, 207, 273, 358
  IOCR 70, 139, 146, 207, 273
  ITOCR xxvii, 70, 72, 79, 82, 90, 92, 94, 99, 101, 104, 106, 108, 112, 113, 118, 139, 143, 146, 148, 149, 163, 207, 208, 263, 271, 283, 284, 287, 291, 293, 301, 303–307, 317, 357, 358, 367, 373
OEM 65, 72
Oliver Heaviside 181, 363
OMIB 126
one-machine infinite bus 126
OPC 185
open platform communications 185
operational change xxvii, 321, 322, 328, 360, 377
operations research 25
OPF 361
opposition-based learning 106, 245
optima 1
optimal power flow 361
optimal power-flow 361
optimal solution 1, 10, 21–23, 25, 28, 51–53, 218, 243, 246, 250, 279, 281, 289, 299, 310, 313, 323, 333, 334, 338, 350, 418
optimization algorithm 1
optimization problem
  constrained optimization problem 12
  constraint satisfaction problem 12
  free optimization problem 12
  no problem 12
optimization technique
  classical 23
  hybrid 46
  meta-heuristic 33

OR 25
original equipment manufacturer 65
over flux relay 61, 66
over frequency relay 61, 66
over voltage relay 61, 66
over-shoot time 62, 148, 164
over-travel delay 62, 92, 148
overhead lines 61
overload current 146, 189, 211, 309, 336, 449
overload factor 118, 122–124, 136, 137, 211

p

P/B relay pair xxiv, 118, 186–188, 201, 202, 204, 205, 210, 216, 218, 222–224, 230, 255, 256, 258, 259, 266, 267, 271, 276, 279, 283, 287–290, 297, 305, 316, 321–325, 329, 350, 374, 377, 399, 402, 411, 413, 437, 449
Pareto front 334, 336, 337, 339, 340, 342, 343
Pareto optimization 333
Pareto point 334
Pareto set 334
particle swarm optimization xxv, 33
PB xxv, 33
peaks and valleys 6, 31
penalized cost function 15, 17–20, 52
penalty function 15
penalty multiplier 15, 17, 18, 231, 246, 311
penalty term 15, 17, 20, 52
per-unit-length 424
performance criteria
  accuracy 15, 49, 82, 86, 89, 113, 173–175, 181, 185, 235, 339, 361, 363, 367, 373, 374, 379
  best solution 21
  complexity xxv, 17, 222, 230, 287
  convergence rate 41, 49, 175, 182, 235
  CPU time xxv, xxvi, 8, 17, 21, 23, 28, 39, 41, 48, 94, 179, 185, 218, 223, 231, 242, 245, 249, 265, 268, 284, 287, 311, 345
  exploitation level 32, 51, 218, 247
  exploration level 19, 29, 32, 41, 43, 51, 242, 245–247, 250, 279, 311, 323
  feasibility 10, 232, 233, 287–289, 304, 360
  mean 21
  median 21
  memory usage 23, 79, 80, 174, 179, 181, 182
  optimality 10, 283, 287, 322, 360
  processing speed xxv, xxvi, 23, 49, 136, 181, 185, 217, 231, 242, 245, 249, 265, 281, 284, 287, 319, 322, 323, 345, 361, 363
  processing time 34, 174, 185, 222, 242, 244, 245, 249, 279, 291
  simplicity 49, 173, 284, 287, 363
  solution quality 18, 40, 41, 185, 223, 231, 242, 244, 271, 287, 322, 323, 329, 361
  standard deviation 21
  worst solution 21
periodic maintenance 62
pilot protection 442
PLC 185, 320
polarizing signal 147
polynomial coefficients 84
polynomial equation 79, 82–85, 87, 89, 90, 93, 99, 113, 114, 163, 330, 375, 380
polynomial regression 375, 380
population biology 34
population size xxvi, 8, 17, 20, 21, 34, 35, 45, 50, 51, 55, 56, 221, 222, 231, 246, 247, 265, 278, 279, 281, 286, 299, 310, 311
population-based algorithm xxv–xxvii, 33, 34, 40, 45, 222, 245, 280
possible search area 222, 223
potential transformer 62
power companies
  ABB 65, 84, 106, 118, 119, 122, 126, 127, 182, 185
  ALSTOM 65, 106, 364, 365
  AREVA 65, 106, 276–278, 286, 299, 313, 372
  CAI 182
  G.E.C. 84
  GE 65, 182, 364
  Schneider Electric 95, 106, 445
  SEL 65, 106
  SIEMENS 65, 182, 313
  Westinghouse 84, 424
power distribution 57
power flow 170–172, 184–186, 190, 201, 204, 320, 323, 360, 361, 363, 387
power flow equations 171, 202–204
power generation 57
power outage 57, 321
power system realization 361
power transmission 57
PQ bus 172, 176

pre-load current 118, 119, 123
pre-processing unit 207, 223, 297
pressure relief relay 61, 67
probabilistic optimization algorithm 33
probabilistic process 32, 40, 277
problem dimension 4, 445
problem-dependent algorithm 33
programmable logic controller 185
programming languages
  C/C++ 23, 190
  Fortran 23
  GNU Octave 26, 179, 188
  JAVA 23
  Julia 23, 188
  MATLAB xxviii, 1, 11, 13, 14, 23, 25–27, 30, 44, 53–56, 77, 85, 91, 96, 102, 114, 115, 120, 122, 125, 132, 133, 138, 157, 166, 167, 177, 179, 188, 190, 205, 221, 224, 228, 229, 231–233, 239, 241, 243–245, 247, 248, 251, 252, 265–268, 274, 276–278, 281, 285, 299, 301, 311, 313, 328, 329, 336, 339, 342, 346, 375
    EST toolbox 182
    MatDyn toolbox 190
    MatEMTP toolbox 182
    MatPower toolbox 182, 190, 204
    PAT toolbox 182
    PSAT toolbox 182
    PST toolbox 182
    Simulink 129, 133–135, 182, 190, 191
    SPS toolbox 182
    VST toolbox 182
  Python 23, 25, 26, 188, 190, 274
    PandaPower library 182, 190
protection system 61
protection zone xxiii, 57, 65, 148
  external fault 65, 66
  in-zone fault 65, 66, 143, 146, 150, 157, 193, 208, 255, 308, 336, 370, 380
  internal fault 65, 66
  out-zone fault 65, 66, 143, 150, 157, 190, 192, 212, 213, 259, 287, 295, 308, 310, 370, 380
protective relays classification 62
  hardware-based xxiii, 62, 65, 82, 273, 293, 303, 309
  software-based 62, 82, 293
PSA 223
pseudocode 17, 42–44, 48, 50
PT 61, 62, 147, 310, 371
PV bus 172, 176, 253

r

R. Storn 45
radial network 71, 139–141, 143, 144, 146, 148, 149, 163, 169, 173, 179, 195, 253, 254, 257, 261, 287, 348
random weight aggregation 334
real-world application 25
recloser 58, 62, 69
rectifier protection 95, 100, 114, 250
reference bus 172, 177, 202
reference current 119
reference signal 147
reference temperature 120
relay comparator 347
relay settings
  current setting multiplier 69
  current tap setting 69, 143–145, 237–239, 250, 251
  multiples of pickup current 70, 79, 97, 145
  multiples of tap current 80, 82, 145
  pickup current setting 69
  pickup setting 69, 82
  plug setting xxiv, 69, 72, 104, 150, 157, 211, 212, 223, 224, 236, 238, 242, 251, 254, 256, 259, 286, 296, 309, 345, 350
  plug setting multiplier 70, 76, 145, 309
  time dial setting 69, 72, 79, 80, 83, 99, 104, 143, 145, 164, 237, 238, 250, 251
  time lever setting 69
  time multiplier setting xxiv, 69, 99, 104, 211, 236, 238, 242, 254, 256, 258, 265, 266, 296, 308, 393, 394
  time setting multiplier 69
relay standard
  European 69, 79, 89, 91, 95, 99, 100, 102, 104, 106, 107, 109, 115, 118, 143, 207, 213, 216, 266, 268, 273–275, 277, 278, 281, 283, 287, 293, 305, 347, 357, 411
  North American xxvi, 70, 79, 93, 95, 96, 99, 100, 102, 104, 106, 107, 111, 115, 118, 143, 164, 207, 212, 213, 217, 231, 273–275, 277, 278, 281, 283, 287, 293, 305, 347, 357, 411

relay technology xxvii, 136, 250, 254, 261, 271, 293, 294, 297, 299, 304, 306, 308, 336
  digital xxiii, xxvii, 62, 65, 82, 112, 263, 273, 293, 295, 299, 303, 306, 309, 310, 347
  electromagnetic 62, 98, 298, 303, 314
  electromechanical xxiii, xxv, xxvii, 62, 65, 84, 89, 90, 94, 98, 112, 113, 148, 164, 217, 251, 263, 272, 273, 285, 286, 293, 295, 298–300, 303–306, 308–316, 336, 347
  numerical xxiii, xxvii, 62, 64, 65, 67, 74, 75, 77, 82, 111–113, 136, 148, 149, 157, 164, 165, 169, 213, 214, 218, 251, 263, 266, 272–278, 280, 283, 285, 286, 293, 295, 299, 300, 303–306, 308–313, 339, 342, 347, 367, 369, 372, 428
    DSP-based 62
    microprocessor-based 62
  solid-state xxiii, 62, 65, 82, 92, 93, 98, 101, 112, 113, 115, 164, 217, 263, 273, 299, 303, 314, 339
  static xxvii, 62, 92, 98, 293, 295, 298–300, 303, 304, 306, 308–316, 347
reliability 57, 61, 66, 94, 148, 190, 253, 290, 335–337, 342, 343, 347, 367
remote terminal unit 185
renewable energy 321, 359–361
reset characteristic 89, 90
resistance-temperature detector xxiii
results tester 228, 232, 233
reverse power relay 61, 66
Robert H. MacArthur 34
Romberg's method 174
Rosenbrock's benchmark function 55
RTD xxiii
RTU 185
RWA 334

S

S-iteration method 174
safety margin 148, 164
Sankey diagram 367, 368
SCADA 185, 323
  microSCADA 185
Schweitzer Engineering Laboratories 106
search space xxv, 9–11, 19, 29, 33, 34, 51, 62, 105, 109, 223, 243, 279, 311, 313, 323, 333
secondary relay 304, 425
sectionalizer 62, 69
security 61, 66, 171, 336, 361
selective time interval xxiii, 148
selectivity xxiv, xxv, 67, 95, 118, 141, 148, 149, 186, 207, 212, 213, 222–224, 230–232, 253–256, 263, 265, 278, 279, 283, 284, 287–289, 295, 297, 303, 308, 310, 348–351, 353–355, 402, 438
  transient 213, 217, 232, 251, 399
sensitivity 18, 23, 62, 67, 174, 181, 367
sequential conic programming 174
series compensation 360
severity level 61
simplicity 67, 253, 303
simulated annealing xxv, 33
single-machine infinite bus 117, 126, 128, 391
single-objective function 6
single-point algorithm 23, 33, 48
single-solution algorithm 33
singular value decomposition 94
slack bus 172, 174–176, 203, 215
smart grid xxvi, 113, 357, 359
SMIB 117, 126, 127, 129–135, 137, 138, 391
soft computing 1
solution features 4
source impedance 70, 143, 428
sparsity 179, 182, 239
speed 67, 68, 77, 99, 118, 136, 164, 211, 213, 261, 310, 336, 339, 380
SR 5
standard
  BS142 99, 101, 103, 139
  G74 193
  GOST R-52735 193
  IEC- 60909, 193
  IEC- 61363, 193
  IEC-255 99
  IEC-602555 99, 100
  IEC-60282 195
  IEC-60781 195
  IEC-608750 369
  IEC-60909 190, 193–195, 197, 202, 206, 435
  IEC-60947 195
  IEC-61363 193–195, 202
  IEC-61660 193
  IEC-62271 194
  IEEE-141 193, 194

standard (*contd.*)
  IEEE-242 194
  IEEE-399 194, 206, 305, 448, 449
  IEEE-946 193
  IEEE-C37 91, 99, 193, 194
  UL-489 193, 194
  VDE-0102 193, 206
  VDE-0103 193
standard points
  close-in 193, 209, 325, 348, 431, 432, 439
  far-end 190, 192, 209, 279, 287, 288, 351, 355, 367, 369, 370, 379, 380, 399, 401, 435
  middle 137, 209, 218, 256–258, 287, 288, 354, 399, 428
  near-end 189, 192, 193, 209, 224, 254, 256, 279, 287–289, 295, 307, 348, 351, 354, 369, 370, 379, 380, 399, 401, 410, 411, 418, 435, 439, 442, 444, 445
  remote-bus 209
  tail-end 209
state estimation 361
statistical software
  MINITAB 375
  R 375
  SAS 375
  SPSS 375
steady-state network 149, 169, 194, 363
steady-state probability 39
steam turbine 171, 364
step-size parameter 48, 50, 56, 104, 310
step-size resolution 43, 54, 85, 95, 112, 117, 164, 296, 299, 304, 311, 313, 345, 380, 413
step-up transformer xxiii, 253, 319, 364, 366
stochastic optimization algorithm 33
stochastic process 32, 244
storage element xxvii, 359, 360
Stott and Alsac 182
sub-algorithm 23, 39, 40, 43, 45, 48, 51, 171, 207, 223, 224, 235, 239, 245–247, 249, 250, 265, 268, 273, 276, 281, 297
sub-transient network 194
supercapacitor 171
supervisory control and data acquisition 185
support vector machine xxvii, 94
support vector regression 94, 319, 330
swing bus 172
swing equation 128, 130–132
switchgears 61
symbolic regression 5
system frequency xxvi, 171, 321, 322, 329
system stability xxvi, 67, 117, 126, 127, 131–133, 136, 190, 211, 296, 299, 308, 361, 367, 387
systematics 34

t

tabu search xxv, 33
Taylor's series 176
TCCCs
  multiple xxvi, 104, 106, 208, 271, 274, 278–281, 283, 287, 288, 290, 291, 293, 294, 304–306, 315, 350, 444, 448, 449
  multistage 263, 284
  optimal 287, 291
  stepwise 358, 359
  unified xxvi, 189, 207, 213, 223, 235, 264, 271, 278, 283, 284, 293, 304, 305, 329, 448, 449
  user-defined 273, 283, 284, 291, 305, 315, 445
TDR 367
Tellegen's theorem 174
temperature rise 119, 121, 124, 126
temperature/frequency-based technique 379
test function 21
TFB 379
the theory of island biography 34
thermal limit xxvi, 117, 118, 211, 296, 308
thermal overload protection 118, 122, 124, 126, 397
thermal relay 118, 119, 121–124, 126, 137
Thumper 367
time delay xxiii, 70, 71, 85, 117, 136, 146–148, 258, 266, 304, 311, 348, 353
time domain reflectometer 367
time-current characteristic curve xxv, 70, 84, 98, 139, 283, 294, 306, 357, 376, 437
time-distance characteristic curve 348
time-temperature curve 122, 125
Tjalling Charles Koopmans 25
tolerance 5
topological change 149, 319, 321–324, 328, 360
traditional optimization algorithm 3, 23
transformation xxvi, 111, 129, 172, 235, 237, 238, 242, 245, 251, 280, 346
transformed current tap setting 239, 251
transformed plug setting 238, 251

transformers 61
transient network 194, 380
transient topology 207, 213–216, 218, 230, 231, 251
trap 11
Trefethen's benchmark function 31
trip 58, 185
two-port network 129, 130

u

ultracapacitor 171
uncertainty xxvi, 261–263, 323, 371, 372, 379
unconstrained function 7
under frequency relay 61, 66
under voltage relay 61, 66
underground cables 61
unified power flow controller 321
uninterruptible power source 319
unit commitment 361
universal functions originator 5, 331
UPFC 321
UPS 319

v

variable
  binary 4, 5
  continuous xxiv, xxv, 4, 5, 14, 17, 148, 160, 207, 217, 218, 281, 285–287, 291, 339, 342, 345, 346, 399, 402, 403, 405, 413, 414, 428, 430, 444
  discrete xxiv, xxv, 4, 5, 14, 43, 44, 53, 62, 143, 148, 160, 207, 212, 217, 218, 222, 232, 236, 243, 274, 285–287, 291, 297, 298, 309–311, 342, 345, 399, 439, 444
  integer 4, 5, 48, 50, 313
  mixed-integer 4, 5
  non-continuous 5
  semi-continuous 342
vector notation form 13
vector optimization 333
violation 7, 15, 17, 20, 222, 223, 226, 228, 229, 279, 287, 304, 311, 316, 439
voltage transformer 62
voltage-controlled bus 172
VT 62

w

watchdog 320
weather condition 321, 322, 363
weighted sum method 334
Wheatstone bridge 367
WSM 334

z

Z-matrix method 173, 418, 455
zone-1 348, 351, 354
zone-2 348–351, 353–355
zone-3 348, 354

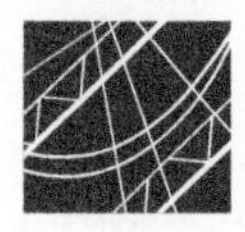

Books in the IEEE Press Series on Power and Engineering

Series Editor: ME El-Hawary, Dalhousie University, Halifax, Nova Scotia, Canada

The mission of IEEE Press Series on Power Engineering is to publish leading-edge books that cover the broad spectrum of current and forward-looking technologies in this fast-moving area. The series attracts highly acclaimed authors from industry/academia to provide accessible coverage of current and emerging topics in power engineering and allied fields. Our target audience includes the power engineering professional who is interested in enhancing their knowledge and perspective in their areas of interest.

1. *Electric Power Systems: Design and Analysis, Revised Printing*
   Mohamed E. El-Hawary
2. *Power System Stability*
   Edward W. Kimbark
3. *Analysis of Faulted Power Systems*
   Paul M. Anderson
4. *Inspection of Large Synchronous Machines: Checklists, Failure Identification, and Troubleshooting*
   Isidor Kerszenbaum
5. *Electric Power Applications of Fuzzy Systems*
   Mohamed E. El-Hawary
6. *Power System Protection*
   Paul M. Anderson
7. *Subsynchronous Resonance in Power Systems*
   Paul M. Anderson, B.L. Agrawal, J.E. Van Ness
8. *Understanding Power Quality Problems: Voltage Sags and Interruptions*
   Math H. Bollen
9. *Analysis of Electric Machinery*
   Paul C. Krause, Oleg Wasynczuk, and S.D. Sudhoff
10. *Power System Control and Stability, Revised Printing*
    Paul M. Anderson, A.A. Fouad
11. *Principles of Electric Machines with Power Electronic Applications*, Second Edition
    Mohamed E. El-Hawary
12. *Pulse Width Modulation for Power Converters: Principles and Practice*
    D. Grahame Holmes and Thomas Lipo
13. *Analysis of Electric Machinery and Drive Systems*, Second Edition
    Paul C. Krause, Oleg Wasynczuk, and S.D. Sudhoff

14. *Risk Assessment for Power Systems: Models, Methods, and Applications*
Wenyuan Li
15. *Optimization Principles: Practical Applications to the Operations of Markets of the Electric Power Industry*
Narayan S. Rau
16. *Electric Economics: Regulation and Deregulation*
Geoffrey Rothwell and Tomas Gomez
17. *Electric Power Systems: Analysis and Control*
Fabio Saccomanno
18. *Electrical Insulation for Rotating Machines: Design, Evaluation, Aging, Testing, and Repair*
Greg C. Stone, Edward A. Boulter, Ian Culbert, and Hussein Dhirani
19. *Signal Processing of Power Quality Disturbances*
Math H. J. Bollen and Irene Y. H. Gu
20. *Instantaneous Power Theory and Applications to Power Conditioning*
Hirofumi Akagi, Edson H. Watanabe and Mauricio Aredes
21. *Maintaining Mission Critical Systems in a 24/7 Environment*
Peter M. Curtis
22. *Elements of Tidal-Electric Engineering*
Robert H. Clark
23. *Handbook of Large Turbo-Generator Operation and Maintenance,* Second Edition
Geoff Klempner and Isidor Kerszenbaum
24. *Introduction to Electrical Power Systems*
Mohamed E. El-Hawary
25. *Modeling and Control of Fuel Cells: Distributed Generation Applications*
M. Hashem Nehrir and Caisheng Wang
26. *Power Distribution System Reliability: Practical Methods and Applications*
Ali A. Chowdhury and Don O. Koval
27. *Introduction to FACTS Controllers: Theory, Modeling, and Applications*
Kalyan K. Sen, Mey Ling Sen
28. *Economic Market Design and Planning for Electric Power Systems*
James Momoh and Lamine Mili
29. *Operation and Control of Electric Energy Processing Systems*
James Momoh and Lamine Mili
30. *Restructured Electric Power Systems: Analysis of Electricity Markets with Equilibrium Models*
Xiao-Ping Zhang
31. *An Introduction to Wavelet Modulated Inverters*
S.A. Saleh and M.A. Rahman

32. *Control of Electric Machine Drive Systems*
Seung-Ki Sul

33. *Probabilistic Transmission System Planning*
Wenyuan Li

34. *Electricity Power Generation: The Changing Dimensions*
Digambar M. Tagare

35. *Electric Distribution Systems*
Abdelhay A. Sallam and Om P. Malik

36. *Practical Lighting Design with LEDs*
Ron Lenk, Carol Lenk

37. *High Voltage and Electrical Insulation Engineering*
Ravindra Arora and Wolfgang Mosch

38. *Maintaining Mission Critical Systems in a 24/7 Environment, Second Edition*
Peter Curtis

39. *Power Conversion and Control of Wind Energy Systems*
Bin Wu, Yongqiang Lang, Navid Zargari, Samir Kouro

40. *Integration of Distributed Generation in the Power System*
Math H. Bollen, Fainan Hassan

41. *Doubly Fed Induction Machine: Modeling and Control for Wind Energy Generation Applications*
Gonzalo Abad, Jesús López, Miguel Rodrigues, Luis Marroyo, and Grzegorz Iwanski

42. *High Voltage Protection for Telecommunications*
Steven W. Blume

43. *Smart Grid: Fundamentals of Design and Analysis*
James Momoh

44. *Electromechanical Motion Devices, Second Edition*
Paul Krause, Oleg Wasynczuk, Steven Pekarek

45. *Electrical Energy Conversion and Transport: An Interactive Computer-Based Approach, Second Edition*
George G. Karady, Keith E. Holbert

46. *ARC Flash Hazard and Analysis and Mitigation*
J.C. Das

47. *Handbook of Electrical Power System Dynamics: Modeling, Stability, and Control*
Mircea Eremia, Mohammad Shahidehpour

48. *Analysis of Electric Machinery and Drive Systems, Third Edition*
Paul C. Krause, Oleg Wasynczuk, S.D. Sudhoff, Steven D. Pekarek

49. *Extruded Cables for High-Voltage Direct-Current Transmission: Advances in Research and Development*
Giovanni Mazzanti, Massimo Marzinotto

50. *Power Magnetic Devices: A Multi-Objective Design Approach*
    S.D. Sudhoff

51. *Risk Assessment of Power Systems: Models, Methods, and Applications, Second Edition*
    Wenyuan Li

52. *Practical Power System Operation*
    Ebrahim Vaahedi

53. *The Selection Process of Biomass Materials for the Production of Bio-Fuels and Co-Firing*
    Najib Altawell

54. *Electrical Insulation for Rotating Machines: Design, Evaluation, Aging, Testing, and Repair, Second Edition*
    Greg C. Stone, Ian Culbert, Edward A. Boulter, and Hussein Dhirani

55. *Principles of Electrical Safety*
    Peter E. Sutherland

56. *Advanced Power Electronics Converters: PWM Converters Processing AC Voltages*
    Euzeli Cipriano dos Santos Jr., Edison Roberto Cabral da Silva

57. *Optimization of Power System Operation, Second Edition*
    Jizhong Zhu

58. *Power System Harmonics and Passive Filter Designs*
    J.C. Das

59. *Digital Control of High-Frequency Switched-Mode Power Converters*
    Luca Corradini, Dragan Maksimoviæ, Paolo Mattavelli, Regan Zane

60. *Industrial Power Distribution, Second Edition*
    Ralph E. Fehr, III

61. *HVDC Grids: For Offshore and Supergrid of the Future*
    Dirk Van Hertem, Oriol Gomis-Bellmunt, Jun Liang

62. *Advanced Solutions in Power Systems: HVDC, FACTS, and Artificial Intelligence*
    Mircea Eremia, Chen-Ching Liu, Abdel-Aty Edris

63. *Operation and Maintenance of Large Turbo-Generators*
    Geoff Klempner, Isidor Kerszenbaum

64. *Electrical Energy Conversion and Transport: An Interactive Computer-Based Approach*
    George G. Karady, Keith E. Holbert

65. *Modeling and High-Performance Control of Electric Machines*
    John Chiasson

66. *Rating of Electric Power Cables in Unfavorable Thermal Environment*
    George J. Anders

67. *Electric Power System Basics for the Nonelectrical Professional*
    Steven W. Blume

68. *Modern Heuristic Optimization Techniques: Theory and Applications to Power Systems*
    Kwang Y. Lee, Mohamed A. El-Sharkawi

69. *Real-Time Stability Assessment in Modern Power System Control Centers*
Savu C. Savulescu

70. *Optimization of Power System Operation*
Jizhong Zhu

71. *Insulators for Icing and Polluted Environments*
Masoud Farzaneh, William A. Chisholm

72. *PID and Predictive Control of Electric Devices and Power Converters Using MATLAB®/Simulink®*
Liuping Wang, Shan Chai, Dae Yoo, Lu Gan, Ki Ng

73. *Power Grid Operation in a Market Environment: Economic Efficiency and Risk Mitigation*
Hong Chen

74. *Electric Power System Basics for Nonelectrical Professional, Second Edition*
Steven W. Blume

75. *Energy Production Systems Engineering*
Thomas Howard Blair

76. *Model Predictive Control of Wind Energy Conversion Systems*
Venkata Yaramasu, Bin Wu

77. *Understanding Symmetrical Components for Power System Modeling*
J.C. Das

78. *High-Power Converters and AC Drives, Second Edition*
Bin Wu, Mehdi Narimani

79. *Current Signature Analysis for Condition Monitoring of Cage Induction Motors: Industrial Application and Case Histories*
William T. Thomson, Ian Culbert

80. *Introduction to Electric Power and Drive Systems*
Paul Krause, Oleg Wasynczuk, Timothy O'Connell, Maher Hasan

81. *Instantaneous Power Theory and Applications to Power Conditioning, Second Edition*
Hirofumi, Edson Hirokazu Watanabe, Mauricio Aredes

82. *Practical Lighting Design with LEDs, Second Edition*
Ron Lenk, Carol Lenk

83. *Introduction to AC Machine Design*
Thomas A. Lipo

84. *Advances in Electric Power and Energy Systems: Load and Price Forecasting*
Mohamed E. El-Hawary

85. *Electricity Markets: Theories and Applications*
Jeremy Lin, Jernando H. Magnago

86. *Multiphysics Simulation by Design for Electrical Machines, Power Electronics and Drives*
Marius Rosu, Ping Zhou, Dingsheng Lin, Dan M. Ionel, Mircea Popescu, Frede Blaabjerg, Vandana Rallabandi, David Staton

87. *Modular Multilevel Converters: Analysis, Control, and Applications*
Sixing Du, Apparao Dekka, Bin Wu, Navid Zargari

88. *Electrical Railway Transportation Systems*
    Morris Brenna, Federica Foiadelli, Dario Zaninelli

89. *Energy Processing and Smart Grid*
    James A. Momoh

90. *Handbook of Large Turbo-Generator Operation and Maintenance, 3rd Edition*
    Geoff Klempner, Isidor Kerszenbaum

91. *Advanced Control of Doubly Fed Induction Generator for Wind Power Systems*
    Dehong Xu, Dr. Frede Blaabjerg, Wenjie Chen, Nan Zhu

92. *Electric Distribution Systems, 2nd Edition*
    Abdelhay A. Sallam, Om P. Malik

93. *Power Electronics in Renewable Energy Systems and Smart Grid: Technology and Applications*
    Bimal K. Bose

94. *Distributed Fiber Optic Sensing and Dynamic Rating of Power Cables*
    Sudhakar Cherukupalli, and George J Anders

95. *Power System and Control and Stability,* Third Edition
    Vijay Vittal, James D. McCalley, Paul M. Anderson, and A. A. Fouad

96. *Electromechanical Motion Devices: Rotating Magnetic Field-Based Analysis and Online Animations,* Third Edition
    Paul Krause, Oleg Wasynczuk, Steven D. Pekarek, and Timothy O'Connell

97. *Applications of Modern Heuristic Optimization Methods in Power and Energy Systems*
    Kwang Y. Lee and Zita A. Vale

98. *Handbook of Large Hydro Generators: Operation and Maintenance*
    Glenn Mottershead, Stefano Bomben, Isidor Kerszenbaum, and Geoff Klempner

99. *Advances in Electric Power and Energy: Static State Estimation*
    Mohamed E. El-hawary

100. *Arc Flash Hazard Analysis and Mitigation,* Second Edition
    J.C. Das

101. *Maintaining Mission Critical Systems in a 24/7 Environment,* Third Edition
    Peter M. Curtis

102. *Real-Time Electromagnetic Transient Simulation of AC-DC Networks*
    Venkata Dinavahi and Ning Lin

103. *Probabilistic Power System Expansion Planning with Renewable Energy Resources and Energy Storage Systems,* First edition,
    Jaeseok Choi and Kwang Y. Lee

104. *Power Magnetic Devices: A Multi-Objective Design Approach*, Second Edition.
    Scott D. Sudhoff.

105. *Optimal Coordination of Power Protective Devices with Illustrative Examples*
    Ali R. Al-Roomi

106. *Alternative Liquid Dielectrics for High Voltage Transformer Insulation Systems: Performance Analysis and Applications*
    Edited by U. Mohan Rao, I. Fofana, and R. Sarathi

Printed and bound by CPI Group (UK) Ltd, Croydon, CR0 4YY